Silicones and Silicone-Modified Materials

ACS SYMPOSIUM SERIES **729**

Silicones and Silicone-Modified Materials

Stephen J. Clarson, EDITOR
University of Cincinnati

John J. Fitzgerald, EDITOR
General Electric Company Silicones

Michael J. Owen, EDITOR
Dow Corning Corporation

Steven D. Smith, EDITOR
Procter and Gamble Corporation

MICHIGAN MOLECULAR INSTITUTE
1910 WEST ST. ANDREWS ROAD
MIDLAND, MICHIGAN 48640

American Chemical Society, Washington, DC

Library of Congress Cataloging-in-Publication Data

Silicones and silicone-modified materials / Stephen J. Clarson, editor ... [et al.].

    p.   cm.—(ACS symposium series, ISSN 0097-6156; 729)

Includes bibliographical references and index.

ISBN 0-8412-3613-5

1. Silicones—Congresses.

I. Clarson, Stephen J., 1959– . II. Series.

QD383.S54 S56   2000
547′.08—dc21   99-48567

The paper used in this publication meets the minimum requirements of American National Standard for Information Sciences—Permanence of Paper for Printer Library Materials, ANSI Z39.48-94 1984.

Copyright © 2000 American Chemical Society

Distributed by Oxford University Press

All Rights Reserved. Reprographic copying beyond that permitted by Sections 107 or 108 of the U.S. Copyright Act is allowed for internal use only, provided that a per-chapter fee of $20.00 plus $0.75 per page is paid to the Copyright Clearance Center, Inc., 222 Rosewood Drive, Danvers, MA 01923, USA. Republication or reproduction for sale of pages in this book is permitted only under license from ACS. Direct these and other permissions requests to ACS Copyright Office, Publications Division, 1155 16th Street, N.W., Washington, DC 20036.

The citation of trade names and/or names of manufacturers in this publication is not to be construed as an endorsement or as approval by ACS of the commercial products or services referenced herein; nor should the mere reference herein to any drawing, specification, chemical process, or other data be regarded as a license or as a conveyance of any right or permission to the holder, reader, or any other person or corporation, to manufacture, reproduce, use, or sell any patented invention or copyrighted work that may in any way be related thereto. Registered names, trademarks, etc., used in this publication, even without specific indication thereof, are not to be considered unprotected by law.

PRINTED IN THE UNITED STATES OF AMERICA

Foreword

THE ACS SYMPOSIUM SERIES was first published in 1974 to provide a mechanism for publishing symposia quickly in book form. The purpose of the series is to publish timely, comprehensive books developed from ACS sponsored symposia based on current scientific research. Occasionally, books are developed from symposia sponsored by other organizations when the topic is of keen interest to the chemistry audience.

Before agreeing to publish a book, the proposed table of contents is reviewed for appropriate and comprehensive coverage and for interest to the audience. Some papers may be excluded in order to better focus the book; others may be added to provide comprehensiveness. When appropriate, overview or introductory chapters are added. Drafts of chapters are peer-reviewed prior to final acceptance or rejection, and manuscripts are prepared in camera-ready format.

As a rule, only original research papers and original review papers are included in the volumes. Verbatim reproductions of previously published papers are not accepted.

ACS BOOKS DEPARTMENT

Contents

Preface   xi

1. **Overview of Siloxane Polymers**   1
   James E. Mark

2. **From Sand to Silicones: An Overview of the Chemistry of Silicones**   11
   Larry N. Lewis

3. **Cationic Ring Opening Polymerization of Cyclotrisiloxanes with Mixed Siloxane Units**   20
   J. Chojnowski, K. Kaźmierski, M. Cypryk, and W. Fortuniak

4. **Hydrogenated and Deuterated Cyclic Poly(dimethylsiloxanes)**   38
   Anthony C. Dagger and J. Anthony Semlyen

5. **Calculation of a Reaction Path for KOH-Catalyzed Ring-Opening Polymerization of Cyclic Siloxanes**   81
   J. D. Kress, P. C. Leung, G. J. Tawa, and P. J. Hay

6. **Mesophase Behavior and Structure of Mesophases in Cyclolinear Polyorganosiloxanes**   98
   Yu. K. Godovsky, N. N. Makarova, and E. V. Matukhina

7. **Photoluminescence of Phenyl- and Methylsubstituted Cyclosiloxanes**   115
   Udo Pernisz, Norbert Auner, and Michael Backer

8. **Synthesis and Characterization of Amino Acid Functional Siloxanes**   128
   J. G. Matisons and A. Provatas

9. **Telechelic Aryl Cyanate Ester Oligosiloxanes: Impact Modifiers for Cyanate Ester Resins**   164
   Steven K. Pollack and Zhidong Fu

10. **UV Curable Silicones from Acryloxymethyldimethylacryloxysilane**   170
    H. K. Chu

11. **Advances in Non-Toxic Silicone Biofouling Release Coatings**   180
    Tim Burnell, John Carpenter, Kathryn Truby,
    Judy Serth-Guzzo, Judith Stein, and Deborah Wiebe

12. **Poly(dimethylsiloxane) Gelation Studies**   194
    R. F. T. Stepto, D. J. R. Taylor, T. Partchuk, and M. Gottlieb

13. NMR Spin–Spin Relaxation Studies of Reinforced
    Poly(dimethylsiloxane) Melts 204
    Terence Cosgrove, Michael J. Turner, Ian Weatherhead,
    Claire Roberts, Tania Garasanin, Randall G. Schmidt,
    Glenn V. Gordon, and Jonathan P. Hannington

14. Synthesis of Organosilicon Linear Polymer
    with Octaorganooctasilsesquioxanes Structural Units
    in Main Chain 214
    N. A. Tebeneva, E. A. Rebrov, and A. M. Muzafarov

15. Halogenated Siloxane-Containing Polymers
    via Hydrosilylation Polymerization 226
    J. Hu and D. Y. Son

16. Radially Layered Poly(amidoamine-organosilicon) Copolymeric
    Dendrimers and Their Networks Containing Controlled
    Hydrophilic and Hydrophobic Nanoscopic Domains 241
    Petar R. Dvornic, Agnes M. de Leuze-Jallouli,
    Michael J. Owen, and Susan V. Perz

17. Organic–Inorganic Hybrid Polymers from Atom Transfer
    Radical Polymerization and Poly(dimethylsiloxane) 270
    Krzysztof Matyjaszewski, Peter J. Miller, Guido Kickelbick,
    Yoshiki Nakagawa, Steven Diamanti, and Cristina Pacis

18. Synthesis and Photoinitiated Cationic Polymerization
    of Monomers Containing the Silsesquioxane Core 284
    James V. Crivello and Ranjit Malik

19. Low Modulus Fluorosiloxane-Based Hydrogels
    for Contact Lens Application 296
    J. Künzler and R. Ozark

20. Fluorosilicones Containing the Perfluorocyclobutane
    Aromatic Ether Linkage 308
    Dennis W. Smith, Jr., Junmin Ji, Sridevi Narayan-Sarathy,
    Robert H. Neilson, and David A. Babb

21. Surface Properties of Thin Film Poly(dimethylsiloxane) 322
    H. She, M. K. Chaudhury, and Michael. J. Owen

22. Aggregation Structure and Surface Properties
    of 18-Nonadecenyltrichlorosilane Monolayer
    and Multilayer Films Prepared by the Langmuir Method ... 332
    Ken Kojio, Atsushi Takahara, and Tisato Kajiyama

23. An Investigation of the Surface Properties and Phase Behavior
    of Poly(dimethylsiloxane-*b*-ethyleneoxide) Multiblock Copolymers   353
    Harri Jukarainen, Stephen J. Clarson, and Jukka V. Seppälä

24. Poly(siloxyethylene glycol) for New Functionality Materials   359
    Yukio Nagasaki and Hidetoshi Aoki

25. Polycarbonate–Polysiloxane-Based Interpenetrating Networks   383
    Sylvie Boileau, Laurent Bouteiller, Riadh Ben Khalifa,
    Yi Liang, and Dominique Teyssié

26. High Strength Silicone–Urethane Copolymers:
    Synthesis and Properties   395
    Emel Yılgör, Ayşen Tulpar, Şebnem Kara, and Iskender Yılgör

27. Silicon-Terminated Telechelic Oligomers by Acyclic Diene
    Metathesis Chemistry   408
    K. R. Brzezinska, K. B. Wagener, and G. T. Burns

28. Hydrogen Bond Interactions and Self-Condensation
    of Silanol-Containing Polymers in Polymer Blends
    and Organic–Inorganic Polymeric Hybrids   419
    Eli M. Pearce, T. K. Kwei, and Shaoxiang Lu

29. A Review of the Ruthenium-Catalyzed Copolymerization
    of Aromatic Ketones and 1,3-Divinyltetramethyldisiloxane:
    Preparation of *alt*-Poly(carbosilane–siloxanes)   433
    William P. Weber, Hongjie Guo, Cindy L. Kepler,
    Timothy M. Londergan, Ping Lu, Jyri Paulsaari,
    Jonathan R. Sargent, Mark A. Tapsak, and Guohong Wang

30. Block Copolymers Containing Silicone and Vinyl Polymer
    Segments by Free Radical Polymerization   445
    D. Graiver, B. Nguyen, F. J. Hamilton, Y. Kim, and H. J. Harwood

31. Synthesis of Polymers Containing Silicon Atoms
    of Regulated Structure   460
    Yusuke Kawakami

32. Investigation of the Synthesis of Bis-(3-Trimethoxy-
    silylpropyl)fumarate   476
    Peter M. Miranda

33. In Situ Formed Siloxane-Silica Filler for Rubbery   485
    Organic Networks
    Libor Matějka

34. **From a Hyperbranched Polyethoxysiloxane Toward Molecular Forms of Silica: A Polymer-Based Approach to the Monitoring of Silica Properties** — 503
    V. V. Kazakova, E. A. Rebrov, V. B. Myakushev, T. V. Strelkova, A. N. Ozerin, L. A. Ozerina, T. B. Chenskaya, S. S. Sheiko, E. Yu. Sharipov, and A. M. Muzafarov

35. **Synthesis, Characterization, and Evaluation of Siloxane-Containing Modifiers for Photocurable Epoxy Coating Formulations** — 516
    Mark D. Soucek, Shaobing Wu, and Srinivasan Chakrapani

36. **TOSPEARL: Silicone Resin for Industrial Applications** — 533
    Robert J. Perry and Mary E. Adams

37. **Oxygen Gas Barrier PET Films Formed by Deposition of Plasma-Polymerized SiOx Films** — 544
    N. Inagaki

INDEXES

Author Index — 561

Subject Index — 563

Preface

In the early planning discussions that led to the American Chemical Society (ACS) "Silicones and Silicone Modified Materials" Symposium, we felt that although many meetings had been held during the 1980s on organosilicon chemistry, no major gathering had been held with the focus specifically on silicones. Our prediction that such a symposium was much needed was clearly proven correct by the quantity and quality of the papers presented and by the attendance and interest from the audience during the four full days of oral presentations and also at the evening poster session in Dallas, Texas.

All major technical symposia are a team effort, so it is a great pleasure to thank Kathleen Havelka and Warren Ford for their help, advice, and encouragement throughout the preparation of the symposium. Kathleen had the patience of a saint with our many requests and "past deadline" updates. Robson Story and his team were their usual efficient and helpful selves in keeping us on track with the ACS Polymer Preprints. Robert Stackman did an outstanding job with coordinated finances and expense claims and deserves our thanks. Dow Corning, Gelest, General Electric, Procter and Gamble, and Wacker kindly provided financial assistance and we thank them for helping to make it possible for many of our presenters to attend this symposium. From my own team at the University of Cincinnati, we thank Mark Van Dyke, Rob Johnston, Jennifer Kearney, Susan Junk, and Melissa Scholle, who each gave invaluable assistance with all the various correspondence in both paper and electronic form for both the symposium and this book.

Anne Wilson of the ACS Books Department has been of terrific help at every stage of the commissioning and preparation of this book. We are fortunate to have the ACS Polymer Preprints for the meeting available for the presenters who were unable to contribute to this ACS Symposium book. We trust, however, that the various chapters from the symposium that are included here are of interest to the global silicones community. For those colleagues not able to join us in Dallas in the Spring of 1998, we are delighted that the ACS has kindly welcomed us back for the Spring Meeting in 2001 in San Diego for what we hope will be another lively gathering of our colleagues from the worldwide silicones community.

STEPHEN J. CLARSON
Department of Materials Science and Engineering
and the Polymer Research Center
College of Engineering
University of Cincinnati
Cincinnati, OH 45221–0012

MICHAEL J. OWEN
Dow Corning Corp.
Mail C041D1, P.O. Box 994
Midland, MD 48686–0994

JOHN J. FITZGERALD
General Electric Company Silicones
260 Hudson River Road, Building 12
Waterford, NY 12188

STEVEN D. SMITH
Procter and Gamble
Miami Valley Laboratories
P.O. Box 538707
Cincinnati, OH 45253–8707

Chapter 1

Overview of Siloxane Polymers

James E. Mark

Department of Chemistry and the Polymer Research Center,
The University of Cincinnati, Cincinnati, OH 45221–0172
(markje@email.uc.edu, jemcom.crs.uc.edu)

This review provides coverage of a variety of polysiloxane homopolymers and copolymers, and some related materials. Specific systems include (i) linear siloxane polymers [-SiRR'O-] (with various alkyl and aryl R and R' side groups), (ii) sesquisiloxane polymers possibly having a ladder structure, (iii) siloxane-silarylene polymers [-Si(CH$_3$)$_2$OSi(CH$_3$)$_2$(C$_6$H$_4$)$_m$-] (where the phenylenes are either <u>meta</u> or <u>para</u>), (iv) silalkylene polymers [-Si(CH$_3$)$_2$(CH$_2$)$_m$-], (v) polysiloxanes of exhanced crystallizability through modifications of chemical and stereochemical structures, (vi) elastomers from water-based emulsions, and (vii) random and block copolymers, and blends of some of the above. Topics of particular importance are preparative techniques, end-linking reactions, and the characterization of the resulting polymers in terms of their structures, flexibilities, transition temperatures, permeabilities, and surface and interfacial properties. Applications of these materials include their uses as high-performance fluids, elastomers, coatings, surface modifiers, separation membranes, soft contact lens, body implants, and controlled-release systems. Also of interest is the use of sol-gel hydrolysis-condensation techniques to convert organosilanes to novel reinforcing fillers within elastomers, and to ceramics modified by the presence of elastomeric domains for the improvement of impact strengths.

Of the semi-inorganic polymers, the siloxane or "silicone" polymers have been studied the most, and are also of the greatest commercial importance (*1-13*). The present review provides an overview of some of these polysiloxanes and related materials, emphasizing their structures, most important and interesting physical properties, and a variety of their applications.

Siloxane-Type Polymers

Preparation. Polymers of the type [-SiRR'O-]$_x$ are generally prepared by a ring-opening polymerization of a trimer or tetramer (*14-19*) where R and R' can be alkyl or aryl and x is the degree of polymerization. In this reaction, macrocyclic species are generally formed to the extent of 10-15 wt %. The lower molecular weight ones are generally stripped from the polymer before it is used in a commercial application. Their presence is also of interest from a more fundamental point of view, in two respects. First, the extent to which they occur can be used as a measure of chain flexibility (20). Second, the separated species can be used to test theoretical predictions of the differences between otherwise identical cyclic and linear molecules (21). In some cases, an end blocker such as YR'SiR$_2$OSiR$_2$R'Y is used to give reactive -OSiR$_2$R'Y chain ends (22). Polymerization of non-symmetrical cyclics gives stereochemically variable polymers [-SiRR'O-] analogous to the totally organic vinyl and vinylidene polymers [-CRR'CH$_2$-]. In principle, it should be possible to prepare them in the same stereoregular forms (isotactic and syndiotactic) which have been achieved in the case of some of their organic counterparts. Work of this type is showing great promise (23-25).

Polymerization of mixtures of monomers can of course be used to obtain random copolymers. They are generally highly irregular, but now in the chemical rather than stereochemical sense. Correspondingly, they generally show little if any crystallizability.

Some topics involving polymerizations and related chemical reactions which were covered at the "Silicones and Silicone Modified Materials" symposium are conversions from silicon itself to semi-inorganics (contribution by Lewis), ring-opening polymerizations (contributions by Chojnowski, Soum, Kress, Jallouli, and Komuro), atom-transfer radical polymerizations (Matyjaszewski), hydrosilation polymerizations (Kaganove, Tronc, Narayan-Sarathy), polymerizations with controlled stereochemistry (Kawakami), condensation polymerizations (Fu), and polysilane syntheses (Newton) (26).

Homopolymers.

Flexibility. The most important siloxane polymer is poly(dimethylsiloxane) (PDMS) [-Si(CH$_3$)$_2$O-] (6,9,11). It is also one of the most flexible chain molecules known, both in the dynamic sense and in the equilbrium sense (20,27-31). Dynamic flexibility refers to a molecule's ability to change spatial arrangements by rotations around its skeletal bonds. The more flexible a chain is in this sense, the more it can be cooled before the chains lose their flexibility and mobility and become glassy. Chains with high dynamic flexibility thus generally have very low glass transition temperatures T_g. Since exposing a polymer to a temperature below its T_g generally causes it to become brittle, low values of T_g can be very advantageous, particularly in the case of fluids and elastomers.

The T_g of PDMS, ~ -125 °C, is the lowest recorded for any common polymer. Two reasons for this extraordinary dynamic flexibility are the unusually long Si-O skeletal bond, and the fact that the oxygen skeletal atoms are not only unencumbered by side groups, they are as small as an atom can be and still have the multi-valency needed to continue a chain structure. Also, the Si-O-Si bond angle of ~143° is much more open than the usual tetrahedral bonding occurring at ~110°. In addition, this bond angle has tremendous deformability. These characteristics also increase the chain's equilibrium flexibility, which is the ability of a chain to be compact when in the form of a random coil. This type of flexibility can have a profound effect on the melting point T_m of a polymer. In this case, it is the origin of the very low T_m (-40 °C) of PDMS.

In this regard, crystallization is very important in the case of elastomers, since crystallites can act as reinforcing agents, particularly if they are strain induced. For this reason, it is of interest to make siloxane-type backbones with increased stiffness, in an attempt to increase the T_m of the polymer. Examples of ways to make a polymer more rigid is to combine two chains into a ladder structure, insert rigid units such as p-phenylene groups into the chain backbone, or add bulky side groups to the backbone.

Insertion of a silphenylene group [-Si(CH$_3$)$_2$C$_6$H$_4$-] into the backbone of the PDMS repeat unit yields either the siloxane meta and para silphenylene polymers (32-38). The T_g of the former polymer is increased to -48 °C, but no crystallinity has been observed to date (*36,37*). Since the repeat unit is symmetric, it should be possible to induce crystallinity by stretching. As expected, the p-silphenylene group has a larger rigidifying effect, increasing T_g to -18 °C, and giving rise to crystallinity with a T_m of 148 °C. The resulting polymer is thus a <u>thermoplastic</u> siloxane. Silarylene polymers having more than one phenylene group in the repeat unit could be of considerable interest because of the various <u>meta,</u> para combinations that could presumably be synthesized. It is intriguing that even some flexible siloxane polymers form mesomorphic (liquid-crystalline) phases (*39,40*).

Unusual polysiloxane structures that were covered at the symposium are stiffened chains (Van Dyke, Zhang, Lauter), cyclics (Semlyen, Dagger), ladders and cages (*41*) (Crivello, Lichtenham, Feher, Rebrov, Carpenter, Rahimian, Haddad), hyperbranched structures (Möller, Herzig, Muzafarov, Vasilenko), dendrimers (*42,43*) (Dvornic, Owen, Vasilenko, Sheiko, Rebrov), and sheets and tubes (Kenney, Katsoulis) (*26*).

Permeability. Siloxane polymers have much higher permeability to gases than most other elastomeric materials (*44*). They have therefore long been of interest for use as gas separation membranes, the goal being to vary the basic siloxane structure to improve selectivity without decreasing permeability. Some of the polymers which have been investigated in a major project (*45*) of this type were: [-Si(CH$_3$)RO-], [-Si(CH$_3$)XO-], [-Si(C$_6$H$_5$)RO-], [-Si(CH$_3$)$_2$(CH$_2$)$_m$-], [Si(CH$_3$)$_2$(CH$_2$)$_m$-Si(CH$_3$)$_2$O-], and [Si(CH$_3$)$_2$(C$_6$H$_4$)$_m$Si(CH$_3$)$_2$O-], where R is typically an n-alkyl group and X is an n-propyl group made polar by substitution of atoms such as Cl or N. Unfortunately, structural changes that increase the selectivity are generally found to decrease the permeability, and vice versa.

Another type of membrane designed as an artificial skin coating for burns also exploits the high permeability of siloxane polymers (*46*). The inner layer of the membrane consists primarily of protein and serves as a template for the regenerative growth of new tissue. The outer layer is a sheet of silicone polymer which not only provides mechanical support, but also permits outward escape of excess moisture while preventing ingress of harmful bacteria.

Soft contact lens prepared from PDMS provide a final example. The oxygen required by the eye for its metabolic processes must be obtained by inward diffusion from the air rather than through blood vessels. PDMS is ideal for such lenses (*46*) because of its high oxygen permeability, but it is too hydrophobic to be adequately wetted by the tears covering the eye. This prevents the lens from feeling right, and can also cause very serious adhesion of the lens to the eye itself. One way to remedy this is to graft a thin layer of a hydrophilic polymer to the inner surface of the lens. Because of the thinness of the coating the high permeability of the PDMS is essentially unaffected.

Some Unusual Properties of Poly(Dimethylsiloxane). Atypically low values are exhibited for the characteristic pressure (*47*) (a corrected internal pressure, which is much used in the study of liquids), the bulk viscosity η, and the temperature coefficient of η. Also, entropies of dilution and excess volumes on mixing PDMS with solvents are much lower than can be accounted for by the Flory Equation of State Theory (*47*) . Finally, as has already been mentioned, PDMS has a surprisingly high permeability.

Although the molecular origin of these unusual properties is still not known definitively, a number of suggestions have been put forward. One involves low intermolecular interactions, and another the very high rotational and oscillatory freedom of the methyl side groups on the polymer. Still others focus on the chain's very irregular cross section (very large at the substituted Si atom and very small at the unsubstituted O atom (*47*)), or packing problems associated with the alternating large and more normal bond angles.

Surface and Interfacial Properties. The polysiloxanes generally have very low surface energies (48,49), and considerable research is underway to measure and control surface and interfacial properties in general. For example, adding fluorine atoms to the side chains on a polysiloxane backbone should have a marked effect in this regard (50).

Unusual properties of polysiloxanes that were covered at the symposium are solubility parameters (Rigby), photoluminescence (Pernisz), formation of mesophases (51-53) (Godovsky), films (Takahara, Inagaki), and surfaces (Wynne, Owen, Kowalewski, Jukarainen) (26).

Reactive Homopolymers.

Types of Reactions. In the typical ring-opening polymerization, reactive hydroxyl groups are automatically placed at the ends of the chains. Substitution reactions carried out on these chain ends can then be used to convert them into other functional groups, and these functionalized polymers can undergo a variety of subsequent reactions. Hydroxyl-terminated chains, for example, can undergo condensation reactions with alkoxysilanes (54). A difunctional alkoxysilane leads to chain extension, and a tri- or tetrafunctional one to network formation. Corresponding addition reactions with di- or triisocyanates represent other possibilities. Similarly, hydrogen-terminated chains can be reacted with molecules having active hydrogen atoms (54). A pair of vinyl or other unsaturated groups could also be joined by their direct reactions with free radicals. Similar end groups can be placed on siloxane chains by the use of an end blocker during polymerization (22). Reactive groups such as vinyls can of course be introduced as side chains by random copolymerizations involving, for example, methylvinylsiloxane trimers or tetramers.

Topics involving functionalized polymers that were covered at the symposium include fluorosilicones (48,50,55) (Narayan-Saratahy), amino acid functionalizations (Matisons), grafts (Priou), hydrosilation reactions (Hu), chain-end functionalizations (Fu, Brzezinska, Miranda), and siloxanes as branches (Kishimoto) (26).

Block Copolymers. One of the most important uses of end-functionalized polymers is the preparation of block copolymers, in part because of the tendency of such copolymers to undergo phase separation into novel morphologies (22,56). The reactions are identical to the chain extensions already mentioned except that the sequences being joined are chemically different. In the case of the $-OSiR_2R'Y$ chain ends, R' is typically $(CH_2)_{3-5}$ and Y can be NH_2, OH, $COOH$, $CH=CH_2$, etc. The siloxane sequences containing these ends have been joined to other polymeric sequences such as carbonates, ureas, urethanes, amides, and imides.

Phase separations, blends, and related subjects (57-64) covered at the symposium involved binodal and spinodal phase separations (Viers), block copolymers (Weber, McGrath, Yilgor, Gravier), blends (Talmon, Krenceski, Singh, Yilgor, Pearce), and interpenetrating networks (65) (Boileau, Wengrovius) (26).

Elastomeric Networks. The networks formed by reacting functionally-terminated siloxane chains with an end linker of functionality three or greater have been extensively used to study molecular aspects of rubberlike elasticity (32,66,67). They are "model", "ideal", or "tailor-made" networks in that a great deal is known about their structures by virtue of the very specific chemical reactions used to synthesize them. For example, in the case of a stoichiometric balance between chain ends and functional groups on the end linker, the critically important molecular weight M_c between cross links is equal to the molecular weight of the chains prior to their end linking. Also, the functionality of the cross links (number of chains emanating from one of them) is simply the functionality of the end-linking agent. Finally, the molecular weight distribution of the network chains is the same as that of the starting polymer, and there should be few if any dangling-chain irregularities.

Since these networks have a known degree of cross linking (as inversely measured by M_c), they can be used to test the molecular theories of rubberlike elasticity, particularly with regard to the possible effects of inter-chain entanglements (*32,66*). Intentionally imperfect networks can also be prepared, by unbalancing the stoichiometry, or by using chains with reactive groups at only one of their ends.

One of the most interesting types of model networks is the bimodal, which consists of very short chains intimately end linked with the much longer chains that are representative of elastomeric materials (*32,66,68-72*). These materials have unusually good elastomeric properties, specifically large values of both the ultimate strength and maximum extensibility. Possibly the short chains contribute primarily to the former, and the long chains primarily to the latter. Also, not only do short chains improve the ultimate properties of elastomers, but long chains improve the impact resistance of the much more heavily cross-linked thermosets.

Some topics involving additives, curing, and reinforcement of elastomers by fillers that were covered at the symposium are: additives (Perry), curing (Tsiang, Singh, Chu, Priou, Wu, Taylor), reinforcement (Osaheni, Okel, Cosgrove, Cohen-Addad, Matejka), and water-based elastomers (Liles, Bowens) (*26*).

Cyclic Trapping. If relatively large PDMS cyclics (*21,73*) are present when linear PDMS chains are end linked, then some of them will be permanently trapped by one or more network chains threading through them (*32,74-76*). Interpretation of the fraction trapped as a function of ring size, using rotational isomeric theory and Monte Carlo simulations, provides very useful information on the spatial configurations of cyclic molecules, and the mobilities of the end-linking chains.

Copolymers.

Random. These materials may be prepared by the copolymerization of a mixture of monomers rather than the homopolymerization of a single type of monomer (*15*). One reason for doing this is to introduce functional species, such as vinyls or hydrogens, along the chain backbone to facilitate cross linking. Another is the introduction of sufficient chain irregularity to make the polymer inherently non-crystallizable.

Block. As already mentioned, the sequential coupling of functionally-terminated chains of different chemical structure can be used to make block copolymers, including those in which one or more of the blocks is a polysiloxane (*77-79*). If the blocks are relatively long, separation into a two-phase system almost invariably occurs. Frequently, one type of block will be in a continuous phase and the other will be dispersed in it in domains having an average size the order of a few hundred angstroms. Such materials can have unique mechanical properties not available from either species when present simply in homopolymeric form. Sometimes, similar properties can be obtained by the simple blending of two or more polymers (*57*).

Applications.

Medical. There are numerous medical applications of siloxane polymers (*46*). Prostheses, artificial organs, facial reconstruction, and catheters, for example, take advantage of the inertness, stability, and pliability of the polysiloxanes. Artificial skin, contact lends, and drug delivery systems utilize their high permeability as well.

Non-Medical. Illustrative non-medical applications are high-performance elastomers, membranes, electrical insulators, water repellents, anti-foaming agents, mold-release agents, adhesives, protective coatings, release control agents for agricultural chemicals, encapsulation media, and hydraulic, heat-transfer, and dielectric fluids (*46*). They are based on the same properties of polysiloxanes just mentioned and also their ability to modify surfaces and interfaces (for example as water repellents, anti-foaming agents, and mold-release agents).

Applications of both types covered at the symposium included hydrogels (Künzler), encapsulants (Samara), tougheners and impact modifiers (Pollack, Kumudinie), surfactants (Davidson), photopatterning and resists (Nagasaki, Harkness, Babitch), release coatings (Gordon), and anti-fouling coatings (Stein) (26).

Silica-Type Materials

Sol-Gel Ceramics for In-Situ Precipitations. A relatively new area involving silicon-containing materials is the hydrolysis and condensation of alkoxysilanes or silicates to give silica (SiO_2) (80). The process is complicated, involving polymerization and branching, but the overall reaction results in production of the desired ceramic-like material. Production of ceramics by this novel route has a variety of advantages. First, much lower temperatures can be used, and higher-purity products obtained. Also, the microstructure of the ceramic can be better controlled, and it is relatively simple to form very thin ceramic coatings. Finally, it is much easier to form ceramics "alloys", using the hydrolysis of a mixture of organometallics, for example silicates and titanates to give a SiO_2 - TiO_2 alloy.

Polymers have now been incorporated in this technology (9,81-108). For example, the same hydrolyses can be carried out within a polymer to generate particles of the ceramic material, typically with an average size of a few hundred angstroms (66,106,107). Considerable reinforcement of elastomers, including PDMS, can be achieved in this way. Because of the nature of this in-situ precipitation, the particles are well dispersed and essentially unagglomerated. The particles are also relatively monodisperse, with almost all of them having diameters in the range 200 - 300 Å. Poly(dimethylsiloxane) has also been reinforced with clay-like materials (109).

Reinforced polysiloxane elastomers from water-based emulsions are also of interest in this regard (110-119).

Polymer-Modified Glasses. If the hydrolyses in silane-polymer systems are carried out using relatively large amounts of silane, then the silica generated can become the continuous phase, with the elastomeric polysiloxane dispersed in it (66,106,107). This approach can be used to obtain relatively tough ceramics of reduced brittleness.

Acknowledgements

It is a pleasure to acknowledge that the author's work in some of these areas has been supported by the National Science Foundation, the Air Force Office of Scientific Research, and the Dow Corning Corporation.

Literature Cited

(1) Noll, W. *Chemistry and Technology of Silicones*; Academic Press: Orlando, FL, 1968.
(2) Bobear, W. J. In *Rubber Technology*; M. Morton, Ed.; Van Nostrand Reinhold: New York, 1973; p. 368.
(3) *Analysis of Silicones*; Smith, A. L., Ed.; John Wiley & Sons: New York, 1974.
(4) Warrick, E. L.; Pierce, O. R.; Polmanteer, K. E.; Saam, J. C. *Rubber Chem. Technol.* **1979**, *52*, 437.
(5) Rochow, E. G. *Silicon and Silicones*; Springer-Verlag: Berlin, 1987.
(6) Zeldin, M.; Wynne, K. J.; Allcock, H. R. *Inorganic and Organometallic Polymers*; American Chemical Society: Washington, DC, 1988.

(7) *Silicon-Based Polymer Science. A Comprehensive Resource*; Zeigler, J. M.; Fearon, F. W. G., Eds.; American Chemical Society: Washington, DC, 1990.
(8) Warrick, E. L. *Forty Years of Firsts. The Recollections of a Dow Corning Pioneer*; McGraw-Hill: New York, 1990.
(9) Mark, J. E. In *Silicon-Based Polymer Science. A Comprehensive Resource*; J. M. Zeigler and F. W. G. Fearon, Eds.; American Chemical Society: Washington, DC, 1990; p. 47.
(10) *The Analytical Chemistry of Silicones*; Smith, A. L., Ed.; John Wiley & Sons: New York, 1991.
(11) Mark, J. E.; Allcock, H. R.; West, R. *Inorganic Polymers*; Prentice Hall: Englewood Cliffs, NJ, 1992.
(12) *Siloxane Polymers*; Clarson, S. J.; Semlyen, J. A., Eds.; Prentice Hall: Englewood Cliffs, 1993.
(13) *Inorganic and Organometallic Polymers II*; Wisian-Neilson, P.; Allcock, H. R.; Wynne, K. J., Eds.; American Chemical Society: Washington, 1994.
(14) *Ring-Opening Polymerization*; Frisch, K. C.; Reegen, S. L., Eds.; Marcel Dekker: New York, 1969.
(15) McGrath, J. E.; Riffle, J. S.; Banthia, A. K.; Yilgor, I.; Wilkes, G. L. In *Initiation of Polymerization*; F. E. Bailey Jr., Ed.; American Chemical Society: Washington, 1983.
(16) *Ring Opening Polymerization*; Ivin, K. J.; Saegusa, T., Eds.; Elsevier: New York, 1984.
(17) *Chain Polymerization*; in *Comprehensive Polymer Science*. Allen, G., Ed.; Pergamon Press: Oxford, 1989; Vol. 3.
(18) Odian, G. *Principles of Polymerization;* Third ed.; Wiley-Interscience: New York, 1991.
(19) Chojnowski, J.; Cypryk, M. In *Polymeric Materials Encyclopedia; Synthesis, Properties, and Applications*; J. C. Salamone, Ed.; CRC Press: Boca Raton, 1996.
(20) Flory, P. J. *Statistical Mechanics of Chain Molecules*; Interscience: New York, 1969.
(21) Semlyen, J. A. In *Siloxane Polymers*; S. J. Clarson and J. A. Semlyen, Eds.; Prentice Hall: Englewood Cliffs, 1993; p. 135.
(22) Yilgor, I.; Riffle, J. S.; McGrath, J. E. In *Reactive Oligomers*; F. W. Harris and H. J. Spinelli, Eds.; American Chemical Society: Washington, 1985; Vol. 282.
(23) Kuo, C.-M.; Saam, J. C.; Taylor, R. B. *Polym. Int.* **1994**, *33*, 187.
(24) Battjes, K. P.; Kuo, C.-M.; Miller, R. L.; Saam, J. C. *Macromolecules* **1995**, *28*, 790.
(25) Clarson, S. J.; Mark, J. E. In *Polymeric Materials Encyclopedia; Synthesis, Properties, and Applications*; J. C. Salamone, Ed.; CRC Press: Boca Raton, 1996.
(26) Preprints of the papers presented at this symposium appeared in *Polymer Preprints* **1998**, *39(1)*, and many appear as full-length articles in this book.
(27) Mark, J. E.; Eisenberg, A.; Graessley, W. W.; Mandelkern, L.; Samulski, E. T.; Koenig, J. L.; Wignall, G. D. *Physical Properties of Polymers;* 2nd ed.; American Chemical Society: Washington, DC, 1993.
(28) Bahar, I.; Zuniga, I.; Dodge, R.; Mattice, W. L. *Macromolecules* **1991**, *24*, 2986.
(29) Bahar, I.; Zuniga, I.; Dodge, R.; Mattice, W. L. *Macromolecules* **1991**, *24*, 2993.
(30) Mattice, W. L.; Suter, U. W. *Conformational Theory of Large Molecules. The Rotational Isomeric State Model in Macromolecular Systems*; Wiley: New York, 1994.

(31) Rehahn, M.; Mattice, W. L.; Suter, U. W. *Adv. Polym. Sci.* **1997**, *131/132*, 1.
(32) Mark, J. E.; Erman, B. *Rubberlike Elasticity. A Molecular Primer*; Wiley-Interscience: New York, 1988.
(33) Dvornic, P. R.; Lenz, R. W. *High Temperature Polysiloxane Elastomers*; Huthig & Wepf: Basel, 1990.
(34) Wang, S.; Mark, J. E. *Comput. Polym. Sci.* **1993**, *3*, 33.
(35) Wang, S.; Mark, J. E. *Polym. Bulletin* **1993**, *31*, 205.
(36) Zhang, R.; Pinhas, A. R.; Mark, J. E. *Macromolecules* **1997**, *30*, 2513.
(37) Zhang, R.; Pinhas, A. R.; Mark, J. E. *Polymer Preprints* **1998**, *39(1)*, 575.
(38) Zhang, R.; Pinhas, A. R.; Mark, J. E. *Polymer Preprints* **1998**, *39(1)*, 607.
(39) Godovsky, Y. K.; Makarova, N. N.; Papkov, V. S.; Kuzmin, N. N. *Makaromol. Chem.* **1985**, *6*, 443.
(40) Friedrich, J.; Rabolt, J. F. *Macromolecules* **1987**, *20*, 1975.
(41) Lichtenhan, J. D. In *Polymeric Materials Encyclopedia; Synthesis, Properties, and Applications*; J. C. Salamone, Ed.; CRC Press: Boca Raton, 1996.
(42) Frechet, J. M. J. *Science* **1994**, *263*, 1710.
(43) Tomalia, D. A. *Sci. Am.* **1995**, *272(5)*, 62.
(44) Stern, S. A.; Krishnakumar, B.; Nadakatti, S. M. In *Physical Properties of Polymers Handbook*; 2nd ed.; J. E. Mark, Ed.; Springer-Verlag: New York, 1996; p. 687.
(45) Reports from the Dow Corning Corporation, Syracuse University, and the University of Cincinnati, under Contract No. 5082-260-0666 from the Gas Research Institute, Chicago, IL 1989.
(46) Arkles, B. *CHEMTECH* **1983**, *13*, 542.
(47) Shih, H.; Flory, P. J. *Macromolecules* **1972**, *5*, 758.
(48) Kobayashi, H.; Owen, M. J. *Macromolecules* **1990**, *23*, 4929.
(49) Owen, M. J. In *Physical Properties of Polymers Handbook*; 2nd ed.; J. E. Mark, Ed.; Springer-Verlag: New York, 1996; p. 669.
(50) Patwardhan, D. V.; Zimmer, H.; Mark, J. E. *J. Inorg. Organomet. Polym.* **1998**, *7*, 93.
(51) *Liquid Crystallinity in Polymers: Principles and Fundamental Properties*; Ciferri, A., Ed.; VCH Publishers: New York, 1991.
(52) Samulski, E. T. In *Physical Properties of Polymers*; 2nd ed.; J. E. Mark, A. Eisenberg, W. W. Graessley, L. Mandelkern, E. T. Samulski, J. L. Koenig and G. D. Wignall, Eds.; American Chemical Society: Washington, DC, 1993; p. 201.
(53) Godovsky, Y. K. In *Polymer Data Handbook* J. E. Mark, Ed.; Oxford University Press: New York, 1998.
(54) Mark, J. E. *Acc. Chem. Res.* **1985**, *18*, 202.
(55) Zhao, Q.; Mark, J. E. *Macromol. Sci., Macromol. Rep.* **1992**, *A(29)*, 221.
(56) Goethals, E. J. *Telechelic Polymers: Synthesis and Applications*; CRC Press: Boca Raton, FL, 1989.
(57) Manson, J. A.; Sperling, L. H. *Polymer Blends and Composites*; Plenum Press: New York, 1976.
(58) *Polymer Blends*; Paul, D. R.; Newman, S., Eds.; Academic Press: New York, 1978; Vols. 1 and 2.
(59) Paul, D. R.; Barlow, J. W.; Keskkula, H. In *Encyclopedia of Polymer Science and Engineering*; H. F. Mark, N. M. Bikales, C. G. Overberger and G. Menges, Eds.; Wiley-Interscience: New York, 1988; Vol. 12.
(60) *Polymer Blends and Composites in Multiphase Systems*; Han, C. D., Ed.; American Chemical Society: Washington, DC, 1984; Vol. 206.
(61) *Multiphase Polymers: Blends and Ionomers*; Utracki, L. A.; Weiss, R. A., Eds.; American Chemical Society: Washington, DC, 1989.

(62) Utracki, L. A. *Polymer Alloys and Blends. Thermodynamics and Rheology*; Hanser Publishers: Munich, 1989.
(63) *Advances in Polymer Blends and Alloys Technology*; Finlayson, K., Ed.; Technomic Publishing Company: Lancaster, PA, 1993; Vol. 4, and preceding volumes in this series.
(64) Lee, W. H. In *Polymer Blends and Alloys*; M. J. Folkes and P. S. Hope, Eds.; Blackie Academic & Professional: London, 1993.
(65) Sperling, L. H. *Interpenetrating Polymer Networks and Related Materials*; Plenum Press: New York, 1981.
(66) Erman, B.; Mark, J. E. *Structures and Properties of Rubberlike Networks*; Oxford University Press: New York, 1997.
(67) Mark, J. E.; Erman, B. In *Polymer Networks*; R. F. T. Stepto, Ed.; Blackie Academic, Chapman & Hall: Glasgow, 1998.
(68) Mark, J. E. In *Elastomers and Rubber Elasticity*; J. E. Mark and J. Lal, Eds.; American Chemical Society: Washington, DC, 1982.
(69) Mark, J. E. *Brit. Polym. J.* **1985**, *17*, 144.
(70) Mark, J. E. *J. Inorg. Organomet. Polym.* **1994**, *4*, 31.
(71) Mark, J. E. *Acc. Chem. Res.* **1994**, *27*, 271.
(72) Mark, J. E. In *Fourth International Conference on Frontiers of Polymers and Advanced Materials*; P. N. Prasad, J. E. Mark, S. H. Kandil and Z. H. Kafafi, Eds.; Plenum: New York, 1997.
(73) *Large Ring Molecules*; Semlyen, J. A., Ed.; John Wiley & Sons: New York, 1996.
(74) de Gennes, P. G. *Scaling Concepts in Polymer Physics*; Cornell University Press: Ithaca, New York, 1979.
(75) Rigbi, Z.; Mark, J. E. *J. Polym. Sci., Polym. Phys. Ed.* **1986**, *24*, 443.
(76) Iwata, K.; Ohtsuki, T. *J. Polym. Sci., Polym. Phys. Ed.* **1993**, *31*, 441.
(77) Noshay, A.; McGrath, J. E. *Block Copolymers: Overview and Critical Survey*; Academic Press: New York, 1977.
(78) Riess, G.; Hurtrez, G.; Bahadur, P. In *Encyclopedia of Polymer Science and Engineering*; H. F. Mark, N. M. Bikales, C. G. Overberger and G. Menges, Eds.; Wiley-Interscience: New York, 1985; Vol. 2.
(79) Yilgor, I.; McGrath, J. E. *Polysiloxane Copolymers/Anionic Polymerization*; Springer Verlag: Berlin, 1988, p. 1.
(80) Brinker, C. J.; Scherer, G. W. *Sol-Gel Science*; Academic Press: New York, 1990.
(81) Schmidt, H.; Wolter, H. *J. Non-Cryst. Solids* **1990**, *121*, 428.
(82) Wilkes, G. L.; Huang, H.-H.; Glaser, R. H. In *Silicon-Based Polymer Science*; J. M. Zeigler and F. W. G. Fearon, Eds.; American Chemical Society: Washington, DC, 1990; Vol. 224; p. 207.
(83) Nass, R.; Arpac, E.; Glaubitt, W.; Schmidt, H. *J. Non-Cryst. Solids* **1990**, *121*, 370.
(84) Schaefer, D. W.; Mark, J. E.; McCarthy, D. W.; Jian, L.; Sun, C.-C.; Farago, B. In *Polymer-Based Molecular Composites*; D. W. Schaefer and J. E. Mark, Eds.; Materials Research Society: Pittsburgh, 1990; Vol. 171; p. 57.
(85) Schmidt, H. In *Better Ceramics Through Chemistry IV*; B. J. J. Zelinski, C. J. Brinker, D. E. Clark and D. R. Ulrich, Eds.; Materials Research Society: Pittsburgh, 1990; Vol. 180; p. 961.
(86) Mark, J. E.; Schaefer, D. W. In *Polymer-Based Molecular Composites*; D. W. Schaefer and J. E. Mark, Eds.; Materials Research Society: Pittsburgh, 1990; Vol. 171; p. 51.
(87) Mark, J. E. *J. Inorg. Organomet. Polym.* **1991**, *1*, 431.
(88) Mark, J. E. *J. Appl. Polym. Sci., Appl. Polym. Symp.* **1992**, *50*, 273.

(89) Schmidt, H. In *Chemical Processing of Advanced Materials*; L. L. Hench and J. K. West, Eds.; Wiley: New York, 1992; p. 727.
(90) Schmidt, H. In *Ultrastructure Processing of Advanced Materials*; D. R. Uhlmann and D. R. Ulrich, Eds.; Wiley: New York, 1992; p. 409.
(91) Mark, J. E. *Angew. Makromol. Chemie* **1992**, *202/203*, 1.
(92) Novak, B. M. *Adv. Mats.* **1993**, *5*, 422.
(93) *Proceedings of the First European Workshop on Hybrid Organic-Inorganic Materials*; Sanchez, C.; Ribot, F., Eds.; Chimie de la Matiere Condensee: Chateau de Bierville, France, 1993.
(94) Clarson, S. J.; Mark, J. E. In *Siloxane Polymers*; S. J. Clarson and J. A. Semlyen, Eds.; Prentice Hall: Englewood Cliffs, 1993; p. 616.
(95) Mark, J. E. In *Frontiers of Polymers and Advanced Materials*; P. N. Prasad, Ed.; Plenum: New York, 1994; p. 403.
(96) Schmidt, H.; Krug, H. In *Inorganic and Organometallic Polymers II*; P. Wisian-Neilson, H. R. Allcock and K. J. Wynne, Eds.; American Chemical Society: Washington, 1994; Vol. 572; p. 183.
(97) Mark, J. E.; Calvert, P. D. *J. Mats. Sci., Part C* **1994**, *1*, 159.
(98) Mark, J. E. In *Diversity into the Next Century*; R. J. Martinez, H. Arris, J. A. Emerson and G. Pike, Eds.; SAMPE: Covina, CA, 1995; Vol. 27.
(99) Mascia, L. *Trends in Polymer Science* **1995**, *3 (2)*, 61.
(100) *Hybrid Organic-Inorganic Composites*; Mark, J. E.; Lee, C. Y.-C.; Bianconi, P. A., Eds.; American Chemical Society: Washington, 1995; Vol. 585.
(101) Mark, J. E. *Macromol. Symp.* **1995**, *93*, 89.
(102) Mackenzie, J. D. In *Hybrid Organic-Inorganic Composites*; J. E. Mark, C. Y.-C. Lee and P. A. Bianconi, Eds.; American Chemical Society: Washington, 1995; Vol. 585; p. 226.
(103) Mark, J. E.; Wang, S.; Ahmad, Z. *Macromol. Symp.* **1995**, *98*, 731.
(104) Mark, J. E. In *Hybrid Organic-Inorganic Composites*; J. E. Mark, C. Y.-C. Lee and P. A. Bianconi, Eds.; American Chemical Society: Washington, 1995; Vol. 585; p. 1.
(105) Wen, J.; Wilkes, G. L. In *Polymeric Materials Encyclopedia: Synthesis, Properties, and Applications*; J. C. Salamone, Ed.; CRC Press: Boca Raton, 1996.
(106) Mark, J. E. *Hetero. Chem. Rev.* **1996**, *3*, 307.
(107) Mark, J. E. *Polym. Eng. Sci.* **1996**, *36*, 2905.
(108) Frisch, H. L.; Mark, J. E. *Chem. Mater.* **1996**, *8*, 1735.
(109) Burnside, S. D.; Giannelis, E. P. *Chem. Mater.* **1995**, *7*, 1597.
(110) Johnson, R. D.; Saam, J. C.; Schmidt, C. M. *U. S. Patent 4,221,688 to the Dow Corning Corporation* **1980**.
(111) Saam, J. C.; Graiver, D.; Baile, M. *Rubber Chem. Technol.* **1981**, *54*, 976.
(112) Saam, J. C. *U. S. Patent 4,244,849 to the Dow Corning Corporation* **1981**.
(113) Graiver, D.; Huebner, D. J.; Saam, J. C. *Rubber Chem. Technol.* **1983**, *56*, 918.
(114) Huebner, D. J.; Saam, J. C. *U. S. Patent 4,567,231 to the Dow Corning Corporation* **1986**.
(115) Liles, D. T. *U. S. Patent 4,962,153 to the Dow Corning Corporation* **1990**.
(116) Liles, D. T.; Lefler, H. V., III, in *18th Water-Borne, Higher-Solids and Powder Coatings Symposium*, New Orleans, 1991.
(117) McCarthy, D. W.; Mark, J. E. *Rubber Chem. Technol.* **1998**, *71*, 000.
(118) Mark, J. E.; McCarthy, D. W. *Rubber Chem. Technol.* **1998**, *71*, 000.
(119) McCarthy, D. W.; Mark, J. E. *Rubber Chem. Technol.* **1998**, *71*, 000.

Chapter 2

From Sand to Silicones: An Overview of the Chemistry of Silicones

Larry N. Lewis

GE Corporate Research and Development Center,
1 Research Circle, Niskayuna, NY 12309

The chemistry of silicones is summarized by following the steps necessary to produce a two-part, platinum-cured silicone containing vinyl-stopped polydimethylsiloxane, Si-H-on-chain siloxane, platinum catalyst and catalyst inhibitor. The process begins with silicon dioxide and follows the steps of conversion to sand to elemental silicon. Silicon is reacted with MeCl to make methylchlorosilanes in the methylchlorosilane reaction (MCS). The products from the MCS reaction are separated by distillation and then hydrolyzed and condensed to make the various siloxane polymers. Polymers with methyl, vinyl or Si-H functionality are made as required for the platinum addition-cured silicone product.

The silicones industry got its start in the late 1930's *(1,2)* and became viable after Rochow's 1940 discovery of the direct process which reacts elemental silicon with MeCl to produce methylchlorosilanes *(3,4)*. This chapter attempts to summarize some of the steps which take place in the process of converting sand into silicones. The "vignette" chosen for this summary is the production of a platinum-cure, so-called addition cured silicone. This brief review will make use of the M, D shorthand wherein an M group is Me$_3$SiO- and a D group is -Me$_2$SiO-. Substituents on silicon other than Me are represented with a superscript so that M^{vi} stands for (H$_2$C=CH)Me$_2$SiO- and D^H stands for -(Me)(H)SiO- *(5)*. Figure 1 summarizes the entire process covered in this review.

Formation of Elemental Silicon

Silica (sand) is reduced in a carbo-electro reduction process to produce chemical grade silicon according to equation 1 *(6)*. The many trace elements present in silicon are either non-reactive or are required for the MCS reaction. Iron apparently has little

© 2000 American Chemical Society

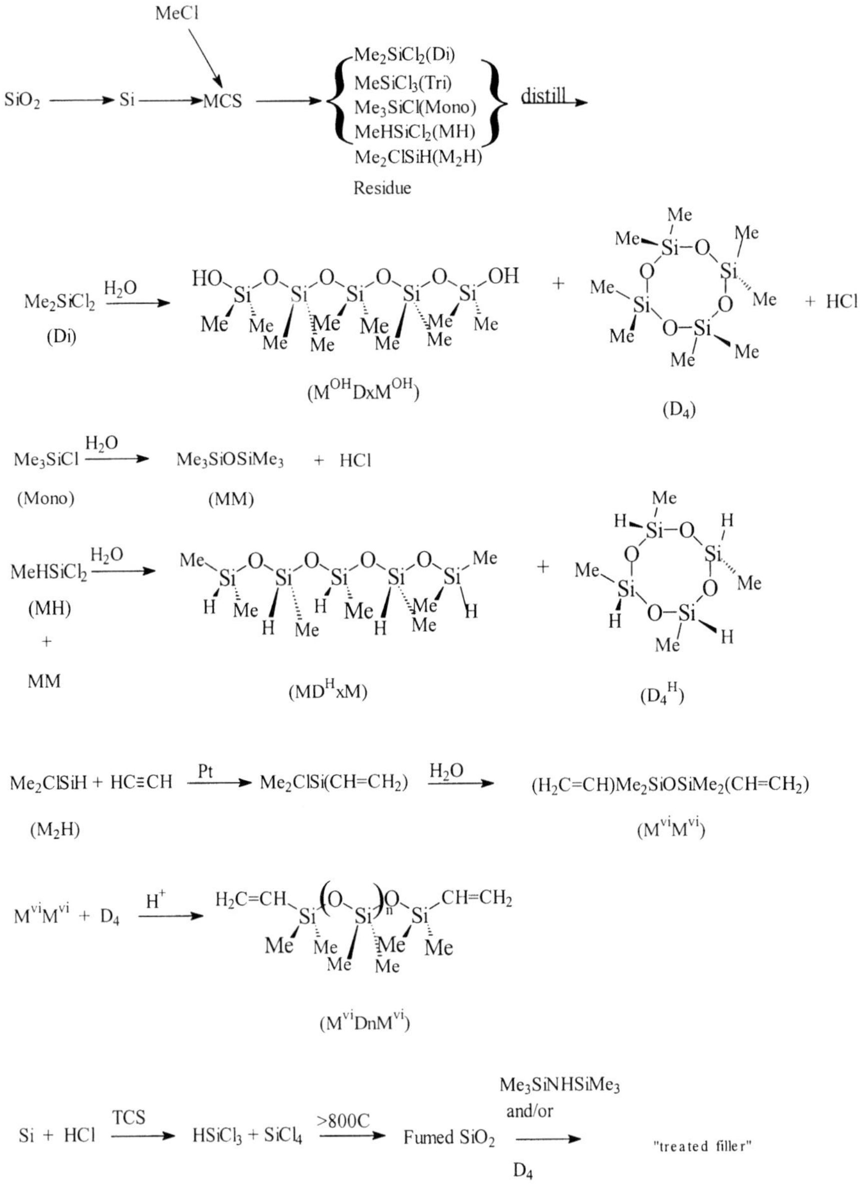

Figure 1. *Overall Scheme for Conversion of Sand to Pt-Curable Silicone Formulation*

effect at its normal levels of 0.5% while aluminum at 0.1 to 0.3% is an essential promoter *(3,4)*. Titanium and calcium are present from 0-200 ppm and may be promoters. Other elements frequently encountered in the ppm level include Cd, Cr, Cu, Ni, P, Pb, Sb, Sn, V, Zn and Zr (3,4). Recent research in the area of silicon made for MCS applications can be found in the proceedings to the meeting held every other year in Norway called, "Silicon for the Chemical Industry" *(7-10)*.

$$SiO_2 + C \xrightarrow[>1200°C]{\text{High voltage}} SiO + SiC \longrightarrow Si + CO \qquad 1$$

The Direct Process or Methyl Chlorosilane Reaction (MCS)

The MCS reaction is shown in equation 2. Typically elemental silicon is reacted

$$Si + MeCl \xrightarrow[\substack{Sn\,(5\text{-}30\text{ ppm}) \\ Al\,(500\text{-}4000\text{ ppm}) \\ 290\text{-}305°C}]{\substack{Cu\,(3\text{-}5\%) \\ Zn\,(400\text{-}2000\text{ ppm})}} \begin{array}{l} Me_2SiCl_2\,(Di,\,75\text{-}90\%) \\ MeSiCl_3\,(Tri,\,5\text{-}10\%) \\ Me_3SiCl\,(Mono,\,1\text{-}5\%) \\ MeHSiCl_2\,(MH,\,0.5\text{-}3\%) \\ Me_2HSiCl\,(M_2H,\,0.1\text{-}1\%) \\ \text{other low boilers }(0.1\text{-}0.5\%) \\ \text{residue }(0.5\text{-}5\%) \end{array} \qquad 2$$

in a fluidized bed reactor in the presence of copper, tin zinc and other promoters *(11)*. Critical factors for the MCS reaction include selectivity for dimethyldichlorosilane (Di), rate of methylchlorosilane production, silicon utilization and spent metal loss. The largest volume polymer product produced is the polydimethylsiloxane polymer (PDMS). PDMS in turn is made from hydrolysis of Di thus selectivity for Di is very important. The MCS reaction is a solid/gas reaction that produces a liquid product mixture. Optimum economic performance is achieved when the silicon utilization is high and the amount of spent metal lost is low.

The mechanism of the MCS reaction has been discussed for over 50 years but detailed understanding is still lacking at the molecular level *(3,4)*. Two interesting and potentially critical pieces of the MCS mechanism puzzle are the form of copper in the reaction and the type of intermediates present. Figure 2 illustrates a mechanism for formation of Di from silicon, copper and methyl chloride.

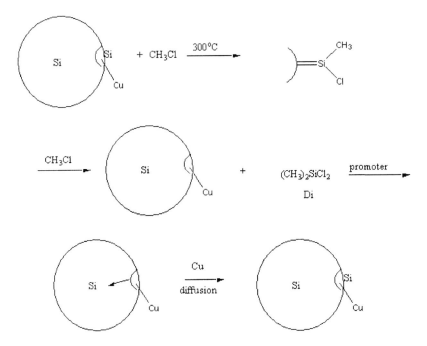

Figure 2. Proposed MCS Mechanism

Cu$_3$Si or "eta-phase" has long been proposed as the active copper species in the MCS reaction (12,13). Recent work has suggested that a silylene intermediate is important in the MCS reaction (14). Activated silicon (as eta-phase) reacts with one equivalent of MeCl to make a copper-bound silylene, Si(Me)Cl. The silylene can then react with a second equivalent of MeCl to form Me$_2$SiCl$_2$.

Siloxane Polymer Formation from Methylchlorosilanes

The product mixture from a typical MCS reaction is subjected to several distillation and isolation steps. The product mixture can be roughly divided into monomers and residue. The monomers are separated from the residue stream by distillation; the residue contains siloxanes and disilanes. Some monomers can be recovered by various redistribution reactions of the residue mixture (15). The individual monomers are separated by distillation where the separation of Di from Tri is difficult. With Di as an example, equation **3** shows the hydrolysis and condensation to form linear and cyclic polysiloxanes. Another useful material is hexamethyldisiloxane (MM) which forms from hydrolysis/condensation of Me$_3$SiCl (mono), equation **4**.

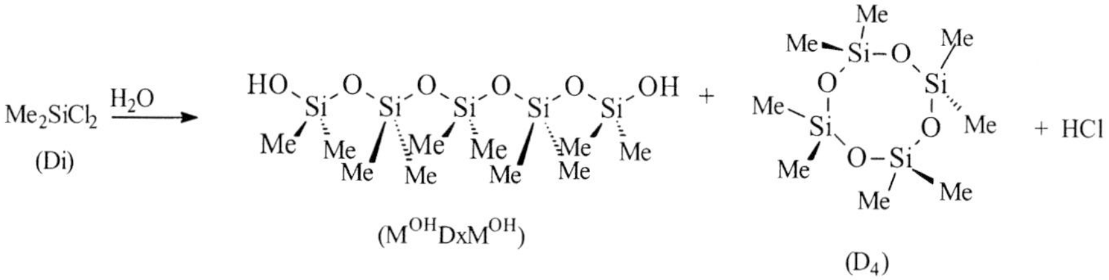

$$\text{Me}_2\text{SiCl}_2 \xrightarrow{\text{H}_2\text{O}} \text{HO-Si(Me)}_2\text{-O-Si(Me)}_2\text{-O-Si(Me)}_2\text{-O-Si(Me)}_2\text{-O-Si(Me)}_2\text{-OH} + (D_4) + \text{HCl}$$
(Di) $(M^{OH}D_xM^{OH})$

3

$$\text{Me}_3\text{SiCl} \xrightarrow{\text{H}_2\text{O}} \text{Me}_3\text{SiOSiMe}_3 + \text{HCl}$$
(Mono)   (MM)

4

The main cyclic product from equation **3** is octamethylcyclotetrasiloxane (D_4) which can be isolated from the product mixture by distillation. Synthesis of polydimethylsilosanes of virtually any degree of polymerization is accomplished by either acid or base catalyzed ring opening polymerization of D_4 with the appropriate concentration of MM as a chain-stopper, equation **5** *(5)*.

The ring-opening polymerization reaction can be used to prepare polymers with other functional groups. An addition-cured formulation uses vinyl-containing and Si-H-containing polymers. Preparation of an Si-H-on-chain polymer from ring-opening polymerization of teramethylcyclotetrasiloxane (D_4^H) and MM is shown in equation 6 and formation of a vinyl-stopped polydimethylsiloxane by ring-opening polymerization of divinyltetramethyldisiloxane and D_4 is shown in equation 7.

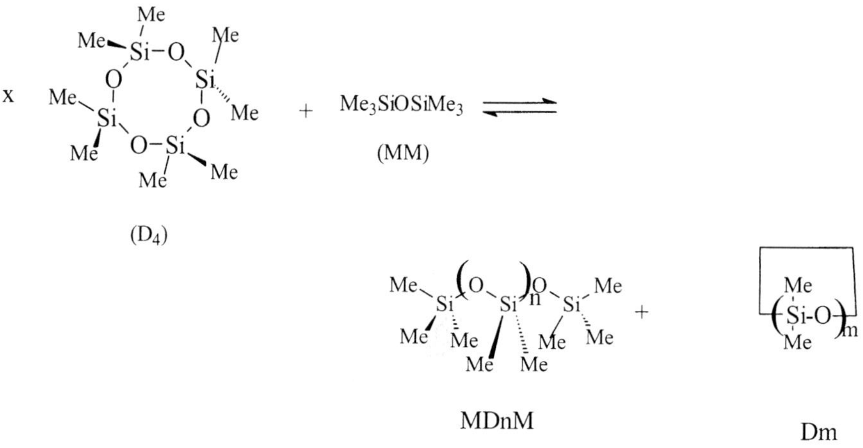

$$x\ (D_4) + \text{Me}_3\text{SiOSiMe}_3 \text{ (MM)} \rightleftharpoons \text{MD}_n\text{M} + D_m$$

5

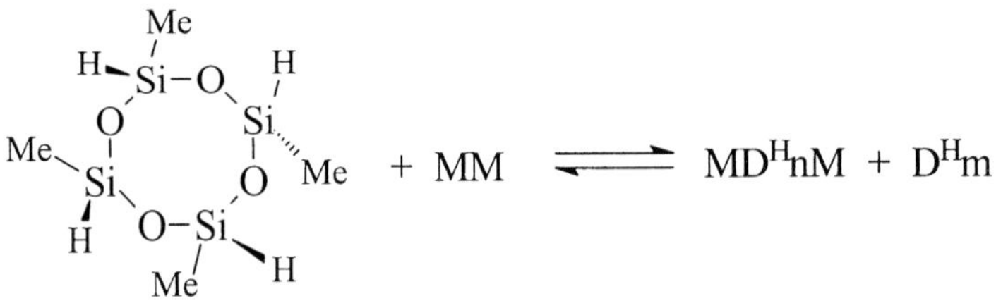

[structure of D_4^H cyclic siloxane] + MM ⇌ $MD^H_nM + D^H_m$

(D_4^H)

6

$(H_2C=CH)Me_2SiOSiMe_2(CH=CH_2)$ + D4 ⇌ $M^{vi}DmM^{vi}$ + Dm

($M^{vi}M^{vi}$)

7

Fillers

The physical properties of cured polysiloxane materials are dramatically influenced by fillers *(1,2)*. So-called non-reinforcing (extenders) and reinforcing fillers are typically used; the most common reinforcing filler is silica. High surface area silica, called fumed silica, is formed by burning the product mixture obtained from the trichlorosilane (TCS) reaction of equation **8**. Only a small amount of untreated

$$Si + HCl \xrightarrow{TCS} HSiCl_3 + SiCl_4$$

8

fumed silica can be added to polydimethylsiloxane polymers. Surface treating is thus carried out in order to improve the blendability of silica with siloxanes. Typical treating agents include Me_3SiCl, $Me_3SiNHSiMe_3$, and D_4. The treatment step is shown in equation **9** where the silanol surface of the fumed silica is modified with trimethylsiloxy groups.

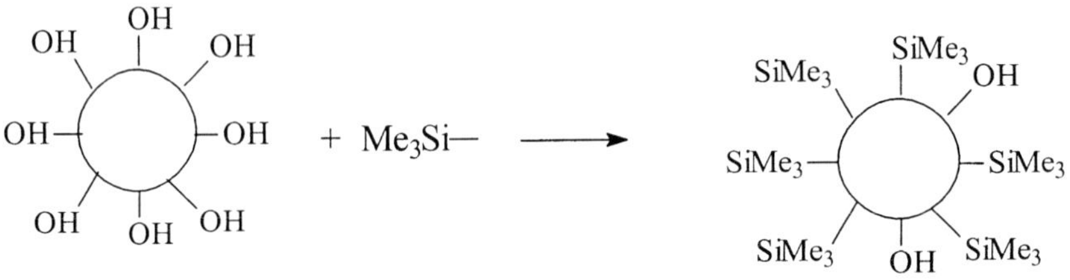

[silica particle with OH groups] + Me_3Si- → [silica particle with $SiMe_3$ and OH groups]

9

Platinum-Catalyzed Addition Cure

The hydrosilylation reaction *(16,17)* is a well known reaction for formation of silicon carbon bonds. When vinyl-containing polysiloxane is reacted with multi-Si-H-containing polysiloxane in the presence of a platinum catalyst, a crosslinked network forms, equation **10**. Platinum is so active for the hydrosilylation reaction that inhibitors are added to moderate the rate of crosslinking *(18)*. A typical platinum catalyst used by industry is the so-called Karstedt catalyst *(19,20)*. A reaction of Karstedt's catalyst with a ligand, L (an inhibitor perhaps), is shown in equation **11**.

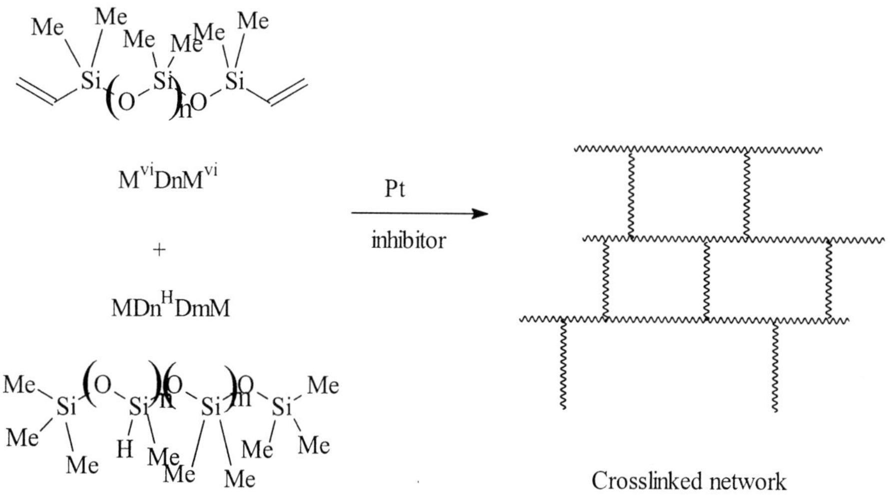

$M^{vi}DnM^{vi}$

+

MDn^HDmM

$\xrightarrow[\text{inhibitor}]{Pt}$ Crosslinked network

10

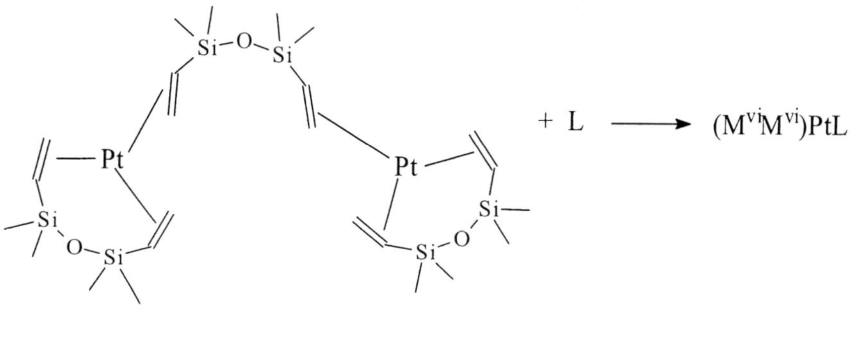

$Pt_2(M^{vi}M^{vi})_3$ + L $\longrightarrow$ $(M^{vi}M^{vi})PtL$

11

A typical low temperature cure inhibitor is dimethyl fumarate which reacts with Karstedt's catalyst to form a platinum-fumarate complex as shown in equation **12** (20).

$$Pt(M^{vi}M^{vi})_x \;+\; 4 \; Me\text{-}O\text{-}C(=O)\text{-}CH=CH\text{-}C(=O)\text{-}O\text{-}Me \;\xrightarrow{C_6D_6}\; \mathbf{2} \qquad \mathbf{12}$$

1   dimethyl fumarate

2 (platinum–fumarate–divinylsiloxane complex)

Other common inhibitors employed for platinum-cured siloxanes include maleates, fumarates, acetylenic alcohols, phosphines and tetramethyltetravinyltetrasiloxane (D_4^{vi}). Other key variables for platinum-cured siloxanes include: The molecular weight of the vinyl-stopped polymer, $M^{vi}DnM^{vi}$; the amount of Si-H in the Si-H-on-chain polymer, $MD^{H}nM$ and the ratio of vinyl to Si-H; the amount of filler and degree of surface treatment, the amount of platinum catalyst and the amount and type of inhibitor.

Summary and Conclusion

With all of the aforementioned variables, hundreds of Pt-curable silicone products are available including: heat-cured rubber, liquid injection moldable products and release coatings. There are also many curable products based on chemistry other than platinum: condensation cure RTV, peroxide-cured and UV-cured epoxy silicones. With other functional groups and/or molecular weights, even more products are possible including: low viscosity fluids/high viscosity gums, alkoxy and acetoxy end groups for condensation cure, amino and epoxy functional groups used for coupling agents, phenyl groups for high temperature applications and fluoro-substituted groups to impart solvent resistance.

Literature Cited

(1) Liebhafsky, H. A.; Liebhafsky, S. S.; Wise, G. *Silicones Under the Monogram*; Wiley Interscience: New York, 1978.

(2) Warrick, E. L., Forty Tears of Firsts; McGraw Hill: New York, 1990.

(3) Kanner, B.; Lewis, K. M. In *Catalyzed Direct Reactions of Silicones*; Lewis, K. M.; Rethwisch, D. G., Eds.; Elsevier Science Publishers B.V.: Amsterdam, 1993; pp. 1-49.

(4) Lewis, L.N., *The Chemistry of Organic Silicon Compounds*, Vol. 2, Rappoport, Z.; Apeloig, Y., Eds, John Wiley, 1998, p. 1581.

(5) Rich, J.; Cella, J.; Lewis, L.; Stein. J.; Singh, N.; Rubinsztajn S.; Wengrovius, J., *Kirk-Othmer Encyclopedia of Chemical Technology*, Fourth Edition, John Wiley: New York, 1997, Vol. 22, pp. 82-142.

(6) Downing, J. H.; Kaiser, R. H.; Wells, J. E. In *Catalyzed Direct Reactions of Silicon*, Lewis, K. M.; Rethwisch, D. G., Eds.; Elsevier: Amsterdam, 1993, p. 67.

(7) *Silicon for the Chemical Industry*, Oye, H. A.; Rong, H., Eds.; Geiranger, Norway, June 16-18, 1992, Institute of Inorganic Chemistry, NTH: Trondheim, Norway, 1992.

(8) *Silicon for the Chemical Industry II*, Oye, H. A.; Rong, H.; Nygaard, L.; Schussler, G.; Tuset, J. Kr., Eds.; Loen, Norway, June 8-10, 1994, Tapir Publishers: Norway, 1994.

(9) *Silicon for the Chemical Industry III*, Oye, H. A.; Rong, H.; Ceccarolli, B.; Nygaard, L.; Tuset, J., Kr., Sandefjord, Norway, June 18-20, 1996, Eds., The Norwegian University of Science and Technology, Trondheim, Norway, 1996.

(10) Lewis, L. N.; Gao, Y.; Bolon, R.; Ravikumar, V.; D'Evelyn, M. In *Silicon for the Chemical Industry IV*, Oye, H. A.; Rong, H. M.; Nygaard, L.; Schussler, G.; Tuset, J. Kr., Eds., Norwegian Univ. of Science & Technology, Trondheim, Norway, 1998, p. 157.

(11) Ward, W. J.; Ritzer, A.; Carroll, K. M.; Flock, J. W. *J. Catal.* 1986, Vol. 100, p. 240.

(12) Voorhoeve, R. J. H. *Organohalosilane: Precursors to Silicones*, Elsevier: New York, 1967.

(13) Floquet, N.; Yilmaz, S.; Falconer, J. L. *J. Catal.* 1994, Vol. 148, p. 348.

(14) Okamoto, M.; Onodera, S.; Okano, T.; Susuki, E.; Ono, Y. *J. Orgnaomet. Chem.* 1997, Vol. 531, pp. 67-71.

(15) Ritzer, A. In *Silicon for the Chemical Industry II*, Oye, H. A.; Rong, H.; Nygaard, L.; Schussler, G.; Tuset, J., Jr., Eds., Loen, Norway, June 8-10, 1994, Tapir Publishers: Norway, 1994, pp. 242-249.

(16) *Comprehensive Handbook on Hydrosilylation*; Marciniec, B., Ed.; Pergamon Press: Oxford, England, 1992.

Chapter 3

Cationic Ring Opening Polymerization of Cyclotrisiloxanes with Mixed Siloxane Units

J. Chojnowski, K. Kaźmierski, M. Cypryk, and W. Fortuniak

Center of Molecular and Macromolecular Studies, Polish Academy of Sciences, Sienkiewicza 112, 90-363 Łódź, Poland

The polymerization of 1,1-diphenyl-3,3,5,5-tetramethylcyclotrisiloxane, **1**, initiated with CF_3SO_3H was studied as a model of cationic polymerization of cyclotrisiloxane with mixed siloxane units to evaluate the chemoselectivity and regioselectivity of this process. The polymerization of **1** occurs in the absence of reactions leading to a randomization of sequences of siloxane units in the polymer chain. Instead, various reactions of end groups accompany this process, leading to the formation of cyclic products and to the broadening of molecular weight distribution. An addition of $CF_3SO_3SiMe_3$ to the polymerization system causes an introducing of the Me_3SiO- group to the end of the polymer chain. Chain transfer to this group occurs intensively, leading to a randomization of Me_3SiOSi and CF_3SO_3Si chain ends. An addition of $CF_3SO_3SiMe_3$ does not decrease the polydispersity and only partially supresses the cyclization. The regioselectivity of the polymerization is rather poor. Monomer **1** is opened between the dimethylsiloxane units and at silicon bound to phenyl groups with an approximate rate ratio 2:1, respectively.

There has been a considerable interest in the controlled synthesis of polysiloxanes of defined structure in connection with the growing use of these polymers in macromolecular engineering. Polysiloxanes are commonly used as segments of block and graft copolymers (*1-5*), which requires a high control of their topology, molecular weight, polydispersity and functionalization. The anionic polymerization of strained-ring cyclotrisiloxanes has, so far, been the most important method of the controlled synthesis of functionalized polysiloxanes. If an appropriate initiator is used, this reaction may approach a living polymerization, leading to a high yield of polymer with a controlled size and functionality of macromolecules, for review see ref. 6,7. In contrast, little effort has been used in examining the kinetically controlled cationic ring opening polymerization as a route for the synthesis of well defined polysiloxanes. It is worth mentioning that the cationic process has often been utilized in the equilibrium polymerization of cyclic siloxanes, as exemplified in refs. 8-10. The cationic polymerization of cyclosiloxanes gives some advantages over the anionic one. In the presence of some initiators, such as strong protic acids, it proceeds fast at room temperature and the initiator may be relatively easily removed from the polymer. Moreover, the reaction may readily be performed with cyclosiloxane monomers bearing functional groups which are not stable under the conditions of the anionic process, such as Si-H and Si-CH$_2$Cl (*8,11*). The recent progress in the understanding of the cationic polymerization of cyclic siloxanes makes a broader application of this reaction fairly feasible as the method of synthesis of well defined siloxane polymers and copolymers (*12,13*).

The purpose of this study is an examination of the chemoselectivity and regioselectivity of the cationic polymerization of cyclotrisiloxanes using a model monomer. We expected that the results would allow to evaluate to what extent the structure of the polymer, obtained by cationic polymerization of cyclotrisiloxanes, could be controlled. Our approach consisted in an investigation of the polymerization of cyclotrisiloxane with mixed siloxane units. The symmetrically substituted 1,1-diphenyl-3,3,5,5,-tetramethylcyclotrisiloxane, **1**, was chosen as a model monomer to avoid the formation of stereoisomers. The same model had previously been used to study the anionic ring-opening polymerization (*14*), what made possible a direct comparison of the selectivity in both cationic and anionic processes. The sequential analysis of products of the polymerization of **1** was believed to give valuable information on the chemo- and regioselectivity of the cationic polymerization of cyclic siloxanes, thus giving a deeper insight into its mechanism which, in many points, is still controversial (*15*).

The polymerization of cyclotrisiloxane with mixed siloxane units was expected to allow for a synthesis of siloxane-siloxane copolymers with a uniform distribution of units. These copolymers are not accessible by the kinetically controlled copolymerization of two cyclic trisiloxane comonomers, which leads to a microsequential order of siloxane units and a gradient arrangement of monomer units along the chain (*16,17*). So far the sequencing in polysiloxane, obtained from cyclotrisiloxane having two kinds of siloxane units, have been studied only for the polymerization in the presence of anionic initiators (*16,18,19*). The use of cationic routes for a controlled synthesis of copolymers with a uniform distribution of units is, therefore, of interest.

Results and Discussions

Chemoselectivity of the Polymerization. Polymerization occurs chemoselectively if chain propagation is not accompanied by any side reactions affecting the structure of resulting macromolecule. Among the undesired processes spoiling the chemoselectivity of polymerization of **1** there could be reactions changing the structure of side groups in the resulting polymer, such as, substitution at phenyl ring or scission of the phenyl group. In particular, the possibility of occurrence of the latter process should be taken into consideration. The phenyl group is known to be relatively easily cleaved from silicon by trifluoromethanesulfonic acid. Matyjaszewski et al. exploited the scission of the phenyl-to-silicon bond by CF$_3$SO$_3$H to functionalize polydiorganosilanes (*20*). The cleavage of phenyl in various structures of silanes was broadly explored by Uhlig (*21*). The reaction of CF_3SO_3H with $Ph_2Si(CD_3)_2$ involving the cleavage of Si-Ph has recently been used as a route to a fully deuterated polydimethylsiloxane (22). However, it should be stressed that the undesired cleavage of phenyl groups seems to be meaningless in the polymerization of **1** under the conditions used in the study. Anyway, that cleavage was registered neither by ^{29}Si NMR nor by the SEC or MALDI TOF analyses. Such a substitution of phenyl would result in a formation of triflate side groups in the polymer chain, which would further lead to branching or cross-linking. A fraction of a higher molecular weight could be expected in this case. However, the size exclusion chromatograms of the polymer have shown a classical, very close to Gaussian, shape even in cases when the polymerization was quenched at a very high monomer conversion (over 90%). The example is given in Figure 1. ^{29}Si NMR spectra do not show any signals which could be attributed to $MeSiO_3$, although the range between -60 ppm to -100 ppm was under close inspection. Evidently the Si-Ph cleavage, even if it occurs here, proceeds much more slowly than the propagation and does not affect the structure of the polymer.

The other possible side reactions affecting the structure of polymer are those leading to the cleavage of the siloxane chains. Three such processes should be considered:
1) Back biting reactions, involving the attack of the propagation center on a siloxane group in the same chain are schematically shown by equation 1. They lead either to the formation of cyclic siloxane or to a rearrangement of the chain fragment (6,23). An "A" denotes the active propagation center.

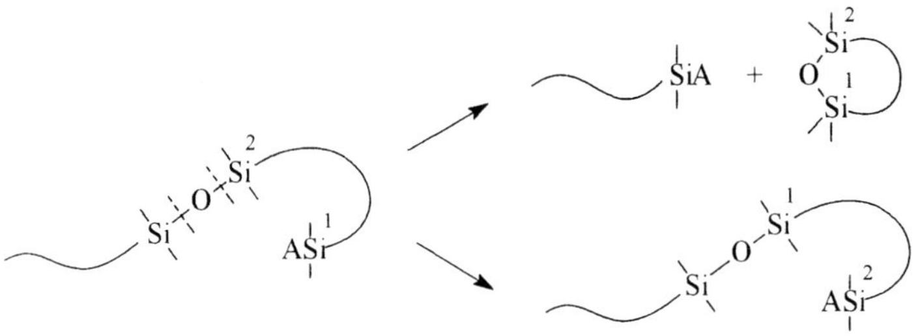

(1)

2) Chain transfer to another chain involving siloxane bond cleavage leads to chain scrambling. It is schematically depicted by equation 2 (6,23).

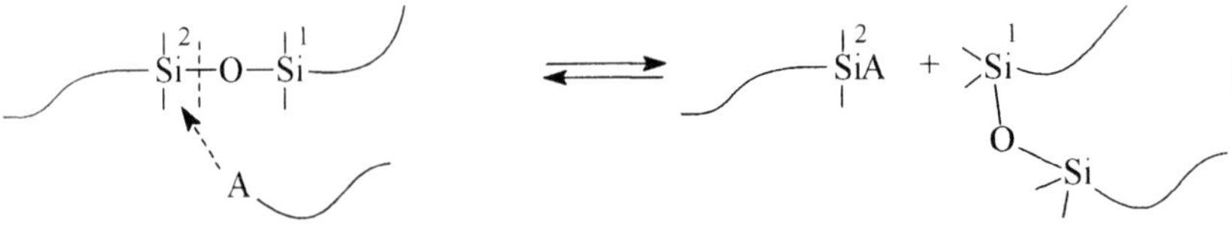

(2)

3) Transfer of the terminal unit to the end of another chain proceeds according to equation 3 (24).

(3) 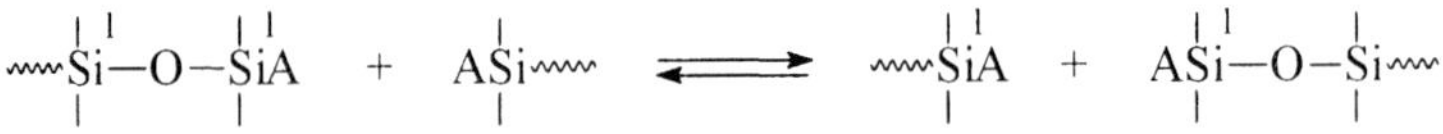 ~~Si—O—SiA + ASi~~ ⇌ ~~SiA + ASi—O—Si~~

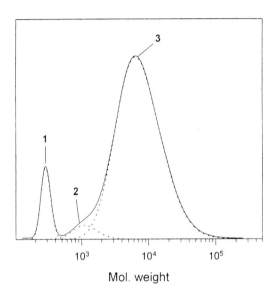

Figure 1. Size exclusion chromatogram of the polymer from the polymerization of monomer **1**, 1.2 mol·dm^{-3}, in n-hexane, at 30°C, in the presence of CF$_3$SO$_3$H, 1·10^{-3} mol·dm^{-3}, and CF$_3$SO$_3$SiMe$_3$, 3·10^{-2} mol·dm^{-3}, quenched at 94% monomer conversion. The peaks correspond to: 1 - cyclic dimer, (Ph$_2$SiO(Me$_2$SiO)$_2$)$_2$; 2 - cyclic oligomers [Ph$_2$SiO(Me$_2$SiO)]$_n$, n>2; 3 - linear polymer. The peaks were separated using computer program based on the Gaussian size distribution among each three populations.

All these three side reactions strongly affect the structure of the copolymer leading to a randomization of sequences, thus studies of sequencing may provide important information about the selectivity of polymerization.

In the chemoselective polymerization of monomer 1 sequencing of siloxane units in the polymer chain depends exclusively on the way in the which monomer is opened and added to the end of the growing chain. There are three non-equivalent places of opening of 1 marked by a, b, c in equation 4, which lead to three different arrangements of siloxane groups in the open chain monomer units.

$$\begin{array}{c} \text{Ph}_2\text{Si(O-)}_2(\text{SiMe}_2\text{O})_2 \end{array} \longrightarrow \begin{cases} \rightarrow \text{SiMe}_2\text{OSiMe}_2\text{OSiPh}_2 \rightarrow \quad a \\ \rightarrow \text{SiPh}_2\text{OSiMe}_2\text{OSiMe}_2 \rightarrow \quad b \\ \rightarrow \text{SiMe}_2\text{OSiPh}_2\text{OSiMe}_2 \rightarrow \quad c \end{cases} \qquad (4)$$

The sequencing depends also on the arrangement in the neighboring monomer unit, which also may be a, b or c. If each combination has an equal probability, then the polymerization occurs chemoselectively as a random (statistical) addition. However, the random addition leads to a specific sequencing of the siloxane units in the polymer chain because each monomer unit, consisting of three siloxane units, enters the polymer chain undivided. Thus, the following ten Ph_2SiO-centered pentads, giving six signals of pentads in ^{29}Si NMR, are forbidden for the chemoselective polymerization: XXXXX, XXXXD+DXXXX, XDXXX+XXXDX, XXXDD+DDXXX, DXXXD, XDXXD+DXXDX (X=Ph_2SiO, D=Me_2SiO). Similarly, in the region of the Me_2SiO unit resonance, four Me_2SiO-centered pentads giving three ^{29}Si NMR signals of pentads, i.e., DDDDD, XXDXX, XXDXD+DXDXX, are forbidden if the polymerization proceeds chemoselectively.

The polymerization of 1 was carried out in n-heptane or in methylene chloride in the presence of CF_3SO_3H or $CF_3SO_3SiMe_3 + CF_3SO_3H$. The process was quenched at a high monomer conversion, in some cases well above 90%. ^{29}Si NMR spectra of an isolated polymers were very similar. A typical spectrum, taken by IGATED technique with addition of $Cr(acac)_3$, is presented in Figure 2. Signals of pentads are well separated. The assignement of pentads was performed on the basis of ref. 16. The range of chemical shifts, where the signals of the "forbidden" Ph_2Si-centered pentads would appear, is marked in the expanded fragment of the spectrum (Figure 2). None of these pentads is observed in the spectrum. Thus, the reactions, causing randomization of sequences of siloxane units in the polymer chain (reactions 1-3), play little role in the mechanism of polymerization of 1.

The above results, together with the known kinetics of the polymerization of hexamethylcyclotrisiloxane (25, 26), the linear increase of molecular weight with monomer conversion (27) in this reaction and the recent proof of the role of tertiary siloxonium ion in this process (12), are in agreement with the simplified mechanism of this reaction represented by equations 5a,b.

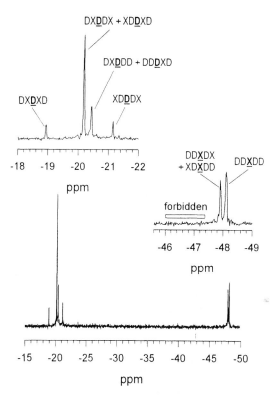

Figure 2. ^{29}Si NMR spectrum of the polymer obtained by polymerization of **1** in the presence of CF_3SO_3H and $CF_3SO_3SiMe_3$ quenched at 94% monomer conversion. The range, where the signals forbidden for the chemoselective process could appear, is marked on the expanded spectrum. X denotes the diphenylsiloxane unit, D denotes dimethylsiloxane unit.

Initiation

$$CF_3SO_3H + O\underset{Si}{\overset{Si}{\diagup\!\!\!\diagdown}} \xrightarrow{CF_3SO_3H} HO\!-\!Si\cdots Si\!-\!OSO_2CF_3 \qquad (5a)$$

Propagation

$$HO\!-\!Si\cdots Si\!-\!OSO_2CF_3 + O\underset{Si}{\overset{Si}{\diagup\!\!\!\diagdown}} \xrightleftharpoons[-CF_3SO_3H]{CF_3SO_3H} HO\!-\!Si\cdots Si\!-\!O^{\oplus}\!\!\underset{Si}{\overset{Si}{\diagup\!\!\!\diagdown}} (CF_3SO_3)_2H^- \qquad (5b)$$

$$\downarrow O\underset{Si}{\overset{Si}{\diagup\!\!\!\diagdown}}$$

$$HO\!-\!Si\cdots SiO\!-\!Si\cdots Si\!-\!OSO_2CF_3 + O\underset{Si}{\overset{Si}{\diagup\!\!\!\diagdown}} \xrightleftharpoons[-CF_3SO_3H]{CF_3SO_3H} HO\!-\!Si\cdots SiO\!-\!Si\cdots Si\!-\!O^{\oplus}\!\!\underset{Si}{\overset{Si}{\diagup\!\!\!\diagdown}} (CF_3SO_3)_2H^-$$

The acid opens the monomer ring leading to the formation of ester and silanol end groups. The end group is reversibly transformed with the participation of acid to the tertiary oxonium ion, which is the active propagation center. The fast equilibrium 5b lies strongly on the side of the ester which creates conditions for a controlled stepwise growth of the polymer chain. However, the data presented in Table I indicate that the above mechanism is oversimplified for the polymerization of monomer **1**. The polydispersity factor is higher than that expected for the controlled polymerization and the process leads to a high yield of cyclic oligomers. The broadening of molecular weight is only partially explained by the continuous initiation connected with the presence of stationary concentration of acids (25,28). The absence of forbidden pentads for the chemoselective process does not permit to exclude side reactions which occur without any fragmentation of monomer units. These reactions may lead to a broadening of molecular weight distribution and to the formation of cyclics. One of such reactions is an intra- or intermolecular condensation of active propagation centers, or the ester end groups, with the nucleophilic OH groups, according to equations 6a,b,c,d.

$$HO\overset{2}{Si}\!\sim\!\overset{1}{Si}O\overset{4}{Si}\!\sim\!\overset{}{Si}O_3SCF_3 + 2\,CF_3SO_3H + \underset{O\!-\!Si}{\overset{}{Si}}\!\overset{O}{\diagup\!\!\!\diagdown}\!\overset{3}{Si} \qquad (6a)$$

$$HO\overset{2}{Si}\!\sim\!\overset{1}{Si}\!\sim\!\overset{3}{Si}O\overset{4}{Si}\!\sim\!\overset{}{Si}O_3SCF_3 + 2\,CF_3SO_3H \qquad (6b)$$

$$\overset{2}{Si}\!\overset{O}{\diagup\!\!\!\diagdown}\!\overset{1}{Si} + \overset{}{Si}\!\overset{O}{\diagup\!\!\!\diagdown}\!\overset{3}{Si} + 2\,CF_3SO_3H \qquad (6c)$$

$$\overset{2}{Si}\!\overset{O}{\diagup\!\!\!\diagdown}\!\overset{3}{Si} + 2\,CF_3SO_3H \qquad (6d)$$

(from intermediate: $HO\overset{2}{Si}\!\sim\!\overset{1}{Si}O^{\oplus}$ with $\overset{3}{Si}\!-\!O$ / $Si\!-\!O$ structure, $HO\!\sim\!\overset{4}{Si}O_3SCF_3$, $(CF_3SO_3)_2H^{\ominus}$)

Table I. Comparison of Polymerization of 1 Initiated by CF_3SO_3H and $CF_3SO_3SiMe_3+CF_3SO_3H$, n-Hexane, 30^0C, $[1]_0 = 1.2$ mol·dm^{-3}

Initiator	Concentration mol·dm^{-3}	$\overline{M}_w$	$\overline{M}_w/\overline{M}_n$	Total Content of Cyclics, %
$CF_3SO_3SiMe_3$ + CF_3SO_3H	$1.0 \cdot 10^{-2}$ $1.0 \cdot 10^{-3}$	54,000	1.8	20 (7)^b
$CF_3SO_3SiMe_3$ + CF_3SO_3H	$3.0 \cdot 10^{-2}$ $1.0 \cdot 10^{-3}$	11,000	1.5	14 (4)^b
$(CF_3SO_3SiMe_2\text{~})_2$ + CF_3SO_3H	$1.5 \cdot 10^{-2}$ $1.0 \cdot 10^{-3}$	42,200	1.7	18 (7)^b
CF_3SO_3H	$1.0 \cdot 10^{-3}$	140.000	2.0	40
CF_3SO_3H	$3.0 \cdot 10^{-3}$	50,000	1.7	38
$CF_3SO_3H^a$	$2.3 \cdot 10^{-3}$	90,000	2.1	35

^{a}Solvent CH_2Cl_2.
^{b}Cyclics except dimer $[Ph_2SiO(Me_2SiO)_2]_2$.

In order to get more information about this process, $Me_3SiO_3SCF_3$, in at least 10-fold excess relative to CF_3SO_3H, was introduced to the polymerization system. The results are shown in Table I. The presence of the ester in excess to the acid introduces the Me_3SiO group at the end of the polysiloxane chain in a concentration approximately equal to the concentration of the ester end group because a direct initiation with $Me_3SiO_3SCF_3$ is possible, according to equation 7. Moreover, the exchange of silanol-ester, according to equation 8, and condensation, according to equations 6 lead to a decrease of the stationary concentration of the SiOH end groups as a result of the addition of $CF_3SO_3SiMe_3$.

$$Me_3SiO_3SCF_3 \xrightarrow[CF_3SO_3H]{monomer} Me_3SiO^{\oplus} \begin{array}{c} Si-O \\ \\ Si-O \end{array} Si \xrightarrow{monomer} \text{propagation} \quad (7)$$
$$(CF_3SO_3)_2H^{\ominus}$$

$$\sim\!\!\sim\!\!\sim\!\!SiOH + CF_3SO_3H \rightleftharpoons \sim\!\!\sim\!\!\sim\!\!SiO_3SCF_3 + H_2O \quad (8)$$

The data presented in Table I show that the presence of the ester reduces the yield of cyclic products. However, even large excess of the ester, relative to the acid, does not prevent the formation of cyclics. In particular, the cyclic dimer is continually formed in a large amounts, which may indicate that the initiation, according to equation 5a, occurs extensively. The introduction of $CF_3SO_3SiMe_3$ does not seem to the narrow molecular weight distribution.

To get information about the end groups present in the growing chains, the MALDI TOF analysis of the polymerization system of **1** in n-hexane was performed. The polymerization was carried out in the presence of an excess of the trimethylsilyl ester at a high $[CF_3SO_3SiMe_3]_0/[CF_3SO_3H]_0$ ratio, which ensured a very small concentration of the SiOH end groups. Thus, the silyl ester groups and the $SiMe_3$ groups should appear at the ends of polymer chains almost exclusively. The reaction was quenched by an addition of a large excess of aqueous Na_2CO_3, which converted all the ester end groups to SiOH groups. The fragment of the spectrum is shown in Figure 3. The spectrum confirms the exclusive formation of chains with an integral number of monomer units. The peaks are attributed to the three types of chains present in the polymerization system, which are distinguished by the structure of end groups, i.e., the chains having the trimethylsilyl group at one chain end and the ester group at the other, the chains having trimethylsilyl groups at both ends and the chains with the ester groups on both sides. They appear in approximate molar ratio 2:1:1, respectively, which corresponds to the statistical distribution of the chain ends. This result indicates that a process leading to fast randomization of chain ends occurs during the polymerization. The process comes from chain transfer to the terminal Me_3SiO siloxane unit according to equation 9.

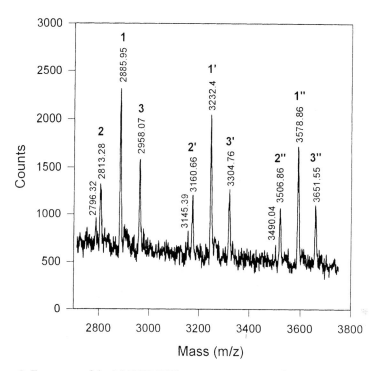

Figure 3. Fragment of the MALDI TOF mass spectrogram of the polymer obtained by the polymerization of **1**. The peaks denoted by numbers correspond to: 1, 1', 1" - $Me_3Si[OSiPh_2(OSiMe_2)_2]_nOH$; 2, 2', 2" - $H[OSiPh_2(OSiMe_2)_2]_nOH$; 3, 3', 3" - $Me_3Si[OSiPh_2(OSiMe_2)_2]_nOSiMe_3$, n = 8, 9, 10, respectively. The SiOH group corresponds to the SiO_3SCF_3 group in the polymerization system. The average masses, denoted over the peaks, contain the mass of Na^+.

$$\begin{aligned}&\text{\textasciitilde O} + \overset{\diagup}{\underset{\diagdown}{\text{Si}}}\overset{-\text{O}}{\underset{-\text{O}}{\diagdown}}\overset{\diagup}{\underset{\diagup}{\text{Si}}} + \text{Me}_3\text{SiOSi}\text{\textasciitilde} \longrightarrow \text{\textasciitilde SiOSi}\text{\textasciitilde} + \text{Me}_3\text{SiO} + \overset{\diagup}{\underset{\diagdown}{\text{Si}}}\overset{-\text{O}}{\underset{-\text{O}}{\diagdown}}\overset{\diagup}{\underset{\diagup}{\text{Si}}}\\ &\qquad\qquad\qquad\qquad\qquad\searrow\\ &\qquad\qquad\qquad\qquad\qquad\text{Me}_3\text{SiOSi}\text{\textasciitilde} + \text{\textasciitilde Si}-\text{O} + \overset{\diagup}{\underset{\diagdown}{\text{Si}}}\overset{-\text{O}}{\underset{-\text{O}}{\diagdown}}\overset{\diagup}{\underset{\diagup}{\text{Si}}}\end{aligned}$$

(9)

Such a transfer was found earlier to extensively occur in the cationic polymerization of hexamethylcyclotrisiloxane with linear oligomers of formula $Me_3Si(OSiMe_2)_nOSiMe_3$, n = 0, 1, 2 (29). The chain transfer constants (k_{tr}/k_p) in benzene (30%) were: 0.29, 0.18 and 0.18 for n=0, 1 and 2, respectively. The chain transfer constant to the chain end in the cationic polymerization of **1** is expected to be of a comparable value.

The unimodal, close to Gaussian, molecular weight distribution of the linear fraction of the polymer obtained by polymerization in the presence of an excess of $Me_3SiO_3SCF_3$ results not from one directional growth of macromolecules, but from a fast randomization of chain ends, according to reactions 5-9.

It is amazing that so large difference of reactivity exists between the cleavage to the Me_3SiO end group and the cleavage of the siloxane bond inside the chain of the polymer obtained from monomer **1**.

Regioselectivity in the Polymerization. The polymerization of **1** would be fully regioselective if the monomer was always opened in the same place. Regardless of the opening site, a, b or c, the polymer would have an alternating arrangement of Ph_2SiO and $(Me_2SiO)_2$ units. Thus, all the Ph_2SiO-centered pentads, except DDXDD, and all the Me_2SiO-centered pentads, except DXDDX and XDDXD, are not allowed in this process. However, the ^{29}Si NMR spectrum presented in Figure 1 reveals that the polymer contains considerable amounts of forbidden pentads XDXDD, DDXDX and DXDXD. Thus, in contrast to a high chemoselectivity, the cationic polymerization of monomer **1** exhibits a poor regioselectivity.

In order to express the regioselectivity in a more quantitative way, the pentad composition was analyzed, according to the second order Markovian statistics (30). That analysis allowed to estimate the probabilities of ring opening P_a, P_b and P_c at three nonequivalent sites: a, b and c, respectively. Those parameters correspond to the relative rates of the monomer opening at the corresponding sites in the ring. The results of the analysis were checked by the Monte Carlo simulation of the pentad content. However, the statistical calculations cannot differentiate between the openings at a and b. Moreover, allowing for a poor precision of the determination of the contribution from various pentads by ^{29}Si NMR the statistical calculations gave several possible solutions (30). Therefore, the identification of terminal triads in the polymer, obtained by quenching the polymerization of **1** at a suitable moment, was necessary. The mathematical analysis in combination with the analysis of terminal triads in the polymer permitted to evaluate P_a, P_b and P_c unequivocally (Table II).

The results indicate that the monomer is mostly opened at the siloxane bond, linking the dimethylsiloxane units, e.g., at site c, as expected. However, a considerable fraction is opened also at site a (*31*). The above result is somewhat strange. Considering the two possible structures of the active propagation center, 10a and 10b, the latter should be disfavoured due to the inductive and steric effects of phenyl groups.

$$\begin{array}{cc}
\underset{a}{\overset{}{\text{Me}_2}} & \underset{b}{\overset{}{\text{Me}_2}} \\
\end{array}\tag{10}$$

The consequence of the formation of the structure 10a is the opening of the ring at site c. Thus, the route c should strongly be preferred and a high regioselectivity could be expected.

The source of the poor regioselectivity may be a more complex mechanism of the cationic polymerization of monomer 1. First, the propagation,, according to equation 5b, is not the sole process of the chain extension. The contribution from other processes may be more significant than expected. Thus, the intermolecular chain coupling, according to equation 6a and 6b, in connection with the fast exchange of the SiOH and SiO_3SCF_3, may affect the sequencing. However, this complication would rather imply that the addition occurs randomly, according to routes a and b, which is not observed. The propagation may also take place by an addition of protonated monomer to the silanol group (*32,33*). However, the contribution of this mechanism should strongly decrease with the introduction of $CF_3SO_3SiMe_3$ to the polymerization system, while the sequencing does not seem to be much affected by the presence of the ester. Moreover, the protonation of 1 should lead to the ring opening of the monomer at site c as well.

Quantum mechanical calculations of the monomer 1, and of its protonated and silylated forms were performed to understand the effect of phenyl groups on the oxonium ion formation. The HF/3-21G* optimized geometries of 1, protonated and silylated at the $Ph_2SiOSiMe_2$ oxygen, are shown in scheme 11. Unexpectedly, the species, protonated at the oxygen atom bound to diphenyl-substituted silicon, is by 2.8 kcal.mol^{-1} lower in energy than that protonated at oxygen linking the two dimethyl-substituted silicon atoms. A similar result was obtained for the two silylated isomers. In this case, the HF/3-21G* energy difference was 1.4 kcal·mol^{-1}. General conclusions of the calculations are that: 1) electron densities on the oxygen atoms of 1 are only slightly modified by the replacement of two methyl by two phenyl groups at one of the silicon atoms; 2) phenyl groups may directly interact with proton and other positively charged groups, attached to siloxane oxygen stabilizing the oxonium ion. This interaction may more than counterbalance the inductive and steric effects of the phenyl substituents destabilizing the positively charged structure. A poor regioselectivity, observed for the cationic polymerization of 1, may be related to these phenomena.

The calculations were performed on the isolated molecules, thus neglecting solvent effects, nevertheless, they point to the structure stabilization resulting from a direct interaction of the phenyl ring with the positively charged center. Such a stabilization of the structures 11a and 11b may affect the reactivity in the polymerization.

a   b   (11)

The regioselectivity in the cationic polymerization of **1** is, on the whole, not worse than that observed in the controlled anionic polymerization of this monomer. The results of triad analysis for some systems of both types of polymerizations of **1** are compared in Table III. The regioselectivity factor is calculated from equation 12, which was derived, assuming that the ring opening predominantly occurs at one site, while openings at other sites during the formation of the polymer chain do not follow directly each other (*30*). This is a crude approximation, however, the comparison, based on the Markov statistics, is also charged with a considerable error.

$$S_r = \frac{3}{200}([DDX] + [XDD]) \qquad (12)$$

Conclusions

The cationic polymerization of **1**, initiated with a strong protic acid, such as trifluoromethanesulfonic acid, proceeds without cleavage of the polymer chain. Thus, processes of back biting, chain scrambling and terminal unit exchange are eliminated, which permits to suppose that cationic polymerization of cyclotrisiloxanes may be used for the synthesis of polysiloxanes with a controlled microstructure of chains. Some recent results seem to confirm this conclusion. Cyclic and macrocyclic products are formed by the end coupling reaction, including the active propagation center (end biting). The addition of $CF_3SO_3SiMe_3$ leads to an efficient reduction of the yield of macrocyclics, but it does not permit to eliminate the formation of lower oligomers, in particular, a cyclic dimer, which was also observed earlier by Sigwalt in the polymerization of hexamethylcyclotrisiloxane (*26,27*). Although presented results are in agreement with the stepwise chain growth mechanism, presented in equations 5b, the molecular weight

Table II. Regioselectivity in Cationic ROP of 1; solvent n-hexane, 30°C, $[1]_0 = 1.2$ mol·dm^{-3}

Initiator	Initial Concentration	P_a	P_b	P_c
CF_3SO_3H	$1.0 \cdot 10^{-3}$	0.24	0.002	0.76
CF_3SO_3H	$3.0 \cdot 10^{-3}$	0.18	0.07	0.75
$CF_3SO_3H^a$	$2.5 \cdot 10^{-3}$	0.42	0.02	0.56
$CF_3SO_3SiMe_3$ + CF_3SO_3H	$1.0 \cdot 10^{-2}$ / $1.0 \cdot 10^{-3}$	0.23	0.01	0.77
$CF_3SO_3SiMe_3$ + CF_3SO_3H	$3.0 \cdot 10^{-2}$ / $1.0 \cdot 10^{-3}$	0.20	0.08	0.72

[a]Solvent CH_2Cl_2.

Table III. Comparison of Triad Contents and Regioselectivity Factor for the Polymerization of 1 in the Presence of Various Cationic and Anionic Initiators

Initiator	Conditions	DDD	XDD+DDX	XDX	XXD+DXX	DXD	XXX	Regioselectivity Factor
CF_3SO_3H	n-hexane, 30°C	7.6	51.6	7.5	0	33.3	0	0.77
$CF_3SO_3SiMe_3$ + CF_3SO_3H	n-hexane, 30°C	6.5	54.4	5.8	0	33.3	0	0.82
$BuMe_2SiOLi^a$	THF, 50°C	9.9	48.2	8.6	0	33.3	0	0.72
~$Me_2SiONMe_4^a$	Toluene, 30°C	13.3	48.2	5.1	6.5	26.8	0	0.72
~Me_2SiOMe_2SiOK + crown 18/6^a	Toluene, 50°C	6.6	56.7	3.4	4.4	28.6	0	0.85

[a]Data taken from ref. 14.

distribution is rather broad because of the permanent initiation and intermolecular chain coupling. An addition of $CF_3SO_3SiMe_3$ permits to control the molecular weight of the polymer, however, it does not seem to change its polydispersity. An extensive chain transfer to the Me_3SiO end group leads to a fast randomization of end groups.

Since the regioselectivity in the ring opening of **1** is poor, the polymer does not have any alternating arrangement of the Ph_2SiO and $(Me_2SiO)_2$ units. The ring is opened mostly between the two dimethylsiloxane unit, although the opening at Ph_2Si grouping also takes place.

Experimental Part

Chemicals. Methylene chloride, CH_2Cl_2, (Fluka, pure) was purified according to standard procedure (*34*). Then it was dried and distilled from CaH_2 to an ampule with a Rotaflo stopcock installed on a high vacuum line and used under dry argon.

n-Hexane, C_6H_{14}, (BDH) was purified according to standard procedure (*34*). Then, it was dried and distilled from metallic Na to an ampule with a Rotaflo stopcock and used under dry argon.

n-Dodecane, $C_{12}H_{26}$, (Aldrich, purity 99+%), was dried by refluxing with CaH_2 and distillation under dry argon to an ampule with a Rotaflo stopcock.

Trifluoromethanosulfonic acid, CF_3SO_3H, (Fluka) was dried and purified by distillation on a high vacuum line to glass bulb with known weight. The bulb was placed in an ampule with a glass hammer and a proper amount of CH_2Cl_2 was distilled to the ampule on a high vacuum line. Then, the glass bulb was broken by a glass hammer. The solution was used with a Hamilton precision syringe under dry argon.

Trimethylsilyl trifluoromethanesulfonate, $CF_3SO_3SiMe_3$, (Aldrich, 99%) dried and purified and then a solution in CH_2Cl_2 was prepared in the same way like the solution of trifluoromethanesulfonic acid in CH_2Cl_2.

Synthesis of Monomer. 1,1-diphenyl-3,3,5,5-tetramethylcyclotrisiloxane, **1**, was prepared by the reaction of 1,3-dichloro-1,1,3,3-tetramethyldisiloxane, with diphenylsilanediol, according to the procedure described earlier (*35*). The monomer was purified by recrystalization from n-heptane and distillation on a high vacuum line (10^{-3} mmHg, bp. 80-85°C).

Polymerization.

Synthesis of Polymers for Sequencing Studies. The polymerization of **1** was carried out under the atmosphere of prepurified argon in a thermostated glass Schlenk type reactor at 30°C. The catalyst solution in CH_2Cl_2 was introduced to the solution of monomer $[\mathbf{1}]_0$=1.2 mol·dm^{-3} in n-hexane and n-dodecane as GC standard, using a Hamilton precision syringe. The reaction was followed by GC analysis of the samples quenched by an excess of Et_3N. The reaction was stopped at the monomer conversion of 60-95%. The polymer solution was washed several times with water and then dried over $CaCl_2$. Cyclics were separated by repetitive dissolving of the polymer in CH_2Cl_2 and precipitating from methanol.

Synthesis of the Polymer for MALDI TOF Experiment. The polymerization of **1** was carried out under the atmosphere of prepurified argon in a thermostated (30°C) glass Schlenk type reactor. The catalyst solution: CF_3SO_3H in CH_2Cl_2 (10^{-3} mol·dm^{-3} in reaction mixture) and $CF_3SO_3SiMe_3$ ($5·10^{-2}$ mol·dm^{-3} in reaction mixture), was introduced to the solution of monomer, $[\mathbf{1}]_0$=1,2 mol·dm^{-3} in n-hexane and n-dodecane as GC standard, using a Hamilton precision syringe. The reaction was stopped at 85 % conversion by adding an excess of $(NH_4)_2CO_3$ in H_2O. The polymer was shaken with a new portion of $(NH_4)_2CO_3$ in water, then dried with $CaCl_2$. All the volatile components were evaporated on a high vacuum line.

Determination of Sequencing in Terminal Monomer Unit. The polymerization of **1** was carried out under the atmosphere of prepurified argon in a thermostated at 30°C glass Schlenk type reactor. Using a precision Hamilton syringe, the catalyst solution CF_3SO_3H in CH_2Cl_2 (10^{-1} mol·dm^{-3} of acid in reaction mixture) was added to monomer solution $[\mathbf{1}]_0$=0.8 mol·dm^{-3} in n-hexane and n-dodecane as GC standard. The reaction was followed by GC analysis. The reaction was stopped by adding an excess of Et_3N. The conversion of monomer was 82%. The polymer solution was washed several times with water and then dried over $CaCl_2$. All the volatile compounds were evaporated on a high vacuum line.

Analytical Procedure. ^{29}Si NMR spectra were taken with Bruker MSL 300 and Bruker DRX 500 Spectrometers. A good resolution for pentads and quantitative integration, as checked on model compounds, was achieved by an addition of Cr(acac)$_3$ and using a gated decoupling technique. The following compounds were integrated as model compounds: monomer **1** and 3,3-diphenyl-1,1,1,5,5,5-hexamethyltrisiloxane. An assignment of signals was made using the results of earlier studies (*16*).

SEC analysis was performed with a LDC Analytical RefractoMonitor IV and an LKB 2150 - HPLC Pump (RIDK) 102, using THF as solvent and PS as standards.

GC analysis was carried out using a Hewlett Packard HP 6890 Series GC system, working with thermal conductivity detector.

The MALDI TOF experiment was done with a Voyager Elite (Per Septive Biosystem), metric: DHB (THF), mode: reflector, laser N_2 337nm, pulse 3μs.

Calculations. The statistical analysis of the copolymer sequencing was performed in terms of Markov chain theory, using the 2nd order Markov chain model. The equations were numerically solved with respect to the conditional probabilities by least squares minimization of the deviations of calculated and measured pentad contributions.

In the Monte Carlo procedure the polymer chain growth was simulated, assuming the conditional probabilities of the three ways of monomer ring opening calculated from Markov analysis.

Ab initio calculations were carried out using standard techniques, as implemented in the Gaussian 94 series of programs (*37*). The equilibrium geometries were calculated at the Hartree-Fock level, using the polarized 3-21G* basis set (*38*).

Acknowledgments

This research was supported by the Committee for Scientific Studies (KBN) grants 3TO9B 08108 and 3TO9A 03015.

Literature Cited

1. Yilgör, I.; Mc Grath, J. E. *Adv. Polym. Sci.* **1988**, *86*, 1.
2. Wagener, K. B.; Zulunga, F.; Wanigatunga, S. *Trends Polymer Sci.* **1996**, *4*, 157.
3. Nakagawa, Y.; Miller, P.; Pacis, C.; Matyjaszewski, K. *ACS Polymer Preprints* **1997**, *38*, 701.
4. Molenberg, A.; Sheiko, S.; Möller, M. *Macromolecules* **1996**, *29*, 3397.
5. Chang, T. C.; Chen, H. B.; Chiu, Y. S. *J. Polym. Sci. Part A* **1996**, *34*, 2613, 3313.
6. Chojnowski, J. In *Siloxane Polymers* Clarson, S. J.; Semlyen, J. A., Eds., Ellis Horwood PTR Prentice Hall: New Jersey, 1993, pp.3-71.
7. Chojnowski, J. *J. Inorg. Organometal. Polym.* **1991**, *1*, 199.
8. Jungst, C. D.; Weber, W. P. *J. Polym. Sci. Part A* **1987**, *25*, 1967.
9. Risch, B. G.; Rodriguez, D. F.; Lyon, K.; Mc Grath, J. E.; Wilkes, G. L. *Polymer J.* **1996**, *37*, 1129.
10. Ziemielis, M.J.; Saam, J.C. *Macromolecules* **1989**, *22*, 2111.
11. Graczyk, T.; Lasocki, Z. *Bull. Acad. Polon. Sci., Ser. Sci. Chim.* **1979**, *27*, 181.
12. Olah, G.A; Li, X.-Y.; Wang, Q.; Rasul, G.; Prakash, G.K.S. *J. Am. Chem. Soc.* **1995**, *117*, 8962.
13. Wang, Q.; Zhang, H.; Prakash, G.K.S.; Hogen-Esch, T.E.; Olah, G.A. *Macromolecules* **1996**, *29*, 6691.
14. Kaźmierski, K.; Cypryk, M.; Chojnowski, J. *Makromol. Chem., Macromol. Symp.* **1998**, *132*, 405.
15. Chojnowski, J. Cypryk, M. In *Polymeric Materials Encyclopedia*; Salamone, J.C., Ed., CRC Press: Boca Raton, 1996, Vol. 2, p. 1682-1695.
16. Kennan, J. J. In *Siloxane Polymers*; Clarson, S. J.; Semlyen, J. A., Eds., Ellis Horwood PTR Prentice Hall: New Jersey, 1993, Chapter 2, p. 72-134.
17. Kobayashi, H.; Nishiumi, W. *Makromol. Chem.* **1993**, *193*, 1403.
18. Baratova, T. N.; Mileshkevich, W. P.; Gurari, V. I. *Vysokomol. Soedin.* **1983**, *125*, 2497.
19. Rózga-Wijas, K.; Chojnowski, J.; Zundel, T.; Boileau, S. *Macromolecules* **1996**, *29*, 2711.
20. Matyjaszewski, K.; Chen, Y. L. *J. Organometal. Chem.* **1988**, *340*, 7.
21. Uhlig, W. *Organometallics* **1994**, *13*, 2843.
22. Dagger, A. G.; Semlyen, J. A. *Polymer* **1998**, *39*, 2621.
23. Sauvet, G.; Lebrun, J. J.; Sigwalt, P. In *Cationic Polymerization and Related Processes*; Goethals, E. J., Ed., Academic Press: London, 1984, p. 237-251.
24. Rubinsztajn, S.; Cypryk, M.; Chojnowski, J. *Macromolecules* **1993**, *26*, 5389.
25. Chojnowski, J. Wilczek, L. *Makromol. Chem.* **1979**, *180*, 117.
26. Nicole, P.; Masure, M.; Sigwalt, P. *Macromol. Chem. Phys.* **1984**, *195*, 2327.

27. Toskas, G.; Besztercey, G.; Moreau, M.; Masure, M. *Macromol. Chem. Phys.* **1995**, *196*, 2715.
28. Sigwalt, P.; Gobin, C.; Nicol, P.; Moreau, M.; Masure, M. *Macromol. Chem. Phys. Macromol. Symp.* **1991**, *42/43*, 229.
29. Chojnowski, J.; Ścibiorek, M. *Makromol. Chem.* **1976**, *177*, 1413.
30. Kaźmierski, K.; Fortuniak, W.; Cypryk, M.; Chojnowski, J. in preparation.
31. Authors are sorry that the values of P_a and P_b in the Table reported in Polymer Preprints (*36*) were interchanged.
32. Wilczek, L.; Rubinsztajn, S.; Chojnowski, J. *Makromol. Chem.* **1986**, *187*, 39.
33. Bischoff, R.; Sigwalt, P. *Macromol. Chem. Phys.*, in press.
34. Perrin, D. D.; Armarego, W. L. F.; Perrin, D. F. *Purification of Laboratory Chemicals*; 2 Edn., Pergamon Press, Oxford, 1980.
35. Mazurek, M.; Ziętera, J.; Sadowska, W.; Chojnowski, J. *Makromol. Chem.* **1980**, *181*, 777.
36. Kaźmierski, K.; Cypryk, M.; Chojnowski, J. *ACS Polymer Preprints* **1998**, *39(1)*, 439.
37. Frisch, M. J.; Trucks, G. W.; et al. *Gaussian 94, Revision C.2*; Gaussian, Inc.: Pittsburgh, PA, 1995.
38. Hehre, W. J.; Radom, L.; Schleyer, P. v. R.; Pople, J. A. *Ab Initio Molecular Orbital Theory*; J. Wiley & Sons: New York, 1986.

Chapter 4

Hydrogenated and Deuterated Cyclic Poly(dimethylsiloxanes)

Anthony C. Dagger and J. Anthony Semlyen

Department of Chemistry, University of York, York, YO10 5DD, United Kingdom

Poly(dimethylsiloxanes) (PDMS) based on the repeat unit $[(CH_3)_2SiO]$ are important commercial polymers with a wide range of applications, for example as adhesives, surfactants, lubricants, sealants, release agents etc. The polymers may be linear, branched or cross-linked into network structures when based on long chain molecules (see Figure 2).

At York, we prepared sharp fractions of the first synthetic cyclic polymers over twenty years ago, following the preparative investigations of Scott [1], Carmichael and his coworkers [2],[3] and Brown and Slusarczuk [4] in cyclic poly(dimethylsiloxane) chemistry. The cyclic polymers that we obtained using preparative gel permeation chromatography (GPC) [5] were narrow molar mass fractions of cyclic PDMS $[(CH_3)_2SiO]_x$ with number average numbers of skeletal bonds up to and beyond 1000. The fractions were obtained from macrocyclic populations recovered from ring-chain equilibration reactions carried out under high dilution conditions. The dispersities of the fractions (as expressed by the ratio of the mass average and number average molar masses, M_w/M_n) were typically about 1.05. The scale of the preparations is illustrated by the fact that a narrow molar mass fraction of cyclics with an average of more than 100 skeletal bonds was obtained on a 30 g scale.

The properties of the novel cyclic PDMS fractions were investigated in joint collaborative work with twenty other University research groups. They included, for example, small angle neutron scattering (SANS) studies (with Imperial College and Bristol), laser light scattering investigations (with Freiburg) and ultrasonic relaxation measurements (with Strathclyde). All these investigations led to a complete characterisation of the cyclic polymers. A further demonstration of their cyclic nature was obtained by entrapping the rings into networks giving the first topological polymers. For example, cyclics with 500 skeletal bonds were entrapped in over 90 % yield (see, for example, Ref [6], [7]).

The corresponding cyclic oligomers and polymers from other polysiloxane systems were also prepared and characterised by our group at York including $[R(CH_3)SiO]_x$, where R = H [8], CH_2=CH [9] and C_6H_5 [10],[11]. We then went on to prepare the first cyclic polymer liquid crystals starting with the cyclic $[H(CH_3)SiO]_x$ materials. For all these studies the corresponding linear PDMS materials or fractions were prepared for purposes of comparison.

In this chapter we show how hydrogenated cyclic PDMS is prepared, characterised and investigated. These studies form the basis of our preparation and characterisation of the first deuterated cyclic polymers [12] [13] [14]. These are per-deuterated PDMS $[(CD_3)_2SiO]_x$ and they were obtained by a modification of the method described by Beltzung and coworkers [15]. Preparative GPC was used to prepare sharp fractions in a similar way to the hydrogenated materials.

Hydrogenated Cyclic PDMS.

When PDMS is produced by acid or base catalysed polymerisation from octamethylcyclotetrasiloxane (D_4) a thermodynamic equilibrium between ring and chain molecules results:

$$-M_y- \leftrightharpoons -M_{y-x}- + M_x \qquad (1)$$

-where $-M_y-$ and $-M_{y-x}-$ represent linear species, M_x represents an x-meric ring species and M represents one monomer unit.

The ring-chain equilibration in PDMS was first investigated by Scott [1] who obtained data giving information on the molecular size distribution of ring and chain species in solution and the undiluted state. This has been followed by the work of Hartung and Camiolo [16] and co-workers [17] [18] [19] [20] extending the investigations to other solvent systems. At this time, kinetic studies of the base-catalysed polymerisation of D_4 by Grubb and Osthoff [21], and Morton and Bostick [22] confirmed that the active species was in fact the silanoate ion (Si-O$^-$ K$^+$).

The understanding of PDMS ring-chain equilibria was much advanced when Brown and Sluzarczuk [4] obtained evidence for the presence of macrocyclic species in the base catalysed reaction. They attempted to characterise their distribution and demonstrated the existence of a continuous macrocyclic population extending to at least D_{400}. The work of Wright and Semlyen [8] [23] [24] has since extended the investigations further, obtaining accurate molar cyclisation equilibrium constants from the concentrations of the cyclic species over a large range of ring sizes through calculations based on the cyclisation theory of Jacobson and Stockmayer [25].

Jacobson And Stockmayer Cyclisation Theory. The cyclic populations in ring-chain equilibrates may be expressed in terms of the molar cyclisation equilibrium constants K_x for the x-meric ring molecules M_x as follows [26]:

$$K_x = \frac{[-M_{y-x}-][M_x]}{[-M_y-]} \quad (2)$$

For a most probable distribution of chain lengths in the linear part of the equilibrate then:

$$K_x = \frac{[M_x]}{p^x} \quad (3)$$

-where p is the extent of reaction of functional groups in the linear polymer. As p is usually close to unity, the K_x values for all but the largest cyclic species may be taken to be equal to the molar concentrations of the cyclic species (in mol dm^{-3}).

A chain undergoing cyclisation must be long enough and flexible enough to obey Gaussian statistics. Kuhn [27] was the first to state that random coil polymer chains would obey the Gaussian expression:

$$W_x(\underline{r}) = (3/2\pi<r_x^2>)^{3/2} \exp(-3r^2/2<r_x^2>) \quad (4)$$

-for the density $W_x(\underline{r})$ of their end-to-end vectors $\underline{r}$, and where $<r_x^2>$ is the mean-square end-to-end distance of an x-meric chain.

Having calculated the fraction of conformations which are suitable for ring formation using Kuhn's Gaussian expression [27], the cyclisation process is then considered as the fixing of one of the chain ends and calculating the density of the distribution that the other end will be at a distance from the fixed atom such that a ring-forming conformation is assumed, i.e. $r = 0$ for the long chain molecule obeying Gaussian statistics:

$$W_x(\underline{0}) = (3/2\pi<r_x^2>)^{3/2} \quad (5)$$

In the Jacobson and Stockmayer cyclisation theory [25] the molar cyclisation equilibrium constants are given by:

$$K_x = (3/2\pi<r_x^2>)^{3/2}(1/N_A\sigma_x) \quad (6)$$

-where N_A is the Avogadro number and σ_x is a symmetry number that represents the number of skeletal bonds of the cyclic that can be opened by the catalyst (2x in the case of PDMS).

The Jacobson and Stockmayer theory [25] gives a simple theoretical expression for the molar cyclisation equilibrium constants for macrocyclics in ring-chain equilibrates. Thus Equation (6) can be used to calculate K_x values provided the corresponding $<r_x^2>$ values are known. Conversely, $<r_x^2>$ values can be obtained by simply measuring K_x values (shown to be equal to molar concentrations).

Figure 1 shows a comparison between the theoretical and experimental data for the macrocyclisation equilibrium constants for PDMS.

Deviations from the theoretical values are observed with the enhanced production of the small cyclics (particularly the tetramer) due to the increased probability of the termini being close together with the correct orientation caused by the favourable conformation of the small molecules [28]. Deviation at high values of x are observed for PDMS in toluene and this has been attributed to excluded volume effects.

Reasons For Studying Ring Macromolecules.

Apart from demonstrating the simple fact that cyclic polymers can be synthesised and isolated, physical studies of such systems allow fundamental investigations into the properties of cyclic macromolecules. Many theoretical predictions have been made with regards to the physical behaviour of a ring compared to that of a corresponding chain. Since PDMS ring-chain equilibrations offer a means of preparing a range of

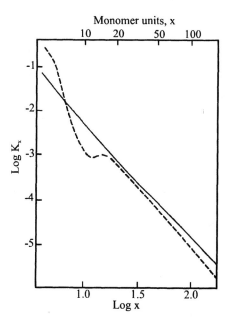

Figure 1: Molar cyclisation equilibrium constants K_x (in mol dm^{-3}) for cyclics $[(CH_3)_2SiO]_x$ in a ring-chain equilibrate in toluene solution (broken line) compared with theoretical values calculated by Flory and Semlyen [28](solid line). (Reproduced from ref. [28]).

ring macromolecules, it is possible to obtain physical experimental data which can be used to determine the validity of these predictions.

It is of interest to note that many natural forms of deoxyribonucleic acid (DNA) have been shown to be circular macromolecules [29]. Furthermore, a detailed knowledge of the properties and molecular conformations of cyclic polymers should lead to a better understanding of the behaviour of linear polymers, as well as giving new insights into the properties of rubbers, gels and other systems where closed loops are known to exist.

Cyclic polymers offer routes to many more different topologies than linear polymers. The number of different architectural structures possible with cyclic polymers are shown in Figure 2 in comparison to those possible with linear polymers.

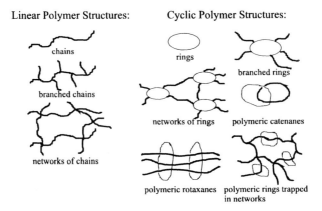

Figure 2: Structural comparison of linear and cyclic polymers.

Additionally, one obvious reason for studying ring macromolecules is to determine if PDMS rings have any particular physical properties which might be industrially applicable. This can only be achieved by building up a detailed picture of the physical properties of PDMS over a range of ring sizes, and defining any major differences which appear with respect to linear PDMS which has already been shown to be exceedingly commercially important.

Preparation Of Hydrogenated Linear And Cyclic PDMS.

As mentioned above, linear PDMS may be prepared by the acid or base catalysed polymerisation of octamethylcyclotetrasiloxane (D_4). A controlled amount of $(CH_3)_3Si$-O-$Si(CH_3)_3$ (hexamethyldisiloxane) is added to cap the reactive end groups resulting in trimethylsilyl end-terminated chains.

Cyclic PDMS is also synthesised by acid or base catalysed polymerisation of D_4, however, the reaction is now carried out at high dilution (and in the absence of any terminating groups). The methods of Brown and Slusarczuk [4] and Chojnowski et al. [30] [31] [32] [33] (see below) have been well established by the group at York and details of both are given below. Both methods establish a ring/chain equilibrium forming both linear and cyclic PDMS. Once equilibrium has been reached the catalyst is neutralised, end groups maybe added to terminate the linear PDMS chains and the cyclics may be separated out.

Brown And Slusarczuk Base Catalysis. The Brown and Slusarczuk [4] method of preparing cyclic PDMS involves an anionic i.e. base catalysed ring-chain equilibrium. The mechanism for the anionic polymerisation of D_4 is shown in Figure 3.

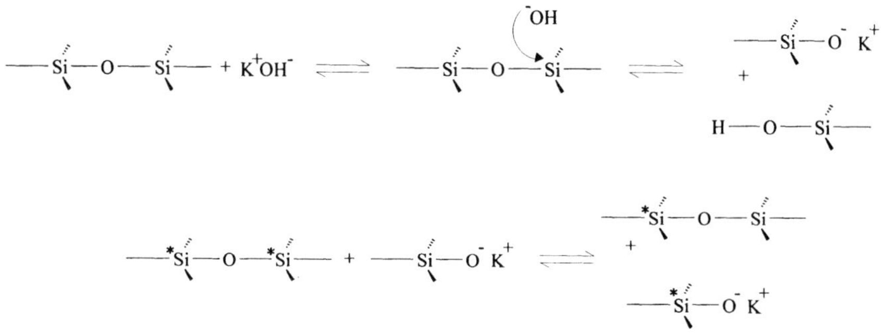

Figure 3: Brown and Slusarczuk reaction mechanism. The * denotes a second siloxane species in the reaction.

Potassium hydroxide is used as the base in the presence of distilled diglyme (2-methoxyethylether). The diglyme complexes the potassium ions and promotes the formation of hydroxide ions. The D_4 starting material must be dried and distilled from calcium hydride prior to use. The reaction is refluxed at 384 K in sodium dried toluene for 14 days under a nitrogen atmosphere. Dry reaction conditions must be maintained since the presence of any water terminates the reaction.

Once the reaction has come to equilibrium, it is quenched by the addition of glacial acetic acid. The cyclic species can then be extracted from the reaction mixture.

Chojnowski Acid Catalysis. The Chojnowksi [30] [31] [32] [33] reaction involves cationic i.e. acid catalysed polymerisation of D_4. The mechanism for the cationic polymerisation is shown in Figure 4. The acid used for the reaction is trifluoromethanesulphonic (triflic) acid. Trimethylsilyltrifluoromethane sulphonate is added to assist the acid as suggested by Lebrun, Sauvet and Sigwalt [34]. The D_4 is dried as in the Brown and Slusarczuk reaction but the dichloromethane (methylene chloride) solvent is not, as in this case a trace amount of water is required for the reaction to proceed. The reaction is carried out at room temperature under a nitrogen atmosphere (to prevent the introduction of excess water) for up to 48hrs.

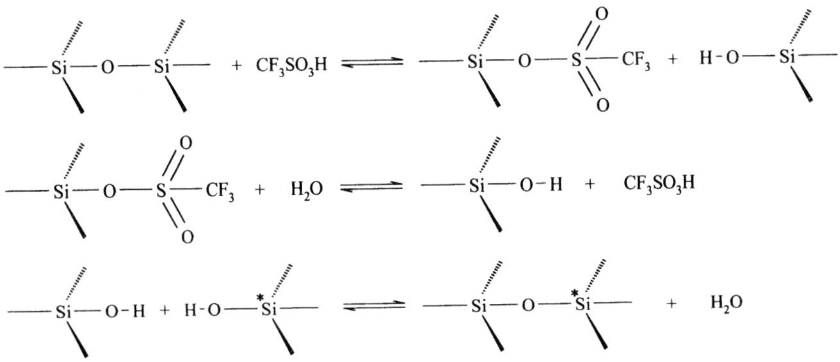

$$\text{—Si—O—Si—} + CF_3SO_3H \rightleftharpoons \text{—Si—O—S(=O)(=O)—CF}_3 + H\text{—O—Si—}$$

$$\text{—Si—O—S(=O)(=O)—CF}_3 + H_2O \rightleftharpoons \text{—Si—O-H} + CF_3SO_3H$$

$$\text{—Si—O-H} + H\text{-O—Si*—} \rightleftharpoons \text{—Si—O—Si*—} + H_2O$$

Figure 4: Chojnowski reaction mechanism. Once again, the * denotes a second siloxane species taking part in the reaction.

Again, once equilibrium has been reached the reaction is quenched with copious amounts of sodium carbonate (to neutralise the acid) and the cyclics may then be extracted.

Both of the above mechanisms are complex processes with polymerisation, depolymerisation, repolymerisation and chain transfer all occurring. Since all the siloxane bonds present in the reaction mixture will stand an equal chance of undergoing reaction whether they are present in D_4, oligomers or polymer a ring/chain equilibrium is established.

Separation Of Linear and Cyclic PDMS. The cyclic PDMS was separated from the linear PDMS by three refluxes at 10% weight by volume in acetone. The cyclics are soluble in acetone, the linears are not and so precipitated out on cooling overnight to room temperature and were removed using a separating funnel. The separated cyclics were analysed by GPC and gas liquid chromatography (GLC) and some of the low cyclic oligomers and D_4 starting material were removed by careful vacuum distillation and washing with methanol (methanol is a good solvent for the smaller cyclic species but a poor solvent for the larger cyclics).

Preparative Gel Permeation Chromatography.

The ability to prepare a series of sharp fractions (i.e. fractions with a dispersity as close to unity as possible) of a variety of molar masses is vital in order to investigate how a number of properties of a polymer scale as a function of chain length [35].

Sharp fractions of cyclic and linear PDMS containing species with up to 100 skeletal bonds can be prepared conveniently by vacuum fractional distillation. Such fractions have dispersities (M_w / M_n) which are typically 1.03 ± 0.02. Larger cyclic and linear PDMS can be fractionated using classical solution fractionation (for example acetone and water mixtures may be used to fractionally precipitate the polymers [36]). This can be very time consuming and broad fractions often result.

GPC may be used to prepare sharp fractions of cyclic and linear PDMS with number average numbers of skeletal bonds in the range $100 < n_n < 1000$. The preparative GPC instrument used for this purpose is described below. The cyclic and linear PDMS fractions obtained this way typically have dispersities (M_w / M_n) of 1.05 ± 0.05, considerably lower than those obtained using conventional fractionation techniques.

Preparative GPC Instrument. Preparative scale GPC units typically involve the use of much larger separation columns, greater solvent volumes and relatively larger amounts of sample than conventional analytical systems. The Preparative GPC instrument used to prepare sharp fractions of polymers was designed and constructed by Mr. D. Sympson at the workshops of the University of York Chemistry Department [5]. A block diagram of the instrument is shown in Figure 5.

Toluene was used as the chromatographic solvent and was delivered from the tank(1) to the still(4) via a constant level device(3). The distilled solvent was pumped through the system by an adjustable pressure pump(5) - the available solvent flow rates being in the range 10 - 180 cm^2 min^{-1} at pressures of up to 100 psi. Adjustable bellows(6) damped the solvent flow and an overpressure release valve(7) was available to control the solvent pressure. The solvent flow was then split into reference and sample streams after passing through a filter(8). The reference stream was only opened at the start of a run and passed through the reference cell of a Waters R4 'Preparative' differential refractometer(12). The sample stream was directed through a 6-port injection valve(9) having a sample loop with nominal capacity of 10 cm^3. The sample loading for each fractionation was 1-2 g of polymer, which was dissolved in ~10 cm^3 toluene. The solution was then filtered using a 0.45 μm filter (Millipore Ltd.) and transferred to the injection loop manually using a 10 cm^3 syringe. The sample stream could then be passed through either column A(10) or column set B(11) or through both in series. Column A was ~120 cm in length and ~5 cm internal diameter. Column set B consisted of two columns each of length ~30 cm and ~5 cm internal diameter in series. All of the columns were packed and supplied by Waters Associates Ltd.. Column A had $^2/_5$ of the column packed with Styragel (cross-linked polystyrene) of nominal porosity 100 nm and the remaining $^3/_5$ was packed with Styragel of nominal porosity 300 nm. Column set B was packed with μ-Styragel 103.

46

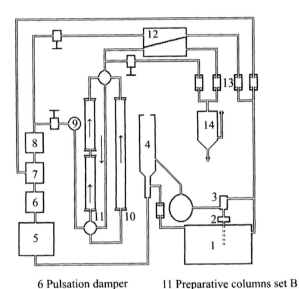

1 Solvent tank	6 Pulsation damper	11 Preparative columns set B
2 Still pump	7 Pressure control valve	12 Differential refractometer with
3 Constant level device	8 Filter	chart recorder
4 Still	9 Sample injection valve	13 Solvent and eluent flow sight glasses
5 Pressure pump	10 Preparative column A	14 Solvent counter and fraction cutter

Figure 5: Schematic of the preparative GPC instrument.

After eluting through the columns, part of the sample flow was directed through the differential refractometer detector, the response being plotted on a Chessell chart recorder. The sample flows were then recombined and the automatic fraction cutter(14) was then used to split the column output volumetrically in to fractions of up to 50 cm^3 (each fraction corresponding to one count) and the fraction cuts were automatically deposited into individual collection containers for subsequent work-up.

"Express train" injections were made when quantities of polymer were fractionated. This technique allowed the injection of a second sample into the system before the complete fractionation of the previous sample, so reducing the time required for several fractionations. A minimum time period was chosen between injections so as to avoid the overlapping of peaks. After a number of fractionation runs the solvent was removed from each fraction by rotary evaporation and the polymer sample recovered.

Characterisation of the Hydrogenated Cyclic PDMS Fractions.

The fractions of cyclic PDMS obtained from the preparative GPC have been analysed using a variety of analytical techniques besides GPC. The results of some of these analyses are given below.

Gas Liquid Chromatography. The lower molecular weight cyclic PDMS fractions have been analysed using a PYE series 104 GLC instrument fitted with a heated katharometer detector. The traces obtained for three fractions are shown in Figures 6, 7 and 8. The traces show that despite being sharp fractions with dispersities $M_w / M_n < 1.05$ the fractions consist of a range of individual ring sizes. Also clear from the traces are the lack of linear siloxanes which have slightly different retention times to the cyclic siloxanes and the lack of other volatile impurities.

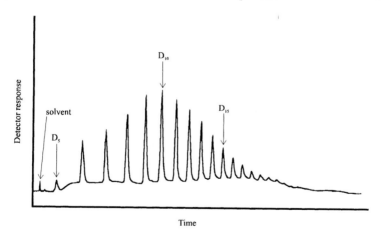

Figure 6: GLC trace of a hydrogenated cyclic PDMS fraction containing rings with 10 to ~42 skeletal bonds.

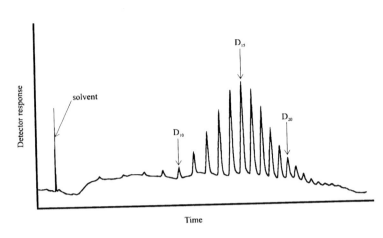

Figure 7: GLC trace of a hydrogenated cyclic PDMS fraction containing rings with 16 to ~52 skeletal bonds.

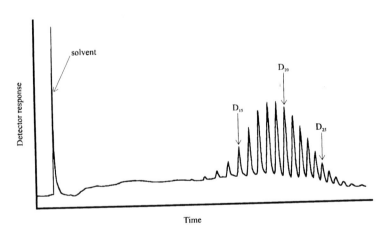

Figure 8: GLC trace of a hydrogenated cyclic PDMS fraction containing rings with 24 to ~60 skeletal bonds.

²⁹Si Nuclear Magnetic Resonance Spectroscopy (NMR). Two of the lower molecular weight cyclic PDMS fractions have been studied using ²⁹Si NMR. The results are shown in Figures 9 and 10. The samples were run in bulk on a Jeol 500 Mhz instrument and show a very interesting effect first observed by Burton et al. [37]. The individual rings can be resolved up to 15 repeat units in a ring. The silicon atoms in a given ring are equivalent, however the different conformations possible in rings of different sizes results in the silicons in the different sized rings experiencing different environments and hence displaying different chemical shifts in the NMR experiment. The minimum chemical shift (σ) occurs for D_8. The peaks then coalesce to the D_{15} peak. Monte-Carlo calculations [38] [39] have shown that marked changes in the shapes of the ring occur up to the D_{15} case, beyond this essentially limiting behaviour has been reached.

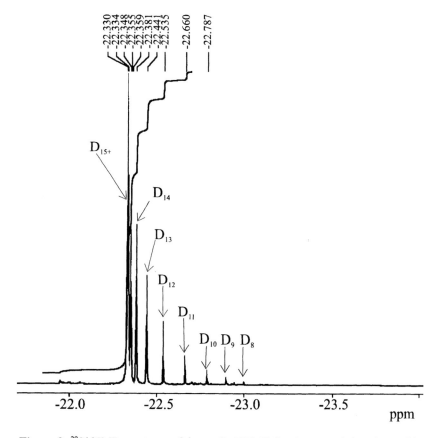

Figure 9: ²⁹Si NMR spectrum of the cyclic PDMS fraction containing rings with 10 to ~42 skeletal bonds.

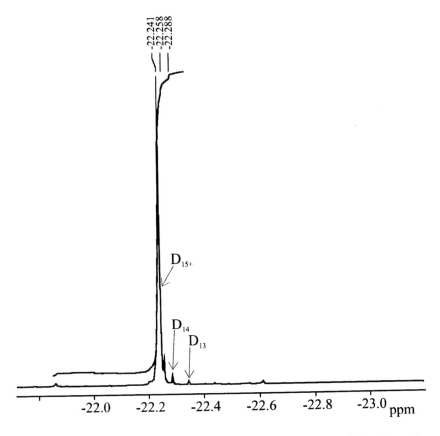

Figure 10: ^{29}Si NMR spectrum of the cyclic PDMS fraction containing ring with 16 to ~52 skeletal bonds.

One further point to note is that comparing the NMR results with the GLC traces for the same fractions it is noticed that the NMR peaks appear to agree quantitatively as well as qualitatively with the GLC peaks. This suggests that the high resolution NMR technique may be useful for analysing the smaller cyclic oligomers (up to and including 15 repeat units).

Low Resolution Electron Ionisation Mass Spectroscopy (EIMS). One of the cyclic PDMS fractions has been analysed using EIMS (see Figure 11). The ionisation pattern again agrees well with the GLC and NMR results above. The clean fragmentation pattern (see Table I) shows the purity of the sample obtained and all of the fragments can be accounted for in terms of rearrangements of the cyclic siloxanes which have been observed previously [40] [41].

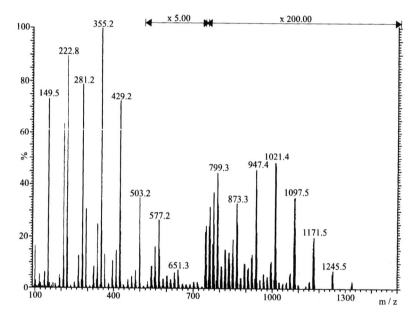

Figure 11: Electron ionisation mass spectrum of a hydrogenated cyclic PDMS fraction.

Table I: Peak assignments for the EIMS spectrum of a hydrogenated cyclic PDMS fraction.

m/z	Composition	Species
281	$Si_4O_4C_7H_{21}$	$(D_4 - CH_3)^+$
355	$Si_5O_5C_9H_{27}$	$(D_5 - CH_3)^+$
429	$Si_6O_6C_{11}H_{33}$	$(D_6 - CH_3)^+$
503	$Si_7O_7C_{13}H_{39}$	$(D_7 - CH_3)^+$
577	$Si_8O_8C_{15}H_{45}$	$(D_8 - CH_3)^+$
651	$Si_9O_9C_{17}H_{51}$	$(D_9 - CH_3)^+$
725	$Si_{10}O_{10}C_{19}H_{57}$	$(D_{10} - CH_3)^+$
799	$Si_{11}O_{11}C_{21}H_{63}$	$(D_{11} - CH_3)^+$
873	$Si_{12}O_{12}C_{23}H_{69}$	$(D_{12} - CH_3)^+$
947	$Si_{13}O_{13}C_{25}H_{75}$	$(D_{13} - CH_3)^+$
1021	$Si_{14}O_{14}C_{27}H_{81}$	$(D_{14} - CH_3)^+$
1095	$Si_{15}O_{15}C_{29}H_{87}$	$(D_{15} - CH_3)^+$
1169	$Si_{16}O_{16}C_{31}H_{93}$	$(D_{16} - CH_3)^+$
1243	$Si_{17}O_{17}C_{33}H_{99}$	$(D_{17} - CH_3)^+$

Investigations of Hydrogenated Cyclic and Linear Poly(dimethylsiloxanes).

As has already been shown, the PDMS ring-chain equilibration provides an excellent route to preparing a wide range of cyclic and linear polymers. To this extent, a large number of studies of both the physical and chemical properties of cyclic PDMS have been carried out and compared with the corresponding linear PDMS. Some examples of these studies are detailed below.

Dimensions Of Rings And Chains In Dilute Solution By Small Angle Neutron Scattering. Theoretical predictions of the z-average mean-square radii of gyration ($<s^2>_{z,c=0}$) for flexible linear and cyclic polymers with the same number of skeletal bonds in dilute solution at the θ-point, which are unperturbed by excluded volume effects, predict a ratio of $<s_l^2>_{z,c=0} / <s_r^2>_{z,c=0} = 2.0$ [42] [43] [44] (where l refers to linear or chain molecules and r refers to ring or cyclic molecules).

Using small angle neutron scattering (SANS) Higgins et al. [45] measured $<s^2>_{z,c=0}$ for four linear and cyclic PDMS polymers of similar molar masses in dilute benzene-d_6 solution at 292 K. A value for the ratio $<s_l^2>_{z,c=0} / <s_r^2>_{z,c=0}$ of 1.9 ± 0.2 in the region $n_z = 500$ (where n_z is the z-average number of bonds) was obtained in good agreement with the theoretically predicted value.

Bulk Viscosity. The bulk viscosities η of sharp fractions of cyclic and linear PDMS have been measured over a wide molar mass range [46] [47] and are shown in Figure 12.

At low molar masses, the viscosities of the cyclics were found to be considerably higher than those of the corresponding linears. The difference between the values decreased with increasing molar mass until a cross-over was observed at $n_n = 100$ (where n_n is the number-average number of bonds). At higher molar masses the ratio η_r / η_l was found to be approximately 0.5 [46] in agreement with the theory of Bueche [48].

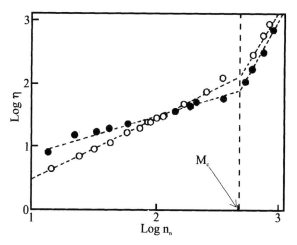

Figure 12: Plot of the logarithm of the viscosity η against the logarithm of the number-average number of skeletal bonds n_n for ring (solid circles) and chain (open circles) PDMS fractions at constant $T - T_g$, where T_g is the glass transition temperature. (Reproduced from ref. [46]).

Values for the critical molar mass for entanglement M_c were found to be ~1.70 x 10^4 g mol^{-1} and ~1.66 x 10^4 g mol^{-1} for ring and chain PDMS respectively. Above this value, the first found for such flexible cyclic molecules, reptation models [49] [50] [51] predict that the motion of linear polymers resemble those of snakes and only the ends of the chains have freedom of motion, moving through 'tubes' created by the entanglements of other molecules. Such models predict that cyclic molecules should be considerably more viscous than the corresponding linear molecules and might even be infinitely viscous. It was found; however, that above M_c the ring polymers are markedly less viscous than the chain polymers [47] and it was concluded that the motion in bulk PDMS does not appear to depend on the movement of chain ends.

Doi [50] has since proposed a tube renewal concept adapted from the melt characteristics of star branched polymers which assumes the motion of rings to be a result of the rearrangement of the tube rather than by the reptation of the polymer down the tube.

Glass Transition Temperature. The thermal behaviour of sharp fractions of linear and cyclic PDMS have been investigated using differential scanning calorimetry [52]. In Figure 13 the glass transition temperatures (Tg) of cyclic and linear PDMS are shown plotted against M_n^{-1}. Linear PDMS shows the expected decrease in Tg with increase in M_n^{-1}. Cyclic PDMS however exhibits the opposite trend. Di Marzio and Guttman [53] have modified the theory of Gibbs and Di Marzio [54] in order to interpret the Tg of these polymers.

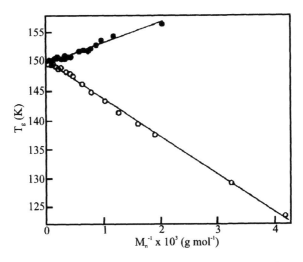

Figure 13: A plot of glass transition temperature (T_g) against reciprocal number-average molar mass (M_n^{-1}) for cyclic (solid circles) and linear (open circles) dimethylsiloxanes. (Reproduced from ref. [52]).

Density and Refractive Index. The densities (ρ) and refractive indices (n_D) of linear and cyclic PDMS have been measured [55] and are shown in Figures 14 and 15 respectively.

There are well defined maxima for the cyclic polymers in both plots at n_n = 22. No such maxima are shown by the chain molecules. As a results of the different bond angles at the silicon and oxygen atoms of the polysiloxane chain, a cyclic molecule with n_n = 22 can adopt a low energy conformation in the all-trans state. Consequently, the ring has a planar disc-shaped character and such rings can pack more efficiently than spherical molecules, giving rise to the increase in density and refractive index.

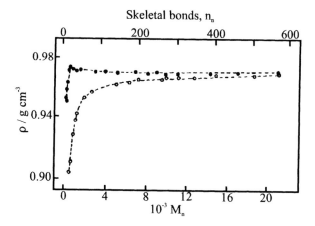

Figure 14: Densities (ρ) of cyclic (solid circles) and linear (open circles) PDMS at 298 K plotted against number-average molar mass. (Reproduced from ref. [55]).

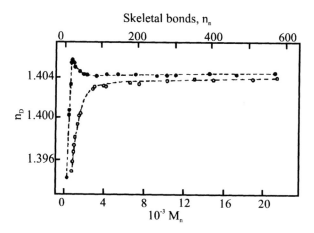

Figure 15: Refractive indices (n_D) of cyclic (solid circles) and linear (open circles) PDMS at 298 K plotted against number-average molar mass. (Reproduced from ref. [55]).

Silicon-29 Nuclear Magnetic Resonance (NMR) Chemical Shift. A Gas Liquid Chromatography (GLC) trace of a cyclic PDMS fraction containing rings - $[(CH_3)_2SiO]_n$ - (D_n) with n = 7 - 19 is shown in Figure 16 [37].

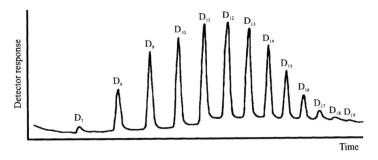

Figure 16: GLC trace for a cyclic fraction containing rings D_x with x = 7-19. (Reproduced from ref. [37]).

The ^{29}Si NMR spectrum for a similar fraction is shown in Figure 17.

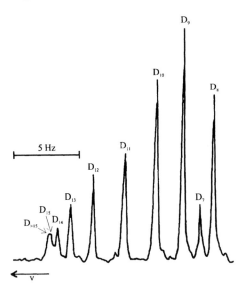

Figure 17: 19.87 MHz ^{29}Si - $\{^1H\}$ NMR spectrum of a cyclic dimethylsiloxane fraction. (Reproduced from ref. [37]).

All silicon atoms in a given ring have the same chemical shift and there are separate resonances for individual rings up to D_{15} where the shifts coalesce to the same value as that for linear PDMS.

Adsorption On Surfaces. The amounts of cyclic and linear PDMS adsorbed on to silica surfaces have been investigated by Fourier transform infrared spectroscopy (FT-IR) [56]. It has been predicted [57] that at low molar masses the adsorption of a cyclic polymer should be greater than that of the corresponding linear polymer, but at high molar masses the reverse should be the case. The predicted crossover was observed (see Figure 18) for PDMS adsorbed onto silica from hexane at ~470 skeletal bonds.

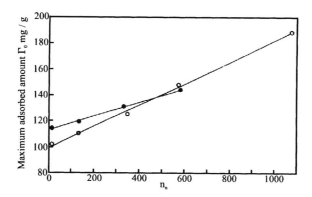

Figure 18: Maximum adsorbed amount Γ_0 plotted as a function of the number average number of skeletal bonds n_n for cyclic (solid circles) and linear (open circles) PDMS adsorbed on silica from hexane. (Reproduced from ref. [56]).

Entrapment Of Rings In To Networks. As discussed earlier, ring molecules are fundamentally different from long chain molecules in that they can form novel topological structures such as catenanes. Investigations have been carried out trapping cyclic polymers into network and rubber structures [58] [59] [60] [61] [62] [63]. The percentage of large rings trapped in to a PDMS network as a function of the number of skeletal bonds n_n is shown in Figure 19.

Deuterated Cyclic PDMS.

Introduction. Deuterated PDMS has previously been obtained from the polymerisation of perdeuterated D_4. This was reported to have been synthesised by Beltzung and co-workers [15] from perdeuterated methanol. The synthesis converts the methanol into methyl iodide via a classical method. The Beltzung reaction scheme is shown in Figure 20:

100
75
50
25
0
Cyclics trapped (%)
0
200
400
600
n_n

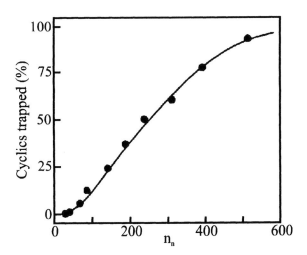

Figure 19: Percentage of cyclic PDMS trapped in a PDMS network, where n_n represents the number of skeletal bonds. (Reproduced from ref. [60]).

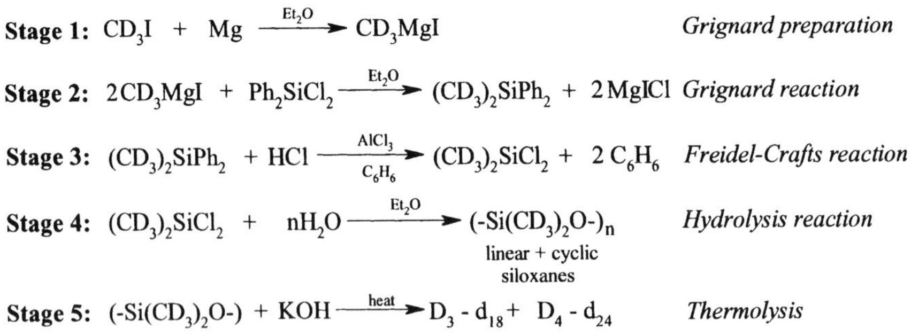

Stage 1: CD_3I + Mg $\xrightarrow{Et_2O}$ CD_3MgI
Grignard preparation
Stage 2: $2CD_3MgI$ + Ph_2SiCl_2 $\xrightarrow{Et_2O}$ $(CD_3)_2SiPh_2$ + 2 MgICl
Grignard reaction
Stage 3: $(CD_3)_2SiPh_2$ + HCl $\xrightarrow[C_6H_6]{AlCl_3}$ $(CD_3)_2SiCl_2$ + 2 C_6H_6
Freidel-Crafts reaction
Stage 4: $(CD_3)_2SiCl_2$ + nH_2O $\xrightarrow{Et_2O}$ $(\text{-}Si(CD_3)_2O\text{-})_n$
linear + cyclic
siloxanes
Hydrolysis reaction
Stage 5: $(\text{-}Si(CD_3)_2O\text{-})$ + KOH $\xrightarrow{heat}$ D_3 - d_{18} + D_4 - d_{24}
Thermolysis

Figure 20: Beltzung outline for per-deuterated D_4 synthesis.

The preparation began by synthesising per-deuterated methyl iodide (CD_3I) from per-deuterated methanol via a classical method. The CD_3I was then reacted with magnesium to produce a Grignard reagent which was subsequently reacted with a dichlorodiphenylsilane to produce bis(trideuteriomethyl)diphenylsilane ((CD_3)$_2SiPh_2$). Freidel Crafts de-arylation was then performed on the silane with benzene and gaseous hydrochloric acid in the presence of aluminium trichloride to remove the phenyl groups and produce the bis(trideuteriomethyl)dichlorosilane ((CD_3)$_2SiCl_2$). This was then hydrolysed under acidic conditions to yield a mixture of linear and cyclic per-deuterated siloxanes. Finally a thermal rectification was made using potassium hydroxide to increase the relative amounts of the cyclic tetramer (the starting material for ring-chain equilibration reactions).

Use Of Triflic Acid To Remove The Phenyl Groups From Dimethyldiphenylsilane.

Background. Investigations into the Beltzung reaction showed that the Friedel-Crafts reaction was not an easy one to achieve under normal laboratory conditions. An alternative method was sought for the removal of the phenyl groups from the dimethyldiphenylsilane intermediate.

A review of the literature soon revealed what appeared to be an alternative route to obtain siloxane oligomers from dimethyldiphenylsilane as shown in Figure 21:

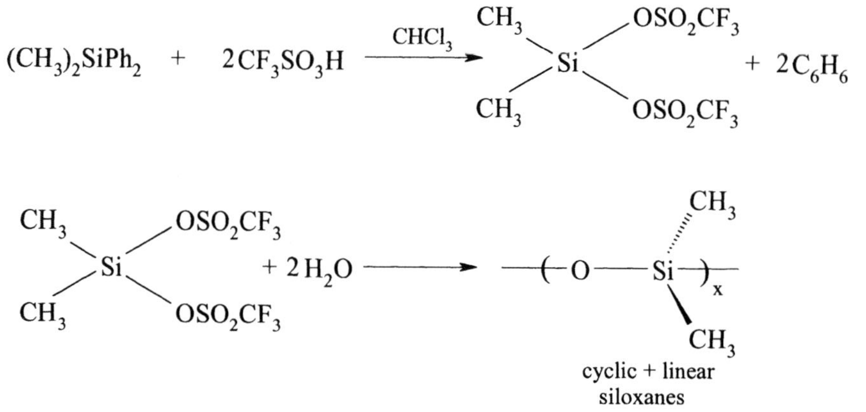

$$(CH_3)_2SiPh_2 + 2CF_3SO_3H \xrightarrow{CHCl_3} (CH_3)_2Si(OSO_2CF_3)_2 + 2C_6H_6$$

$$(CH_3)_2Si(OSO_2CF_3)_2 + 2H_2O \longrightarrow \text{—}(O\text{—}Si(CH_3)_2)_x\text{—}$$
cyclic + linear siloxanes

Figure 21: Proposed use of triflic acid to produce siloxane oligomers.

This reaction had not previously been attempted, however, work carried out by Matyazewski and co-workers [64] [65] showed that the removal of the phenyl groups from dimethyldiphenylsilane could be readily achieved at room temperature with two equivalents of triflic acid (Figure 22) and Uhlig [66] had shown that the presence of water leads to siloxane production (Figure 23) - an unwanted side-product in that particular investigation.

$$2\,CF_3SO_3H + PhSi(CH_3)_2Si(CH_3)_2Ph \longrightarrow CF_3SO_3Si(CH_3)_2Si(CH_3)_2OSO_2CF_3 + 2\,PhH$$

Figure 22: Matyjaszewski and coworkers preparation of silyl triflates from silanes and oligosilanes with phenyl groups.

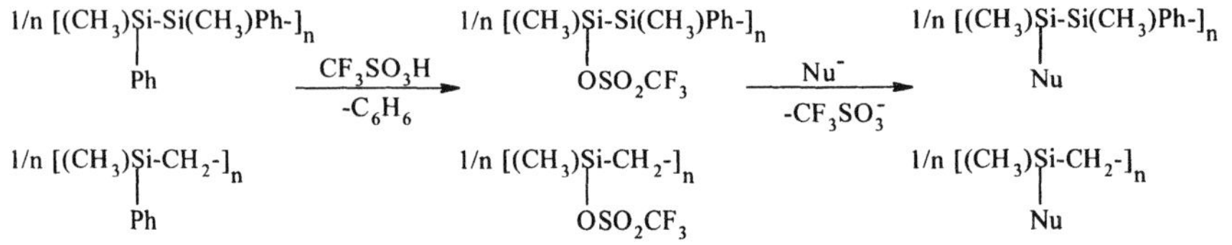

$$1/n\,[(CH_3)\underset{Ph}{\overset{|}{Si}}\text{-}Si(CH_3)Ph\text{-}]_n \xrightarrow[-C_6H_6]{CF_3SO_3H} 1/n\,[(CH_3)\underset{OSO_2CF_3}{\overset{|}{Si}}\text{-}Si(CH_3)Ph\text{-}]_n \xrightarrow[-CF_3SO_3^-]{Nu^-} 1/n\,[(CH_3)\underset{Nu}{\overset{|}{Si}}\text{-}Si(CH_3)Ph\text{-}]_n$$

$$1/n\,[(CH_3)\underset{Ph}{\overset{|}{Si}}\text{-}CH_2\text{-}]_n \qquad 1/n\,[(CH_3)\underset{OSO_2CF_3}{\overset{|}{Si}}\text{-}CH_2\text{-}]_n \qquad 1/n\,[(CH_3)\underset{Nu}{\overset{|}{Si}}\text{-}CH_2\text{-}]_n$$

Figure 23: Examples of Uhlig's work with novel organosilicon polymers.

This chemistry was thoroughly investigated in order to produce perdeuterated siloxane materials.

Preparation of Deuterated Cyclic PDMS.

Introduction. The method described here firstly details a route to the per-deuterated small cyclic dimethylsiloxanes from CD_3I [12], and then a route to both linear and cyclic deuterated PDMS [13] [14], based on the investigations in to the use of triflic acid as a means of removing the phenyl groups from the intermediate dimethyldiphenylsilane. Details for the preparation of the deuterated intermediate bis(trideuteriomethyl)diphenylsilane (($CD_3)_2SiPh_2$) are not given here but may be found elsewhere [12]. The synthesis reported here is more convenient to carry out, requires less purification of reagents and produces per-deuterated dimethylsiloxanes in an improved yield.

Experimental. The first range of per-deuterated cyclics have been achieved in improved yield via the route shown in Figure 24 from per-deuterated methyl iodide [12].

$$CD_3I + Mg \rightarrow CD_3MgI$$

$$2\,CD_3MgI + Ph_2SiCl_2 \rightarrow (CD_3)_2SiPh_2 + 2\,MgICl$$

$$n(CD_3)_2SiPh_2 + 2n\,CF_3SO_3H + nH_2O \rightarrow \underset{\text{cyclic dimethyl siloxanes}}{(Si(CD_3)_2O)_n} + \text{other products}$$

Figure 24: Convenient synthesis of per-deuterated dimethylsiloxanes.

The per-deuterated cyclic dimethylsiloxanes produced in these reactions were to be further polymerised in various ways to yield mixtures of larger cyclic and linear polymers which could be separated and fractionated using preparative GPC in a recognised fashion.

Preparation Of Small Cyclic Per-Deuterated Dimethylsiloxanes. Investigations have shown triflic acid to be capable of removing phenyl groups from dimethyldiphenylsilane and polymerisation of the hydrolysis products is possible in situ. The reaction was now carried out using bis(trideuteriomethyl)diphenylsilane prepared from CD_3I via a Grignard to produce per-deuterated dimethylsiloxanes.

The overall reaction required that 2.66×10^{-1} mol (57.95 g) of $(CD_3)_2SiPh_2$ were reacted with 5.32×10^{-1} mol (79.78 g) of triflic acid in the presence of 2.66×10^{-1} mol (4.79 g) of water.

The silyl triflate intermediates proposed in the reaction mechanism (see below) have been isolated and characterised previously [64] [67] and were reported to be extremely reactive, so for reasons of safety, the reaction was achieved in three stages. The procedure for each stage was as follows: the silane was diluted in dichloromethane at a ratio of 1 : 5 (w : v), one third of the required amount of water was added and the mixture stirred vigorously. One third of the amount of triflic acid was then added extremely cautiously at room temperature and allowed to react for two hours. The reaction was then cooled to <5°C before 200 mls of an approximately 1 M aqueous sodium carbonate solution was added slowly to neutralise the acid. The mixture was allowed to come to room temperature and once clear and colurless, the two layers obtained were separated, the dichloromethane layer was washed with deionised water and evaporated down under reduced pressure to remove the chlorinated solvent.

The progress of the triflic acid reaction was monitored using GLC. This technique is a rapid method of studying the volatile silanes and siloxanes being produced [68]. The product was then reacted further by an identical procedure, analysed again by GLC and the process repeated once more. This resulted in 18.24 g of a clear colourless oil. This was analysed by ^1H, ^{13}C and ^{29}Si Nuclear Magnetic Resonance (NMR) spectroscopy, InfraRed (IR) absorption, Gas Chromatography / Electron Ionisation - Mass Spectometry (GC/EI-MS) and by Gas Chromatography / Chemical Ionisation - Mass Spectrometry (GC/CI-MS)(see below for further details of the analyses).

The traces obtained from the GLC analysis of the reaction (Figure 25) clearly show the production of what have been identified, through the use of GC/EI-MS, as linear phenyl terminated poly(dimethylsiloxane) oligomers (Ph-$(Si(CD_3)_2O)_n$-$Si(CD_3)_2Ph$). The subsequent removal of the terminal phenyl groups upon further reaction with triflic acid was seen to result in a rearrangement to small cyclic oligomers of per-deuterated dimethylsiloxane.

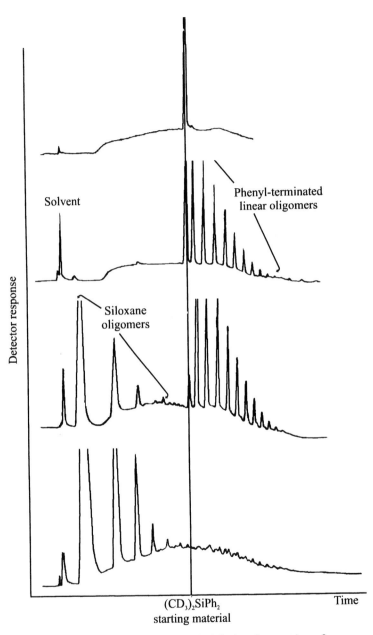

Figure 25: GLC traces recorded during the reaction of bis(trideuteriomethyl)diphenyl silane with triflic acid.

Analysis Of The Small Per-Deuterated Dimethylsiloxanes.

The structure and purity of the per-deuterated dimethylsiloxanes were confirmed by various analytical techniques described below:

1H NMR. A small sample of the product was analysed in dilute CDCl$_3$ solution spiked with a known amount of dichloromethane using a Jeol 270 MHz NMR instrument (see Figure 26). As expected for a fully deuterated dimethylsiloxane compound, the ^{1}H spectrum showed virtually no signals from the product, however the minute signals at approximately 0 ppm allowed the percentage deuteration to be calculated when the integrations were compared with a similar spectrum recorded with hydrogenated PDMS.

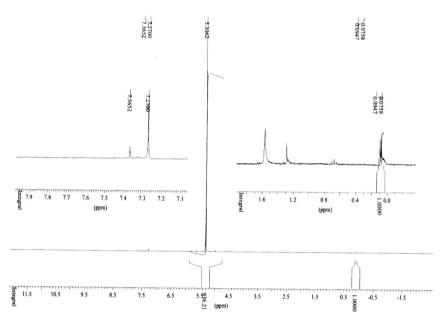

Figure 26: ^{1}H NMR spectrum of the deuterated product spiked with dichloromethane.

The ^{1}H spectrum of the deuterated product shows the almost total deuteration due to the absence of peaks at approximately 0 ppm. All of the phenyl groups have been completely removed from the silane and the hydrolysis has produced deuterated PDMS of high purity. The minute peaks observed at 7.27 and 7.37 ppm are due to residual solvents (CHCl$_3$ and C$_6$H$_6$ respectively). The large peak at 5.31 ppm is due to the dichloromethane used as a reference for calculating the percentage deuteration in comparison with a similar spectrum recorded with the hydrogenated cyclic tetramer (D$_4$).

^{13}C NMR. Again as expected for deuterated dimethylsiloxane compounds virtually no signals were seen to arise from the product in the ^{13}C spectrum.

^{29}Si NMR. ^{29}Si NMR has been identified as another convenient technique for characterising small cyclic poly(dimethylsiloxanes) [37]. The technique allows individual ring species to be identified upto fifteen repeat units in size i.e. D_{15}. A sample of the per-deuterated dimethylsiloxane was analysed undiluted using a Bruker 500 Mhz instrument and the spectrum is shown (Figure 27). Although the spectrum was not locked to any signal as no solvent was present (and so chemical shifts should not be taken as absolute but are shown as a guide), four clearly distinguishable peaks are present. In comparison to the peaks observed with protonated cyclics [37], it is believed that these peaks are due to the tetrameric (-18.6 ppm), pentameric (-21 ppm), hexameric (-21.7 ppm) and septameric (-21.9 ppm) cyclic species. Higher cyclics may well be present but could not be reliably identified due to the broad peaks observed. The broad peaks (typically with half maximum height peak widths of 10 Hz compared to 0.3 Hz for the protonated cyclics) are a result of silicon/deuterium coupling these may be resolved by running a silicon/deuterium de-coupled spectrum at high resolution to confirm the presence of the higher cyclics.

IR. A small sample of the oil produced from the triflic acid reaction was anlysed as a neat fluid between sodium chloride discs using a Perkin-Elmer FT-1000 Fourier Transform instrument. The resulting spectrum (Figure 28) shows a characteristic broad peak of siloxane asymmetric Si-O-Si stretches at 1076 cm^{-1} [69]. The C-D

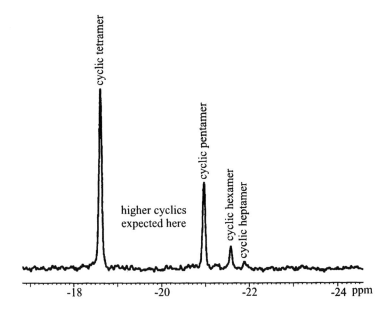

Figure 27: [29]Si spectrum of the deuterated small cyclic dimethylsiloxanes.

stretching band is clearly visible at 2217 cm⁻¹, whilst the sharp peak at 1013 cm⁻¹ can be assigned to that of the symmetric CD_3 deformation. A spectrum of the hydrogenated cyclic tetramer (D_4) was obtained in the same fashion and is shown in Figure 29 for reference. The siloxane asymmetric stretch is visible at 1070 cm⁻¹. The C-H stretching bands, absent from the deuterated spectrum, are clearly visible at 2905 and 2963 cm⁻¹. Also visible is a sharp band due to the symmetric CH_3 deformation at 1261 cm⁻¹ corresponding to the expected shift in frequency of approximately 1.4 times that with deuterium.

Gas Chromatography/Electron Ionisation Mass Spectroscopic Analysis Of The Small Per-Deuterated Dimethylsiloxane Cyclics. A sample of the D-PDMS oil was analysed using GC/EI-MS. This technique separates the volatile siloxane oligomers present and produces a mass spectrum of each component [68] [70]. An Autospec double focussing magnetic sector instrument was used linked to an HP 5890 series II GC fitted with capillary columns. The GC trace obtained (Figure 30) clearly shows the high purity of the sample. Minute peaks observed at 142 and 1009 counts being the only impurities present. The EI-MS spectra were obtained for all of the dimethylsiloxane peaks from the tetramer to the nonamer inclusive (Figures 31 and 32 show those of the hexamer and the nonamer respectively). All of the spectra show a common fragmentation pattern. The most noticeable fragmentation is the facile loss of one CD_3 group leading to a $(M-CD_3)^+$ ion in all cases. Another common loss corresponds to a mass of 100, this corresponds to the loss of $Si(CD_3)_4$ from the cyclic $(M-CD_3)^+$ species and this fragmentation has been accounted for previously in terms of a rearrangement of the cyclic siloxane [40] [41]. A table of the common peaks observed in the EI-MS spectra is given in Table II.

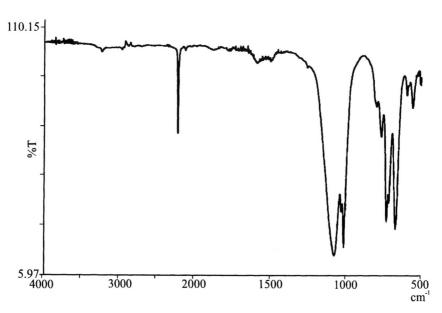

Figure 28: IR spectrum of the deuterated small cyclic dimethylsiloxanes.

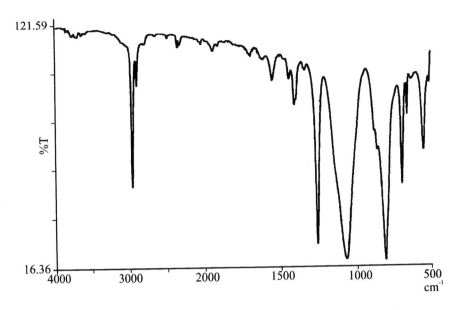

Figure 29: IR spectrum of the hydrogenated cyclic tetramer (D_4) - for reference.

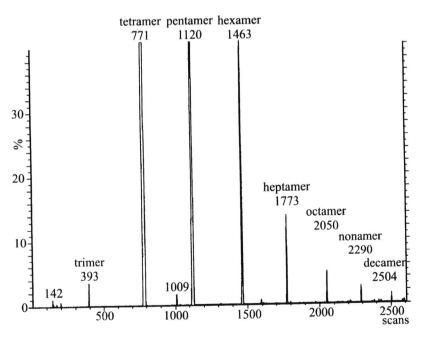

Figure 30: GC trace of the deuterated small cyclic dimethylsiloxanes.

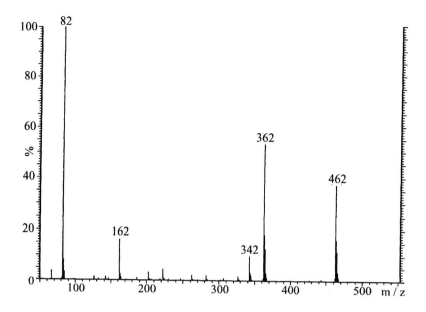

Figure 31: GC/EI-MS of the per-deuterated cyclic dimethylsiloxane hexamer.

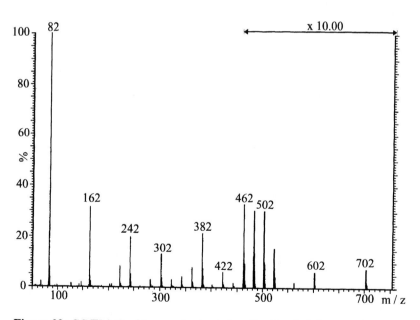

Figure 32: GC/EI-MS of the per-deuterated cyclic dimethylsiloxane nonamer.

Table II: Peak assignments for the GC/EI-MS spectra of the small per-deuterated cyclic dimethylsiloxanes.

m/z	Composition	Species
82	SiC_3D_9	$(Si(CD_3)_3)^+$
142	$Si_2O_2C_3D_9$	$(dimer - CD_3)^+$
162	$Si_2OC_5D_{15}$	$(Si(CD_3)_2OSi(CD_3)_3)^+$
202	$Si_3O_4C_3D_9$	$(tetramer - CD_3$ and $- SiCD_3)_4)^+$
222	$Si_3O_3C_5D_{15}$	$(trimer - CD_3)^+$
242	$Si_3O_2C_7D_{21}$	$(Si(CD_3)_2OSi(CD_3)_2OSi(CD_3)_3)^+$
282	$Si_4O_5C_5D_{15}$	$(pentamer - CD_3$ and $- Si(CD_3)_4)^+$
302	$Si_4O_4C_7D_{21}$	$(tetramer - CD_3)^+$
322	$Si_4O_3C_9D_{27}$	$(Si(CD_3)_2O[Si(CD_3)_2O]_2Si(CD_3)_3)^+$
362	$Si_5O_6C_7D_{21}$	$(hexamer - CD_3$ and $- Si(CD_3)_4)^+$
382	$Si_5O_5C_9D_{27}$	$(pentamer - CD_3)^+$
402	$Si_5O_4C_{11}D_{33}$	$(Si(CD_3)_2O[Si(CD_3)_2O]_3Si(CD_3)_3)^+$
422	$Si_6O_8C_7D_{21}$	$(octamer - CD_3, - Si(CD_3)_4$ and $- Si(CD_3)_4)^+$
442	$Si_6O_7C_9D_{27}$	$(septamer - CD_3$ and $- Si(CD_3)_4)^+$
462	$Si_6O_6C_{11}D_{33}$	$(hexamer - CD_3)^+$
482	$Si_6O_5C_{13}D_{39}$	$(Si(CD_3)_2O[Si(CD_3)_2O]_4Si(CD_3)_3)^+$
502	$Si_7O_9C_9D_{27}$	$(nonamer - CD_3, - Si(CD_3)_4$ and $- Si(CD_3)_4)^+$
522	$Si_7O_8C_{11}D_{33}$	$(octamer - CD_3$ and $- Si(CD_3)_4)^+$
542	$Si_7O_7C_{13}D_{39}$	$(septamer - CD_3)^+$
602	$Si_8O_9C_{13}D_{39}$	$(nonamer - CD_3$ and $- Si(CD_3)_4)^+$
622	$Si_8O_8C_{15}D_{45}$	$(octamer - CD_3)^+$
702	$Si_9O_9C_{17}D_{51}$	$(nonamer - CD_3)^+$

Clearly all the EI sepctra show no evidence for the presence of protonated PDMS despite the normal exposure of the deuterated dimethylsiloxanes to typical laboratory methodology - glassware, solvents etc.. The presence of the peak m/z 82 is in good agreement with a time-of-flight secondary ionisation mass spectrometry (TOFSIMS) study of fully deuterated PDMS carried out by Zhang and co-workers [71], as well as results obtained previously for the corresponding small hydrogenated PDMS cyclics [40] [41], and is thought to be due to a number of different rearrangements including a rearrangement of the skeletal -[$(CD_3)_2SiO$]- backbone of the hydroxyl- terminated deuterated siloxane $HO[(CD_3)_2SiO]_nH$, with a subsequent methyl shift to give -$Si(CD_3)_3$.

To try and confirm the presence of any linear hydroxyl- terminated deuterated siloxanes, which are not easily resolved from the cyclic siloxanes by GLC, the technique of GC/CI-MS was applied to the sample. This technique has been reported to clearly show the presence of hydroxyl end-groups in the analogous hydrogenated PDMS [70].

Gas Chromatography/Ammonia Chemical Ionisation Mass Spectroscopic Analysis Of The Small Per-Deuterated Dimethylsiloxane Cyclics. The CI spectra were also recorded of the small perdeuterated dimethylsiloxane cyclics using the same instrument as previously with ammonia. Three competing processes are known to occur with ammonia in the case of PDMS [70]. Quasimolecular ions are formed either by hydrogen or ammonium ion addition as well as subsequent trideuteriomethyl loss. Ions can be observed at M +18, M + 2(18 - 19), M + 1, and M - 18. Each of these processes are observed for the cyclics upto the decamer and spectra were obtained for the pentamer up to the decamer inclusive (Figures 33 and 34 show those of the hexamer and the nonamer respectively).

It has been reported that silanol terminated linear PDMS produces the $(M + NH_4)^+$ ion in ammonia CI [70]. Whilst the EI spectra of an hydroxyl terminated per-deuterated linear PDMS would appear indistinguishable from that of the cyclic per-deuterated PDMS of the same chain length, the facile trideutriomethyl loss followed by water ejection from the linear diol would result in the same $(M + NH_4)^+$ ion as the trideuteriomethyl loss from the cyclic. This would result in a series of quasimolecular peaks at 18 m/z higher for the linear than the corresponding cyclic. No such peaks were observed in this study and hence the amount of hydroxyl terminated per-deuterated linear PDMS produced may be taken as negligible. A table of the common quasimolecular peaks observed for the per-deuterated cyclic dimethylsiloxanes in the CI-MS spectra is given in Table III.

Preparation of Larger Per-Deuterated Siloxanes.

The successful preparation of the small per-deuterated cyclic dimethylsiloxanes now meant that larger cyclics and linears could in theory be prepared in a recognised fashion from the small rings via either the Brown and Sluzarczuk [4] or the Chojnowski [30] [31] [32] [33] methods. The investigations in to the triflic acid reaction with dimethyldiphenylsilane, however, showed the possibility that the reaction might be catalytic and so this was attempted with deuterated materials and is described below.

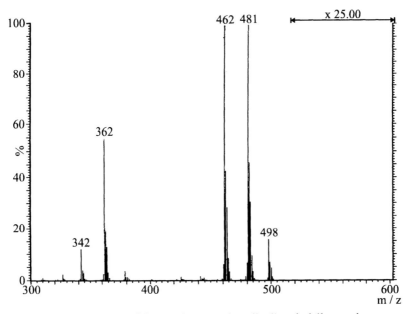

Figure 33: GC/CI-MS of the per-deuterated cyclic dimethylsiloxane hexamer.

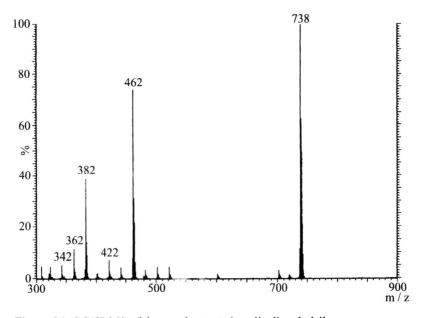

Figure 34: GC/CI-MS of the per-deuterated cyclic dimethylsiloxane nonamer.

Table III: Peak assignments for the GC/CI-MS spectra of the small per-deuterated cyclic dimethylsiloxanes.

m/z	Composition	Species
382	$Si_5O_5C_9D_{27}$	$(pentamer - CD_3)^+$
401	$Si_5O_5C_{10}D_{30}H$	$(pentamer + H)^+$
418	$Si_5O_5NC_{10}D_{30}H_4$	$(pentamer + NH_4)^+$
462	$Si_6O_6C_{11}D_{33}$	$(hexamer - CD_3)^+$
481	$Si_6O_6C_{12}D_{36}H$	$(hexamer + H)^+$
498	$Si_6O_6NC_{12}D_{36}H_4$	$(hexamer + NH_4)^+$
542	$Si_7O_7C_{13}D_{39}$	$(septamer - CD_3)^+$
561	$Si_7O_7C_{14}D_{42}H$	$(septamer + H)^+$
578	$Si_7O_7NC_{14}D_{42}H_4$	$(septamer + NH_4)^+$
622	$Si_8O_8C_{15}D_{45}$	$(octamer - CD_3)^+$
641	$Si_8O_8C_{16}D_{48}H$	$(octamer + H)^+$
658	$Si_8O_8NC_{16}D_{48}H_4$	$(octamer + NH_4)^+$
702	$Si_9O_9C_{17}D_{51}$	$(nonamer - CD_3)^+$
721	$Si_9O_9C_{18}D_{54}H$	$(nonamer + H)^+$
738	$Si_9O_9NC_{18}D_{54}H_4$	$(nonamer + NH_4)^+$

Catalytic Preparation Of Both Linear And Cyclic Per-Deuterated PDMS. As proposed, the probability that the triflic acid reaction with the bis(trideuteriomethyl)diphenylsilane was catalytic was extremely high and has been confirmed in the first catalytic preparation of both linear and cyclic per-deuterated dimethlysiloxanes from the same reaction [13] [14].

The $(CD_3)_2SiPh_2$ was prepared from trideuteriomethyliodide (CD_3I) as detailed elsewhere [12]. 145.0 g (0.665 mol) of the silane was dissolved in 220 ml of CH_2Cl_2 under a nitrogen atmosphere. 20.0 g (0.133 mol) of triflic acid was slowly added over ½hr and the solution gently stirred at room temperature. The concentrations were chosen such that complete conversion of the silane to siloxanes would result in an overall siloxane concentration of ~230 g l^{-1} shown by Brown and Slusarczuk [4] (in the case of PDMS in toluene) and by Lebrun and coworkers [34] (in the case of PDMS in dichloromethane) to yield good amounts of cyclic PDMS in dilute solution conditions. The amount of triflic acid added was one tenth of the stoichiometric amount required to replace the phenyl groups with triflate groups as carried out in earlier investigations in to this reaction [12]. Once the triflic acid was completely added, one drop of de-ionised water was added and the reaction continued to be gently stirred at room temperature under nitrogen.

The reaction was monitored using GLC, and as observed previously, was seen to produce a series of phenyl terminated linear siloxane oligomers over time. Once the distribution of these oligomers was seen not to change with time a further drop of de-ionised water was added to the reaction. Eventually, the smaller siloxanes were produced, and the distribution of phenyl terminated linear oligomers was seen to reduce. Further evidence enabling a mechanism to be proposed came from the observation of turbidity in the reaction mixture after the addition of the water which was seen to clear after some time. This turbidity has been reported previously by Lebrun and co-workers [34] as a result of the water released on condensation of silanol groups produced in a ring/chain equilibration of hexamethylcyclotrisiloxane (D_3) with triflic acid in dichloromethane (such silanol groups would result here from the rapid hydrolysis of the highly reactive silyl triflates produced).

The cycle of dropwise addition of water and monitoring of the reaction by GLC until no further change in the distribution of products occurred, was continued for 10 days. (Once all of the phenyl groups have been removed from the reaction this becomes essentially a ring-chain equilibration as identified by Chojnowski and co-workers [30] [31] [32] [33]).

Once equilibrated, a copious amount of sodium hydrogen carbonate ($NaHCO_3$) was added to neutralise the triflic acid. The dichloromethane solution was then washed several times with de-ionised water before being evaporated down under reduced pressure.

This resulted in 45.9 g of a clear colourless viscous oil containing a mixture of cyclic and linear polymers. The small cyclic per-deuterated dimethylsiloxanes were carefully distilled off this residue. The cyclics and high molar mass linear polymer in the residue were separated in a recognised fashion by precipitating the linear polymer from a hot solution of the equilibrate in acetone (10% w/v), allowing the solution to cool to room temperature and stand for 24hrs. The supernatant solution containing the cyclics was decanted from the viscous linear polymer and the acetone removed by evaporation under reduced pressure yielding 6.5g of polydisperse cyclic polymer.

Fractionation of Deuterated Cyclic PDMS.

The polydisperse per-deuterated cyclic PDMS recovered from the catalytic reaction was fractionated in to twelve sharp fractions using the preparative GPC instrument detailed earlier [5]. A section of the preparative trace is shown in Figure 35. The fractions were analysed using analytical GPC calibrated with cyclic PDMS standards and the results of the analysis are shown in Table IV.

The molar masses of the cyclic fractions show a typical spread for a ring/chain equilibration reaction with ring sizes ranging from $n_n = 45$ (number-average number of bonds in a ring) up to $n_n = 605$ for the highest fraction obtained. The dispersities (M_w/M_n) of the samples were also typically low, all being less than 1.2.

The purity of the deuterated cyclic PDMS fractions were verified using infrared absorption, mass spectrometry and ^1H, ^{13}C and ^{29}Si nuclear magnetic resonance spectroscopies.

Proposal Of A Reaction Mechanism.

Trifluoromethanesulfonic (triflic) acid has previously been reported to selectively remove phenyl groups from methylphenylsilanes under mild conditions [64] [65], produce siloxanes in the presence of water [66], as well as catalyse the polymerisation of cyclic PDMS [30] [31] [32] [33] [34].

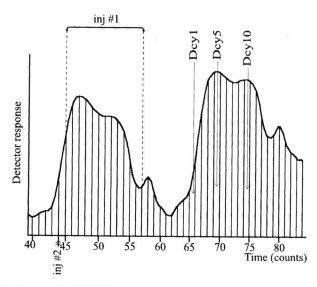

Figure 35: Preparative GPC chromatogram recorded during the fractionation of the deuterated cyclic PDMS.

Table IV: Data from the fractionation of the deuterated cyclic PDMS.

Fraction	Yield g	Weight average molar mass (M_w) g mol^{-1}	Number average molar mass (M_n) g mol^{-1}	Dispersity (M_w/M_n)	Number average number of bonds (N_n)
1	0.03	25500	24200	1.05	605
2	0.19	21200	20100	1.05	503
3	0.42	18400	17400	1.06	435
4	0.53	15100	14000	1.08	350
5	0.63	12400	11400	1.09	285
6	0.64	10300	9300	1.11	233
7	0.60	8000	6900	1.16	173
8	0.55	6200	5400	1.15	135
9	0.52	4400	3900	1.13	98
10	0.48	3300	2900	1.14	73
11	0.39	2600	2200	1.18	55
12	0.22	2000	1800	1.11	45

In light of the findings of our work so far, it is suggested that the acid removes the phenyl group from the silane producing the silyl triflate ester (I) and di-ester (II) as intermediates (Figure 36). The silyl triflates have been previously identified as valuable reagents in organosilicon chemistry [72] and are readily hydrolysed to silanol functionalities in the presence of water which is an excellent nucleophile under the reaction conditions. The silanols will readily self condense or condense with the silanoate ester groups to yield siloxane linkages and regenerate the triflic acid in an analogous process to that proposed by Chojnowski and co-workers for the polymerisation of D_4 with triflic acid.

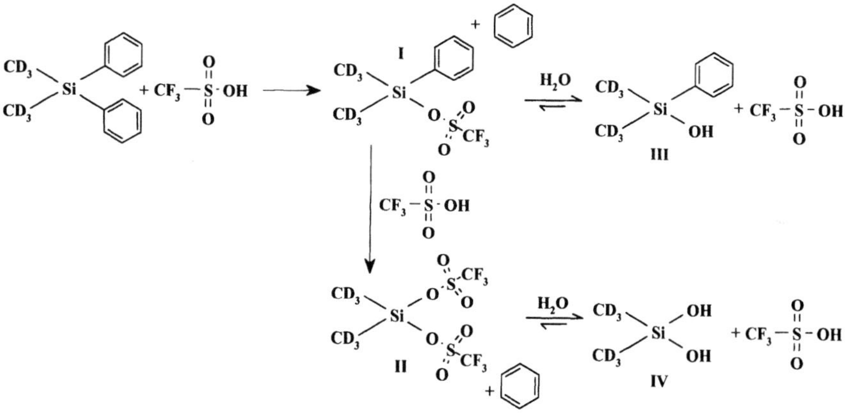

Figure 36: Production of silyl triflates and silanol intermediates.

The production of phenyl terminated dimethylsiloxane oligomers can be accounted for via the reactions shown in Figure 37. Reactions A and B show how silanol self condensation and silanol/silanoate ester condensation respectively yield the diphenyltetra(trideuteriomethyl)disiloxane oligomer (V). Reaction C shows an example of how condensation of larger silanols and silanoate esters may lead to the higher oligomers. An example of the rearrangement to cyclic (VIII) or linear hydroxyl-terminated (IX) dimethylsiloxanes (depending on reaction conditions) is given in Figure 38. Again it is noted that the reactions regenerate the triflic acid reagent and therefore account for the fact that catalytic amounts of the acid are able to achieve the same reaction over a longer time scale.

Summary.

Narrow fractions of hydrogenated cyclic poly(dimethylsiloxanes) have been prepared in good yield. A comprehensive series of analytical techniques have displayed the purity of the products and demonstrated the isolation of cyclic polymers.

The further work carried out investigating the effects of dilution and triflic acid concentration on the reaction with bis(trideuteriomethyl)diphenylsilane has optimised a catalytic route to both linear and cyclic per-deuterated PDMS of high purity in good yields enabling the first narrow fractions of deuterated cyclic polymers to be prepared.

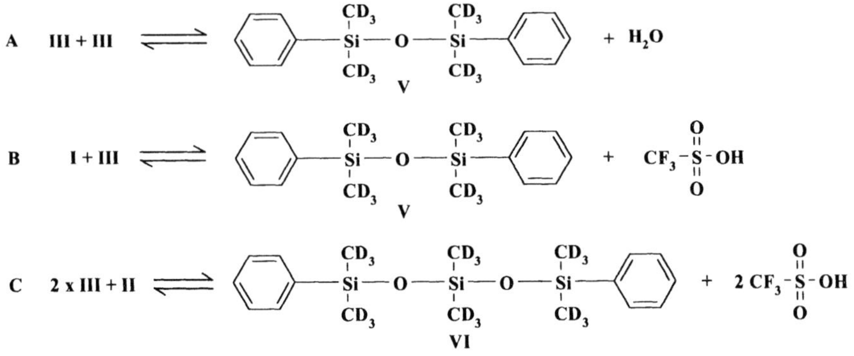

Figure 37: Condensations leading to phenyl-terminated linear oligomers.

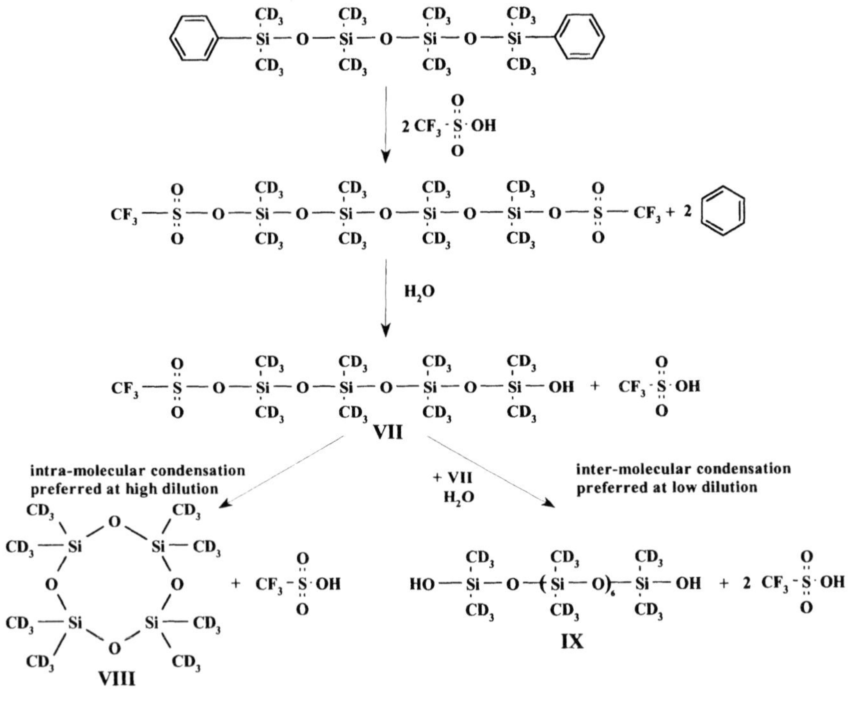

Figure 38: Further reaction of phenyl-terminated oligomers leading to cyclic or linear siloxane production depending on reaction conditions.

The reaction is also expected to provide a convenient route to well controlled random hydogenated/deuterated PDMS copolymers with specific amounts of labelling for neutron scattering studies. This may be achieved by blending the required amounts of $(CH_3)_2SiPh_2$ and $(CD_3)_2SiPh_2$ prior to reacting with a catalytic amount of triflic acid.

The identification of a viable route which yields sufficient quantities of per-deuterated cyclic polymers to allow fractionation in to very sharp fractions, now enables us for the first time to carry out neutron scattering experiments investigating the conformations and properties of cyclic and linear polymers in chemically identical blends.

The conformational and dynamic properties of cyclic polymers have been the subject of considerable interest over a number of years [73] [74]. The new synthetic procedure developed for the preparation of per-deuterated cyclic and linear poly(dimethylsiloxane) (PDMS) polymers [12] [13] [14] now offers unique possibilities for neutron scattering studies. For the first time, it has allowed sufficient deuterated materials to be available for fractionation using our preparative GPC instrument. The resulting narrow fractions are ideally suited for both static and dynamic studies of cyclic and linear polymers. This range incorporates the critical molar mass for entanglement in PDMS.

Despite the extensive literature which is available on the behaviour of cyclic polymers in solution, the static properties and indeed the dynamic properties of ring polymers in the melt remain largely unexplored. As recently pointed out by Müller, Wittmer and Cates [75] "no experimental study of the radius of gyration of rings in the melt appears to have been made. A good understanding of the static properties of the system is of course an indispensable starting point for a reasonable description of the dynamics, so this is unfortunate."

Small angle neutron scattering (SANS) techniques are extremely powerful methods for studying the conformations of polymers. SANS is the only probe of the polymer conformation in the bulk state, relying on the large difference between the scattering lengths of hydrogen and deuterium.

Only two limited SANS studies of linear PDMS have been reported in the literature. Lapp et al. [76] carried out SANS experiments on an asymmetric mixture of h-PDMS (M_w=14600) and d-PDMS (M_w =267000) at low concentration of deuterated chains (0.9 and 6%) whereas Beaucage et al. [77] have recently extended these experiments to symmetric 50/50 h-PDMS/d-PDMS blends with M_w ranging from 15000 to 300000 g/mol.

In collaboration with Heriot-Watt University (Edinburgh), the University of Surrey and Imperial College, we have carried out a series of SANS experiments which have begun to investigate a wide range of linear and cyclic PDMS polymers in chemically identical blends using the materials described in this chapter. The preliminary results obtained are extremely interesting. They will be compared with current theories, computer simulations and the results of previous studies in dilute solution and are to be reported in the near future [78].

Acknowledgments.

ACD would like to thank Dow Corning (Barry, South Wales) for the financial funding of this project and useful discussions. Figures adapted from research papers published by JAS and his co-workers are referenced in the legends. They were published in the journal *Polymer* and are reproduced by permission of Butterworth-Heinemann Ltd.

Literature Cited.

1. Scott, D.W., *J. Amer. Chem. Soc.*, **68**, 2294 (1946).
2. Carmichael, J.B. and Winger, R., *J. Polym. Sci.*, **A3**, 971 (1965).
3. Carmichael, J.B., Gordon, D.J. and Isakson, F.J., *J. Phys. Chem.*, **71**, 2011 (1967).
4. Brown, J.F. and Slusarczuk, G.M.J., *J. Amer. Chem. Soc.*, **87**, 931 (1965).
5. Dodgson, K. Sympson, D. and Semlyen, J.A., *Polymer*, **19**, 1285 (1978).
6. Clarson, S.J. and Semlyen, J.A., (Eds.), *"Siloxane Polymers"*, PTR Prentice Hall, New Jersey (1993).
7. Semlyen, J.A. and coworkers, Studies of Cyclic and Linear Poly(dimethylsiloxanes), Parts 1-34, Polymer (1977-98).
8. Wright, P.V. and Semlyen, J.A., *Polymer*, **11**, 462 (1970).
9. Formoy, T.R. and Semlyen, J.A., *Polymer Comm.*, **30**, 86 (1989).
10. Beevers, M.S. and Semlyen, J.A., *Polymer*, **12**, 373 (1971).
11. Clarson, S.J. and Semlyen, J.A., *Polymer*, **27**, 1633 (1986).
12. Dagger, A.C. and Semlyen, J.A., *Polymer*, **39**, 2621 (1998).
13. Dagger, A.C. and Semlyen, J.A., *Polym. Prepr. (Am. Chem. Soc., Div. Polym. Chem.)*, **39**, 579 (1998).
14. Dagger, A.C. and Semlyen, J.A., *Polym. Commun.*, in press.
15. Beltzung, M., Picot, C., Rempp, P. and Herz, J., *Macromolecules*, **15**, 1584 (1982).
16. Hartung, H.A. and Camiolo, S.M., Division of Polymer Chemistry, 141st National Meeting of the American Chemical Society, Washington, D.C. (March 1962).
17. Carmichael, J.B. and Kinsinger, J.B., *Canad. J. Chem.*, **42**, 1996 (1964).
18. Carmichael, J.B. and Winger, R., *J. Polym. Sci.*, **A, 3**, 971 (1965).
19. Carmichael, J.B., Gordon, D.J. and Isakson, F.J., *J. Phys. Chem.*, **71**, 2011 (1967).
20. Carmichael, J.B., *J. Macromol. Chem.*, **1**, 207 (1966).
21. Grubb, W.T. and Osthoff, R.C., *J. Amer. Chem. Soc.*, **77**, 1405 (1955).
22. Morton, M. and Bostick, E.E., *J. Polym. Sci.*, **A2**, 523 (1964).
23. Wright, P.V., D. Phil. Thesis, University of York (1970).
24. Wright, P.V., *J. Polym. Sci.: Polym. Phys. Ed.*, **11**, 51 (1973).
25. Jacobson, H. and Stockmayer, W.H., *J. Chem. Phys.*, **18**, 1600 (1950).
26. Flory, P.J., *"Principles Of Polymer Chemistry"*, Cornell University Press, Ithaca (1953).
27. Kuhn, W., *Kolloid-Z*, **68**, 2 (1934).
28. Flory, P.J. and Semlyen, J.A., *J. Amer. Chem. Soc.*, **88**, 3209 (1966).
29. Wang, J.C., *"Circular DNA"*, Chapter 7 of Semlyen, J.A. (Ed.), *"Cyclic Polymers"*, Elsevier Applied Science Publishers Ltd., London (1986).
30. Chojnowski, J., Mazurek, M., Ścibiorek, M. and Wilczek, L., *Die Makromolekulare Chemie*, **175**, 3299 (1974).

31. Chojnowski, J. and Ścibiorek, M., *Makromol. Chem.*, **177**, 1413 (1976).
32. Chojnowski, J., Ścibiorek, M. and Kowalski, J., *Makromol. Chem.*, **178**, 1351 (1977).
33. Chojnowski, J. and Wilczek, L., *Makromol. Chem.*, **180**, 117 (1979).
34. Lebrun, J.-J., Sauvet, G. and Sigwalt, P., *Makromol. Chem., Rapid Commun.*, **3**, 757 (1982).
35. De Gennes, P.G., *"Scaling Concepts In Polymer Physics"*, Cornell University Press, Ithaca (1979).
36. Dodgson, K. and Semlyen, J.A., *Polymer*, **18**, 1265 (1977).
37. Burton, B.J., Harris, R.K., Dodgson, K., Pellow, C.J. and Semlyen, J.A., *Polym. Commun.*, **24**, 278 (1983).
38. Edwards, C.J.C., Rigby, D., Stepto, R.F.T. and Semlyen, J.A., *Polymer*, **24**, 395 (1983).
39. Edwards, C.J.C., Rigby, D., Stepto, R.F.T., Dodgson, K. and Semlyen, J.A., *Polymer*, **24**, 391 (1983).
40. VandenHeuvel, W.J.A., Smith, J.L., Firestone, R.A. and Beck, J.L., *Analytical Letters*, **5**, 285 (1972).
41. Pickering, G.R., Olliff, C.J. and Rutt, K.J., *Organic Mass Spectrometry*, **10**, 1035 (1975).
42. Kramers, H.A., *J. Chem. Phys.*, **14**, 415 (1946).
43. Zimm, B.H. and Stockmayer, W.H., *J. Chem. Phys.*, **17**, 1301 (1949).
44. Casassa, E.F., *J. Polym. Sci.*, **A**, **3**, 605 (1965).
45. Higgins, J.S., Dodgson, K. and Semlyen, J.A., *Polymer*, **20**, 552 (1979).
46. Dodgson, K., Bannister, D.J. and Semlyen, J.A., *Polymer*, **22**, 663 (1980).
47. Orrah, D.J., Semlyen, J.A. and Ross-Murphy, S.B., *Polymer*, **29**, 1452 (1988).
48. Bueche, F., *J. Chem. Phys.*, **40**, 84 (1964).
49. Doi, M. and Edwards, S.F., *"The Theory Of Polymer Dynamics"*, Oxford University Press, Oxford (1986).
50. Doi, M., *J. Polm. Sci.: Polym Phys. Ed.*, **21**, 665 (1983).
51. Graessley, W.W., *Farad. Symp.*, **18**, 7 (1983).
52. Clarson, S.J., Dodgson, K. and Semlyen, J.A., *Polymer*, **26**, 930 (1985).
53. Di Marzio, E.A. and Guttman, C.M., *Macromolecules*, **20**, 1403 (1987).
54. Gibbs, J.H. and Di Marzio, E.A., *J. Chem. Phys.*, **28**, 373 (1958).
55. Bannister, D.J. and Semlyen, J.A., *Polymer*, **22**, 377 (1981).
56. Patel, A., Cosgrove, T. and Semlyen, J.A., *Polymer*, **32**, 1313 (1991).
57. van Lent, B., Scheutjens, J.M.H.M. and Cosgrove, T., *Macromolecules*, **20**, 366 (1987).
58. Garrido, L., Mark, J.E., Clarson, S.J. and Semlyen, J.A., *Polym. Commun.*, **26**, 53 (1985).
59. Garrido, L., Mark, J.E., Clarson, S.J. and Semlyen, J.A., *Polym. Commun.*, **26**, 55 (1985).
60. Clarson, S.J., Mark, J.E. and Semlyen, J.A., *Polym. Commun.*, **27**, 244 (1986).
61. Clarson, S.J., Mark, J.E. and Semlyen, J.A., *Polym. Commun.*, **28**, 151 (1987).
62. Fyvie, T.J., Frisch, H.L., Semlyen, J.A., Clarson, S.J. and Mark, J.E., *J. Polym. Sci.: Polym. Chem.*, **9**, 2503 (1987).

63. Huang, W., Frisch, H.L., Hua, Y. and Semlyen, J.A., *J. Polym. Sci.: Polym. Chem.*, **28**, 1807 (1990).
64. Matyjaszewski, K. and Chen, Y.L., *Journal Of Organometallic Chem.*, **340**, 7 (1988).
65. Ruehl, K.E. and Matjaszewski, K., *Journal Of Organometallic Chem.*, **410**, 1 (1991).
66. Uhlig, W., *Organometallics*, **13**, 2843 (1994).
67. Howells, R.D. and Mc Cown, J.D., *Chemical Reviews*, **77**, No. 1, 69 (1977).
68. Steinmeyer, R.D. and Becker, M.A., *"Chromatographic Methods."* Chapter 10 of Smith, A.L., (Ed.), *"The Analytical Chemistry of Silicones"*, Chemical Analysis Vol. 112, John Wiley & Sons, Inc., New York (1991).
69. Lipp, E.D. and Smith, A.L., *"Infrared, Raman, Near-Infrared, and Ultraviolet Spectroscopy"* Chapter 11 of Smith, A.L., (Ed.), *"The Analytical Chemistry of Silicones"*, Chemical Analysis Vol. 112, John Wiley & Sons, Inc., New York (1991).
70. Moore, J.A., *"Mass Spectrometry."* Chapter 13 of Smith, A.L., (Ed.), *"The Analytical Chemistry of Silicones"*, Chemical Analysis Vol. 112, John Wiley & Sons, Inc., New York (1991).
71. Zhang, X.K., Stuart, J.O., Clarson, S.J., Sabata, A. and Beaucage, G., *Macromolecules*, **27**, 5229 (1994).
72. Emde, H., Domsch, D., Feger, H., Frick, U., Götz, A., Hergott, H., Hoffmann, K., Kober, W., Krägeloh, K., Oesterle, T., Steppan, W., West, W. and Simchen, G., *Synthesis*, 1 (1982).
73. Semlyen, J.A., (Ed.), *"Cyclic Polymers"*, Elsevier Applied Science, London (1986).
74. Semlyen, J.A., (Ed.), *"Large Ring Molecules"*, J. Wiley & Sons, Chichester (1996).
75. Müller, M., Wittmer, J.P. and Cates, M.E., *Phys. Rev. E*, **53**, 5063 (1996).
76. Lapp, A., Picot, C. and Benoît, H., *Macromolecules*, **18**, 2437 (1985).
77. Beaucage, G., Sukumaran, S., Clarson, S.J., Kent, M.S. and Schaefer, D.W., *Macromolecules*, **29**, 8349 (1996).
78. Dagger, A.C., Arrighi, V., Shenton, M.J., Clarson, S.J. and Semlyen, J.A., in preparation.

Chapter 5

Calculation of a Reaction Path for KOH-Catalyzed Ring-Opening Polymerization of Cyclic Siloxanes

J. D. Kress[1], P. C. Leung[2,†], G. J. Tawa[1,3], and P. J. Hay[1]

[1]Theoretical Division, Los Alamos National Laboratory, Group T-12, Mail Stop B268, Bikini Atoll Road, TA-3 SM-30, Los Alamos, NM 87545
[2]Technical Computing Department, 3M Corporation, St. Paul, MN 55144-1000

As evidenced by the many contributions to this symposium, siloxanes represent a class of semi-organic polymers of great importance to industry. Traditionally, polydimethylsiloxane can be prepared by base-catalyzed ring-opening of cyclic dimethyl siloxanes. *Ab initio* electronic calculations were conducted to examine a gas-phase reaction path for the KOH catalyzed ring-opening polymerization of octamethylcyclotetrasiloxane (D_4) and hexamethylcyclotrisiloxane (D_3). Two different stable KOH-D_4 adduct structures were found for the initial step along the D_4 reaction path: (1) a multidendate interaction between the K atom and the four O atoms in the D_4 ring, and; (2) a side-on addition complex between KOH and a Si-O bond in the ring, where the Si atom is five-fold coordinated. Continuing from the addition complex, the D_4 reaction path leads to five-fold coordinated Si atom transition state, then to a stable insertion product (KOH inserts into the ring). Retention of stereochemistry about the five-fold coordinated Si atom is maintained in the formation and cleavage of the Si-O bonds. The relative stability of a ring-opened product $HO[Si(CH_3)_2O]_4K$ is also considered. The KOH-D_4 results are compared with previous calculations for the KOH-D_3 reaction path. Adducts between D_3 and other K-containing bases larger than KOH were studied to determine whether KOH is a realistic model for the chain propagation step. The energy along the D_3 reaction path was also modeled both in a moderately polar solvent (tetrahydrofuran, THF). The solvation energy was calculated using a recent implementation of an electrostatic model, where the solute molecule is placed in a non-spherical cavity in a dielectric continuum. The effect of basis set and electron correlation on the gas-phase energy and solvation energy was studied.

[3]Current address: Frederick Cancer Research and Development Center, National Cancer Institute, Building 430, Miller Drive, Frederick, MD 21702-1201.

†Deceased.

© 2000 American Chemical Society

Introduction

The importance of poly(dialkylsiloxane)s as inorganic backbone polymers is well established.[1,2] The synthesis of linear siloxane polymers can be divided into two general classes:[3] polycondensation of bifunctional siloxanes and ring-opening polymerization of cyclic oligosiloxanes. The mechanism of the latter initiated (catalyzed) by a base (anionic polymerization) is the focus of this work. A shorthand notation is useful for referring to the cyclic siloxanes. The $(CH_3)_2SiO$ unit will be represented as D, *i.e.*, the cyclic tetramer octamethylcyclotetrasiloxane is denoted as D_4. The base-catalyzed reaction involves a catalyst MX, where M is an alkali metal or ammonium cation and X is a counterion. The siloxane ring-opening polymerization process of interest is composed of an initiation reaction

$$D_n + MX \rightleftharpoons [X - (SiMe_2 - O-)_{n-1}SiMe_2 - O]^- M^+ \qquad (1)$$

followed by a chain propagation step,

$$[X-(SiMe_2-O-)_m SiMe_2-O]^- + D_n \to [X-(SiMe_2-O-)_{m+n} SiMe_2-O]^- \qquad (2)$$

where $[X-(SiMe_2-O-)_m SiMe_2-O]^-$ is the growing species (denoted P later) and Me is a methyl (CH_3) group. For $n > 3$, (1) is represented as reversible and the polymerization is controlled by thermodynamics.[1,2] But for $n = 3$ the reversibility is greatly suppressed due to ring strain in D_3 (absent for rings with $n > 3$) and the polymerization is controlled by kinetics.[1,2]

According to one proposed mechanism,[3,4] a cation (*e.g.* K^+ or Li^+) forms a complex with the growing species and D_n. Rearrangement of the complexes occurs, resulting in extension of the siloxane polymer chain by the reaction

$$[P - SiMe_2 - O]^- \ldots M^+ \ldots D_n \to [P - (SiMe_2 - O-)_n SiMe_2 - O]^- \ldots M^+. \qquad (3)$$

A multidentate structure between the cation and the O atoms in the D_n ring has been proposed[4] as the transition state in order to explain the trend in the rate of reaction for n=5 to 9. The multidentate structure is akin to the interaction of a cation with polydentate ether ligands in crown ethers. Other studies, however, suggest a mechanism in which the anion attacks D_n directly (2) and the metal ion is not involved in the ring-opening step.[5,1,2] For the anion mechanism, a pentavalent (five-fold coordinated) Si atom intermediate was proposed.[5] Structural entropy changes in various conformations of ligand attachment to trigonal bipyramidal structures were used to explain the observed trend in reactivity for D_6 and D_7. The level of association between the cation and the growing species depends on the polarity of the solvent used in the process.

Disiloxane $(SiH_3)O(SiH_3)$ has been the subject of many electronic structure calculations.[6-9] The Si-O-Si bond angle in disiloxane is very flexible and thus the geometry is a challenge to predict correctly. Molecular mechanics force fields based on *ab initio* calculations on disiloxane and other small molecules have been used to study[10] the torsional barrier in hexamethyldisiloxane and structural conformations of D_4. *Ab initio* calculations on several conformations of polysiloxane $(H_2SiO)_n$ for $n = 3$, 4 and 5 have also been reported.[11] Calculations on small siloxane molecules have also been used as models for zeolites.[12]

The present work focuses on the thermodynamically controlled reaction $KOH + D_4 \rightarrow HO[Si(CH_3)_2O]_3K$. Previous calculations[13] on the kinetically controlled reaction $KOH + D_3 \rightarrow HO[Si(CH_3)_2O]_3K$ are summarized and compared with the $KOH + D_4$ reaction. For this reaction, KOH acts as the catalyst. Adducts between D_3 and other K-containing bases larger than KOH were studied to determine whether KOH is a realistic model for the chain propagation step. The energy along the $KOH + D_3$ reaction path was also modeled both in a moderately polar solvent (tetrahydrofuran, THF). The solvation energy was calculated using a recent implementation of a electrostatic model, where the solute molecule is placed in a non-spherical cavity in a dielectric continuum.

Computational Method

Electronic Structure. Calculations were performed using the GAUSSIAN 92 and GAUSSIAN 94 programs.[14] The Hartree-Fock (HF) optimized geometries were determined using techniques described in Ref. 14. The choice of basis sets resulted from a balance between quantitative accuracy and computational effort. To this end a modest STO-3G basis was used throughout for carbon and hydrogen atoms (unless specified otherwise) and the LANL2DZ effective core potential[15] (ECP) was used throughout for potassium (unless specified otherwise). For silicon and oxygen atoms, 3-21G, 3-21G(d), 6-31G*, and 3-21G(2d) bases were examined and discussed[13] for D_3. For the 3-21G(d) and 6-31G* bases, one polarization d-orbital per Si and O atom was used. For the 3-21G(2d) basis, the two (double) polarization d-orbitals per Si and O was used. The effect of basis set on the conformation for D_3 and D_4 has been studied (Table I). For the D_3 results,[13] the 3-21G basis yields a Si-O bond distance, r(Si-O), that is about 0.04 Å too long, a Si-O-Si angle, a(Si-O-Si), that is about 6° too large, and, a O-Si-O angle, a(O-Si-O), that is 5° too small, compared to experiment.[17] For the single [3-21G(d)] and double [3-21G(2d)] polarization basis sets, r(Si-O) is within 0.01 Å and the angles are within 1° or better compared to experiment. In contrast to D_3, the D_4 ring is very flexible. In a molecular mechanics study[10], eight different non-planar conformations were found within 4 kcal/mol of the S_4 (four-fold symmetric) ground-state. Nicholas et al.[16] find that a double polarization basis is needed to obtain a reasonable non-linear Si-O-Si angle, a(Si-O-Si), for disiloxane, which is very flexible and is thus difficult to converge. Thus, the equilibrium value of a(Si-O-Si) for D_4 is very sensitive with respect to basis set (Table I). The 3-21G basis predicts the wrong conformation with respect to experiment, a(Si-O-Si) is too large by 16°, and r(Si-O) that is 0.03 Å too long. The double polarization basis 3-21G(2d) does predict the correct conformation and the error in the floppy angle a(Si-O-Si) is reduced to 5°. Both the 3-21G and 6-31G* bases were used in the D_3 reaction path studies and the 3-21G basis was used for the D_4 study. Although more accurate, the use

Table I. Optimized geometries for the D_3 and D_4 rings. (Adapted from ref. 13.)

Si,O basis set	Conformation	Si-O bond distance (Å)	Si-O-Si bond angle (°)	O-Si-O bond angle (°)
D_3				
3-21G	Planar	1.672	137.1	102.9
3-21G(d)	Planar	1.642	133.1	106.9
3-21G(2d)	Planar	1.635	132.6	107.4
exp't (Ref. 15)	Planar	1.635±0.002	131.6±0.4	107.8±0.7
D_4				
3-21G	Planar	1.652	161.1	108.9
3-21G(d)	Boat-Boat	1.630	154.3	109.3
3-21G(2d)	S_4	1.632	140.0	108.9
exp't (Ref. 15)	S_4	1.622±0.003	144.8±1.2	110.0±1.8

of the 3-21G(2d) basis to determine the reaction path is too computationally expensive.

Solvation Model. First, the quantum mechanical electrostatic potential is calculated on a grid surrounding the solute molecule for each optimized geometry along the gas-phase reaction path. An electrostatic potential (ESP) fit[18] is then performed to obtain a set of atom-centered charges. Next a molecular volume, composed of a union of spheres centered at each atom of the solute, is constructed around the solute molecule. Each sphere has a radius about equal to the van der Waals radius of the atom upon which it is centered[13]. The solvent is assumed to reside outside the molecular surface and is characterized as a dielectric continuum with a static dielectric constant ϵ. To represent tetrahydrofuran (THF), a moderately polar solvent used in polymerization, $\epsilon=8.2$ (Ref. 19). The region inside of the molecular surface is assigned $\epsilon = 1$. The set of atom charges determined from the electronic structure calculation is then used in the governing macroscopic Poisson equation which is solved[20] to obtain the response of the solvent to the solute charge density. The classical electrostatic solvation energy $\Delta G(Solv)$ is calculated from this response and the potential of mean force is given by $W = E(Electronic) + \Delta G(Solv)$.

Results

Reaction paths for the KOH catalyzed ring-opening of D_4 and D_3 in the gas-phase were constructed. The calculated 3-21G structures of a side-on addition complex (AC) and a multidendate (MD) adduct are shown in Fig. 1. The structures along a D_4 reaction path continuing from the AC are shown in Fig. 2. (The atoms are numbered for reference in Figs. 1 and 2.) Similar structures for KOH-D_3 adducts and a KOH-D_3 reaction path were shown in Ref. 13. The 3-21G gas-phase energies for the KOH-D_4 and KOH-D_3 reaction paths are shown in Fig. 3. The zero of energy for each path is defined as the energy for the AC structure. For $r > 0$ ($r = 0$ is the addition complex structure), the reaction coordinate r is the Si(7)-O(8) bond distance in D_4. For $r < 0$, r is not specific, but is only representative of going downhill in energy from R (reactant) to AC (left to right) or from R to MD (right to left). The specific structures along the reaction path are discussed in the next section. In Fig. 4, the structures of the KOH-D_3 and KO$-$Si$(CH_3)_2-$O$-$Si$(CH_3)_2-$OH (KOM$_2$) addition complexes are shown. KOM$_2$ represents a more realistic model of the attacking end of the growing polymer compared to using KOH. In Table II, the binding energies of addition complexes using various molecules to model the growing end of the polymer are given. Solvation energies along the D_3 reaction coordinate were obtained by calculating the potential of mean force using the dielectric continuum model[20] described above. The effect of electron correlation on the energy was calculated with Möller Plesset 2nd order perturbation theory (MP2) at the HF geometry. In Fig. 5, the Hartree-Fock (HF) gas-phase reaction path energy and the corresponding HF and MP2 potential of mean force (solvated energy) is plotted. The method used to calculate the energy is specified. In all cases, the HF method was used to optimize the geometry.

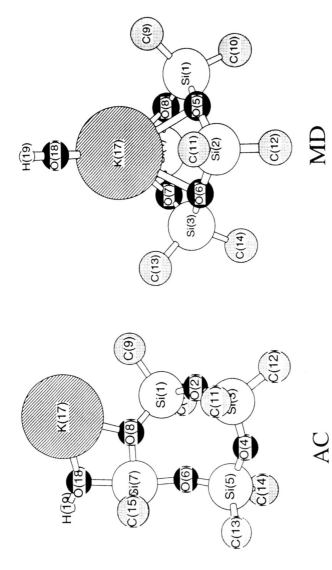

Figure 1. Structures of adducts of KOH and D$_4$. (a) Addition complex (AC). (b) Multidendate structure (MD). The black, gray, and, striped balls are O, C, and K atoms, respectively. The white balls are Si atoms and the H atom on KOH. H atoms on methyl groups are omitted for clarity. C atoms behind the plane of the paper are hidden by out-of-plane C atoms in the addition complex.

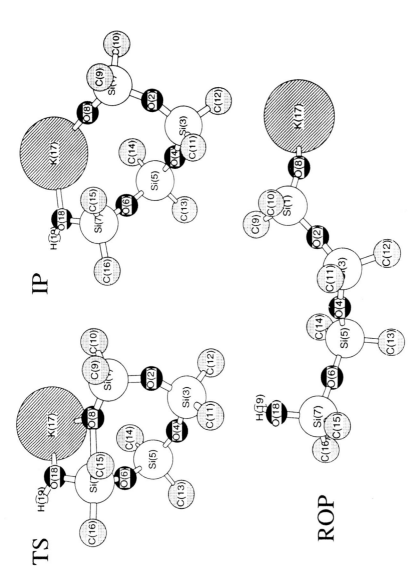

Figure 2. Structures of critical points along the reaction path for KOH catalyzed ring-opening polymerization of D_4. **TS**. Transition state. **IP**. Insertion product. **ROP**. Ring-opened product. H atoms on methyl groups are omitted for clarity. C atoms behind the plane of the paper are hidden by out-of-plane C atoms in **R** and **AC**.

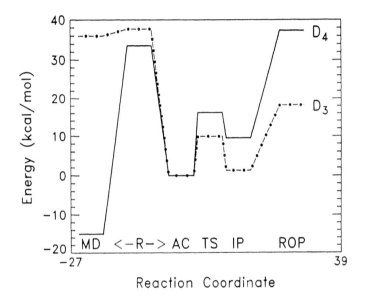

Figure 3. The gas-phase reaction path of KOH catalyzed ring-opening polymerization of D_4 (solid line) and D_3 (dash-dot line) calculated with the HF method and 3-21G basis set. The structures along the reaction coordinate for D_4 are defined in Figs. 1 and 2. The addition complex (AC) is defined as the zero of energy for both cases. The reaction coordinate is not to scale and is only representative. (Adapted from. ref. 13.)

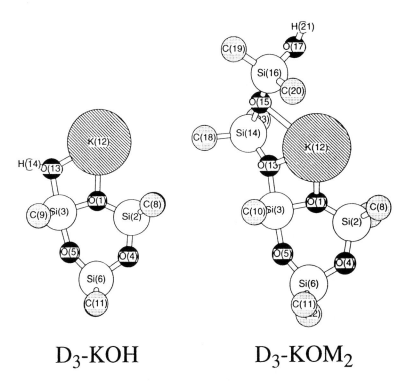

Figure 4. Structures of addition complexes with D₃. (a) KOH. (b) KO–Si(CH₃)₂–O–Si(CH₃)₂–OH (KOM₂). The black, gray, and, striped balls are O, C, and K atoms, respectively. The white balls are Si atoms and the H atom on KOH. H atoms on methyl groups are omitted for clarity. C atoms behind the plane of the paper are hidden by out-of-plane C atoms in the addition complex.

Table II. Thermochemistry of addition reactions to siloxane rings. 3-21G basis set for Si and O atoms.

Ring	Reactant	Binding Energy (kcal/mol)
D_3	KOH	-38
D_3	$KOSi(CH_3)_2OH$ (KOM_1)	-38
D_3	$KOSi(CH_3)_2OSi(CH_3)_2OH$ (KOM_2)	-38
D_3	$KOSi(CH_3)_3$ (KO-tb)	-33
D_4	KOH	-34

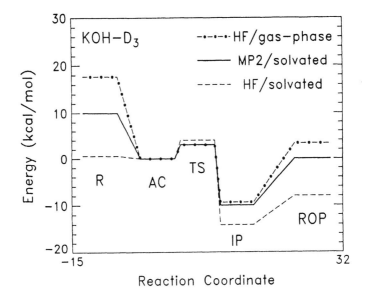

Figure 5. The reaction path of KOH catalyzed ring-opening polymerization of D_3. HF method, gas-phase (dash-dot line); HF method, potential of mean force, solvated (dashed line). MP2 method, potential of mean force, solvated (solid line). 6-31G* basis sets are used for Si and O for both the energy and geometry calculations. A STO-3G basis set is used for the CH_3 groups and the LANL2DZ ECP is used for K. The structures along the reaction coordinate are defined in Ref. 13. The addition complex (AC) is defined as the zero of energy for both cases. The reaction coordinate is not to scale and is only representative. (Adapted from. ref. 13.)

Discussion

Gas-Phase Calculations. At the starting point R in Fig. 3 are the reactants, KOH and a D_4 molecule at infinite separation. The calculated[13] and experimental[17] conformation for D_3 is planar. Whereas the D_3 ring is strained, the higher D_n (n> 3) rings are quite floppy. As discussed above, the calculated ground-state conformation of D_4 is very sensitive to the basis set. For the 3-21G reaction path considered here, the predicted ground-state structure (planar) is different than the experimental conformation (S_4). The energy differences between these different conformations is negligible compared to the interaction energy between KOH and D_4 and therefore of minor concern in the study of the reaction path. The D_4 ring is flexible enough to end up in the conformation that maximizes the binding with KOH.

Proceeding from the reactants (R), two different KOH-D_4 adduct structures have been found for the initial step along the reaction path. One structure is a multidendate (MD) interaction between the K atom and the four O atoms in the D_4 ring that has been previously been proposed[4] as the transition state for this reaction. This structure (Figure 1, multidendate structure, MD), is akin to the interaction of a cation with polydentate ether ligands in crown ethers. This direction along the reaction path is represented in Fig. 3 as going to the left from R. The other structure is a side-on addition complex (AC), where atom O(42) in the KOH molecule forms a bond with atom Si(7) and the atom K(41) forms bonds with ring atom O(8) and KOH atom O(42). This is shown in Fig. 1 (structure AC). This direction along the reaction path is represented in Fig. 3 as going to the right from R. The atom Si(8) is 5-fold coordinated with atom O(42) (KOH) and a ring atom O(6) in the axial positions and the two methyl groups and another ring atom [O(8)] in the equatorial positions. The KOH molecule is not in the plane of the ring, and the ring itself is not perfectly planar. For KOH-D_3, the AC structure (Fig. 4a) also contains[13] a 5-fold coordinated Si atom with a similar trigonal bipyramidal arrangement of O atoms and methyl groups. The AC structure for KOH-D_3 is more stable than the MD structure by 36 kcal/mol (Fig. 3). In contrast, the MD structure for KOH-D_4 is more stable that the AC structure by 15 kcal/mol (Fig. 3). For D_3, the MD structure is less stable than the AC structure due to the steric hindrance of the methyl groups on the Si atoms which the shield the K atom interacting with the O atoms in the ring. This is not true for D_4 which is floppy (no ring strain), where the methyl groups can move away from the center of the ring (Fig. 1).

To this point the reaction path from the reactants to the addition complex has been downhill. To continue, a reaction coordinate r is defined to be the Si-O bond stretch involving the 5-coordinated atom Si(7) and atom O(8) in the ring of the addition complex. (The reaction path of the subsequent ring-opening of D_4 starting from the MD structure has not been studied.) The reaction coordinate was determined manually in increments of typically $\Delta r=0.2$ Å starting at the addition complex geometry (defined as r=0). The Si(7)-O(8) distance in the D_4 AC is 1.73 Å (3-21G) and almost identical to the corresponding Si-O distance (1.75 Å) in the D_3 addition complex. Every coordinate was optimized in the system except the reaction coordinate. The initial geometry in the optimization was taken from the result of the previous step. A barrier (transition state, TS) of 16 kcal/mol is found at r=0.8 Å for D_4, whereas a similar TS structure of 10 kcal/mol is found[13] at r=0.6 Å for D_3. Note that for the larger (better) 6-31G* basis, the D_3 TS has a lower barrier of 3 kcal/mol at r=0.2 Å. The Si(7)-O(8) distance in the D_4 TS is 2.53 Å and the comparable Si-O distance in the D_3 TS is 2.35 Å (3-21G basis) and 1.94 Å (6-31G* basis), respectively.

The nature of 5-coordinate Si species, either as stable forms or reactive intermediates, has been the subject of numerous recent experimental and theoretical studies.[23-25] X-ray crystal structures have been reported for a variety of 5-coordinate Si compounds, of which the closest forms to the present siloxane systems include the bicyclics $Si(O-C_2-O)_2R^-$ and $Si(O-C_2-O)_2X^-$, where R and X denote alkyl and halide ligands, respectively, and C_2 denotes intervening aliphatic or aromatic carbon linkages in the bidendate rings. A few anionic structures with pentaoxy ligands $Si(O-C_2-O)_2(OR)^-$ have also been reported. The 5-coordinate species we observe in the present calculations for both D_3 and D_4 would be of the form $Si(OR)_3(CH_3)_2^-$ where, as shown in AC in Fig. 1 for D_4. Atoms O(6) and O(42) form the apices and atom O(8) forms one equatorial position of a trigonal bipyramid (TBP); the two methyl groups C(33) and C(34) form the other two equatorial positions. The calculated angles for this structure are: O_{ax}-Si-O_{ax} (173.5°), O_{ax}-Si-O_{eq} (90.4° and 88.5°), and C_{eq}-Si-O_{eq} (116.2° and 120.8°), where ax=axial and eq=equatorial. Going from the addition complex to the transition state (Figs. 1 and 2), the equatorial atom O(8) becomes axial and the two axial atoms [O(6) and O(42)] become equatorial. Also, one of the equatorial methyl groups C(34) becomes axial. Experimental structures for 5-coordinate Si compounds span the range along the coordinate between TBP and rectangular pyramid (RP). As R becomes progressively less bulky for the pentaoxy $Si(O-C_2-O)_2(OR)^-$ series,[25] the structures change from TBP to RP, while the $Si(O-C_2-O)_2X^-$ species tend more towards TBP. Other pentacoordinate structures have been invoked in sol-gel synthesis[26] and zeolite formation[27] from silica-containing precursors. Recent evidence has also been reported on 5-coordinate cationic structures in acid-catalyzed siloxane polymerization,[28] as contrasted to anionic forms we find here in base-catalyzed siloxane polymerization.

The mechanism in Figs. 1 and 2 corresponds to the situation where the incoming OH^- group associated with K^+ occupies an axial position of the TBP and the leaving O-Si group of the D_3 ring occupies an equatorial position. In this situation, retention of stereochemistry about the Si is maintained in the formation and cleavage of the Si-O bonds, as has been discussed earlier.[23-25] Inversion stereochemistry results in the standard S_N2 chemistry involved in C chemistry, where both entering and leaving groups occupy the axial positions of the TBP *transition state*. In the present case of OH attack on siloxane, the TBP form is a *stable point* along the reaction coordinate.

While we have not examined the case of more bulky substituents R, as in $[(R)(CH_3)Si-O]_n$ rings, the result that the stereochemistry is retained at the Si atom in D_3 where addition occurs is consistent with recent experimental studies of chain configurations. Kuo et al. observed[29] that the anionic polymerization of 1,3,5-trimethyl-1,3,5,tris(3',3',3'-trifluoropropyl)cyclotrisiloxane with lithium silanoate proceeds with roughly equal probability of forming meso or racemic configurations relative to the configuration of the reacting chain end; a pentavalent transition state for the polymerization was also proposed. Compared to the experiment, a significant difference in our results arises from the fact that we find both a stable 5-coordinate addition complex (AC) as well as a subsequent 5-coordinate transition state (TS) in the addition step.

The next step on the reaction path in Fig. 3, the minimum between r=2.6 and 3.0 Å, corresponds to the insertion of KOH into the D_4 ring, where atom K(41) is inserted between the Si(7)-O(8) and O(42)-H groups. This is designated as a insertion product (Fig. 2, structure IP). A similar IP structure at r=2.0 Å was found[13] for D_3.

The final point in Fig. 3 corresponds to the ring-opened product (Fig. 2, structure ROP). Because optimizing a floppy molecule which has many local minima is difficult, if not impossible, this energy was calculated by fixing r and optimizing all other coordinates. This yields a nearly "linear" ring-opened product for $r = 7.8Å$, with a 3-21G energy of about +3 kcal/mol relative to the reactants. Since there is no ring strain in D_4, this energy difference is the amount of energy it takes to break the K-OH and Si-O bonds and form the K-OSi and HO-Si bonds. The energy difference between reactants and ring-opened products for D_3 also includes a contribution due to the release of ring strain (estimated[13] to be about 15 and 11 kcal/mol for the 3-21G and 3-21G(d) basis sets, respectively).

The present results for both D_3 and D_4 suggest that the insertion product is more stable than the ring-opened product. However, this energy difference can be an artifact of limitations in the gas-phase models. The K cation in the ring-opened product can be considered coordination unsaturated. In solution, the K atom can interact and coordinate with Si-O moieties in unreacted D_3 and PDMS in the environment.

For D_3, the manually-optimized transition state structure was refined[13] using gradient optimization procedures[21] and Quadratic Synchronous Transit[22] (QST) techniques; force constant matrix calculations were also performed to classify various points along the reaction path. For D_3, all positive force constants, and thus, proper local minima, were found for the reactants, addition complex, and insertion product; a single negative force constant was found for the 3-21G TS structures, thus verifying a proper transition state. The manually-optimized transition state has not been refined nor has force constant matrix analyses been performed for the KOH-D_4 reaction path.

The general trend of the variation of geometries and atom charges along the D_3 reaction path was summarized in Ref [13]. The change in bond angles between AC and TS was discussed above in terms of a trigonal bipyramidal structure about the the 5-fold coordinated Si atom. The bond between this Si atom and the KOH O atom shrinks by about 0.2 Å between the AC and TS. The distance between the K and O atoms from KOH gradually increases as the reaction path is traversed from R to IP. For the reaction path Si-O bond there is less charge separation for the AC compared to both R and TS. The charge on O atom from KOH increases in increments of about 0.2 going from R to AC to TS. However, the change in the K atom charge along the reaction path is negligible.

Adducts between D_3 and other molecules larger than KOH were studied to determine whether KOH is a realistic model for the chain propagation step (2). In Table II, the binding energies of two different silanoates KOM_1 and KOM_2 and of $KOSi(CH_3)_3$ (KO-tb) to D_3 are presented. As in the D_4 reaction path studies described above, the 3-21G basis set was used for the Si and O atoms. The binding energy of KOM_1 and KOM_2 is the same as the KOH binding energy to within 1 kcal/mol or less. The binding energy of KO-tb is about 5 kcal/mol weaker relative to the KOH binding energy to D_3, and comparable to the KOH-D_4 binding energy. Therefore, with respect to the energetics of the adducts, KOH seems to be a realistic model to describe the addition complex. In Fig. 4, the structures of the $KOH-D_3$ and KOM_2-D_3 are shown. Both structures exhibit the 5-fold coordinated Si atom described above, with the KOH O atom and a ring O atom forming the apices of a trigonal bipyramidal structure. The K-O and Si-O bond distances in the adduct and rings are similar. Additionally, the K atom in KOM_2 adduct forms a multidendate interaction with the middle O atom in KOM_2 as well as with the O atom from the K-O-Si- group and a O atom from the ring. This structure is an example of back-biting, where the K atom can attack other O atoms in the chain.

Solvation Calculations. The Hartree-Fock (HF) potential of mean force in the solvent (Fig. 5) is similar to the HF gas phase potential energy curve between the AC and ROP structures. Both curves exhibit a barrier between the stable addition complex and the insertion product. In the gas-phase, the transition state is lower than the reactants. But, in the reactants are stabilized more by the solvent relative to the transition state with the result that the potential of mean force for the transition state is higher than the reactants. In general, the R and ROP are stabilized more by the solvent, relative to the AC, TS, and IP, since the R and ROP structures are more open. The openness exposes the Si and O atoms in the ring more to the solvent and, thus, $\Delta G(solv)$ is more negative. To provide a quantitative estimate of the solvated reaction path including electron correlation, MP2 potentials of mean force (Fig. 5) were generated by adding $\Delta G(solv)$ calculated with a HF/6-31G* wavefunction to the MP2/6-31G* gas-phase electronic energy. The MP2 reactants are 10 kcal/mol higher than the AC. The MP2 barrier (TS) is +3 kcal/mol relative to the AC. The MP2 products (ROP) are 10 kcal/mol below the MP2 reactants energy and are 3 kcal/mol below the MP2 barrier. The insertion product is the most stable state for all three cases in Fig. 3.

Basis Sets. The effect of basis set and electron correlation along the $KOH-D_3$ gas-phase reaction path was discussed int Ref. 13. Relative to the addition complex energy, electron correlation (MP2 calculation) increases the reactants energy and decreases the barrier to products. The better basis (6-31G*) has a more dramatic effect relative to the 3-21G energies (Fig. 3); the reactants energy is reduced from 38 to 18 kcal/mol and the difference between the barrier energies is reduced from 10 to 3 kcal/mol relative to the 3-21G values. The general effect of the 6-31G* relative to the 3-21G basis set along the solvated reaction path is to stablize (lower) the potential of mean force relative to the AC energy. By comparing the HF/3-21G gas-phase and MP2/6-31G* solvation calculations for the KOH-D_3 reaction (including a correction for ring strain), a very crude estimate was derived to correct the HF/3-21G gas-phase calculation for the KOH-D_4 reaction path. This crude estimate yields for the MP2/6-31G* solvation energy 6, 0, 9, −1, and 12 kcal/mol for the KOH-D_4 reaction path structures R, AC, TS, IP, and ROP, respectively. In comparison the MP2/6-31G* solvation reaction path (Fig. 5) for D_3, the D_4 the TS, IP, and ROP are higher in energy.

The effect of basis set superposition errors[30] (BSSE) on the relative energies between R (the KOH and D_3 reactants) and AC (the addition complex) was examined[13] for various basis sets for the KOH-D_3 reaction path. This BSSE analysis was somewhat inconclusive since typically such estimates are used for weakly-bound complexes, such as hydrogen-bonded species, where the two moieties retain their molecular indentity. The effect of BSSE has not been examined for the KOH-D_4 reaction path.

Conclusion

Ab initio electronic calculations were conducted to examine a gas-phase reaction path for the KOH catalyzed ring-opening polymerization of octamethylcyclotetrasiloxane (D_4) and hexamethylcyclotrisiloxane (D_3). Two different stable KOH-D_4 adduct structures were found for the initial step along the D_4 reaction path: (1) a multidendate (Fig. 1, structure **MD**) interaction between the K atom and the four O atoms in the D_4 ring, and; (2) a side-on addition complex (Fig. 1, structure **AC**) between KOH and a Si-O bond in the ring, where the Si atom is five-fold coordinated. Continuing from the addition complex, the D_4 reaction path leads to a five-fold coordinated Si atom transition state (Fig. 2, structure **TS**), then to a stable insertion product (Fig. 2, structure **IP**). Retention of stereochemistry about the five-fold coordinated Si atom is maintained in the formation and cleavage of the Si-O bonds. The relative stability of a ring-opened product (Fig. 2, structure **ROP**) is also considered. The KOH-D_4 results are compared with previous calculations for the KOH-D_3 reaction path. Adducts between D_3 and other K-containing bases larger than KOH were studied to determine whether KOH is a realistic model for the chain propagation step. The energy along the D_3 reaction path was also modeled in a moderately polar solvent (tetrahydrofuran, THF). The solvation energy was calculated using a recent implementation[20] of an electrostatic model, where the solute molecule is placed in a non-spherical cavity in a dielectric continuum. The effect of basis set and electron correlation on the gas-phase energy and solvation energy was studied. The addition of electronic correlation (Moller-Plesset 2nd order perturbation theory) to the 6-31G* gas-phase reaction path destabilizes (increases) the reactants and ring-opened product energies relative to the addition complex; however, the transition state relative energy change is negligible.

Finally, we note that the solvation model does not include the direct interaction of the electronic structure of a solvent molecule with the solute. For example, an O atom on a THF molecule could participate in multidentate bonding with the K atom and an O atom in the D_3 ring. Such solvation models that involve a few solvent molecules explicitly coupled with a dielectric continuum description may be an avenue for future work.

Acknowlegments. We thank Stefan Klemm, Smarajit Mitra, David Misemer, Richard Martin, Lawerence Pratt, John Blair, Shih-Hung Chou, and Kris Zaklika for many valuable discussions and thank Holman Brand for assistance with the graphics. This work was supported by a Cooperative Research and Development Agreement (CRADA) "Properties of Polymers and Organic Dye Molecules" between the 3M Corporation and Regents of the University of California/Los Alamos National Laboratory (LANL). The work at LANL is carried out under the auspices of the U. S. Department of Energy, Contract No. W-740-ENG-36.

Literature Cited

(1) Saam, J. C. In *Silicon-Based Polymer Science: A Comprehensive Resource*; Ziegler, J. M., Gordon Fearon, F. W., Eds.; Am. Chem. Soc.: Washington, DC, 1990; pp 71-90.
(2) Wright, P. V. In *Ring Opening Polymerization*; Ivin, K. J., Saegusa, T., Eds.; Elsevier: New York, 1984; pp 1055-1133.
(3) Chojnowski, J. In *Siloxane Polymers*; Clarson, S. J., Semylen, J. A., Eds; Prentice Hall: Englewood Cliffs, NJ, 1993; pp 1-71; Chojnowski, J. In *Silicon Chemistry*; Corey, J. Y., Corey, E. R., Gaspar, P. P., Eds.; Ellis Horwood: New York, 1988; pp 297-306.
(4) Mazurek, M.; Chojnowski, J. *Makromol. Chem.* **1977**, *178*, 1005.
(5) Laita, Z.; Jelinek, M.; *Vysokomol. Soed.* **1962**, *4*, 1739. [*Polym. Sci. USSR (Eng. Transl.)* **1965**, *4*, 535.]
(6) Grigoras, S.; Lane, T. H. *J. Comp. Chem.* **1987**, *8*, 84.
(7) Shambayati, S.; Blake, J. F.; Wierschke, S. G.; Jorgensen, W. L.; Schreiber, S. L. *J. Am. Chem. Soc.* **1990**, *112*, 697.
(8) Luke, B. T. *J. Phys. Chem.* **1993**, *97*, 7505.
(9) Koput, J.; *J. Phys. Chem.* **1995**, *99*, 15874. Nicholas, J.B.; Feyereisen, M. J.; *Chem. Phys.* **1995**, *103*, 8031. Csonka, G.I.; Reffy, J.; *Theochem* **1995**, *332*, 187. Baer, M.R.; Sauer, J.; *Chem. Phys. Lett.* **1994**, *226*, 405.
(10) Grigoras, S.; Lane, T. H. *J. Comp. Chem.* **1988**, *9*, 25.
(11) Kudo, T.; Hashimoto, F; Gordon, M.S.; *J. Comput. Chem.* **1996**, *17*, 1163.
(12) Sauer, J.; *Chem. Rev.* **1989**, *89*, 199.
(13) Kress, J.; Leung, P. C.; Tawa, G. J.; Hay, P. J. *J. Am. Chem. Soc.* **1997**, *119*, 1954.
(14) Frisch, M. *et al.*; *GAUSSIAN 92, Revision A*; Gaussian Inc.: Pittsburgh, PA, 1992. Frisch, M. *et al.*; *GAUSSIAN 94, Revision D.3*; Gaussian Inc.: Pittsburgh, PA, 1995. *J. Am. Chem. Soc.* **1992**, *114*, 8191.
(15) Hay, P. J.; Wadt, W. R. *J. Chem. Phys.* **1985**, *82*, 299.

(16) Nicholas, J. B.; Winans, R. E.; Harrison, R. J.; Iton, L. E.; Curtiss, L. A.; Hopfinger, A. J. *J. Phys. Chem.* **1992**, *96*, 10247.
(17) Oberhammer, H.; Zeil, W.; Fogarasi, G.; *J. Molecular Structure* **1973**, *18*, 309.
(18) Breneman, C. M.; Wiberg, K. B.; *J. Comp. Chem.* **1990**, *11*, 361.
(19) Holland, R. S.; Smyth, C. P.; *J. Phys. Chem.* **1955**, *59*, 1088.
(20) Tawa, G. J.; Pratt, L. R. In *Structure and reactivity in aqueous solution: Characterization of chemical and biological systems*; Cramer, C. J.; Truhlar, D. G., Eds.; ACS Symposium Series 568; ACS: Washington DC, 1984; p 60. Tawa, G. J.; Pratt, L. R. *J. Am. Chem. Soc.* **1995**, *117*, 1625; Corcelli, S. A.; Kress, J. D.; Pratt, L. R.; Tawa, G. J. *BIOCOMPUTING. Proceedings of the 1996 Pacific Symposium*; Hunter L.; Klein, T. E., Eds.; World Scientific: Singapore, 1995, pp. 142-159. Pratt, L. R.; Tawa, G. J.; Hummer, G.; Garcia, A. E.; Corcelli, S. A. *Int. J. Quantum Chem.* **1997** *64*, 121. Tawa, G. J.; Martin, R. L.; Pratt, L. R.; Russo, T. V. *J. Phys. Chem.* **1996** *100*, 1515.
(21) Schlegel, H. B.; *J. Comp. Chem.* **1982** *3*, 214.
(22) Halgren, T. A.; Lipscomb, W. N.; *Chem. Phys. Lett.* **1977** *49*, 225. Peng, C.; Schlegel, H. B.; *Israeli J. Chem.* **1994** *33*, 449.
(23) Holmes, R. R.; *Chem. Rev.* **1990**, *90*, 17.
(24) Corriu, R. J. P.; Young, J. C. In *The Chemistry of Organic Silicon Compounds*; Patai, S., Rappoport, Z., Eds.; John Wiley and Sons: Chichester, 1989; pp 1241-1288.
(25) Holmes, R.R.; Day, R.O.; Payne, J.S. *Phosphorus, Sulfur, Silicon* **1989**, *42*, 1.
(26) Iler, R. K. *The Chemistry of Silica*; Wiley: New York, 1979.
(27) Herreros, B.; Carr, S. W.; Klinowski, J. *Science* **1994**, *263*, 1585. Herreros, B.; Klinowsky, J. *J. Phys. Chem.* **1994**, *99*, 1025.
(28) Olah, G.A.; Li, X.-Y.; Wang, Q.; Rasul, G.; Surya Prakash, G.K. *J. Am. Chem. Soc.* **1995**, *117*, 8962.
(29) Kuo, C.-M.; Saam, J. C; Taylor, R. B. *Polymer International* **1994**, *33*, 187.
(30) Frisch, M. J.; Del Bene, J. E.; Binkley, J. S.; Schaefer, III, H. F.; *J. Chem. Phys.* **1986** *84*, 2279.

Chapter 6

Mesophase Behavior and Structure of Mesophases in Cyclolinear Polyorganosiloxanes

Yu. K. Godovsky[1], N. N. Makarova[2], and E. V. Matukhina[3]

[1]Department of Polymer Materials, Karpov Institute of Physical Chemistry, Vorontsovo Pole 10, Moscow 103064, Russia
[2]Nesmeyanov Institute of Organo-Element Compounds, Russian Academy of Sciences, Vavilov str. 28, Moscow 117813, Russia
[3]Moscow State Pedagogical University, Malaya Pirogovskaya 1, Moscow 119882, Russia

The phase behavior and structure of the mesophases in cyclolinear polyorganosiloxanes (POCS[1]) with siloxane cycles of various dimensions bonded either by only oxygen atoms or by flexible spacers is considered. It is shown that the 2D-columnar ordering in the mesophase is not typical for all POCS. Formation of a novel mesophase with 1D-order has been discovered for a series of POCS containing cyclohexasiloxane chain fragments substituted with methyl groups (PMCS-6). The most probable arrangement of the PMCS-6 molecules in the mesophase is a layer packing formed by monomolecular layers stacked with good long-range order. Within a layer the cyclohexasiloxane fragments are arranged with a strong tendency to lie on their flat sides along the planes of the layer. The interlayer periodicity being connected with the cyclohexasiloxane thickness only is completly non-sensitive to the chemical structure of PMCS-6. The temperature region of 1D-mesophase existence, however, can be directly regulated by the chemical structure of PMCS-6.

During the last decade a new family of non-mesogenic mesophase polymers – cyclolinear polyorganosiloxanes (POCS) - consisting of linear chains of repeated siloxane cycles of various dimensions bonded either by only oxygen atoms or by flexible spacers was developed (1–5). The general formula of this class of polymers is

$$\left[\begin{array}{c} \\ -Si \\ R^* \end{array} \begin{array}{c} O[(R)_2SiO]_n \\ \\ O[(R)_2SiO]_m \end{array} \begin{array}{c} R^* \\ Si-O-[(R)_2SiO]_l \\ \end{array} \right]_p$$

[1] The another abbreviation, i.e. CLPOS, has been used in the previous works.

where R, R* are alkyl or aryl substitutes, R= R* or R ≠ R*; m =1, 2; n=1, 2; l ≥ 0. The number of the repeat cycles **p** in the macromolecules may range from several to several hundreds.

Similar to linear poly(di-n-alkylsiloxane)s (*4,7*), polyphosphazenes (*4,8*) and polysilanes (*9*), POCSs are able to form columnar liquid crystalline phases (mesophases). Typical POCS with alkyl substitutes are rather flexible. However, in spite of their flexibility, the temperature interval of the stability of POCSs mesophases are surprisingly large (up to three hundred degrees). They are a strong function of the cycle size, the length and nature of organic side chains, the length and nature of spacers, the local tacticity in silsesquioxane fragments and the degree of polymerization (*1–6*).

The classification of polymer mesophases have been considered in many reports (*9-10*). On analyzing the published data there arises much controversy in the identification. In order to avoid confusion in terminology it is necessary to define the basic terms relating to polymer columnar mesophases before further discussion. Columnar polymer systems have both a correlation of the centers of gravity and molecular orientation, but have mesomorphic properties due to the conformational disorder both of the polymer back-bone and side chains. The structural unit of the polymeric columnar mesophase is a macromolecule. In columnar phases macromolecules form regular 2D-periodic arrays. The two-dimensional symmetry of the column packing and the parameters of two-dimensional lattice are strongly dependent on the form and dimensions of the cross-section of a polymer molecule.

It should be noted that 2D-mesophases formed by long side chains in a series of comb-like polymers (*7*) can not be attributed to columnar mesophases due to the lack of properties inherent to columnar mesophases. By the same reason sanidic mesophases (*12*) including 2D-ordered one are not to be classified as columnar ones.

The columnar phases are usually classified according to the symmetry of two-dimensional lattice. Thus, the columnar phases can be subdivided into hexagonal C_h or rectangular C_r (*13*).

Initially we assumed that the columnar type of ordering is a characteristic structure of all mesomorphic POCSs. However, quite recently we recognized that the mesophase structure of a series of POCS cannot be described in terms of the conventional classification used for organic mesophase polymers including 2D-columnar types. In this paper we will discuss the phase behavior and structure of the mesophases in POCS compounds. In particular, the discussion will focus on the effects of chemical structure of both the siloxane backbone and the organic side groups on the phase transitions, the ability to form mesophase structures, the temperature range of mesophase existence and the structural evidence indicating the difference between the columnar and monolayered-type mesophases.

Experimental

The thermal transitions were recorded with a different scanning calorimetry (DSC) Perkin-Elmer DSC-7 according to a conventional procedure. The wide-angle X-ray procedure was performed with filtered CuKα radiation using DRON-3M diffractometer with an asymmetric focusing monochromator (a bent quartz crystal), equipped with a heating and a cooling camera wherein the temperature was automatically regulated (±1°). Diffraction patterns were recorded in transmission mode.

Results and Discussion

2-Dimensional Columnar Mesophase. A series of atactic noncrystallizable POCS homopolymers comprising tetra-[1] and pentasiloxane[2] cyclic fragments [POCS-4 and POCS-5] substituted with ethyl, propyl, phenyl groups[3] were observed to display columnar mesomorphic behavior as well as their linear analogues, polydi-n-alkylsiloxanes (6). In several cases PECS-4 and PECS-5 copolymers containing linear diethylsiloxane units as comonomer are able to form columnar mesophase too (14). Moreover, contrary to polydimethylsiloxane (PDMS), which reveals no mesomorphic behavior, formation of columnar mesophase has been observed in POCS-4, with methyl side groups when its chains are enriched by *trans*-units (15).

X-ray and DSC data on temperature transitions of POCS-4 and -5 (with or without linear siloxane spacers) are summarized in Table I. Noncrystallizable mesomorphic POCS-4 and -5 and their mesomorphic copolymers transform into the columnar mesomorphic glass below the glass transition temperature T_g. Transition to mesomorphic glass does not affect the ordering in columnar mesophase. As one can see from Table I, the temperature interval of the stability of the mesophase depends strongly on the cycle size, the length and nature of organic side chains, the length of spacers and the local tacticity in silsesquioxane fragments.

Columnar Polymesomorphism of POCSs. Depending on the dimensions and the nature of both the side chains and cyclosiloxane moieties, the POCS compounds form C_h or/and C_r columnar phases.

The majority of noncrystallizable atactic POCS-4 and -5 exhibit only one columnar mesomorphic modification (mesophase I). The X-ray pattern of all POCS compounds in mesophase I show a single intense reflection, whose angular position ($2\theta_m$) depends on the type of polymer (Figure 1).

Comparative study of the structural characteristics of the mesophase I has shown the identical hexagonal type of ordering for all POCS in mesophase I (14). Figures 2 and 3 illustrate the strong tendencies toward an increase in d_m as the size of side groups or siloxane cycle increases and a decrease in d_m as a function of the length of the flexible spacer. This corresponds to the hexagonal columnar ordering in mesophase I. The changes in slopes in the plot of d_m vs. T (Figure 2 and 3) correlate with the glass transition seen in the DSC data (Table I). So, the difference in slopes can be attributed to the effect of changes in thermal expansion.

---

[1] Poly[oxy(organocyclotetrasiloxane-2,6-diyl)]s and
[2] Poly[oxy(ethylcyclopentasiloxane-2,6-diyl)],
[3] Hereafter, organic substituents will be designated as follows: O, organic; A, alkyl; M, methyl; Et, ethyl; Pr, propyl; Ph, phenyl.

Table I. Effect of Chemical Structure on the Temperature Transitions of Polyorganosiloxanes exhibiting 2D-columnar Mesophase

MACROMOLECULES	R	ℓ	$T_g,°C$	$T_m,°C$	$T_{II \to I},°C$	$T_i,°C$
1. POCS-4	CH_3	0	-55	70	95	108
	C_2H_5	0	-110	—	—	275
	C_2H_5	1	-112	—	—	130
	C_2H_5	2	-131	—	—	70
	C_2H_5	3	-129	—	—	—
	C_3H_7	0	-55	—	—	325
	C_6H_5	0	68	—	180	> 400
2. POCS-5	C_2H_5	0	-110	—	10	40
	C_2H_5	2	-130	—	—	34
3. LINEAR ANALOGUES						
PDES	C_2H_5		-139	-62	10	53
PDPS	C_3H_7		-110	65	—	207
PDPhS	C_6H_5		49	210	—	> 400

T_m — melting point, T_i —isotropization temperature, $T_{II \to I}$ —transition from the low temperature mesophase to the high temperature one.

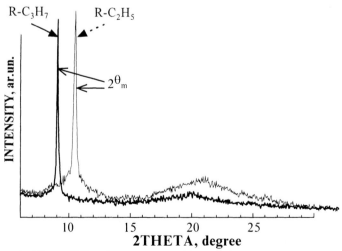

Figure 1. X-ray diffraction patterns of POCS homopolymers containing cyclotetrasiloxane fragments in mesomorphic state with 2D-hexagonal molecular packing.

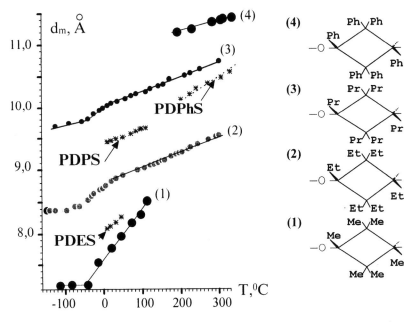

Figure 2. Effect of the chemical structure of organic side groups on the structural parameter d_m of 2D-columnar mesophase.

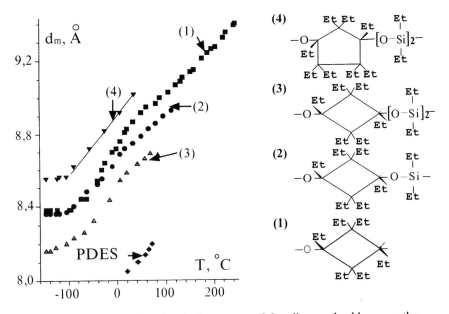

Figure 3. Effect of the chemical structure of the siloxane backbone on the structural parameter d_m of 2D-columnar mesophase.

Interesting enough is the analogy of the phase and structural behavior of two POCS, namely PECS-4 and PPCS-4, and H-bonded columnar mesomorphic analogues built of the molecules of 1,3-dioxitetraalkyldisiloxanes with the general formula: $HO-Si(C_nH_{2n+1})_2-O-Si(C_nH_{2n+1})_2-OH$ with $n=2,3$ (*16*). The latter consists of dimeric associates stacked one on top of the other. In both cases the hexagonal columnar mesophase is observed, the parameter **a** of the hexagonal lattice as well as the behavior of the dependencies **a** vs. **T** for H-bonded systems to be well in accordance with that for the covalent-bonded ones. The only significant difference is the width of the temperature range of the mesophase existence, which is of course much narrower in the case of the H-bonded analogues.

The reversible polymesomorphic transition within the columnar phases has been detected in several POCS homopolymers. In PMCS-4 (*15*) and PPhCS-4 (*17*), the hexagonal columnar phase has been observed at higher temperatures and the rectangular one at lower temperatures. On heating a tentative phase sequence can be set up as follows $C_r \rightarrow C_h$. Note that the limiting factor in realization of polymesomorphic transitions in PMCS-4 and PPhCS-4 is quite different. As to PMCS-4, there is a critical ratio between *trans*- and *cis*-units providing both crystallization and realization of a mesomorphic state, i.e. the polymesomorphic transition is a part of the multistage transition 'crystal $\rightarrow C_r \rightarrow C_h$'. As to PPhCS-4, the Ph-groups mobility appears to be a factor responsible for its mesophase behavior.

Furthermore, polymesomorphism of the columnar phases are inherent to atactic noncrystallizable PECS-5 too (*18*). On heating, a polymesomorphic transition resulting from the appearance of the higher-symmetry rectangular lattice takes place, the tentative phase sequence can be set up as follows $C_{r1} \rightarrow C_{r2}$.

As follows from our studies, the tendency toward ordering in columnar POCS manifests itself at two interrelated levels, inter- and intramolecular ones. On the one hand, the cyclic fragments tend to realize conformations with the smallest intramolecular distances. (Figure 4a). On the other hand, this conformation ensures such an arrangement of the side groups when their interaction can be sufficiently strong to affect the intramolecular ordering and to stabilize a certain rotational configuration. While in organic polymers the isomers are stabilized only by Van der Waals repulsion of atoms or radicals, in POCS this mechanism is complemented by a factor related to the peculiarity of POCS macromolecules: the different chemical nature of the backbone and side chains. As in the linear analogues, the side groups in POCS will tend to form an organic shell for the siloxane chain.

The data presented for POCS as well as data for mesomorphic linear polyorganosiloxanes are visual evidence of inherent possibilities for the controlled molecular design of 2D-columnar systems. Indeed, the temperature interval of the stability as well as the structural parameters of the 2D-columnar mesophase are a strong function of the cycle size, the length and nature of organic side chains, the length of spacers, the local tacticity in silsesquioxane fragments and the degree of polymerization.

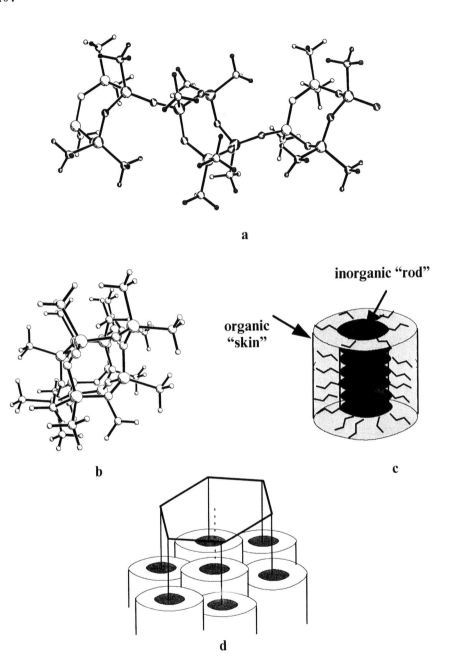

Figure 4. Self-organization of POCS-4 and POCS-5 into a 2D-columnar mesophase. Computer simulation of molecular conformation of PMCS-4 (a, b). Ordering on molecular level (c). Ordering of cycles on intermolecular level (d).

1-Dimensional Monolayered Mesophase. General characterization of the phase behavior of homopolymers[4], copolymers and oligomers[5] with decamethylhexasiloxane cycles (PMCS-6) obtained by means of DSC and X-ray studies is summarized in Tables II - III. A comparison of this data with corresponding data presented in Table I reveals marked differences between the factors responsible for mesophase behavior in PMCS-6 and POCS, which exhibited 2D-columnar mesophase.

First, as seen from Table II, the changes in microtacticity of PMCS-6 homopolymer up to its atactic type have no influence on the ability to form a mesophase. Quite a different situation is observed for atactic methyl-substituted POCS-4 and POCS-5. The latter polymers, like their linear analogue, polydimethylsiloxane, are unable to form mesomorphic structures. As mentioned above, the formation of the columnar mesophase has been observed in POCS-4 with methyl side groups when its chains were enriched by *trans*-units (*15*) and in POCS-4 and -5 (*14, 17*) with side chains longer than methyl. Moreover, the replacement of the tetrasiloxane cycles in macromolecules of methyl-substituted POCS by the hexasiloxane ones results in a quite unexpected remarkable rise of the thermal stability of the mesophase.

Second, the decrease in the degree of polymerization of POCS with the 2D-columnar ordering below p=20 leads to the lack of mesomorphic properties. In contrast, as one can see from Table III, PMCS-6 oligomers exhibit mesomorphic behavior even when p=5.

Moreover, the mesomorphic state of PMCS-6 is characterized by several uncommon and specific features of the structural behavior inherent exclusively to this class of POCS. Let us consider this aspect in detail. The X-ray pattern of all PMCS-6 in mesophase show one narrow well-pronounced reflection at the angular position $2\theta_m$ and two amorphous galo: the first galo with maximum at $2\theta_{a1}$ is in the angle range $2\theta=9°-12°$, the halo is in the range $2\theta=16°-25°$ (Figure 5).

Noncrystallizable mesomorphic PMCS-6s transform into mesomorphic glasses below the glass transition temperature T_g. The transition to the mesomorphic glass does not affect the ordering in the mesophase but is accompanied by the remarkable lowering of the slopes in the plot of d_m vs. T due to the change in the thermal expansion (Figure 6). The slope in the plot of d_m vs. T at temperatures above the glass transition suggests that the long-range order in the mesophase exists in the base plane, i.e. the mesophase is characterized by the lack of the order along the axes of molecular chains. In general, at first sight, X-ray patterns of PMCS-6 in mesophase as well as their temperature transformation seems to be similar to that typical of a 2D-hexagonal columnar packing of POCS (Figure 1). However, a pronounced difference between the mesophase structural data behavior of PMCS-6 and that of 2D-columnar hexagonal POCS reveals. In contrast to POCS-4 and -5, the angular position of the characteristic mesophase reflection ($2\theta_m$) for PMCS-6 happened to be completely non-sensitive to the chemical structure of the macromolecules. On varying chemical structure of PMCS-6 one can only detect the changes in the temperature region of mesophase existence (Table II-III) and in the amorphous

---
[4] Poly[oxy(methylcyclohexasiloxane-2,8-diyl)]s
[5] Olygo[oxy(methylcyclohexasiloxane-2,8-diyl)]s

Table II. Effect of Chemical Structure on the Temperature Transitions of POCS with Decamethylcyclohexasiloxane Fragments (PMCS-6).

MACROMOLECULE		$T_g,°C$	$T_m,°C$	$T_{II \to 1},°C$	$T_i,°C$
1. HOMOPOLYMERS					
	trans-	−91	−50 - −55	105 - 110	291 - 312
	cis-	−91	—	90 - 100	232 - 253
	atactic-	−91	—	—	240 - 93
2. COPOLYMERS					
	m=1	−93	0 - 5	80 - 90(*)	90 - 110
	m=2	−96	−20 - −16	—	—
	m=3	−95	—	—	—
	trans-	−71	—	0 - 25(*)	120 - 150
	cis-	−72	−20 - 5	—	50 - 85
T- hexamethyltetrasiloxane cycle					
3. BLOCK-COPOLYMERS					
	p=3, m=1	−90	—	130 - 140	200 - 210
	p=3, m=3	−100	—	—	90 - 100
	p=3, m=5	−98	—	—	—
	p=5, m=1	−95	—	150 - 170	240 - 250
	p=5, m=3	−92	—	140 - 150	200 - 210
	p=5, m=5	−93	—	110 - 130	180 - 90

For notation of temperatures see Table I.

Table III. Characteristics of Mesophase Oligomers (p=5) with General Formula:

End group -X	$\Sigma \Delta V_i^E, A^3(*)$	$\beta^C(**)$	$T_g,°C$	$T_i, °C$
- H	13.4	0.993	−73	120 - 135
- Si (CH$_3$)$_3$	182.3	0.915	−91	95 - 115
-Si(CH$_3$)$_2$ - (CH$_2$)$_2$- CN	256.1	0.885	−90	170
-Si(CH$_3$)$_2$ - C$_4$H$_9$	284.9	0.874	−96	110 - 120
-Si(CH$_3$)$_2$ - C$_6$H$_{13}$	353.3	0.848	−95	110 - 116
-Si(CH$_3$)$_2$ (CH$_2$)$_4$OCOC$_6$H$_{13}$	546.6	0.783	−86	90
-Si(CH$_3$)$_2$ -(CH$_2$)$_4$OCOC$_6$H$_5$	492.6	0.800	−79	115 - 120

* Van der Vaals volume of two end groups
** Volume fraction of central cyclic fragments
For notation of temperatures see Table I.

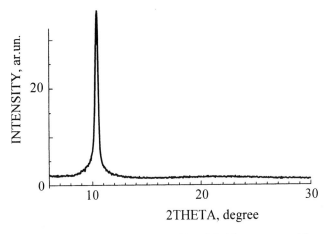

Figure 5. X-ray diffraction patterns of PMCS-6 in mesomorphic state.

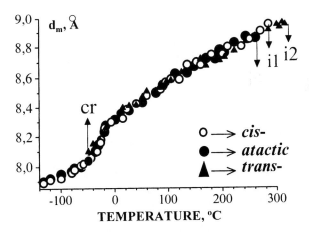

Figure 6. Temperature dependencies of interlayer distance d_m for PMCS-6 homopolymers with various microtacticity. The arrows indicate the isotropization temperature for *cis*-(i1), *atactic*- (i2) and *trans*- (i3) microtacticity and the melting temperature for the *trans*-microtacticity (cr).

scattering distribution, in particular, in the angular position of the maximum of the first amorphous galo $2\theta_{a1}$ (Figure 5).

The whole set of the experimental data leads to the conclusion that the replacement of rigid tetrasiloxane cycles in macromolecules of methyl-substituted POCS by the flexible hexasiloxane ones results in qualitative transformations in the structural organization of the mesomorphic state.

Taking into account the properties of the parameter d_m, which is completely non-sensitive to the chemical structure of the macromolecules, one can conclude that the mesophase of PMCS-6 is characterized by one-dimensional (1D-) positional long-range order. It suggests that all experimental evidence obtained for mesophase behavior of PMCS-6 can be described exclusively within the limits of a 1D-structural model. This conclusion is confirmed by the examination of the X-ray diffraction patterns of non-crystallizible atactic PMCS-6 homopolymer obtained at temperatures of about 10-40° higher than the glass transition temperature. These patterns show a relatively low intensity maximum at 2θ' in addition to the intense principle mesophase reflection at $2\theta_m$. The ratio d_m/d' is equal to 2, confirming the formation of a 1D-ordered structure.

To explain the reasons for the mesomorphic behavior of PMCS-6 one has to conclude that there is an anisotropy of the intermolecular forces in the mesophase of PMCS-6 and that a structural unit of 1D-mesophase is a supramolecular element.

Macromolecules of PMCS-6 do not comprise rigid moieties (mesogens) with an essentially high geometrical shape anisotropy and specific groups that are able to create the anisotropy of the intermolecular interactions. Hence, one may suggest that it is the amphiphilic constitution of the PMCS-6 macromolecules that has to play the dominant role in the mesophase formation. It means that the main driving forces for the mesophase development in PMCS-6 are similar to that for the 2D-columnar mesophase POCS systems. In connection with this a question arises. What are the principle differences between two classes of POCS, i.e. between 2D-columnar systems (POCS-4 and -5) and PMCS-6 that are responsible for the decrease in the dimensionality of mesophase long-range order ? Note that the volume fraction of the organic side groups $\beta^{al}=\Sigma\Delta V_i^{al}/\Sigma\Delta V_i$ for PMCS-6 ($\beta^{al}=0.44$) are comparable with the correspondent value for the 2D-columnar PMCS-4 ($\beta^{al}=0.47$).

As mentioned above, the formation and stabilization of 2D-columnar structures by the macromolecules of POCS including PMCS-4 are caused by the tendency to nanoscale phase separation on inter- and intramolecular levels. The presence of the two rigid moieties in their macromolecules ensures a stability of the intra-molecular correlations of the cyclic fragments creating the molecular conformational state where the siloxane chain are enveloped by the organic side groups (Figure 4).

To clarify the main reason for the formation of 1D-mesophase in PMCS-6 one has to compare the molecular conformation of PMCS-6 with that of the PMCS-4. This comparison reveals significant differences. In contrast to macromolecules of PMCS-4 where the cycles are located at a certain angle with respect to the main axis (Figure 4), the cycles in the macromolecules of PMCS-6 are arranged presumably parallel to the main axis (Figure 7a). This feature of molecular conformation of PMCS-6 resulting from the high flexibility of hexasiloxane cycles sharply reduces the possibility of the cycles to correlate on the molecular level.

It is important to emphasize that there is a uniform chain fragment for all PMCS-6 that determines the ability of the molecules to form mesophase structures as well as the value of the structural parameter of the mesophase d_m. This feature can be inherent only to a decamethylhexasiloxane cycle. The d_m value near T_g is close to the thickness of the cycle. Schematically a decamethylhexasiloxane cycle can be represented as an inorganic "disk" with an organic "tire". It seems that it is the geometrical shape anisotropy (although non-pronounced) of a decamethylhexasiloxane cycle in combination with the anisotropy of intermolecular interactions resulted from the specific feature of the organic distribution around the siloxane cycle (Figure 7b) that creates the conditions for the self-ordering of PMCS-6 macromolecules into the mesomorphic structure.

Model of structural organization. As follows from the properties of d_m parameter, the supramolecular unit responsible for the mesophase behavior must be uniform for all PMCS-6. A monomolecular layer is the only possible type of suprastructure that meets these requirements. The d_m value corresponds to the thickness of a decamethylhexasiloxane cycle. So, within a layer the cyclohexasiloxane fragments are arranged with a strong tendency to lie on their flat sides along the planes of the layer. The molecular arrangement characterized by the existence of an equal-spaced (with periodicity equal to d_m value) monomolecular layers (Figure 7c) is the only model providing an adequate explanation of the experimental evidence obtained for the PMCS-6's mesophase behavior. Indeed, when the density of the cycles in a monomolecular layer is high enough the value of an interlayer d_m-spacing should be independent of the chemical structure. As for the intralayer diffraction scattering, in general, it is an amorphous one. However, the distribution of its intensity, in particular the angular position of the maximum of the first amorphous galo $2\theta_{a1}$, is a function of the chemical structure of PMCS-6.

Hence, the mesophase of PMCS-6 may be classified as a novel type of mesomorphic structures. Moreover, we also believe that the direct consequence of such a structural organization of the mesophase is a striking ability of the PMCS-6 macromolecules to spread at the air-water interface into Langmuir monolayers and their self-organized collapse via transformation to multilayers (*19-21*).

The structural model of 1D-mesophase explains adequately the whole set of experimental data obtained for PMCS-6 mesophase including the polymesomorphic phenomena observed for a majority of PMCS-6.

1D-Monolayer Polymesomorphism. DSC scans show that a number of PMCS-6 exhibit a polymesomorphic transition II→ I within the temperature region of 1D-mesophase existence (Table II). As a rule for atactic non-crystallizible PMCS-6, the X-ray diffraction patterns obtained below $T_{II \to I}$ reveal no additional Bragg reflections besides the principal reflex at $2\theta_m$. While the parameter d_m shows systematic linear increase with the temperature due to the thermal expansion, the slope in the plot $2\theta_{a1}$ vs. T experiences a change just near $T_{II \to I}$. A similar

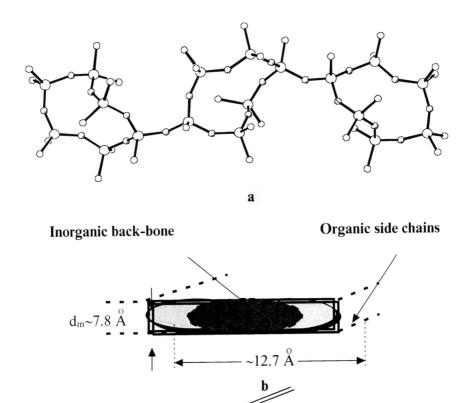

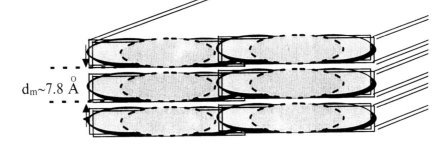

Figure 7. Self-assembling of PMCS-6 into a 1D-monolayered mesophase. Molecular conformation (a). Cross-section of molecule (b). 1D-packing of monomolecular layers (c).

transformation of the X-ray diffraction pattern has been observed for PMCS-6 under the other governing factors.

Firstly, the varying of the homopolymer microtacticity affects both the value $2\theta_{al}$ and the slopes in the plots $2\theta_{al}$ vs. T without any influence on the principle mesophase parameter d_m and the plot d_m vs. T. Secondly, the transitions 'mesophase → mesophase glass' exhibited as a rule by atactic non-crystallizible PMCS-6 also give rise to the changes of the slopes in the plots $2\theta_{al}$ vs. T. Note, that the intralayer amorphous scattering makes its significant contribution to the integral intensity of the first diffusive galo, the angular position of its maximum $2\theta_{al}$ to be dependent on the molecular structure of 1D-system.

Taking into account all these experimental regularities, one can reasonably believe, that the change in the slope in the plots $2\theta_{al}$ vs. T observed at the poly-mesomorphic transition temperatures $T_{II\to I}$ is the result of the thermotropic changes in molecular conformations and in the segment mobility, which lead to the transformation of the intralayer short-range order only.

The transformation of long-range order during the polymesomorphic transition is observed only for two high stereoregular PMCS-6 (Table II). Below $T_{II\to I}$ the X-ray diffraction patterns indicate one or four very weak Bragg reflections in addition to the intense principle one at $2\theta_m$. This experimental evidence allow us to conclude that the cooling below $T_{II\to I}$ has not induced any changes in the 1D- monolayer packing with interlayer d_m-spacing. However, there appears a second level of order below $T_{II\to I}$, i.e. the long-range order within monolayers themselves. It is quite evident that the rise in the degree of order in the monolayer is directly connected with the stereoregularity of the molecules forming the mesophase.

Mesophase oligomers and block-copolymers. To clarify the role of the length of macromolecules in the mesophase behavior, a series of cyclolinear oligomers and block copolymers were synthesized and studied. The results obtained are summarized in Tables II, III. It was established that PMCS-6 oligomers with the degree of polymerization p=3 are not able to form a mesophase. The mesophase behavior similar to the advanced homopolymers occurs only for oligomers with the degree of polymerization p≥5. One can see that the mesophase ordering in oligomers is controlled by both the amphiphilic molecular constitution and geometrical factor, i.e. aspect ratio of the cyclic block.

The minimal aspect ratio of the cyclic oligomer block essential for the mesophase formation is quite comparable with the corresponding values for classical organic LC compounds. The temperature range of the mesophase existence expanses considerably with further increase of the length of the cyclic part of oligomers due to the sharp rise of the isotropisation temperature. The structural organization of the mesophase appears to be non-sensitive to the chemical nature of the end groups. We believe that this property occurs due to a remarkable increase in the ratio between the van der Waals volumes of the central core and that of the end groups. For example, for PMCS-6 oligomers, this ratio exceeds one order of magnitude, while in classical LC compounds the ratio of these volumes are quite comparable. Hence, taking into account typical LC properties of PMCS-6 oligomers one can consider the central core composed of five siloxane cycles to be a peculiar moiety responsible for the mesophase behavior. Therefore, the corresponding block-copolymers with very long flexible spacers between the moieties should exhibit mesomorphic behavior. Table III shows that block-copolymers with p=5 strongly obey this requirement.

Thus, the data obtained for mesomorphic oligomers and block-copolymers based on PMCS-6 show the duality of their mesomorphic properties. It means that these systems exhibit some properties characteristic of classical LC compounds simultaneously with the features inherent to PMCS-6 systems.

Conclusions

The analysis of the effects of chemical structure both the siloxane backbone and the organic side groups of cyclolinear polyorganosiloxanes (POCS) on the phase transitions, the ability to form mesophase structures, the temperature range of mesophase existence and the structural evidence has been carried out.

It was shown that POCS comprising tetra- and pentasiloxane cyclic fragments substituted with ethyl, propyl, phenyl groups display columnar mesomorphic behavior as well as their linear analogues, polydi-n-alkylsiloxanes. Self-organization of POCSs into a 2D-columnar mesophase implies self-coordinated ordering on intra- and intermolecular levels. Siloxane cycle fragments tend to correlate on molecular level packing one on top of the other. The ordering of siloxane cycles within a molecule enables organic side substituents to wrap around the siloxane chain. This arrangement of organic chains corresponds to their best overlapping. On the other hand intramolecular cycle ordering itself gains stability due to the interactions between organic groups on inter- and intramolecular levels. The temperature interval of the stability as well as the structural parameters of the 2D-columnar mesophase are a strong function of the cycle size, the length and nature of organic side chains, the length of spacers, the local tacticity in silsesquioxane fragments and the degree of polymerization

The most probable arrangement of the molecules of POCSs comprising decamethylhexasiloxane cyclic fragments (PMCS-6) in the mesophase is a layer packing formed by monomolecular layers stacked with a good long-range order. The chemical structure of PMCS-6 does not influence the main structural parameter of 1D-mesophase d_m being connected with the stacking periodicity related to the layer thickness. Within a layer the cyclohexasiloxane fragments are arranged with a strong tendency to lie on their flat sides along the planes of the layer. However, the temperature region of 1D-mesophase existence is a function of the chemical structure of PMCS-6.

Self-organization of PMCS-6 into the 1D-mesophase is directly associated with formation of supramolecular elements – monomolecular layers. This phenomenon is caused by conformational peculiarities of PMCS-6 molecules that outcome from the properties of cyclohexasiloxane fragments. Contrary to rigid tetra- and pentasiloxane rings, flexible hexasiloxane rings have no ability to correlate within a molecule. As a result, the siloxane PMCS-6 chain is unable to wrap itself with organic side groups. However, amphiphilic constitution and anisotropic form of hexasiloxane ring promote nanoscale separation exclusively on the intermolecular level.

The structure of PMCS-6 mesophase cannot be described in terms of the conventional classification used for organic mesophase polymers. From our viewpoint, PMCS-6 are the novel class of mesomorphic systems.

Acknowledgments

This material is based upon work supported by Russian Foundation for Basic Research (grant 96-03-32497). One of the authors (Yu.K.G.) would like to thank Alexander von Humboldt Foundation for the Research Award.

Literature Cited

1. Makarova, N.N.; Godovsky, Yu. K. *Progr. Polym. Sci,* **1997**, *22*, 1001.
2. Makarova, N.N.; Godovsky, Yu. K.; Lavrukhin, B.D. *Polymer Science (Russia),* **1995**, *A37*, 375.
3. Godovsky, Yu. K.; Makarova, N.N. *Phyl.Trans.R.Soc.Lond.A,* **1994**, *384*, 45.
4. Godovsky, Yu. K.; Papkov, V.S. *Adv.Polym.Sci,* **1989**, *88*,129.
5. Godovsky, Yu. K.; Makarova, N.N.; Kuzmin N.N. *Makromol.Chem. Makromol. Symp.,* **1988**, *188*, 119.
6. Godovsky, Yu. K.; Makarova, N.N.;. Matukhina, E.V *Polym.Prepr. ACS,* **1998**, *39*, 485.
7. Molenberg, A.; Moeller, M.; Sautter, E. *Progr.Polym.Sci.*, **1997**, *22*, 1133
8. Magill, J.H. *J. of Inorganic and Organometallic Polymers*, **1992**, *2*, 213.
9. Weber, P.; Guillon, D.; Scoulious, A.; Miller, R.D. *J. Phys. (Paris),* 1989, *50*, 793.
10. Definition of basic terms relating to low mass and polymer liquid crystals. IUPAC Commission IV.1., **1994**.
11. Ungar, G. *Polymer*, **1993**, *34*, 2050.
12. Herrmann-Schonherr, O.; Wendorff, J.H.; Kreuder, W.; Ringsdorf, H. *Makromol. Chem., Rapid Commun.* **1986**, *7*, 97.
13. Sauer, T. *Macromolecules*, **1993**, *26*, 2057.
14. Makarova, N.N.; Petrova, I.M.; Matukhina, E.V.; Godovsky, Yu. K.; Lavrukhin, B.D. *Polymer Science (Russia),* **1997**, *A39*, 1078.
15. Matukhina, E.V.; Boda, E.E.; Timofeeva, T.V.; Godovsky, Yu. K.; Makarova, N.N.; Petrova, I.M.; Lavruchin, B.D. *Polymer Science (Russia),* **1996**, *A38*, 1020.
16. Makarova, N.N.; Kuzmin, N.N.; Godovsky, Yu. K.; Matukhina, E.V. *Dokl. Akad. Nauk SSSR,* **1988**, *300*, 372.
17. Matukhina, E.V.; Godovsky, Yu. K.; Makarova, N.N.; Petrova, I.M. *Polymer Science (Russia),* **1995**, *A37*, 1021.
18. Makarova, N.N.; Matukhina, E.V.; Godovsky, Yu. K.; Lavrukhin, B.D. *Polymer Science (Russia),* **1992**, *B33*, 56.

19. Sautter, E.; Belousov, S.I.; Pechhold, W.; Makarova, N.N.; Godovsky, Yu. K. *Polymer Science (Russia)*, **1996**, *A38*, 1020.
20. Fang, J.; Dennin, M.; Knobler, C.M.; Godovsky, Yu. K.; Makarova, N.N.; Yokoyama, H. *J.Phys.Chem.B.*, **1997**, *101*, 3147.
21. Godovsky, Yu. K.; Fang, J.; Knobler, C.M.; Makarova, N.N., *Polym.Prepr. ACS*, **1998**, *39*, 543.

Chapter 7

Photoluminescence of Phenyl- and Methylsubstituted Cyclosiloxanes

Udo Pernisz[1], Norbert Auner[2], and Michael Backer[3]

[1]Dow Corning Corporation, Central Research and Development,
Midland, MI 48686–0994
[2]Institut für Anorganische Chemie, J. W. Goethe Universität,
Marie-Curie-Strasse 11, D–60439, Frankfurt, Germany
[3]Institut für Anorganische und Allgemeine Chemie,
Humboldt Universität, Berlin, Germany

The observation of strong blue photoluminescence from several different types of phenylated Si-containing molecules upon excitation with UV light was investigated by measuring the emission and excitation spectra of the luminescence as well as the time dependence of the phosphorescence. A range of variously substituted, Si-containing cyclic compounds was synthesized and analyzed in order to identify the origin of the photoluminescence and to determine the role of the substituents in the effect. The two classes of compounds studied were 2,3-diphenyl-substituted silacyclobutenes and stereoregularly-built phenylated cyclosiloxanes.

The excitation with UV light of phenyl-containing Si compounds such as 2,3-diphenylsilacyclobutene **1** and siloxanes of type **2** results in the emission of strong, blue photoluminescence from the solid material. It can be assumed that the origin of the photoluminescence is essentially associated with the π-electron system of the substituents at the C and/or Si since that corresponding class of compounds commonly shows this effect although for small molecules such as benzene or stilbene the emission usually occurs in the UV[1]. The fact that the photoluminescence can be observed in the visible gave rise to a series of investigations aimed at understanding the role of the Si atom when it is directly bound to such an aromatic moiety with the assumption that the Si shifts the emission from the UV into the visible part of the spectrum.

The questions of the effect that the Si atom has on the electronic structure of aromatic molecules can be expanded to include the effect of other substituents on the Si, including different π-conjugated substituents. A second part of this question regards the presence of the Si-O- group or, generally, the effect of linear or cyclic siloxane configurations with which the aromatic part interacts via the Si atom to which it is bound. The cyclosiloxanes are of interest in the wider context of the photoluminescent properties of silica materials containing various defects, including mechanical ones (strained bond angles) and chemical ones (non-bridging oxygen defects, and Si radicals due to dangling bonds).

A series of suitable molecules covering the range of interest was investigated by measuring the emission and excitation spectra of the luminescence and phosphorescence components as well as the time dependence of the latter effect.

Experimental Details

The compounds investigated fall in two classes. The first one is based on a silacyclobutene subunit which is obtained through cycloaddition reaction of dichloroneopentylsilene to tolane (Tolane CycloAdduct, TCA). Starting from 1,1-dichloro-2,3-diphenyl-1-silacyclo-2-butene **1** (TCA), a whole series of organo-substituted derivatives is available by usual organometallic routes. Utilizing the silicon dichlorofunctionality and the outstanding chemical and thermal stability of **1**, this silacyclobutene building block is easily incorporated into cyclosiloxanes yielding silaspirocyclic compounds.

The second class deals with stereoregularly-built phenylated trimethylsiloxy-cyclosiloxanes of various ring sizes of which **2** shows the cyclotetrasiloxane as an

example. (The phenyl group on the front Si is given as Ph, the trimethylsiloxo group on the back Si was omitted for clarity.) This group of compounds, [PhSi(OSiMe$_3$)O]$_n$, n = 4, 6, 8, 12, was synthesized following the literature.[2]

Tolane Cycloadduct (TCA) Derivatives. TCA, or 1,1-dichloro-2,3-diphenyl-4-neopentyl-1-silacyclo-2-butene, **1**, is preparatively easily accessible[3] by starting from equimolar amounts of trichlorovinylsilane, tolane, and lithium-*tert*-butyl, forming crystalline **1** in nearly quantitative yield. The preparation of alkyl and aryl derivatives is performed by substituting the Cl on the Si employing the usual organometallic synthetic route, e.g., by use of Grignard reagents or Li organyls. The silanediol **3** is obtained quantitatively from **1** in a controlled hydrolysis. Thermolysis of **3** between 80 °C and 110 °C upon reflux in toluene solution leads to the condensation product 1,3-siloxanediol **4**. This solid was characterized by x-ray analysis after re-crystallization from toluene.

[Scheme showing conversion of compound 1 (dichlorosilacyclobutene with Ph, Ph, and neopentyl substituents) with H₂O / 0 °C / NaHCO₃ to compound 3 (dihydroxy analog), then to compound 4 (a disiloxanediol dimer), with loss of H₂O and ΔT.]

TCA-containing Cyclosiloxanes. From co-condensation of **4** with dichlorofunctional silanes such as dichlorodimethyl- or -diphenylsilane, such as R_2SiCl_2, R = Me, Ph, the bisilacyclobutene-substituted cyclic trisiloxanes as exemplified by **5** are obtained. Their monocrystalline forms were characterized by x-ray analysis.

[Reaction scheme: compound 4 + Me_2SiCl_2 → compound 5 (cyclic trisiloxane with two silacyclobutene units and a Me₂Si group).]

Based on the extraordinary chemical and thermal stability, the 1,1-dichloro-2,3-diphenyl-4-neopentyl-1-silacyclobutene subunit can be easily incorporated into dimethylsiloxane chains by co-condensation reactions with α,ω-oligosiloxanediols. Thus, condensation of **1** with 1,1,3,3-tetramethyl-1,3-disiloxanediol, **6**, generates the discrete spirocycle 1-(2′,3′-diphenyl-4′-neopentyl-1-silacyclo-2′-butene)-3,3,5,5-tetramethylcyclotrisiloxane, **8**, while reaction with the phenyl-substituted siloxane analog, **7**, yields the phenylcyclotrisiloxane **9**.

Co-condensation of **4** with 1,1,3,3,5,5-hexaphenyl-1,5-trisiloxanediol even generates the corresponding 8-membered cycle which was also characterized by single crystal x-ray analysis. Photoluminescence studies of this compound will be reported elsewhere.

Instrumentation. The initial observations of strong blue photoluminescence, i.e., light with wavelength $\lambda > 400$ nm, in the compounds discussed were made visually by irradiating the samples with light from a pulsed nitrogen laser at a wavelength $\lambda = 337$ nm and a beam energy of *ca.* 200 μJ per pulse.

[Scheme showing compound 1 (dichlorosilacyclobutene with Ph groups) + disiloxane diol (R = Me 6, R = Ph 7) reacting in Et₂O/NR₃ to give cyclic product 8 (R = Me) and 9 (R = Ph).]

The photoluminescence spectra were obtained with a SPEX Fluorolog 2 instrument (Jobin/Yvon) with excitation and emission monochromators equipped with single gratings of 1200 lines/mm, blazed at wavelengths $\lambda = 330$ nm and $\lambda = 500$ nm, respectively. Both monochromators had a focal length of $f = 0.22$ m; at the commonly selected slit widths of 0.25 mm, the spectral bandwidth of the instrument was typically 1 nm. The powdered samples were inserted into the center tube (having ca. 3 mm inner diameter) of a quartz dewar with which measurements could be performed both at room temperature and at the temperature of liquid nitrogen (77 K).

Phosphorescence excitation and emission were obtained by using a flash lamp of 3 µs duration (at half maximum) for excitation of the sample at select wavelengths and recording the emission intensity at longer wavelengths after an adjustable delay time during an appropriately set time window (photon counting). Similarly, phosphorescence intensity decay time measurements were performed by varying the delay time at fixed excitation and emission wavelengths.

Results and Discussion

Photoluminescence of Silacyclobutene Derivatives. The visual observation of blue photoluminescence is confirmed by the spectral distribution of the emitted light measured under continuous excitation at a wavelength of $\lambda = 320$ nm. Figure 1 shows the luminescence emission spectra of the four TCA derivatives investigated: ethyl-TCA, methyl-TCA, hydroxyl-TCA, and phenylethynyl-TCA. Both Cl are replaced with the same substituent for each compound. The measurements were made at room temperature using the samples in powder form, filled into thin quartz tubes. The four samples have very similar emission characteristics with the maximum located at around 400 nm; there is only a small effect from the substitutent on its position. This indicates that the photoluminescence is associated with the cis-stilbene part of the silacyclobutene. However, the emission intensity is strongly affected by the substituent. Most notably, the phenylethynyl group appears to quench the photoluminescence emission, an effect also observed with polysilanes.[4] Another factor in the emission process appears to be the ring structure itself through which the Si atom is attached to the cis-stilbene group as is observed when comparing **1** to a linear analog of comparable electronic configuration such as trimethylsilyl-cis-stilbene which does not show observable photoluminescence upon excitation with light of 337 nm. In contrast, one observes strong blue emission from the triphenylsilyl-*cis*-stilbene under these conditions, confirming the previously stated role of the Si atom in lowering the energy levels of aromatic compounds bound to it.

While there is no discernible effect of the O atom from the hydroxyl groups at the Si atom on the photoluminescence emission maximum, compared to the C of the methyl and ethyl groups, the oxygen bridge between the Si atoms of the bis-TCA-disiloxanediol configuration, **4**, appears to shift the maximum of the emission to longer wavelengths as shown in Figure 2. This observation raises the question of the effects that contribute to the shift of the emission from the phenyl groups in the *cis*-stilbene moiety of the TCA unit. Since the oligosiloxanediols as well as the TCA-cyclosiloxanes can be readily prepared it is of interest to understand the various possible contributions to the photoluminescence of these Si-containing compounds.

Photoluminescence of Cyclosiloxane. For comparison, the photoluminescence emission spectra of the silaspirocycles **8** and **9** are shown in Figure 3; both exhibit the strong blue emission seen in the other parent silacyclobutenes ($\lambda > 400$ nm). Again it is noted that the methyl-substituted siloxane ring has a higher emission intensity than the phenyl-containing ring. This is interesting because the perphenylated cyclosiloxane (commercially available) is itself also a very strong blue-luminescing material, see Figure 4, while the permethylated cyclosiloxanes do not exhibit any similar photoluminescence, a statement equally true for siloxane rings containing any substituent other than one with a π–electron system. Notably, the photoluminescence intensity increases with the number of the phenyl substitutents as they replace the methyl groups at the silicon. Thus, we found a series of decreasing photoluminescence intensity for the cyclotrisiloxane D_3 compounds as $D_3^{Ph2} > D_2^{Ph2}D^{Me2} > D^{Ph2}D_2^{Me2}$; D_3^{Me2} shows no photoluminescence. Another conclusion that can be drawn from the fact that non-aromatic substitution on the Si atom in a siloxane ring does not give rise to (visible) photoluminescence is that the effect – at least at the observed very high intensity – has its origin not in the siloxane ring itself but in the Si-Ph bond.

The experimental findings presented in the preceding material lead to the question of the mechanism by which the UV photoluminescence emission usually observed in the far UV with most aromatic compounds is shifted towards the blue and into the visible part of the spectrum. An analogous effect was observed for the organic molecule phenylazo-*tert*-butyl in which the ionization energy is lowered by 0.5 eV when the C on the azo group is replaced by Si, that is, when the *tert*-butyl is replaced by trimethylsilyl.[5] This is explained by an inductive effect resulting in considerable electron donation from the Si substituent. In an attempt to analyze the role played by the Si atom that is attached to an aromatic moiety such as a phenyl group, in the conversion of the absorbed photon energy to visible luminescence, the series of stereoregular phenylcyclosiloxanes (of which the four-membered molecule is shown as **2**) of different ring sizes was investigated.

Effect of Ring Size of Phenyl-substituted Cyclosiloxanes. The stereoregularly built phenylated cyclosiloxanes (all-cis and all-trans) with trimethylsiloxy groups attached opposite each phenyl group, **2**, has been described in the literature and was

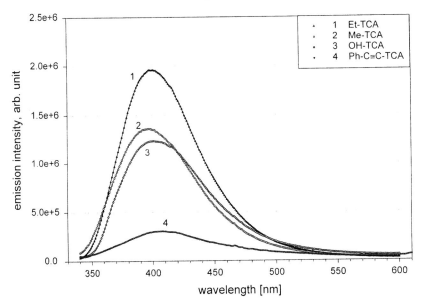

Figure 1. Photoluminescence emission spectra at room temperature of differently silicon-substituted tolane cycloadduct (TCA) compounds with general fomula 1,1-R_2-2,3-diphenyl-4-neopentyl-1-silacyclo-2-butene where R = ethyl (trace 1), R = methyl (trace 2), R = hydroxyl (trace 3), and R = phenylethynyl or $C_6H_5C\equiv C-$ (trace 4). Excitation wavelength 320 nm, spectral bandwidth 1 nm.

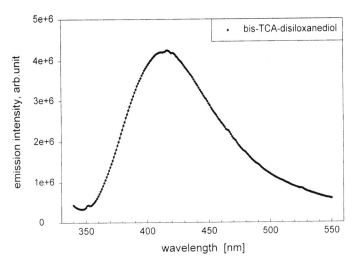

Figure 2. Photoluminescence spectrum of the bis-TCA-disiloxanediol, **4**, at room temperature, as a solid. Excitation wavelength 320 nm, spectral bandwidth 2 nm.

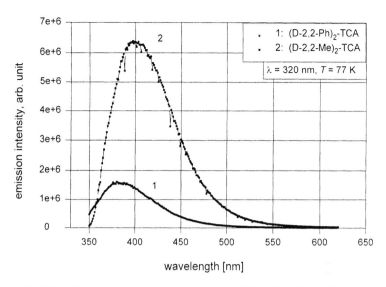

Figure 3. Photoluminescence emission spectra of functionalized silaspirocycles (D-2,2-R)$_2$-TCA or 1-(2′,3′-diphenyl-4′-neopentyl-1-silacyclo-2′-butene)-3,3,5,5-tetra-R-cyclotrisiloxane, **8** and **9**. Lower trace (1): R = phenyl (Ph); upper trace (2): R = methyl (Me). Measured at liquid nitrogen temperature, T = 77 K, excitation wavelength 320 nm, spectral bandwidth 1 nm.

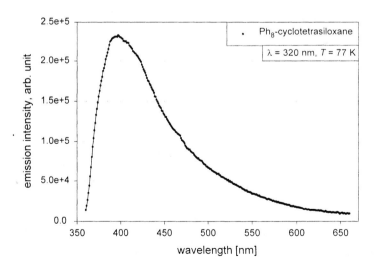

Figure 4. Photoluminescence emission spectrum of octaphenylcyclotetrasiloxane solid, at liquid nitrogen temperature. Excitation wavelength 320 nm, spectral bandwidth *ca.* 8 nm.

accordingly re-synthesized. The four cyclics prepared were the four-, six-, eight-, and twelve-membered rings; in the context of this work, they will be referred to as D4, D6, D8, and D12, using D as an abbreviation for the functionalized Si-O- group in the siloxane chain. For comparison, a commercially available small linear compound, tetraphenyldisiloxanediol, was also included in the photoluminescence study. Although not a cyclic, it is referred to as D2 in the following.

Since the phenyl groups in the stereoregular cyclosiloxanes chosen for this investigation all lie on the same side of the siloxane ring, it appears possible that an effect of energy transfer between the phenyl groups is observed in the photoluminescence emission, either by direct interaction between the π-electrons of the aromatic rings, or mediated by the siloxane groups separating adjacent phenyl groups that are attached to the Si atoms. In addition to an interaction between the pendent groups, to the extent that they participate in the internal energy exchange and transfer from the absorption to the emission process, one could also expect an effect of the ring size itself on the photoluminescence characteristics.

Figure 5 shows the photoluminescence emission spectra of the four stereoregular phenylcyclosiloxanes, D4 ... D12, together with the linear tetraphenyl-siloxanediol, D2. The graph demonstrates that the ring structure has a pronounced effect on the spectral characteristics of the photoluminescence, however, it is not immediately obvious in what way the cyclosiloxane size controls each spectrum. One notes the very strong emission above 370 nm of the D2; this feature recurs in the other molecules with decreasing intensity although there is a D8-D6 reversal in this sequence. (The short-wavelength flank of all spectra is due to a second-order filter with an absorption edge at 370 nm.) The strong wide peak around 460 nm of the D8 ring is surprising.

Figure 6 which shows the luminescence excitation spectra does not clarify the picture although one can observe the same rank ordering in the emission intensity (at a wavelength of 505 nm) as with the emission spectra shown in Figure 5 except for the D2 molecule having the lowest value. This appears plausible considering that if interaction effects were essentially localized between two adjacent aromatic groups the behavior of the large ring should be very similar to that of the short linear molecule.

The other common feature of the spectra in Figure 6 is the sequence of small peaks and shoulders which can be analyzed into at least two vibrational bands, one having an energy separation of 0.36 eV (2930 cm^{-1}), and the other separated by 0.13 eV (1050 cm^{-1}). Two more line series are discernible but were not further identified and assigned. The two major series are ascribed to the C-H stretch vibration on the phenyl ring, and the Si-O-Si vibration from the siloxane ring, according to IR data in the literature.[6] The appearance of the Si-O vibronic structure indicates that the siloxane ring participates in the energy relaxation processes between the absorption and emission of photons in these compounds.

While the measurements of the photoluminescence intensity under steady-state excitation as discussed above did not reveal a systematic influence from the cyclosiloxane rings, it was possible to observe such an effect in the characteristics of the phosphorescence emission. Two types of experiments were performed in this mode, namely recording the emission intensity after a fixed delay time ($\Delta t = 50$ μs) at a fixed emission wavelength ($\lambda_{ms} = 505$ nm) as a function of excitation wavelength,

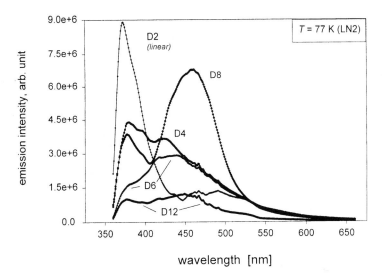

Figure 5. Photoluminescence emission spectra of the stereoregular phenylcyclosiloxanes, [PhSi(OSiMe$_3$)O]$_n$, n = 4, 6, 8, 12, together with the linear tetraphenyl-1,3-siloxanediol, measured (in powder form) at the temperature of liquid nitrogen, T = 77 K. Excitation wavelength 280 nm, spectral bandwidth $ca.$ 2 nm, edge filter (370 nm). D2: linear; D4: n=4; D6: n=6; D8: n=8; D12: n=12.

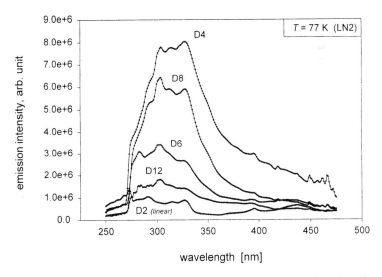

Figure 6. Photoluminescence excitation spectra of the stereoregular phenylcyclosiloxanes, [PhSi(OSiMe$_3$)O]$_n$, n = 4, 6, 8, 12, together with the linear tetraphenyl-1,3-siloxanediol, measured (in powder form), at the temperature of liquid nitrogen, T = 77 K. Emission wavelength 505 nm, spectral bandwidth $ca.$ 1 nm. D2: linear; D4: n=4; D6: n=6; D8: n=8; D12: n=12.

and measuring the time dependence of the phosphorescence emission at a fixed excitation wavelength (λ_{xc} = 320 nm) at two emission wavelengths (λ_{ms} = 380 nm and λ_{ms} = 458 nm).

Figure 7 shows the phosphorescence excitation spectra of the five compounds. Intensity is no longer an obvious distinction, except for the D12 molecule for which the data are scaled up by a factor of 10, and for D2 which drops off in emission towards long wavelengths. However, the characteristic features in each spectrum can be sorted according to the symmetry of the siloxane ring: most obviously, D4 and D8 fall into one group, and D6 with D12 fall into another one; D2 shares most of the peaks common to both groups with more uniform intensity distribution. It thus appears that 3-fold and 4-fold symmetry of the siloxane ring emphasizes different features in the (vibronic) structure of the long-lived phosphorescence of phenylated siloxane compounds, and that the disiloxanediol constitutes the building block which exhibits the basic interactions by which the absorbed photon energy relaxes into the emitting state from which the energy is released at long times.

This model was confirmed with a measurement of the phosphorescence time dependence which is shown in Figure 8. Two features can be observed. The first one is a build-up of the delayed emission intensity for about 50 µs during which time the energy from molecular state that absorbs 3.87 eV photons (320 nm) is pumped into the emitting state at 3.26 eV (380 nm, cf. Figure 5) at a higher rate than it emits energy. The second one is the ordering of the spectra by the appearance of a second such peak for the higher-symmetry molecules, i.e., D2, D4, and D8, which is missing from the traces of the molecules with 3-fold symmetry, i.e., D6, and D12. It is remarkable that D2 shares the double-peak feature with the molecules of 4-fold symmetry; this suggests that the 3-fold symmetry of the siloxane ring suppresses a second electronic state to which the phenyl-Si system has otherwise access, and from which long-lived delayed emission can occur.

It is also noted that the D6 molecule shows a substantial emission with much longer time constant than the comparatively rapid decay observed between 30 µs and 100 µs for the other molecules. This rapid decay which applies to all five species, is characterized by a time constant between 10 µs and 20 µs, somewhat dependent on the molecule, while for times t > 70 µs the D6 molecule shows an additional exponential phosphorescence decay characterized by a time constant of 0.25 ms. The origin of the electronic state from which this emission occurs or the mechanism by which it is created are currently not understood.

Summary and Conclusions. The photoluminescence emission obtained from π-electron systems such as stilbene or benzene is red-shifted into the blue part of the visible spectrum if a Si atom is attached to the aromatic moiety. Additional substitutions on the Si atom affect the spectral features of the molecule as demonstrated with the diphenylsilacyclobutene compounds **1** or **3** and their derivatives (tolane cycloadducts or TCA). Diol derivatives of this molecule that were also investigated include the cyclosiloxanes **5**, and **8** or **9** (spirocyclics).

With a related class of compounds, the phenylated cyclosiloxanes, the effect of siloxane ring size on the spectral characteristics of the blue photoluminescence was investigated in more detail. It was shown that the cyclosiloxane symmetry is the controlling parameter for the spectral distribution of the phosphorescence intensity as

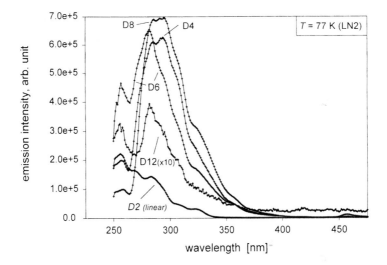

Figure 7. Phosphorescence excitation spectra of the stereoregular phenylcyclosiloxanes, [PhSi(OSiMe$_3$)O]$_n$, n = 4, 6, 8, 12, together with the linear tetraphenyl-1,3-siloxanediol, measured (in powder form), at the temperature of liquid nitrogen, T = 77 K. Emission wavelength 505 nm, spectral bandwidth *ca.* 8 nm, edge filter (370 nm). Excitation flash duration (at half maximum) 3 µs, delay time (after flash trigger) before sampling 50 µs, sampling window size 10 ms, cumulative emission from 10 flashes per data point.

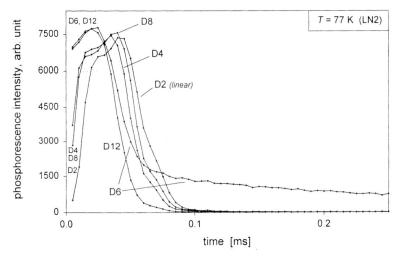

Figure 8. Phosphorescence time dependence of the emission from the stereo-regular phenylcyclosiloxanes, [PhSi(OSiMe$_3$)O]$_n$, n = 4, 6, 8, 12, together with the linear tetraphenyl-1,3-siloxanediol, measured (in powder form), at the temperature of liquid nitrogen, T = 77 K. Excitation wavelength 320 nm, emission wavelength 380 nm, spectral bandwidth ca. 8 nm, edge filter (370 nm). Excitation flash duration (at half maximum) 3 µs, delay time increment 5 µs, sampling window size 10 µs, cumulative emission from 10 flashes per data point.

well as for the temporal behavior of the emission. The specific features in the phosphorescence excitation spectrum and the phosphorescence emission time dependence that are modified by the size and symmetry of the siloxane ring are also present in the linear phenylsiloxanediol which indicates that the molecular structure responsible for the observed effects consists of two phenyl groups each attached to a Si atom in a siloxane configuration. From this observation, as well as from the fact that the Si–O vibronic signature is observed in the excitation spectrum of each molecule with phenylsiloxane structure, it is concluded that the siloxane unit participates in the energy absorption, relaxation, and emission processes in such compounds. This is remarkable since the siloxane moiety itself does not exhibit a comparable photoluminescence phenomenon as demonstrated by octamethylcyclotetrasiloxane in comparison to octaphenylcyclosiloxane. Thus the effect of the Si atom bonded to the phenyl group on the shift of the photoluminescence from the UV into the blue part of the visible spectrum is further modified by the interaction with the siloxane bridges between the phenyl groups.

Acknowledgements

The experimental skills of A.A. Hart and F.N. Noble who carried out most of the photoluminescence measurements reported here are gratefully acknowledged.

References

1. Jaffé, H.H., Orchin, M. *Theory and Applications of Ultraviolet Spectroscopy*; John Wiley and Sons, New York, 1992.
2. Shchegolikhina, O.I., Igonin, V.A., Molodtsova, Yu.A., Pozdniakova, Yu.A., Shdanov, A.A., Strelkova, T.V., Lindeman, S.V., *J. Organomet. Chem.* **1998**, 562(1-2), 141-51.
3. Auner, N., Seidenschwarz, C., Seewald, N., Herdtweck, E. *Angew. Chem.* **1991**, 103, 1172; Int. Ed. Engl. **1991**, 30, 1151.
4. Horn, Keith A.; Grossman, Robert B.; Thorne, Jonathan R. G.; Whitenack, Anne A., *J. Am. Chem. Soc.* (1989), 111(13), 4809-21.
5. Bock, H., Wittel, K., Veith, M., Wiberg, N., *J. Am.Chem.Soc.* **1976**, 98, 109-14.
6. Lipp, E.D., Smith, A.L., in: *The Analytical Chemistry of Silicones* (A.L. Smith, ed.), Chemical Analysis 112; John Wiley & Sons: New York, 1991; Chapter 11, pp 305.

Chapter 8

Synthesis and Characterization of Amino Acid Functional Siloxanes

J. G. Matisons and A. Provatas[1]

Polymer Science Group. Ian Wark Research Institute, University of South Australia, Mawson Lakes 5095, Adelaide, South Australia, Australia

The preparation of polymers containing pendant or terminal amino acids, peptides, and polypeptides has in the past been limited to organic polymers, and so has neglected inorganic polymers such as siloxanes. A novel series of amino acid functional siloxanes, suitable for use in personal care products, have been prepared and characterized for the first time. Such products can be synthesized by a three–step procedure. The preparation of amino acid siloxanes has now been achieved by hydrosilylation of an unsaturated functional amino acid with three hydrido functional siloxanes: tetramethyldisiloxane, containing terminal Si–H groups, poly(methylhydrogen)siloxane (DP = 33) and poly[(methylhydrogen)-co-(dimethyl)]siloxane (DP = 176), both of which contain pendant Si–H groups. Amino acid functionalized siloxanes were prepared and characterized in a three-step process. Firstly, allylcarboxy alanine tert–butyl ester is prepared by the dicyclohexylcarbodiimide coupling of 3–butenoic acid to alanine tert–butyl ester. Subsequent hydrosilylation by telechelic tetramethyldisiloxane or poly(methylhydrogen)siloxane and deprotection of tert–butyl ester yields the desired amino acid siloxane diadducts or polymers in high yields, typically as viscous oils.

Siloxanes have been widely used for many years as surfactants because of their high surface energy and activity as well as their stability towards heat, chemicals and UV radiation. The specific surface free energy of siloxanes (surface tension of ~21 mN/m) is significantly lower than that of most hydrocarbons, which means that they will not only be positively adsorbed at hydrocarbon surfaces, but also that siloxanes lower the surface tension of their solutions (1). This property makes siloxanes very useful as surfactants.

[1]Current address: Weapons System Division, Defence Science and Technology Organisation, P. O. Box 1500, Salisbury, 5095, Adelaide, South Australia 5108, Australia.

Siloxane surfactant properties stem from the electronic and structural properties of the Si-O and Si-C bonds, which impart to the surfactant its specific physical, chemical and mechanical properties. The Si-O bonds act to reduce steric conflicts between methyl groups on neighboring silicon atoms, by permitting unhindered rotation about the Si-O and Si-C bonds. Furthermore, the partial ionic character of the Si-O bond itself allows distortion of the large bond angle at oxygen to further relieve any additional steric hindrance. This freedom of rotation about the Si-O and Si-C bonds gives ideal screening of the polar Si-O-Si backbone by the non polar methyl groups, giving siloxane polymers excellent film forming properties. These unique surfactant properties have been investigated using siloxanes containing ionically charged hydrophilic groups (cationic and anionic species) (2). Our interest resides in the adsorption of cationic α-amino-β-hydroxy functional siloxanes, which have a variety of applications associated with the reactivity of the hydrophilic amino groups, attached to the inert hydrophobic siloxane backbone.

The preparation of polymers containing pendant or terminal amino acids, peptides, and polypeptides has in the past been limited to organic polymers, and so has neglected inorganic polymers such as the siloxanes. Organic polymer backbones incorporating pendant amino acids or proteins have been prepared using polymethacrylic acids (3–6), polymethacrylic amides (7), polyurethane amides (8) and poly(alkylene phosphates) (9). Such functionalized polymers are subsequently used in complexation studies with multivalent metal ions and in conformational studies of polymer behavior in aqueous solution. Pseudo-poly(amino acids) are one of the newest classes of polymers investigated as potential implantable, degradable materials in medical applications.

Research impetus on the preparation of amino acid functional siloxane polymers has stemmed from the growing needs of polymeric medical implants currently in use (10). Such polymeric materials must be inert, non-thrombogenic (i.e. prevent blood clotting on an artificial implant) and have a long working life in the body before replacement. By combining the hydrophobic polymeric properties of siloxanes (being biologically inert, having high as well as low temperature stability, and chemical resistance), with hydrophilic, pendant amino acid groups, biomedical copolymers having the required functionality necessary to use in biomedical implant applications could be produced.

One of the first published reports investigating the synthesis of amino acid functionalized siloxanes used end-to-end coupling of suitably terminated polymers (11). The resultant copolymers did have amino acid groups, but not directly associated with the siloxane backbone. Interest in such materials gradually waned, largely as a result of the synthetic difficulties encountered. In 1982, a new, facile route to the preparation of terminal amino acid siloxanes was published (12), utilizing primary amino-terminated siloxanes to initiate the polymerization of N-carboxy-α-amino acid anhydrides. This synthetic route became the primary strategy for preparing amino acid functionalized siloxanes in reasonable yields (13).

Amino acids bearing unsaturated vinyl or allyl groups can be grafted onto pendant hydridosiloxanes by hydrosilylation (14,15). The hydrosilylation reaction involves the addition of Si-H functional compounds to unsaturated organic compounds in the presence of a transition metal catalyst, usually platinum. Several amino acid functionalized siloxanes have been prepared this way; first by synthesizing allylcarboxy derivatives of the tertiary butyl amide amino acid prior to hydrosilylation onto Si-H siloxane backbones (*16*). Such derivatized backbones were then used as chiral stationary phases (CSP's) in gas chromatography for separating amino acid enantiomers, as well as following the racemization occurring during peptide and protein hydrolysis (*16*). The chiral recognition from using such CSP's results from the diastereomeric, hydrogen bonding association between the amino acid functional groups (*17*). Chiral stationary phases are now commercially available, and can be modified to suit particular physicochemical requirements. The incorporation of various organic segments into the siloxane backbone [e.g. a $(CH_2)_3$ organic spacer group (*18*) or a rigid cyclohexyl spacer group (*19*)], has markedly shortened chromatographic retention times without any apparent loss in resolution for separating enantiomeric amino acid derivatives.

Existing separations of amino acid enantiomers, using commercially available stationary phases such as L-valine *tert*-butylamide bound to polydimethylsiloxane (Chirasil-Val; Figure 1) (*16*), L-valine-*S*-1-(α-phenylethyl)-amide bound to polydimethylsiloxane (*20*) and octakis(3-O-butyryl-2,6-di-O-pentyl)-γ-cyclodextrin (*21*) are well documented. Lohmiller *et al.* (*22*) used a series of L-(valine) - *tert*-butylamide (n = 1-4) groups bound to poly (β-methyl)-siloxy-α-methylpropionic acid copolymers to separate the enantiomers of N-TFA-amino acid esters (where N-TFA is the protecting group, N-trifluoroacetic acid). The ^1H NMR and thermodynamic data in this study revealed that chiral recognition is most effective when short pendant chains are used (n = 1). Ôi *et al.* (*23*) recently prepared a whole series of tripeptide derivatives capable of excellent enantiomeric separation of a wide variety of racemic compounds including alcohols, amines, amino alcohols, carboxylic acids, hydroxy acids as well as amino acids. A tripeptide, using a modified *s*-triazine derivative of L-valyl-L-valyl-L-valine isopropyl ester bonded to an amino siloxane oil and N-(2-aminoethyl)-3-aminopropyl silica gel (*24*) was most effective in separating most racemic compounds.

Our aim was to prepare amino acid functional siloxanes, suitable not only for use as either biomedical implant or chromatographic materials, but also suitable for textile and personal care applications. Significantly, we sought to prepare such materials not only in high yield using readily available reagents, but also in the absence of any side reactions, as both the adsorption characteristics and surface properties of the resulting materials were then to be the subject of further research. Characterization of the synthesized polymers using three different siloxane backbones was completed, prior to their examination as surface-active agents on E-glass and textile fibers (Scheme 1).

Experimental

The following reagents were at least laboratory grade, and used as supplied: acetic acid, allyl glycidyl ether, 1,3-dicyclohexylcarbodiimide, hexachloroplatinic acid, trifluoroacetic acid, vinyl acetic acid (Aldrich); bentonite clay, celite, citric acid, magnesium sulfate, and sodium hydrogen carbonate (ACE Chemicals); morpholine (Fluka Buchs); the amino acid derivatives N-*tert*-BOC-L-alanine, N-*tert*-BOC-glycine, N-*tert*-BOC-leucine, N-*tert*-BOC-phenylalanine, N-*tert*-BOC-valine, β-alanine *tert*-butyl ester·HCl, glycine *tert*-butyl ester·HCl, leucine *tert*-butyl ester·HCl, phenylalanine *tert*-butyl ester·HCl, valine *tert*-butyl ester·HCl all from Sigma; 1,3-Bis-(3-amino propyl)tetramethyldisiloxane (Shin Etsu). Dry toluene was distilled over sodium wire

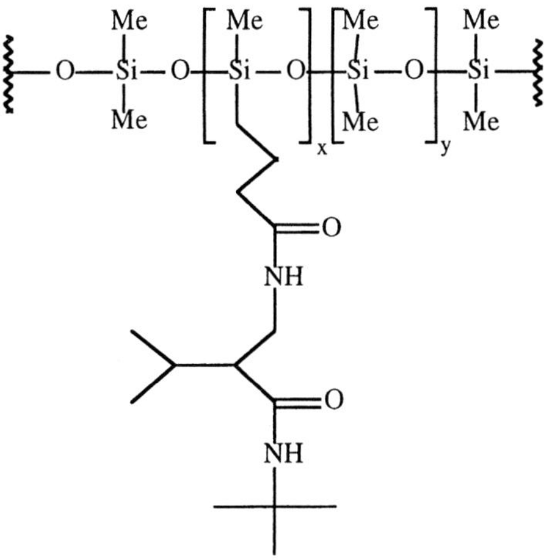

Figure 1. Chirasil-Val Stationary Phase.

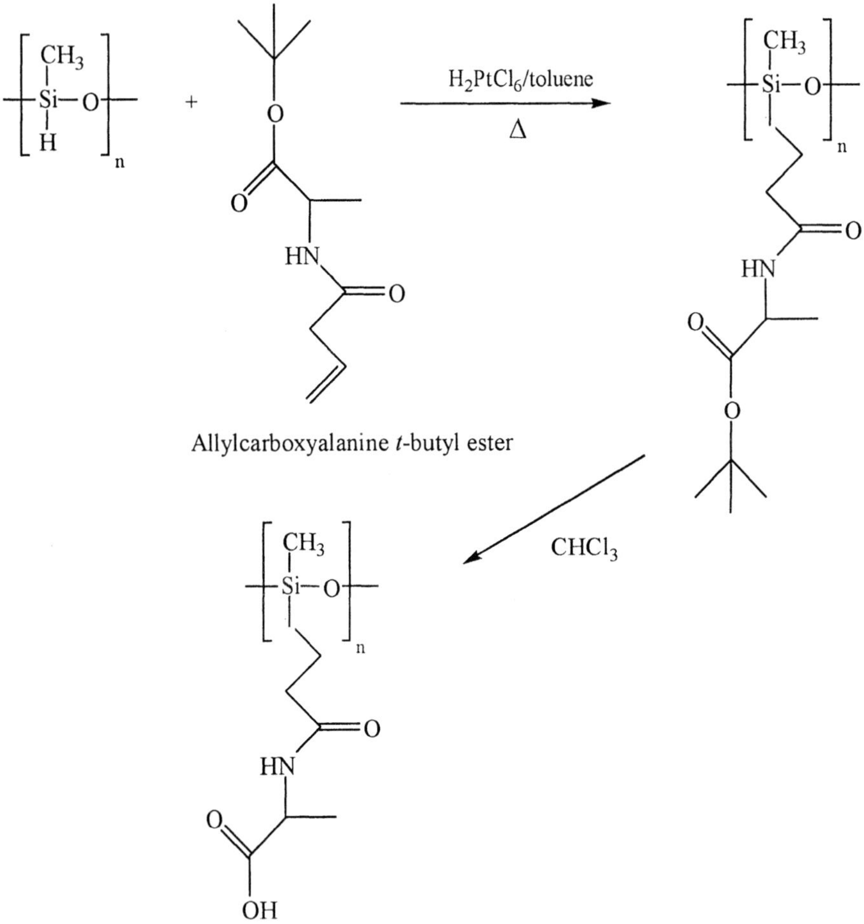

Allylcarboxyalanine *t*-butyl ester

Scheme 1 Synthesis of Amino Acid Disiloxanes.

(bp 110°C) and stored over 4Å molecular sieves. Diethyl ether was dried by distilling over sodium wire with benzophenone once a violet color had developed indicating the formation of the benzophenone ketyl radical. Chloroform and dichloromethane were dried over calcium chloride for 24 hrs before distilling over calcium sulfate and storing over 4Å molecular sieves (25).

Fourier Transform-Infra Red spectroscopy was carried out with the use of a BIORAD Model FTS65 FT-IR spectrometer using NaCl plates. Spectra were obtained over the wave number range from 700 cm^{-1} to 4000 cm^{-1} (resolution of 2 cm^{-1}) with the use of a MCT liquid nitrogen cooled detector with coaddition of 64 scans. Nuclear Magnetic Resonance analyses was conducted with a Varian Gemini Fourier Transform NMR spectrometer (200 MHz) and associated software. All spectra obtained with use of CDCl$_3$ (Cambridge Isotope Laboratories) as solvent and internal standard (^{1}H NMR δ = 7.26 ppm and ^{13}C NMR δ = 77.00 ppm) unless stated otherwise. Number of transients for ^{1}H and ^{13}C NMR spectra was generally 16 and 2000 respectively. All 2D NMR experiments were performed using standard software provided by the manufacturer (Version 6.2). ^{29}Si NMR experiments carried out on Varian 300 MHz with trichloromethylsilane as reference. Homonuclear ^{1}H-^{1}H correlated spectra were measured using the COSY pulse sequence. The spectral width was varied between 750-1000 Hz. Generally, number of transients (NT) and number of increments (NI) was 160 and 300 respectively. The heteronuclear ^{13}C-^{1}H chemical shift correlated spectra were measured with the HETCOR pulse sequence. The spectral width in the ^{13}C dimension varied between 10000 and 15000 Hz with NT = 256 and NI = 64. The decoupler frequency and spectral width in the 2D spectrum were set to match those used in the measurements of the corresponding 1D spectrum.

TGA analyses were performed using a high resolution TA 2950 thermogravimetric analyzer at 10°C/min. DSC analyses were preformed using a TA 2920 differential scanning calorimeter in closed aluminum pans under nitrogen at 20°C/min. Elemental analyses were obtained from the microanalytical service of the Department of Chemistry at the University of Queensland.

Preparation of Poly[(methylhydrogen)-*co* -(dimethyl)] Siloxanes. A mixture of poly(hydrogenmethyl)siloxane, (DP = 33, 68 g, 0.0314 mol), octamethylcyclotetrasiloxane (500 g, 1.69 mol), hexamethyldisiloxane (2.24 g, 0.014 mol) and bentonite clay (18 g, 1.5% wt/wt bentonite clay/reagents) was stirred at reflux (100°C) for 16 hrs. After cooling to room temperature the mixture was filtered through a 10 mm polypropylene filter bag. The colorless clear oil (89.0%) was then dried in vacuo to remove any excess octamethylcyclotetramethylsiloxane. Characterization of the product was carried out by NMR and FT-IR. ^{1}H NMR (ppm) δ = 0.1 (s, SiCH$_3$), 4.7 (s, SiH); ^{13}C NMR (ppm) δ = 0.1 (SiCH$_3$); ^{29}Si NMR (ppm) δ = -21.6 (s, [-OSiMe$_2$-]); FT-IR (cm^{-1}) v = 2968 (m), 2161 (s), 1455 (m), 1266 (s), 1100 (s).

Preparation of Allylcarboxy Functionalized Amino *tert*-Butyl Ester Acids (1 – 5). 3-Butenoic acid (1 g, 11.6 mmol) was added dropwise to a solution of dicyclohexylcarbodiimide (1.19 g, 5.8 mmol) in dry dichloromethane (50 mL) at 0°C (ice bath). Subsequent stirring for 1 h at room temperature produced a white precipitate (dicyclohexylurea), which was removed by filtration, before the specific amino acid *tert*-

butyl ester was added to solution (i.e. alanine 1.05 g; glycine 1.00 g; leucine 1.20 g; phenyl alanine 1.49 g; valine 1.10 g; all 5.8 mmol). Morpholine was added till a pH of 7 was reached, and the resulting reaction mixture left to stir for 24 hrs. The product was then extracted by dichloromethane (100 mL), washed with dilute hydrochloric acid (5%, 3 × 50 mL), sodium hydrogen carbonate (5%, 3 × 50 mL), and finally with water (3 × 50 mL) before drying over magnesium sulfate. Recrystallization from dichloromethane-pentane, gave yields between 84 to 95% (Table I).

Product 1 (Allylcarboxy alanine *tert*-butyl ester; white crystals, yield 92%)

^{1}H NMR (ppm) δ = 0.9 (db, NHCH(CH_3)), 1.5 (s, C(CH$_3$)$_3$), 2.4 (tr, NHCH(R)), 3.1 (db, CHCH_3), 3.4 (m, CHCH$_3$), 3.6 (db, CH$_2$=CHCH_2), 5.1 (m, CH$_2$=CH), 5.9 (m, CH_2=CH);

^{13}C NMR (ppm) δ = 18.3 (NHCH(CH$_3$)), 28.4 (C(CH$_3$)$_3$), 35.4 (CHCH$_3$), 41.9 (CHCH$_3$), 46.4 (CH$_2$=CHCH$_2$), 80.8 (C(CH$_3$)$_3$), 118.2 (CH$_2$=CH), 131.5 (CH$_2$=CH), 174.3 (COOC(CH$_3$)$_3$), 174.2 (CONHCH(CH$_3$));

FT-IR (cm^{-1}) ν = 3328 (m), 3069 (m), 2983 (m), 2917 (m), 2855 (m), 1739 (s), 1660 (s), 1135 (m).

Analysis. Calcd. for C$_{11}$H$_{20}$NO$_3$: C, 57.64%; H, 8.73%; N 12.27%. Found: C, 57.44%; H, 8.98%; N 11.56%.

Product 2 (Allylcarboxy glycine *tert*-butyl ester; orange crystals, yield 95%)

^{1}H NMR (ppm) δ = 1.5 (s, C(CH$_3$)$_3$), 2.1 (m, NHCH$_2$), 3.8 (db, CH$_2$=CHCH_2), 5.1 (m, CH$_2$=CH), 5.9 (m, CH_2=CH);

^{13}C NMR (ppm) δ = 24.0 (C(CH$_3$)$_3$), 45.7 (CH$_2$=CHCH$_2$), 48.9 (NHCH$_2$), 80.9 (C(CH$_3$)$_3$), 118.9 (m, CH$_2$=CH), 131.8 (m, CH$_2$=CH), 170.8 (COOC(CH$_3$)$_3$), 174.2 (CONHCH$_2$);

FT-IR (cm^{-1}) ν = 3308 (m), 3086 (m), 2965 (m), 2918 (m), 2848 (m), 1741 (s), 1657 (s), 1139 (s).

Analysis. Calcd. for C$_{10}$H$_{18}$NO$_3$: C, 56.6%; H, 8.53%; N 9.29%. Found: C, 56.07%; H, 8.41%; N 13.08%.

Product 3 (Allylcarboxy leucine *tert*-butyl ester; white crystals, yield 91%)

^{1}H NMR (ppm) δ = 0.9 (db, CH(CH_3)$_2$), 1.4 (s, C(CH$_3$)$_3$), 1.6 (m, CH(CH$_3$)$_2$), 1.9 (m, NHCH(R)), 3.0 (db, NHCH(R)), 3.3 (tr, CH_2CH(CH$_3$)$_2$), 4.6 (m, CH$_2$=CHCH_2), 5.2 (m, CH$_2$=CH), 6.0 (m, CH_2=CH);

^{13}C NMR (ppm) δ = 22.3, 23.1 (CH(CH$_3$)$_2$), 24.0 (C(CH$_3$)$_3$), 45.7 (CH(CH$_3$)$_2$), 45.9 (CH$_2$=CHCH$_2$), 54.3 (CH$_2$CH(CH$_3$)$_2$), 56.9 (NHCH(R)), 80.2 (C(CH$_3$)$_3$), 117.8 (m, CH$_2$=CH), 131.8 (m, CH$_2$=CH), 169.3 (COOC(CH$_3$)$_3$), 174.2 (CONHCH(R));

FT-IR (cm^{-1}) ν = 3310 (m), 3066 (m), 2963 (s), 2910 (s), 2849 (m), 1790 (s), 1746 (s), 1646 (s), 1093 (s).

Analysis. Calcd. for C$_{14}$H$_{26}$NO$_3$: C, 62.22%; H, 9.63%; N 10.37%. Found: C, 61.05%; H, 10.09%; N 11.27%.

Product 4 (Allylcarboxy phenylalanine *tert*-butyl ester; orange crystals, yield 95%)

^{1}H NMR (ppm) δ = 1.4 (s, C(CH$_3$)$_3$), 2.8 (m, NHCH(R)), 3.1 (db, NHCH(R)), 3.4 (db, CH$_2$=CHCH_2), 3.6 (db, CH(CH_2Ph)), 5.1 (m, CH$_2$=CH), 5.9 (m, CH_2=CH), 7.2 (m, Ph);

^{13}C NMR (ppm) δ = 27.9 (C(CH$_3$)$_3$), 41.5 (NHCH(CH$_2$Ph)), 45.7 (CH$_2$=CHCH$_2$), 56.9 (CH$_2$Ph), 80.2 (C(CH$_3$)$_3$), 117.8 (m, CH$_2$=CH), 126.7, 128.5, 129.6, 137.9 (Ph), 131.8 (m, CH$_2$=CH), 169.3 (COOC(CH$_3$)$_3$), 174.2 (CONH(R));

FT-IR (cm^{-1}) ν = 3380 (w), 3329 (w), 3308 (m), 3092 (m), 2973 (m), 2937 (m), 2864 (m), 1736 (s), 1659 (s), 1187 (s), 1129 (s).

Analysis. Calcd. for C$_{17}$H$_{24}$NO$_3$: C, 67.10%; H, 7.89%; N 9.21%. Found: C, 67.4%; H, 8.76%; N 9.62%.

Product 5 (Allylcarboxy valine *tert*-butyl ester; white crystals, yield 84%)

^1H NMR (ppm) δ = 0.8 (tr, CH(CH$_3$)$_2$), 1.2 (s, C(CH$_3$)$_3$), 1.9 (m, NHCH(R)), 2.8 (db, NHCH(R)), 3.6 (db, CH(CH$_3$)$_2$), 4.2 (db, CH$_2$=CHCH$_2$), 5.7 (m, CH$_2$=CH), 6.5 (d, CH$_2$=CH);

^{13}C NMR (ppm) δ = 18.4 (CH(CH$_3$)$_2$), 27.9 (C(CH$_3$)$_3$), 41.6 (CH$_2$=CHCH$_2$), 45.6 (CH(CH$_3$)$_2$), 57.2 (NHCH(R)), 81.8 (C(CH$_3$)$_3$), 119.5 (m, CH$_2$=CH), 131.2 (m, CH$_2$=CH), 169.7 (COOC(CH$_3$)$_3$), 170.4 (CONHCH(R));

FT-IR (cm^{-1}) ν = 3337 (m), 3290 (m), 3074 (m), 2964 (s), 2917 (s), 2845 (s), 1752 (m), 1663 (s), 1192 (s).

Analysis. Calcd. for C$_{13}$H$_{24}$NO$_3$: C, 60.94%; H, 9.38%; N 10.93%. Found: C, 59.96%; H, 9.69%; N 11.39%.

α,ω–Amino Acid Functional Siloxanes Containing Tetramethyldisiloxane (6 – 10).

To a stirred solution of allylcarboxy amino acid *tert*-butyl ester (i.e. **1** 0.75 g; **2** 0.90 g; **3** 1.08 g; **4** 1.05 g; **5** 1.23 g) in 50 mL toluene, is slowly added a 0.5 molar amount of tetramethyldisiloxane under nitrogen (Scheme 1). Several drops of a 1% hexachloroplatinic acid solution in tetrahydrofuran is added, and the mixture is left to reflux, until ^1H NMR or FTIR indicates a complete loss of Si-H and allyl resonances. The solution is then allowed to cool, and activated charcoal is stirred into the mixture for 2 hrs before filtering the resulting slurry through celite, washing with toluene (2 × 10 mL) and drying under vacuum. The product is then washed with saturated citric acid (3 × 50 mL), sodium hydrogen carbonate (5%, 3 × 50 mL) and water (3 × 50 mL) before drying over magnesium sulfate, filtering and drying under vacuum to yield the disiloxane products **(6 – 10)**.

Siloxane 6 (1,3-Bis-(3-alanine *tert*-butyl ester propyl)tetramethyldisiloxane; white oil, yield 85%)

^1H NMR (ppm) δ = 0.1 (s, SiCH$_3$), 0.6 (m, CH$_2$Si), 0.9 (db, NHCH(CH$_3$)), 1.6 (m, CH$_2$CH$_2$Si), 1.6 (db, NHCH(CH$_3$)), 1.7 (s, C(CH$_3$)$_3$), 1.9 (m, NHCH(CH$_3$)), 4.2 (CH$_2$CH$_2$CH$_2$Si);

^{13}C NMR (ppm) δ = 0.1 (SiCH$_3$), 14.3 (CH$_2$Si), 19.7 (CH$_2$CH$_2$Si), 22.9 (NHCH(CH$_3$)), 28.1 (C(CH$_3$)$_3$), 42.3 (CH$_2$CH$_2$CH$_2$Si), 56.9 (NHCH(CH$_3$)), 80.2 (C(CH$_3$)$_3$), 169.3 (CONHCH(CH$_3$)), 174.2 (COOC(CH$_3$)$_3$));

^{29}Si NMR (ppm) δ = 8.1 (s, [-OSiMe$_2$(CH$_2$)$_3$CONH-]);

FT-IR (cm^{-1}) ν = 3304 (m), 3064 (m), 2963 (m), 2922 (m), 2841 (m), 1735 (s), 1661 (s), 1409 (m), 1261 (s), 1051 (s);

Analysis. Calcd. for C$_{30}$H$_{64}$Si$_2$N$_2$O$_7$: C, 58.06%; H, 10.32%; N 4.51%. Found: C, 59.26%; H, 10.78%; N 4.86%.

Siloxane 7 (1,3-Bis-(3-glycine *tert*-butyl ester propyl)tetramethyldisiloxane; orange oil, yield 87%)

^{1}H NMR (ppm) δ = 0.1 (s, SiCH$_3$), 0.6 (m, CH$_2$Si), 0.9 (db, NHCH_2), 1.5 (s, C(CH$_3$)$_3$), 1.7 (m, CH_2CH$_2$Si), 1.9 (m, NHCH$_2$), 2.4 (CH_2CH$_2$CH$_2$Si), 2.8 (db, NHCH_2);

^{13}C NMR (ppm) δ = 0.1 (SiCH$_3$), 14.3 (CH$_2$Si), 19.7 (CH$_2$CH$_2$Si), 21.4 (SiCH(CH$_3$)CH$_2$), 28.1 (C(CH$_3$)$_3$), 42.3 (CH$_2$CH$_2$CH$_2$Si), 56.9 (NHCH$_2$), 81.2 (C(CH$_3$)$_3$), 169.4 (CONHCH$_2$), 170.2 (COOC(CH$_3$)$_3$);

^{29}Si NMR (ppm) δ = 7.9 (s, [-OSiMe$_2$(CH$_2$)$_3$CONH-]);

FT-IR (cm^{-1}) ν = 3310 (m), 2963 (m), 2912 (m), 2842 (m), 1733 (s), 1658 (s), 1410 (m), 1263 (s), 1086 (s);

Analysis. Calcd. for C$_{28}$H$_{60}$Si$_2$N$_2$O$_7$: C, 56.76%; H, 10.13%; N 4.73%. Found: C, 57.26%; H, 9.83%; N 4.7%.

Siloxane 8 (1,3-Bis-(3-leucine *tert*-butyl ester propyl)tetramethyldisiloxane; orange oil, yield 92%)

^{1}H NMR (ppm) δ = 0.1 (s, SiCH$_3$), 0.5 (m, CH$_2$Si), 0.8 (db, CH(CH_3)$_2$), 1.2 (m, CH(CH$_3$)$_2$), 1.3 (s, C(CH$_3$)$_3$), 1.5 (m, CH_2CH$_2$Si), 1.6 (m, NHCH(R)), 2.1 (CH_2CH$_2$CH$_2$Si), 2.8 (db, NHCH(R)), 3.3 (tr, CH_2CH(CH$_3$)$_2$);

^{13}C NMR (ppm) δ = 0.1 (SiCH$_3$), 14.3 (CH$_2$Si), 19.7 (CH$_2$CH$_2$Si), 22.6, 23.4 (CH(CH$_3$)$_2$), 28.1 (C(CH$_3$)$_3$), 42.3 (CH$_2$CH$_2$CH$_2$Si), 45.8 (CH(CH$_3$)$_2$), 55.1 (CH$_2$CH(CH$_3$)$_2$), 56.9 (NHCH(R)), 81.2 (C(CH$_3$)$_3$), 170.1 (CONHCH(R)), 173.2 (COOC(CH$_3$)$_3$);

^{29}Si NMR (ppm) δ = 8.0 (s, [-OSiMe$_2$(CH$_2$)$_3$CONH-]);

FT-IR (cm^{-1}) ν = 3310 (m), 3061 (m), 2963 (m), 2922 (m), 2843 (m), 1738 (s), 1662 (s), 1418 (m), 1265 (s), 1096 (s), 1053 (s);

Analysis. Calcd. for C$_{38}$H$_{78}$Si$_2$N$_2$O$_7$: C, 62.46%; H, 10.69%; N 3.84%. Found: C, 62.32%; H, 10.64%; N 4.01%.

Siloxane 9 (1,3-Bis-(3-phenylalanine *tert*-butyl ester propyl)tetramethyldisiloxane; orange oil, yield 91%)

^{1}H NMR (ppm) δ = 0.1 (s, SiCH$_3$), 0.6 (m, CH$_2$Si), 1.6 (m, CH_2CH$_2$Si), 1.7 (s, C(CH$_3$)$_3$), 1.9 (m, NHCH(R)), 2.3 (s, CH_2Ph), 2.8 (db, NHCH(R)), 3.4 (CH_2CH$_2$CH$_2$Si), 7.2 (m, Ph);

^{13}C NMR (ppm) δ = 0.1 (SiCH$_3$), 14.3 (CH$_2$Si), 19.7 (CH$_2$CH$_2$Si), 28.1 (C(CH$_3$)$_3$), 42.3 (CH$_2$CH$_2$CH$_2$Si), 41.9 (CH$_2$Ph), 56.9 (NHCH(R)), 80.2 (C(CH$_3$)$_3$), 126.7, 128.6, 129.6, 137.9 (Ph), 169.3 (CONH(R)), 174.2 (COOC(CH$_3$)$_3$);

^{29}Si NMR (ppm) δ = 7.72 (s, [-OSiMe$_2$(CH$_2$)$_3$CONH-]);

FT-IR (cm^{-1}) ν = 3310 (m), 3061 (m), 2963 (s), 2919 (s), 2845 (m), 1742 (s), 1660 (s), 1419 (m), 1263 (s), 1073 (s).

Analysis. Calcd. for C$_{38}$H$_{74}$Si$_2$N$_2$O$_7$: C, 61.2%; H, 8.6%; N 7.52%. Found: C, 61.26%; H, 8.68%; N 7.7%.

Siloxane 10 (1,3-Bis-(3-valine *tert*-butyl ester propyl)tetramethyldisiloxane; orange oil, yield 75%)

^{1}H NMR (ppm) δ = 0.1 (s, SiCH$_3$), 0.5 (m, CH$_2$Si), 0.8 (tr, CH(CH_3)$_2$), 1.4 (s, C(CH$_3$)$_3$), 1.6 (m, CH_2CH$_2$Si), 2.1 (m, NHCH(R)), 2.3 (m, CH(CH$_3$)$_2$), 3.4 (CH_2CH$_2$CH$_2$Si);

^{13}C NMR (ppm) δ = 0.1 (SiCH$_3$), 14.3 (CH$_2$Si), 17.8 (CH(CH$_3$)$_2$), 19.7 (CH$_2$CH$_2$Si), 28.1 (C(CH$_3$)$_3$), 31.5 (CH$_2$CH$_2$Si), 46.1 (CH(CH$_3$)$_2$), 57.1 (NHCH(R)), 81.9 (C(CH$_3$)$_3$), 169.9 (CONHCH((R)), 173.2 (COOC(CH$_3$)$_3$);

^{29}Si NMR (ppm) δ = 8.0 (s, [-OSiMe$_2$(CH$_2$)$_3$CONH-]);

FT-IR (cm^{-1}) ν = 3310 (m), 3061 (m), 2963 (m), 2922 (m), 2841 (m), 1732 (m), 1660 (m), 1419 (m), 1263 (m), 1052 (m);

Analysis. Calcd. for C$_{36}$H$_{74}$Si$_2$N$_2$O$_7$: C, 61.54%; H, 10.54%; N 3.98%. Found: C, 61.43%; H, 10.11%; N 4.21%.

Polymers Containing Poly(methylhydrogen)siloxane (PMHS), (11 – 15). To a stirred solution of allylcarboxy amino acid *tert*-butyl ester (i.e. **1** 1.00 g; **2** 1.00 g; **3** 0.95 g; **4** 0.95 g; **5** 0.95 g) in 50 mL toluene, is slowly added a 0.5 molar amount of poly(methylhydrogen)siloxane (DP = 33) under nitrogen (Scheme 1). Several drops of a 1% hexachloroplatinic acid solution in tetrahydrofuran is added, and the mixture is left to reflux, until ^{1}H NMR indicates a complete loss of Si-H and allyl resonances. The solution is then allowed to cool, and activated charcoal is stirred into the mixture for 2 hrs, before filtering the resulting slurry through celite, washing with toluene (2 × 10 mL) and drying under vacuum. The product is then washed with saturated citric acid (3 × 50 mL), sodium hydrogen carbonate (5%, 3 × 50 mL) and water (3 × 50 mL) before drying over magnesium sulfate, filtering and drying under vacuum to yield the polymeric products **(11 – 15)**.

Siloxane 11 (Poly-(3-alanine *tert*-butyl ester propyl)methylsiloxane; white oil, yield 89%)

^{1}H NMR (ppm) δ = 0.1 (s, SiCH$_3$), 0.6 (m, CH$_2$Si), 0.9 (db, NHCH(CH_3)), 1.6 (m, CH_2CH$_2$Si), 1.6 (db, NHCCH(R)), 1.7 (s, C(CH$_3$)$_3$), 1.9 (m, NHCH(R)), 4.0 (CH_2CH$_2$CH$_2$Si);

^{13}C NMR (ppm) δ = 0.1 (SiCH$_3$), 14.3 (CH$_2$Si), 19.7 (CH$_2$CH$_2$Si), 22.5 (NHCH(CH$_3$)), 28.1 (C(CH$_3$)$_3$), 42.3 (CH$_2$CH$_2$CH$_2$Si), 56.9 (NHCH(R)), 80.2 (C(CH$_3$)$_3$), 169.3 (CONH(R)), 174.2 (COOC(CH$_3$)$_3$);

^{29}Si NMR (ppm) δ = 8.1 (s, [-OSiMe$_2$(CH$_2$)$_3$CONH-]);

FT-IR (cm^{-1}) ν = 3304 (m), 3064 (m), 2963 (m), 2922 (m), 2841 (m), 1735 (s), 1661 (s), 1409 (m), 1261 (s), 1051 (s).

Siloxane 12 (Poly-(3-glycine *tert*-butyl ester propyl)methylsiloxane; orange oil, yield 77%)

^{1}H NMR (ppm) δ = 0.1 (s, SiCH$_3$), 0.6 (m, CH$_2$Si), 0.9 (db, SiCH(CH_3)CH$_2$), 1.5 (s, C(CH$_3$)$_3$), 1.7 (m, CH_2CH$_2$Si), 1.9 (m, NHCH$_2$), 2.4 (CH_2CH$_2$CH$_2$Si), 2.8 (db, NHCH_2);

^{13}C NMR (ppm) δ = 0.1 (SiCH$_3$), 14.3 (CH$_2$Si), 19.7 (CH$_2$CH$_2$Si), 21.4 (SiCH(CH$_3$)CH$_2$), 28.1 (C(CH$_3$)$_3$), 42.3 (CH$_2$CH$_2$CH$_2$Si), 56.9 (NHCH$_2$), 81.2 (C(CH$_3$)$_3$), 169.4 (CONH(R)), 170.2 (COOC(CH$_3$)$_3$);

^{29}Si NMR (ppm) δ = 8.0 (s, [-OSiMe$_2$(CH$_2$)$_3$CONH-]);

FT-IR (cm^{-1}) ν = 3310 (m), 2963 (m), 2912 (m), 2842 (m), 1733 (s), 1658 (s), 1410 (m), 1263 (s), 1086 (s).

Siloxane 13 (Poly-(3-leucine *tert*-butyl ester propyl)methylsiloxane; orange oil, yield 90%)

^{1}H NMR (ppm) δ = 0.1 (s, SiCH$_3$), 0.5 (m, CH$_2$Si), 0.8 (db, CH(CH_3)$_2$), 1.2 (m, CH(CH$_3$)$_2$), 1.3 (s, C(CH$_3$)$_3$), 1.5 (m, CH_2CH$_2$Si), 1.6 (m, NHCH(R)), 2.1 (CH_2CH$_2$CH$_2$Si), 2.8 (db, NHCH(R)), 3.3 (tr, CH_2CH(R));

^{13}C NMR (ppm) δ = 0.1 (SiCH$_3$), 14.3 (CH$_2$Si), 19.7 (CH$_2$CH$_2$Si), 22.6, 23.4 (CH(CH$_3$)$_2$), 28.1 (C(CH$_3$)$_3$), 42.3 (CH$_2$CH$_2$CH$_2$Si), 45.8 (CH(CH$_3$)$_2$), 55.1 (CH$_2$CH(CH$_3$)$_2$), 56.9 (NHCH(R)), 81.2 (C(CH$_3$)$_3$), 170.1 (CONHCH(R)), 173.2 (COOC(CH$_3$)$_3$);

^{29}Si NMR (ppm) δ = 8.0 (s, [-OSiMe$_2$(CH$_2$)$_3$CONH-]);

FT-IR (cm^{-1}) ν = 3310 (m), 3061 (m), 2963 (m), 2922 (m), 2843 (m), 1738 (s), 1662 (s), 1418 (m), 1265 (s), 1096 (s), 1053 (s).

Siloxane 14 (Poly-(3-phenylalanine *tert*-butyl ester propyl)methylsiloxane; orange oil, yield 89%)

^{1}H NMR (ppm) δ = 0.1 (s, SiCH$_3$), 0.6 (m, CH$_2$Si), 1.6 (m, CH$_2$CH$_2$Si), 1.7 (s, C(CH$_3$)$_3$), 1.9 (m, NHCH(R)), 2.3 (s, CH$_2$Ph), 2.4 (CH$_2$CH$_2$CH$_2$Si), 2.8 (db, NHCH(R)), 7.2 (m, Ph);

^{13}C NMR (ppm) δ = 0.1 (SiCH$_3$), 14.3 (CH$_2$Si), 19.7 (CH$_2$CH$_2$Si), 28.1 (C(CH$_3$)$_3$), 42.3 (CH$_2$CH$_2$CH$_2$Si), 56.9 (NHCH(R)), 41.9 (CH$_2$Ph), 80.2 (C(CH$_3$)$_3$), 126.7, 128.6, 129.6, 137.9 (Ph), 169.3 (CONHCH(R)), 174.2 (COOC(CH$_3$)$_3$);

^{29}Si NMR (ppm) δ = 7.9 (s, [-OSiMe$_2$(CH$_2$)$_3$CONH-]);

FT-IR (cm^{-1}) ν = 3310 (m), 3061 (m), 2963 (s), 2919 (s), 2845 (m), 1742 (s), 1660 (s), 1419 (m), 1263 (s), 1073 (s).

Siloxane 15 (Poly-(3-valine *tert*-butyl ester propyl)methylsiloxane; white oil, yield 78%)

^{1}H NMR (ppm) δ = 0.1 (s, SiCH$_3$), 0.5 (m, CH$_2$Si), 0.8 (tr, CH(CH$_3$)$_2$), 1.4 (s, C(CH$_3$)$_3$), 1.6 (m, CH$_2$CH$_2$Si), 2.1 (m, NHCH(R)), 2.3 (m, CH(CH$_3$)$_2$), 3.6 (CH$_2$CH$_2$CH$_2$Si);

^{13}C NMR (ppm) δ = 0.1 (SiCH$_3$), 14.3 (CH$_2$Si), 17.4 CH(CH$_3$)$_2$), 19.4 (CH$_2$CH$_2$Si), 28.1 (C(CH$_3$)$_3$), 31.3 (CH$_2$CH$_2$CH$_2$Si), 46.4 (CH(CH$_3$)$_2$), 57.9 (NHCH(R)), 81.2 (C(CH$_3$)$_3$), 168.9 (CONHCH(R)), 173.2 (COOC(CH$_3$)$_3$);

^{29}Si NMR (ppm) δ = 8.0 (s, [-OSiMe$_2$(CH$_2$)$_3$CONH-]);

FT-IR (cm^{-1}) ν = 3304 (m), 3051 (m), 2963 (s), 2925 (s), 2841 (m), 1732 (m), 1660 (s), 1410 (m), 1262 (s), 1071 (s).

Polymers Containing Poly[(methylhydrogen)-*co*-(dimethyl)] Siloxane (16 – 20).
To a stirred solution of each specific allylcarboxy amino acid *tert*-butyl ester (i.e. 0.5 g; 1.45-1.97 mmols) in 50 mL toluene, is slowly added a 0.5 molar amount of poly[(methylhydrogen)-*co*-(dimethyl)]siloxane (DP = 23) under nitrogen. Several drops of a 1% hexachloroplatinic acid solution in tetrahydrofuran is then added, and the mixture is left to reflux, until the ^{1}H NMR indicates a complete loss of Si-H and allyl resonances. The solution is then allowed to cool, and activated charcoal is stirred into the mixture for 2 hrs, before filtering the resulting slurry through celite, washing with toluene (2 × 10 mL) and drying under vacuum. The product is then washed with saturated citric acid, (3 × 50 mL) dilute sodium hydrogen carbonate (5%, 3 × 50 mL) and water (3 × 50 mL) before drying over magnesium sulfate, filtering and drying under vacuum to yield the polymeric siloxane products **(16 – 20)**.

Siloxane 16 (Poly-[(3-alanine *tert*-butyl ester propyl)-*co*-(dimethyl)]siloxane; white oil, yield 84%)

^{1}H NMR (ppm) δ = 0.1 (s, SiCH$_3$), 0.6 (m, CH$_2$Si), 0.9 (db, NHCH(CH$_3$)), 1.6 (m, CH$_2$CH$_2$Si), 1.6 (db, NHCH(CH$_3$)), 1.7 (s, C(CH$_3$)$_3$), 1.9 (m, NHCH(CH$_3$)), 4.0 (CH$_2$CH$_2$CH$_2$Si);

^{13}C NMR (ppm) δ = 0.1 (SiCH$_3$), 14.6 (CH$_2$Si), 19.6 (CH$_2$CH$_2$Si), 22.4 (NHCH(CH$_3$)), 28.1 (C(CH$_3$)$_3$), 42.3 (CH$_2$CH$_2$CH$_2$Si), 56.4 (NHCH(R)), 81.2 (C(CH$_3$)$_3$), 170.5 (CONH(CH$_3$)), 173.2 (COOC(CH$_3$)$_3$);

^{29}Si NMR (ppm) δ = 8.1 (s, [-OSiMe$_2$(CH$_2$)$_3$CONH-]).

FT-IR (cm^{-1}) ν = 3317 (m), 3068 (m), 2965 (m), 2912 (m), 2846 (m), 1732 (s), 1665 (s), 1413 (m), 1263 (s), 1056 (s).

Siloxane 17 (Poly-[(3-glycine *tert*-butyl ester propyl)-*co*-(dimethyl)]siloxane; white oil, yield 83%)

^{1}H NMR (ppm) δ = 0.1 (s, SiCH$_3$), 0.6 (m, CH$_2$Si), 0.9 (db, SiCH(CH$_3$)CH$_2$), 1.5 (s, C(CH$_3$)$_3$), 1.7 (m, CH$_2$CH$_2$Si), 1.9 (m, NHCH$_2$), 2.4 (CH$_2$CH$_2$CH$_2$Si), 2.8 (db, NHCH$_2$);

^{13}C NMR (ppm) δ = 0.1 (SiCH$_3$), 14.3 (CH$_2$Si), 19.7 (CH$_2$CH$_2$Si), 21.4 (SiCH(CH$_3$)CH$_2$), 28.1 (C(CH$_3$)$_3$), 42.3 (CH$_2$CH$_2$CH$_2$Si), 56.9 (NHCH$_2$), 81.2 (C(CH$_3$)$_3$), 169.4 (CONHCH$_2$), 170.2 (COOC(CH$_3$)$_3$);

^{29}Si NMR (ppm) δ = 8.0 (s, [-OSiMe$_2$(CH$_2$)$_3$CONH-]);

FT-IR (cm^{-1}) ν = 3310 (m), 2963 (m), 2912 (m), 2843 (m), 1732 (s), 1649 (s), 1411 (m), 1263 (s), 1079 (s).

Siloxane 18 (Poly-[(3-leucine *tert*-butyl ester propyl)-*co*-(dimethyl)]siloxane; white oil, yield 72%)

^{1}H NMR (ppm) δ = 0.1 (s, SiCH$_3$), 0.5 (m, CH$_2$Si), 0.8 (db, CH(CH$_3$)$_2$), 1.2 (m, CH(CH$_3$)$_2$), 1.3 (s, C(CH$_3$)$_3$), 1.5 (m, CH$_2$CH$_2$Si), 1.6 (m, NHCH(R)), 2.0 (CH$_2$CH$_2$CH$_2$Si), 2.8 (db, NHCH(R), 3.3 (tr, CH$_2$CH(CH$_3$)$_2$);

^{13}C NMR (ppm) δ = 0.1 (SiCH$_3$), 14.5 (CH$_2$Si), 19.9 (CH$_2$CH$_2$Si), 22.4, 23.1 (CH(CH$_3$)$_2$), 28.7 (C(CH$_3$)$_3$), 42.2 (CH$_2$CH$_2$CH$_2$Si), 45.87 (CH(CH$_3$)$_2$), 55.1 (CH$_2$CH(CH$_3$)$_2$), 55.9 (NHCH(R)), 81.2 (C(CH$_3$)$_3$), 170.1 (CONHCH(R)), 173.2 (COOC(CH$_3$)$_3$);

^{29}Si NMR (ppm) δ = 8.0 (s, [-OSiMe$_2$(CH$_2$)$_3$CONH-]);

FT-IR (cm^{-1}) ν = 3298 (m), 3056 (m), 2963 (m), 2912 (m), 2840 (m), 1743 (s), 1661 (s), 1418 (m), 1265 (s), 1096 (s), 1053 (s).

Siloxane 19 (Poly-(3-phenylalanine *tert*-butyl ester propyl)-*co*-(dimethyl)]siloxane; orange oil, yield 75%)

^{1}H NMR (ppm) δ = 0.1 (s, SiCH$_3$), 0.6 (m, CH$_2$Si), 1.6 (m, CH$_2$CH$_2$Si), 1.7 (s, C(CH$_3$)$_3$), 1.9 (m, NHCH(R)), 2.3 (s, CH$_2$Ph), 2.4 (CH$_2$CH$_2$CH$_2$Si), 2.8 (db, NHCH(R)), 7.2 (m, Ph);

^{13}C NMR (ppm) δ = 0.1 (SiCH$_3$), 14.3 (CH$_2$Si), 19.7 (CH$_2$CH$_2$Si), 28.1 (C(CH$_3$)$_3$), 42.3 (CH$_2$CH$_2$CH$_2$Si), 56.9 (NHCH(R)), 41.9 (CH$_2$Ph), 80.2 (C(CH$_3$)$_3$), 126.7, 128.6, 129.6, 137.9 (Ph), 169.3 (CONHCH(R)), 174.2 (COOC(CH$_3$)$_3$);

^{29}Si NMR (ppm) δ = 7.9 (s, [-OSiMe$_2$(CH$_2$)$_3$CONH-]);

FT-IR (cm^{-1}) ν = 3290 (m), 3072 (m), 2968 (s), 2914 (s), 2839 (m), 1739 (s), 1665 (s), 1419 (m), 1262 (s), 1058 (s).

Siloxane 20 (Poly-[(3-valine *tert*-butyl ester propyl)-*co*-(dimethyl)]siloxane; white oil, yield 77%)

^{1}H NMR (ppm) δ = 0.1 (s, SiCH$_3$), 0.5 (m, CH$_2$Si), 0.8 (tr, CH(CH$_3$)$_2$), 1.4 (s, C(CH$_3$)$_3$), 1.6 (m, CH$_2$CH$_2$Si), 2.1 (m, NHCH(R)), 2.3 (m, CH(CH$_3$)$_2$), 3.6 (CH$_2$CH$_2$CH$_2$Si);

^{13}C NMR (ppm) δ = 0.1 (SiCH3), 14.3 (CH2Si), 17.4 CH(CH3)2), 19.4 (CH2CH2Si), 28.1 (C(CH3)3), 31.3 (CH2CH2CH2Si), 46.4 (CH(CH3)2), 57.9 (NHCH(R)), 81.2 (C(CH3)3), 168.8 (CONHCH(R)), 173.7 (COOC(CH3)3);
^{29}Si NMR (ppm) δ = 8.0 (s, [-OSiMe2(CH2)3CONH-]);
FT-IR (cm^{-1}) ν = 3309 (m), 3055 (m), 2964 (s), 2915 (s), 2847 (m), 1736 (m), 1662 (s), 1415 (m), 1265 (s), 1061 (s).

Deprotection of *Tert*-Butyl Ester Groups (21 – 35). A two-fold excess of trifluoroacetic acid was slowly added to a solution of the amino acid *tert*-butyl ester siloxanes (6 – 20) in chloroform (10 mL), and the resulting solutions left to stir for 4-6 days at room temperature. The product was extracted by diethyl ether (100 mL), washed with hydrochloric acid (5%, 3 × 50 mL), sodium hydrogen carbonate (5%, 3 × 50 mL) and water (3 × 50 mL), before drying over magnesium, filtering and drying under vacuum to yield the resultant polymers (**21 – 35**, Table I).

Table I. Quantities and Yields for Amino Acid Siloxanes Obtained After Deprotection of *Tert*-Butyl Ester Groups by Trifluoroacetic Acid.

Deprotected Siloxanes	Reagents (mmol)	TFA	Yield (%)	Product Appearance	FT-IR (cm^{-1}) $^νNH^{3+}$	$^νC=O$
21	0.69 (**6**)*	1.38	61	clear, viscous oil	3312	1743
22	0.75 (**7**)	1.5	49	white, viscous oil	3323	1738
23	0.49 (**8**)	0.98	54	white solid	3359	1746
24	0.91 (**9**)	1.82	70	white solid	3327	1748
25	0.53 (**10**)	1.06	39	white, viscous oil	3379	1739
26	0.69 (**11**)	1.38	48	clear, viscous oil	3387	1746
27	0.42 (**12**)	0.84	67	white, viscous oil	3319	1751
28	0.49 (**13**)	0.98	59	white, viscous oil	3365	1759
29	0.91 (**14**)	1.82	57	white, viscous oil	3348	1737
30	0.97 (**15**)	1.94	66	white, viscous oil	3326	1748
31	0.76 (**16**)	1.56	71	clear, viscous oil	3384	1743
32	1.05 (**17**)	2.11	57	white, viscous oil	3375	1746
33	1.09 (**18**)	2.18	49	white, viscous oil	3391	1745
34	0.75 (**19**)	1.5	69	white, viscous oil	3383	1732
35	0.9 (**20**)	1.80	62	white, viscous oil	3324	1739

* Bold numbers in brackets represent starting siloxanes employed for the deprotection reaction

Results and Discussion

The reactivity of three different siloxanes was investigated. Tetramethyldisiloxane, having α,ω–telechelic Si-H groups, is the simplest siloxane used, and is available commercially. Chemical microanalysis could be used in this case to confirm the elemental composition of the final product. Accurate microanalysis is meaningless with the other larger siloxanes, due to their polydispersity. Poly(methylhydrogen)siloxane (DP = 33; Figure 2) having ~ 33 pendant Si-H groups along the siloxane backbone, was also available commercially. It was used to investigate the efficiency of replacing all the pendant Si-H groups with amino acid side groups. Often hydrosilylation onto such a siloxane backbone leaves residual Si-H groups, generating more than one reaction product and leaving the resulting complex mixture unsuitable for further kinetic studies of adsorption onto various surfaces. We also prepared poly[(methylhydrogen)-co-(dimethyl)]siloxane, $Me_3SiO[Me(H)SiO]_{23}[Me_2SiO]_{150}$, to assess the reactivity of the pendant Si-H groups to the synthetic regime, when surrounded by inert Me_2SiO groups in a higher molecular weight, significantly more viscous, siloxane polymer.

Kantor et al. (26) first prepared a poly[(methylhydrogen)-co-(dimethyl)siloxane, such as $Me_3SiO-[Me(H)SiO]_{23}[Me_2SiO]_{150}$, in 1954. We prepared this polymer by heating poly (methylhydrogen siloxane) (DP = 33), octamethylcyclotetrasiloxane and hexamethyldisiloxane at 100°C in the presence of acid treated bentonite clay. The average molecular weight (M_w) of the final siloxane was 12,300 g/mol (by GPC). In the ^1H NMR, the ratio of Si-Me protons to Si-H protons was 60:1, indicating a copolymer containing 150 Si-Me groups and 23 Si-H groups. The FTIR shows the characteristic Si-H absorption at $v = 2160$ cm^{-1}.

The synthetic route was chosen to prepare amino acid functional polymers of defined structure and molecular weight without side reactions occurring. Scheme 1 shows a three step procedure to prepare what is essentially an amino acid functional disiloxane adduct, or α,ω–telechelic product, of low molecular weight. This can then be contrasted with the amino acid functional poly[(methylhydrogen)-co-(dimethyl)]siloxane adducts also prepared, having considerably higher molecular weights.

Five representative amino acids were chosen to evaluate the efficacy and generality of the synthetic procedure: alanine, glycine, leucine, phenylalanine and valine, (where R = -CH$_3$, -H, -CH$_2$CH(CH$_3$)$_2$, -CH$_2$C$_6$H$_5$ and -CH(CH$_3$)$_2$, respectively). A gradual increase in the aliphatic nature of R groups (i.e. R = -CH$_3$, alanine; -CH(CH$_3$)$_2$, valine; -CH$_2$CH(CH$_3$)$_2$, leucine), was selected to observe the structural or morphological differences associated with increasing carbon content (this systematic change in R groups was considered important in projected future studies examining the kinetics of adsorption of the derivatized siloxane products onto various surfaces). To minimize racemization and maintain chiral integrity, all amino acids used were L stereoisomers.

The final step does involve the deprotection of either tert-butyl ester or N-tert-BOC groups from the parent amino acid, by addition of trifluoroacetic acid in chloroform. Trifluoroacetic acid is a strong acid capable of breaking down the siloxane backbone (27), so it was decided to first test the stability of the polydimethylsiloxane under reaction conditions in the presence of the acid. Polydimethylsiloxane (0.5 g, 200 fluid, Dow Corning) was dissolved in 10 mL chloroform and excess trifluoroacetic acid

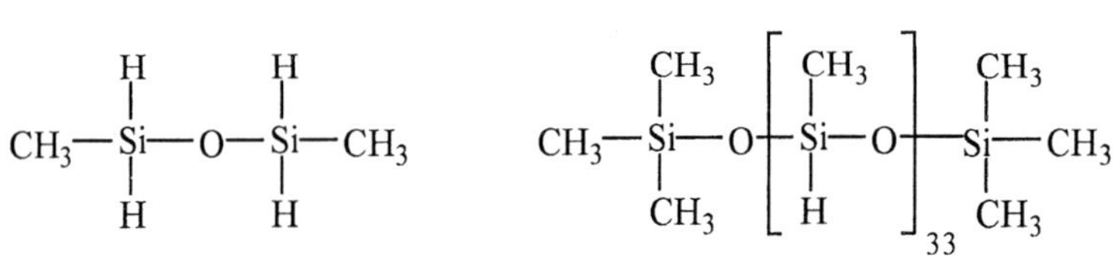

Figure 2. Tetramethyldisiloxane and Poly(methylhydrogen)siloxane.

was added and left to stir for 4 days. The mixture was then extracted with diethyl ether (2 × 50 mL) and water (3 × 10 mL) before drying over magnesium sulfate, filtering and drying in vacuo to give a clear oil which was characterized by NMR.

Preparation of Amino Acid Functionalized Siloxanes. Amino acids having protected carboxylic acid groups (by esterification with *tert*-butyl groups), were coupled to three different siloxane backbones (Scheme 1). The first reaction involves the preparation of allylcarboxy functional amino acid *tert*-butyl esters using established synthetic peptide chemistry (*28*), where dicyclohexylcarbodiimide (DCC) binds 3-butenoic acid to the amino acid *tert*-butyl ester. Since Sheehan and Hess reported their findings on DCC in 1955 (*29*), dicyclohexylcarbodiimide has become an important coupling reagent in amino acid chemistry. DCC readily forms crystalline salts with many acids, producing the dicyclohexylurea side product (DCU), which can be simply recovered by filtration. Mixing both the amino and carboxy reactants with DCC at 0°C for one hour, and then filtering the resulting white dicyclohexylurea precipitate, completes the coupling reaction. Subsequent work-up of the solution by washing with dilute hydrochloric acid, followed by dilute sodium hydrogen carbonate and water, then storing over magnesium sulfate, before final filtering and drying under vacuum isolates the product. Recrystallization in this case, from dichloromethane-pentane yields the final pure products (**1 – 5**). Yields for the five allylcarboxy amino acids are high (84-96%). Both glycine and phenylalanine analogs formed orange crystals, whereas the remaining three amino acid analogs formed white crystals. Coloration is known in extended hydrogen bonded array structures, although alanine compounds (including functionalized polymers, *vide infra*) almost always form white colored products. Elemental analysis confirms composition of all five products.

Characterization of the allylcarboxy functional amino acid *tert*-butyl esters was completed by FTIR, ^{1}H, ^{13}C and ^{29}Si NMR, as well as thermal analysis. The FT-IR absorbances between 3086 and 3066 cm^{-1} for the five allylcarboxy amino acids is typical of allyl containing organics with further absorbances at 1736 cm^{-1} and 1659 cm^{-1} due to the ester vibration (ν_{COO}) of the amino acid and the amide carbonyl (ν_{CONH}), respectively (*30*).

Both ^{1}H and ^{13}C NMR spectra (Figures 3A and 3B) show the characteristic allyl resonances (^{1}H: δ = 5–6 ppm, ^{13}C: 117–132 ppm) and *tert*-butyl resonances of the protecting groups (^{1}H: δ = 1.5–1.7 ppm, ^{13}C: 24 and 81 ppm). Assignment of all resonances in the ^{1}H and ^{13}C NMR spectra of the allylcarboxy functional amino acids was difficult, and the application of 2D NMR experiments was necessary to define each resonance. Previous characterization attempts on similar compounds, the allylcarboxy-amino acids protected by *tert*-butylamide (*17*), as opposed to the *tert*-butyl ester protected amino acids prepared here, only reported general NMR and FTIR characteristics, with no precise structural assignments of all resonances. Such spectral information can be obtained however, using the 2D NMR pulse sequences, COSY and HETCOR. The COSY 2D NMR experiment elucidates the nearest neighbor proton coupling, and so permits all proton resonances to be assigned (Figure 4). For instance, starting from a known resonance, such as the phenyl group at δ = 7.2 ppm, and linking to the cross coupling peaks (off diagonal peaks) reveals the exact position of the neighboring methylene protons **f**, attached to the phenyl ring. Similarly, following the

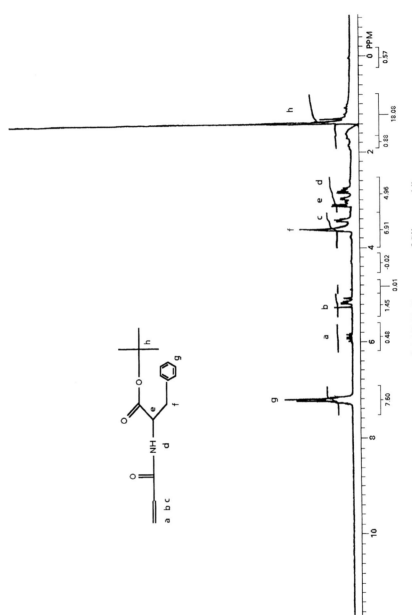

Figure 3 A) The 1H NMR Spectrum of Siloxane (**4**).

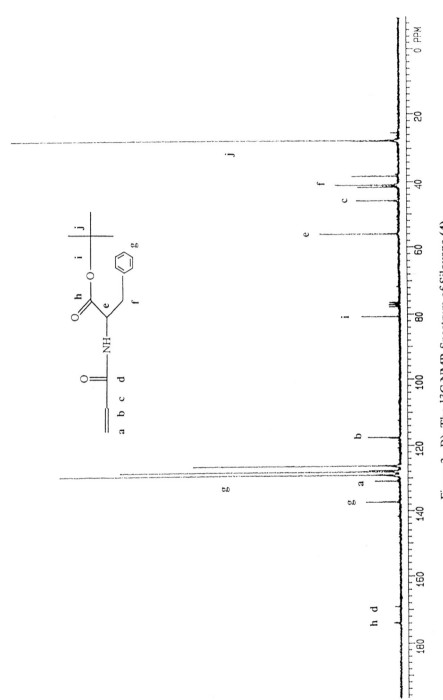

Figure 3 B) The ^{13}C NMR Spectrum of Siloxane (**4**).

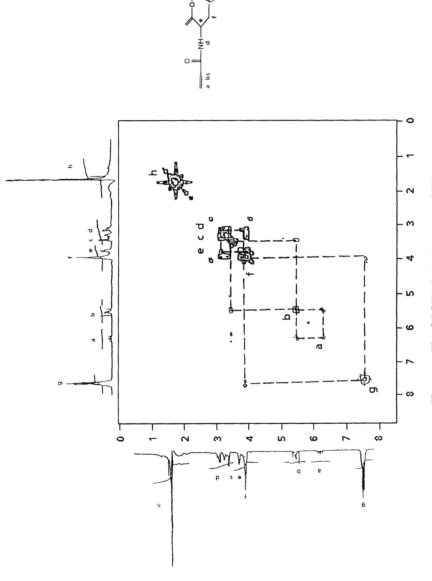

Figure 4 The COSY NMR Spectrum of Siloxane (4).

cross peaks further from resonance **f** reveals the methine proton **e** at δ = 3.4 ppm, as the next neighboring atom in the molecule. Following this procedure (using the nearest neighbor coupling cross peaks), there is now no difficulty in assigning all the resonances of each of the allylcarboxy amino acids prepared **(1 – 5)**.

Performing HETCOR 2D NMR experiments on all five allylcarboxy amino acids reveals the correlation between proton resonances and their corresponding carbon resonances (the heteronuclear species), and so permits the definitive assignment of all the carbon resonances. Figure 5 shows a typical HETCOR spectrum performed on one of the products, the allylcarboxy phenylalanine *tert*-butyl ester **(4)**, where the x-axis corresponds to the ^{1}H NMR spectrum, and the y-axis corresponds to the ^{13}C NMR spectrum. By simply linking the proton resonances on the x-axis, to the corresponding contour peaks of the carbon resonances on the y-axis, complete spectral characterization of all carbon resonances in each product **(1 – 5)** is possible [all the proton resonances have been assigned accurately by COSY]. For example, the ^{13}C NMR resonance at δ = 24 ppm (Figure 6) arises from the methyl groups of the *tert*-butyl ester (Figure 3B), while the resonance at 80.2 ppm (Figure 5) belongs to the quaternary carbon of the *tert*-butyl group (Figure 3B).

The second reaction in Scheme 1, involves the hydrosilylation of allylcarboxy amino acid *tert*-butyl esters **(1 – 5)**, to tetramethyldisiloxane in toluene, by using a small amount of chloroplatinic acid in tetrahydrofuran as the catalyst. The desired telechelic *tert*-butyl ester amino acid functionalized siloxanes **(6 – 10)**, are produced in high yields (75-95%). The products are isolated by washing with saturated citric acid, dilute sodium hydrogen carbonate and water before drying over magnesium sulfate. Saturated citric acid was used, as stronger mineral acids are known to rapidly attack *tert*-butyl ester groups (*21*); particularly *tert*-butyloxycarbonyl groups. Citric acid is a weak acid (initial pK_a 3.1) and does not cause any deprotection of *tert*-butyl groups (*29*).

It was thought that increasing the hydrocarbon content of the R group (i.e. from alanine, -CH$_3$; to valine, -CH(CH$_3$)$_2$; to leucine, -CH$_2$CH(CH$_3$)$_2$) would give rise to decreasing solubility, especially when coupled to a polydimethylsiloxane backbone. Kania *et al.* (*12*) reported the insolubility of block copolymers, comprising of both polydimethylsiloxane and polypeptide blocks; such as siloxane-poly–DL-valine block copolymers that were insoluble across a wide range of solvents. Polyvaline is known to be soluble only in solvents such as dichloroacetic acid and polyphosphoric acid (*30*) although some slight chloroform solubility exists (*31*). The polypeptides used in such previous studies are not the single, small molecules (the amino acids) used in this study. The large molecular size of polypeptides promotes their insolubility in many solvents. No solubility problems were encountered by us in any of the amino acid functional polymers, even in those siloxanes containing many pendant amino acid groups.

The hydrosilylation reaction was monitored by both the disappearance of Si-H resonances in the ^{1}H NMR spectrum (δ = 4.7 ppm, Figure 6A) and the corresponding absorbance in the FTIR spectrum (v_{Si-H} = 2163 cm^{-1}). FTIR analysis also shows typical siloxane vibrations at 3360 cm^{-1} ($v_{NH/OH}$), 1730 cm^{-1} ($v_{C=O}$), 1640 cm^{-1} (v_{CONH}), 1405 cm^{-1} (v_{Si-CH2}), 1263 cm^{-1} (v_{Si-CH3}) and at 1060 cm^{-1} (v_{C-O}). Furthermore, the concomitant disappearance of allyl resonances at δ = 5.1 and 5.9 ppm (Figure 6A), with the formation of new resonances at δ = 0.6 (CH$_2$C*H*$_2$Si), 1.6 (C*H*$_2$CH$_2$Si), 2.4 ppm (C*H*$_2$CH$_2$CH$_2$Si), associated with the silylpropyl group,

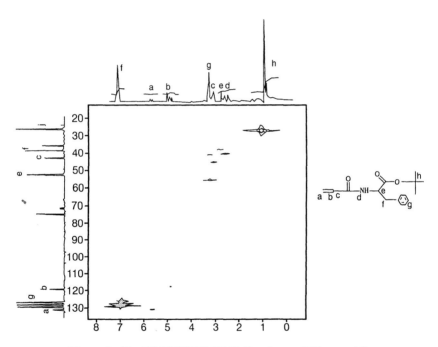

Figure 5 The 2D HETCOR NMR Spectrum of Siloxane **(4)**.

(CH$_2$)$_3$Si, clearly indicates hydrosilylation is complete. Similarly, silylpropyl resonances appear in the ^{13}C NMR spectrum (δ = 14.3, 19.7 and 42.3 ppm; Figure 6B), with loss of all allyl resonances.

Using the NMR spectra in Figure 6, and following previous spectral assignment of allylcarboxy amino acids *tert*-butyl esters (by using the COSY spectrum, Figure 7), the assignment of all resonances in the alanine functionalized siloxane (**6**), is possible. Identifying all resonances in the ^{1}H and ^{13}C NMR spectra of the siloxane functional, protected-amino acid product (**6**) was still difficult, and the extended application of 2D NMR experiments was necessary to assign each resonance. The 2D COSY NMR spectrum (Figure 7), reveals the nearest neighbor coupling, permitting the accurate assignment all the proton resonances. For example, the peaks in the region of δ = 1.3 – 1.5 ppm are congested and difficult to assign. An expanded COSY experiment clarifies this region so that the resonance for the methine proton, **f**, is found at 1.5 ppm, while the resonance for the methylene proton near the silicon atom, **c**, also occurs at 1.5 ppm, with the nearby resonance of the *tert*-butyl peak, **h**, at 1.3 ppm (Figure 7). All the ^{13}C NMR resonances for the product (**6**) were assigned using the HETCOR experiment previously described. Likewise the proton and carbon resonances for all the products (**6 – 10**) can be similarly assigned. Furthermore, to confirm the NH peak assignment, deuterium exchange of these labile amino hydrogen atoms with D$_2$O was examined. Deuterium exchange causes the shift of the amino hydrogen peaks in the ^{1}H NMR, permitting such resonances to be readily identified.

^{29}Si NMR analysis reveals the presence of a single characteristic resonance between δ = 7 – 8 ppm belonging to the -OSiMe$_2$(CH$_2$)$_3$CONH- group (Figure 8). These ^{29}Si NMR results are consistent with published literature values for amino functional siloxanes (*32*). Silicon NMR can also be used to gauge the progress of the hydrosilylation reaction, through the disappearance of Si-H functional species at δ = – 5.0 ppm and the subsequent growth of peaks between 7 and 8 ppm. Finally, elemental analysis confirms the elemental composition of the telechelic products (**6 – 10**).

Generally, hydrosilylation reactions produce two adducts, α (minor product, -SiCH(CH)$_3$CH$_2$-) and β (major product, -SiCH$_2$CH$_2$CH$_2$-), Figure 9. The preparation of the minor α product is termed Markovnikov addition while the preparation of the major β product is termed anti-Markovnikov addition. Once the ^{1}H NMR spectrum has been assigned correctly, it is possible to determine the extent of Markovnikov / anti-Markovnikov addition. For instance, if x is defined as the number of -SiCH(CH)$_3$- groups associated with Markovnikov additions, the formula for determining x is given by (*33*):-

$$\frac{4(m-x)}{3x} = \frac{\int CH_2CH_2 \text{ protons from } ^1H \text{ NMR}}{\int CH_3 \text{ protons from } ^1H \text{ NMR}}$$

where m = 2 for tetramethyldisiloxane, 23 for poly[(methylhydrogen)-*co*-(dimethyl)] siloxane and 33 for poly(methylhydrogen)siloxane. Typically, only a minor amount of α product (Markovnikov addition) was observed in all hydrosilylation products (**6 – 10**), with anti-Markovnikov addition (β adduct) dominating (Table II).

150

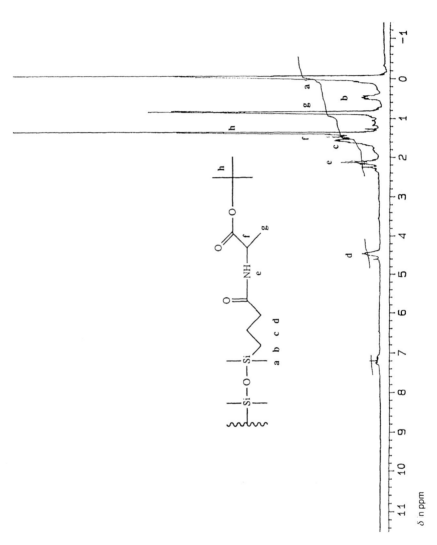

Figure 6 A) The ^{1}H NMR Spectrum of Siloxane (6).

151

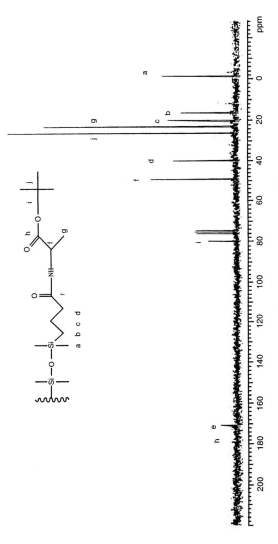

Figure 6 B) The 13C NMR Spectrum of Siloxane (6).

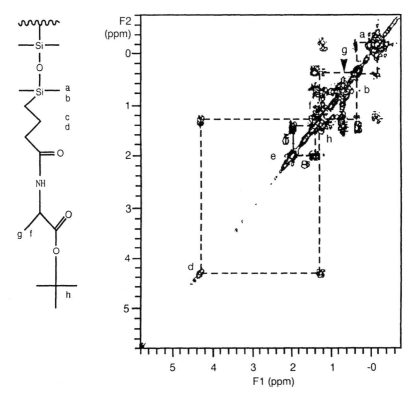

Figure 7 The 2D COSY NMR Spectrum of Siloxane **(6)**.

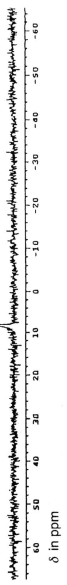

Figure 8 The ^{29}Si NMR Spectrum of Siloxane (**6**).

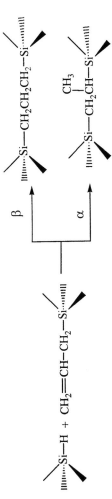

Figure 9 Typical Hydrosilylation.

Table II. The % α and β Adduct for Siloxanes (6 – 20) {Ratio of CH_2CH_2 vs. $CHCH_3$}.

Siloxane No.	$(m-x)/x$ [a]	%α adduct [b]	%β adduct
6 (Ala.)	3.6/0.7	19	81
7 (Gly.)	3.9/0.6	16	84
8 (Leu.)	1.6/0.15	6.5	93.5
9 (Phe.)	–[c]	–	100
10 (Val.)	2.3/0.2	8.7	91.3
11 (Ala.)	3.53/0.56	15.7	84.3
12 (Gly.)	3.8/0.63	16.6	83.4
13 (Leu.)	1.9/0.17	8.9	91.1
14 (Phe.)	–	–	100
15 (Val.)	1.3/0.05	3.8	96.2
16 (Ala.)	2.9/0.42	14.5	85.5
17 (Gly.)	2.86/0.4	13.9	86.1
18 (Leu.)	1.8/0.34	18.8	81.2
19 (Phe.)	–	–	100
20 (Val.)	1.4/0.02	1.3	98.7

[a] calculated from proton NMR integral
[b] % α adduct = $100x/m$
[c] Amount of α adduct is too small to integrate accurately (i.e. less than 0.5%).

Interestingly, a correlation exists between the steric size of the R groups associated with the amino acids and % α adduct formed. As the size of the R group increases, the amount of α adduct formed decreases, illustrating that the larger R groups (such as valine), permit less steric crowding than the smaller alanine or glycine analogs (R = CH_3 or H, respectively) (*34*). The constrained geometry of the large phenyl group of the phenylalanine *tert*-butyl ester, imposes considerable steric constraints on its product siloxane (**9**), resulting in the more sterically demanding α adduct being formed in low yield. The stereochemical reproducibility in forming the alanine and phenylalanine functional siloxanes (**6** and **9**) was evaluated by repeating each reaction three times and examining the α adduct amounts using ^{1}H NMR. In all reactions examined the value for x is ±1.

Hydrosilylation reactions are far from straightforward; and side reactions such as isomerization (*35*), hydrogenation (*36*), decomposition (*37*) of the associated unsaturated compounds, or scission of the siloxane bonds resulting in gel formation (*38*) can occur. Side reactions involving the pendant Si-H groups of either poly(methylhydrogen)siloxane or poly[(methylhydrogen)-*co*-(dimethyl)]siloxane polymers can also lead to a redistribution process among the siloxane chains, with evolution of gaseous silane by-products (*39*). Another possible side reaction involves hydrosilylation of Si-H groups to the carbonyl groups present in the starting reagents (*40*). For instance, Boileau *et al.* (*41*) systematically investigated the side reactions that arise from the hydrosilylation of poly(methylhydrogen)siloxane with allylcarbamates, and found branched polymers in the ^{29}Si NMR spectra.

It is possible for the allyl carboxy amino acid *tert*-butyl ester, during hydrosilylation, to produce a number of side products, where competitive hydrosilylation of Si-H group to the carbonyl group or the allyl group may occur. It must be remembered however, that hydrosilylation occurs faster than most side reactions. Then because the allyl addition proceeds much faster, the hydrosilylation selectivity can be controlled simply by using an excess of the allyl compound (in this way all the Si-H is consumed before it can react to any significant extent with the carbonyl groups). The excess allyl amino acid is then removed in vacuo after hydrosilylation is complete. Furthermore, such side reactions are unlikely to be occurring in the synthesis of our amino acid siloxanes, since no branched siloxane structures were observed in the ^{29}Si NMR spectra; such branched structures occurring between –55 to –65 ppm.

Comparing the % α adducts (Markovnikov addition products) formed using the larger siloxane backbones to the tetramethyldisiloxane analogs prepared above, reveals similar trends. Again as the R group of the amino acid siloxanes is increased, the α adduct (Markovnikov addition) decreases; with steric constraints resulting in the larger R groups, like phenylalanine, having very little Markovnikov addition (Table II).

The final step of Scheme 1, involves the deprotection of the *tert*-butyl group to produce the product, amino acid tetramethyldisiloxanes. Such transformations were carried out using an excess of trifluoroacetic acid in chloroform. The *tert*-butyl group resonances in the ^{1}H NMR spectrum of siloxane **(24)** (δ = 1.5 - 1.7 ppm C(CH_3)$_3$, Figure 10A) disappear as the reaction proceeds (Figure 10B) with the concomitant appearance of the carboxylic acid resonance at δ = 10.9 ppm (shown in the expanded region of Figure 10B). After the reaction is complete, the product is extracted using dilute hydrochloric acid, dilute sodium hydrogen carbonate and water before drying. Elemental analysis confirmed the elemental composition of the final products.

FTIR analysis of the unprotected telechelic siloxanes shows typical siloxane vibrations (i.e. νSi-CH$_3$ 1263 cm^{-1} and νSi-CH$_2$ 1420 cm^{-1}) concomitant with carboxylic acid vibrations (νCOOH 3350 cm^{-1} and νC=O 1710 cm^{-1}). The decrease in intensity of the C-H vibrations at 2960 to 2850 cm^{-1} corresponds to the loss of the *tert*-butyoxycarbonyl group of **(24)**. The vibrations at 3060 cm^{-1} are characteristic of the aromatic C-H stretching vibrations of the phenyl group in phenylalanine.

The poly[amino acid *tert*-butyl ester] siloxanes were also prepared, using either poly(methylhydrogen)siloxane **(11–15)** or poly[(methylhydrogen)-*co*-(dimethyl)]siloxane **(16 – 20)**, in good yields (78–90% and 77–84%, respectively). Deprotection of these siloxane polymers results in viscous white colored oils **(26 – 35)** forming in quite high yields. Spectral characterization was carried out on these polymers

(NMR and FTIR) similar to that for the α,ω-telechelic tetramethyldisiloxane analogs **(21 – 25)**. Additionally, mass spectroscopy provided detailed molecular mass and fragmentation patterns. For example, the poly-(3-phenylalanine, *tert*-butyl ester, propyl methyl)siloxane **(34)**, mass spectrum shows fragmentation patterns typical for polydimethylsiloxanes, i.e. m/e 74 ($SiMe_2$) and phenyl ring fragmentations, together with the repeated loss of CH_3SiOR groups (m/e = 367), where the R group corresponds to the organofunctional side chain (phenyl).

Thermal Analysis of Amino Acid Functional Siloxanes. For all amino acid siloxanes prepared, thermogravimetric analysis revealed similar weight loss patterns. Typically, a rapid weight loss (~50%) occurs above 200°C, followed by another smaller weight loss around 350°C (Figure 11). The derivatized siloxanes are not expected to display high thermal stability, because of the large organic content inherent to the polymeric side chains, and consistent with published literature values of similar siloxane organic polymers *(41)*. Yilgör *et al.* *(42)* did prepare siloxane-urea copolymers which also revealed initial weight losses around 300°C, and which by 400°C did lose more than 50% of their initial weight. On the other hand, they also prepared siloxane-imide copolymers, possessing much higher thermal stability, where initial weight loss occurs above 400°C, and only 15% total weight loss is seen even at 500°C. Such high thermal stability was attributed to the increased hydrogen bonding of the polyimide structures, believed to be responsible for improved mechanical properties in these copolymers. Even though our amino acid siloxanes **(21 – 35)** should similarly have a high degree of hydrogen bonding as a result of their pendant amino acid groups, no increased thermal stability is seen.

Differential scanning calorimetry (DSC) is also applied to functionalized siloxanes to evaluate their intrinsic thermal properties, particularly the glass transition temperature, T_g. All siloxanes were cooled down to sub-ambient temperatures of –150°C (liquid nitrogen cooled) and then slowly heated at a rate of 10°C/min to 220°C in the DSC. Samples were accurately weighed and placed in hermetically sealed aluminum pans. The glass transition temperatures for the functionalized siloxanes **(21 – 35)**, was in the range –81 to –38°C (Table III), and was often difficult to detect. Such difficulties can be attributed to the amino acid functional groups promoting widespread intramolecular and intermolecular hydrogen bonding in polymers. Such hydrogen bonding restricts main chain rotation, and often, particularly in natural polymers like polysaccharides, no T_g is seen until decomposition of the main chain occurs, as intra- and inter-molecular hydrogen bonds stabilize the structure *(43,44)*. Above the T_g, the siloxanes undergo melting (T_m) leading to decomposition peaks at temperatures above 160°C *(45)*. A typical DSC trace for siloxane **(23)**, the leucine diadduct, is shown in Figure 12, with a T_g of –74°C and a T_m of 193°C.

Summary

The formation of amino acid siloxanes, containing a free pendant carboxylic acid group, and free from side reactions, or the formation of minor products, was accomplished for the first time. Five amino acids (alanine, glycine, leucine, phenylalanine and valine) were

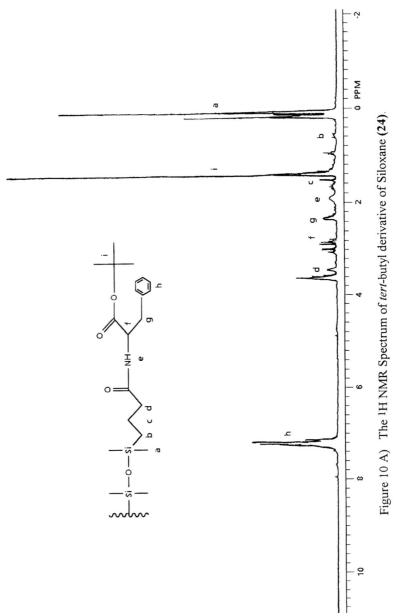

Figure 10 A) The ^1H NMR Spectrum of *tert*-butyl derivative of Siloxane (24).

159

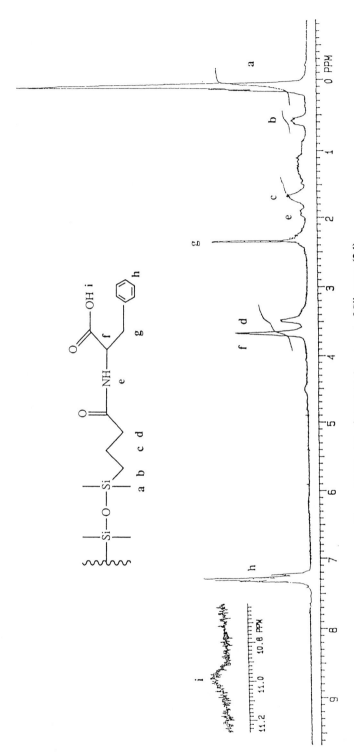

Figure 10 B) The ¹H NMR Spectrum of Siloxane (**24**).

160

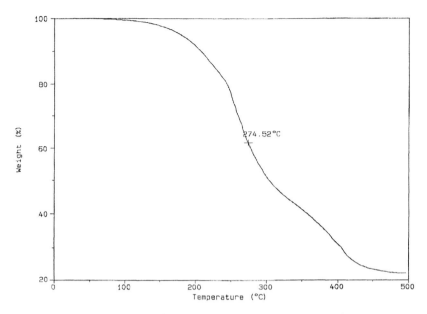

Figure 11 TGA of Siloxane **(25)** [heating rate 10°C/min].

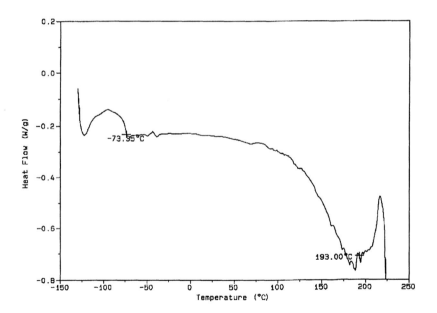

Figure 12 DSC Trace of Siloxane **(23)**.

Table III DSC Results for Amino Acid Functional Siloxanes.

Siloxane	Backbone	Amino Acid	T_g, °C	T_m, °C
21	TMDS[a]	Alanine	−78.1 m[b]	138.2 m
22		Glycine	−57.9 m	−
23		Leucine	−74.0 m	193 m
24		Phenylalanine	−45.0 m	−
25		Valine	−45.4 m	165.1 m, 168.9 m
27	D33	Glycine	−38.7 m	158.4 s
29		Phenylalanine	−70.0 m	165.1 s, 168.0 s
31	S3	Alanine	−80.1 m	−
33		Phenylalanine	−77.3 m	−

[a] TMDS = tetramethyldisiloxane, D33 = poly(methylhydrogen)siloxane and S3 = poly[(methylhydrogen)-co-(dimethyl)]siloxane.
[b] peak intensities: m = medium; s = strong; w = weak.

systematically coupled to three different siloxane backbones: tetramethyldisiloxane, poly(methylhydrogen)siloxane and poly[(methylhydrogen)-co-(dimethyl)]siloxane in order to evaluate molecular weight effects on reaction efficiency. All products were formed in high yields and characterized by FTIR, NMR (^{1}H, ^{13}C, ^{29}Si and where needed 2D techniques, COSY and HETCOR), elemental analysis and thermal analysis (TGA and DSC).

Systematic investigations of the adsorption of such amino acid functional siloxanes on various surfaces where molecular effects (including steric considerations), surface molecular orientation and coordination, can now be undertaken. Such adsorption behavior intrinsic to biomolecular activity will be the subject of future reports.

References

1. Kendrick, T. C.; Parbhoo, B.; White, J. W. In *The Silicon-Heteroatom Bond*; Patai, S. and Rappoport, Z., Eds.; John Wiley and Sons: New York, 1991; Chapter 3.
2. Kollmeier, B. J. *Tenside Surf. Det.* **1986**, *26*, 26.
3. Morcellet, M.; Loucheux, C.; Daoust, H. *Macromolecules*, **1982**, *15*, 890.
4. Lekchiri, A.; Morcellet-Sauvage, J.; Morcellet, M. *Macromolecules*, **1987**, *20*, 49.
5. Methenitis, C.; Morcellet, J.; Pneumatikakis, G.; Morcellet, M. *Macromolecules*, **1994**, *27*, 1455.

6. Pneumatikakis, G.; Pitsikalis, M.; Morcellet, J.; Morcellet, M. *J. Polym. Sci., Polym. Chem.* **1995**, *33*, 2233.
7. Morcellet-Sauvage, J.; Morcellet, M.; Loucheux, C. *Polym. Bull.* **1983**, *10*, 473.
8. Kartvelishivili, T.; Kvintradze, A.; Katsarava, R. *Macromol. Chem. Phys.* **1996**, *197*, 249.
9. Penczek, S.; Kaluzy'nski, K.; Baran, J. In *Macromolecules*; Kahovec, J., Ed.; VSP: Netherlands, 1993, 231.
10. Kumaki, T.; Sisido, M.; Imanishi, Y. *J. Biomed. Mat. Res.* **1985**, *19*, 785.
11. Chow, S.; Byck, J. S. US Patent 3 562 353, 1971.
12. Kania, C. M.; Nabizadeh, H.; McPhillimy, D. G.; Patsiga, R. A. *J. Appl. Polym. Sci.* **1982**, *27*, 139.
13. Yoon, K. S.; Sung, Y. K.; Kim, S. W. *Pollimo*, **1995**, *19*, 139.
14. Teyssié, D.; Caille, J.; Quentin, J.; Yu, J.; Boileau, S. *Silicones in Coatings*, 2nd Conf.; Brussels: PRA, London, 1996.
15. Frank, H.; Nicholson, G. J; Bayer, E. *J. Chromatogr. Sci.* **1977**, *15*, 174.
16. Celma, C.; Giralt, E. *J. Chromatogr.* **1991**, *562*, 447.
17. Feibush, B.; Gil-Av, E. *Tetrahedron* **1970**, *26*, 1361.
18. Koppenhoefer, B.; Mühleck, U.; Walser, M.; Lohmiller, K. *J. Chromatog. Sci.* **1995**, *33*, 217.
19. Frank, H.; Nicholson, G. J; Bayer, E. In *Amino Acid Analysis by Gas Chromatography*, Vol 2.; Gehrke, C. W.; Kuo, K. C. T.; Zumwalt, R. W., Eds.; CRC: Boca Raton, FL, 1987.
20. Zimmerman, G.; Haas, W.; Faasch, H.; Schmalle, H.; König, W. A. *Liebigs Ann. Chem.* **1985**, 2165.
21. König, W. A.; Krebber, R.; Mischnick, P. *J. High Resolut. Chromatogr.* **1989**, *12*, 732.
22. Lohmiller, K.; Bayer, E.; Koppenhoefer, B. *J. Chromatogr.* **1993**, *634*, 65.
23. Ôi, N.; Kitahara, H.; Matsushita, Y.; Kisu, N. *J. Chromatogr.* **1996**, *722*, 229.
24. Ôi, N.; Nagase, M.; Swada, Y. *J. Chromatogr.* **1993**, *292*, 427.
25. Aramarego, W. L.; Perin, D. D. *Purification of Laboratory Chemicals*, 3rd ed.; Pergamon: Oxford; U.K., 1988.
26. Kantor, S. W.; Grubb, W. T.; Osthoff, R. C. *J. Am. Chem. Soc.* **1954**, *76*, 5190.
27. Noll, W. *Chemistry and Technology of Silicones*; Academic Press: New York, 1968.
28. Sheehan, J. C.; Hess, G.P. *J. Am. Chem. Soc.* **1955**, *77*, 1067.
29. Jones, J. *Amino Acid and Peptide Synthesis*; Davies, S. G., Ed.; Oxford Press: Oxford, 1994, pp17-18.
30. Bamford, C. H.; Elliot, A.; Hanby, W. E. *Synthetic Polypeptides*; Academic; New York, 1956; p325.
31. Coleman, D.; Farthing, A. C. *J. Chem. Soc.* **1950**, 3218.
32. Taylor, R. B.; Parbhoo, B.; Fillmore, D. M. In *The Analytical Chemistry of Silicones*; Smith, A. L., Ed.; Wiley: New York, 1991; Chapter 12.
33. Britcher, L. G.; Kehoe, D. C.; Matisons, J. G.; Swincer, A. G. *Macromolecules*, **1995**, *28*, 3110.
34. Britcher, L. G. Ph.D. Thesis, University of South Australia, 1996.
35. Carless, H. A. J.; Haywood, D. J. *J. Chem. Soc., Chem. Commun.* **1980**, 980.

36. Lestel, L.; Cheradame, H.; Boileau, S. *Polymer*, **1990**, *31*, 1155.
37. Yu, J. M.; Teyssié, D.; Boileau, S. *Polym. Bull.* **1992**, *28*, 435.
38. Torrès, G. Ph.D. Thesis, Université P. et M. Curie, Paris, 1986.
39. Spinu, M.; McGrath, J. E. *J. Polym. Sci. Part A: Polym. Chem.* **1991**, *29*, 657.
40. Marciniec, B. *Comprehensive Handbook on Hydrosilylation*; Pergamon: Oxford, 1992.
41. Yu, J. M.; Teyssié, D.; Boileau, S. *J. Polym. Sci.: Part A: Polym. Chem.* **1993**, *31*, 2373.
42. Yilgör, I.; Lew, B.; Steckle, W. P. Jr.; Riffle, J. S.; Tyagi, D.; Wilkes, G. L.; McGrath, J. E. *Polym. Prepr., Am. Chem. Soc. Div.* **1983**, *24*, 35.
43. Bair, H. E. In *Assignment of the Glass Transition*; Seyler, R. J., Ed.; ASTM: Philadelphia, 1994; p52.
44. Bershtein V. A.; Egorov, V. M. *Differential Scanning Calorimetry of Polymers: Physics, Chemistry, Analysis, Technology;* Ellis Harwood: New York, 1994; Sec. 4.7.
45. Hatakeyama, T.; Quinn, F. X. *Thermal Analysis Fundamentals and Applications to Polymer Science*; John Wiley & Sons, New York, 1994; Chapter 5.

Chapter 9

Telechelic Aryl Cyanate Ester Oligosiloxanes: Impact Modifiers for Cyanate Ester Resins

Steven K. Pollack[1] and Zhidong Fu[2]

[1]Department of Chemistry and the Polymer Science and Engineering Program, Howard University, Washington, DC 20059
[2]Colgate-Palmolive, 909 River Road, P.O. Box 1343, Piscataway, NJ 08855

We describe the synthesis of monomers and characterization of polymers and blends derived from telechelic siloxanes terminated with dimethyl(4-cyanatophenyl)silyl units. These are synthesized by base-catalyzed equilibration of THP-protected 1,1,4,4-tetramethyl-1,4-bis(4-hydroxyphenyl)disiloxane with octamethylcyclotretrasiloxane (D_4) and octaphenylcyclotetrasiloxane (D_4") followed by deprotection and reaction with cyanogen bromide. Homopolymerization of the telechelic siloxane cyanate esters produce cross-linked materials whose glass transition temperatures range from 277°C(disiloxane) to -120°C(dimethyl-10-mer). Blending with commercial novolac based-cyanate ester resins (Primaset PT-30) produces a single phase material for the disiloxane and two-phased materials for the higher oligomers. The completely dispersed rubber phase has an average domain size of 20 microns.

Cyanate ester resins (CERs) have found extensive use in applications where low-dielectric, high thermal stability materials are needed[1]. Polymerization occurs by the cyclotrimerization of aryl cyanate ester groups

164 © 2000 American Chemical Society

These applications include printed wire circuit boards (PCWBs), and radomes. Recently, these resins have been proposed as materials for use as the matrix in structural composites used in civilian air transport. In this application, the organic materials must exhibit low flammability as well as having the appropriate strength and toughness for this type of application. CERs, by themselves possess appropriate flammability properties, but are too brittle for use alone as matrix materials for structure composites. However, their toughness can be greatly enhanced by the introduction of modifiers such as thermoplastics or elastomers. Siloxane based elastomers provide both the requisite improvement in toughness as well as having excellent flame retardancy. To that end, there have been efforts to develop siloxane based impact modifiers for CERs. Previous workers have described approaches utilizing either amine or alcohol terminated siloxanes as impact modifiers[2]. There is also a patent issued for the development of allyl(2-phenylcyanate) terminated siloxanes[3]. In the former case, there is concern that the cross-linking reaction may introduce polar functionality into these materials causing excessive moisture absorbance. In the latter materials, the added hydrocarbon component of the allyl group is a potential problem in terms of combustibility of the siloxane. We have focused our attention on developing a family of telechelic siloxanes terminated with p-cyanatophenyl groups.

Experimental

NMR The nuclear magnetic resonance spectra were obtained using a GE-NMR QE-300 model spectrometer with chloroform-d ($CDCl_3$) as solvent and TMS or $CDCl_3$ was used as internal reference. ^1H NMR operated at 300.6 MHz and ^{13}C NMR operated at 75.2 MHz.

IR The infrared spectra were recorded on Perkin-Elmer 1600 series FT-IR spectrophotometer. Liquid samples were placed on NaCl plates. Solid samples were ground and mixed with KBr power. Rubbery samples were ground in liquid nitrogen and mixed with KBr powder. The mixtures were then pressed in a hand KBr press to give a transparent thin film.

SEM Scanning electron micrographs (SEMs) were obtained using an R.J. Lee Inc. Personnel SEM. Micrographs were obtained from uncoated samples using the low pressure mode of operation. Elemental composition of the various phases in the materials was analyzed using energy dispersive X-ray analysis (EDXA). Images were obtained utilizing a backscatter detector thus contrast is due primarily to difference in atomic number.

Thermal Analysis Differential scanning calorimetry was performed using a Perkin-Elmer Pyrus 1 DSC. Scan rates were 20 °C/min and carried out under a helium atmosphere. Thermogravimetric analysis was carried out using a Perkin-Elmer TGA-7 interfaced to a Pyrus Data Station. Analyses were run at 20 °C/min scan rates of 20 cc/min flow of either nitrogen or air (as indicated).

Synthesis of bis[1,3-p-tetrahydropyranyloxyphenyl]-1,1,3,3-tetramethyldi-siloxane (5): Solid tetramethylammonium hydroxide (0.30 g) was added to 34.0 g of **4** (synthesized via a modification of the method described by McGrath and co-workers[4]). The mixture was placed in high vacuum (0.2 mmHg) while stirring for three hours. When the reaction was completed, the 3500-3000 cm^{-1} broad peak disappeared. The resulting product is a white solid (29.9 g, 88 % yield). IR (neat, cm^{-1}): 2948 (strong, CH), 2875 (strong, CH), 1594 (strong, Ar), 1564, 1500 (strong, SiAr), 1467, 1454, 1441, 1388, 1356, 1239 (strong, SiMe), 1202, 1180, 1117 (strong SiO), 1038 (broad, Si-O-Si), 966, 921, 872, 840, 791, 704, 642, 565 (SiO) cm^{-1}. ^1H NMR ($CDCl_3$): 7.44 (d,4H), 7.03 (d, 4H), 5.44 (t, 2H), 3.6-3.9 (m, 4H), 1.5-2.1 (m, 12H), 0.29 (s, 11H). ^{13}C NMR ($CDCl_3$): 158.0 (Ar), 134.4 (Ar), 132.0 (Ar), 115.7 (Ar), 96.9 (THP), 62.0 (THP), 30.3 (THP), 25.2 (THP), 18.7 (THP), 0.9 (SiMe).

Synthesis of α,ω-(p-tetrahydropyranyloxyphenyl)-oligo-(dimethylsiloxane) (7): Calculated amounts of **5**, D_4, and 1 w % of tetramethylammonium hydroxide and a stirring bar were placed in a flame dried, three-neck round bottomed flask equipped with an addition funnel and condenser. The mixture was maintained at 80 to 90 °C for two to three hours while purging with nitrogen. After equilibration, the mixture was cooled down to room temperature and pumped on a high vacuum line for 2 hours to remove residual D_4. ^{1}H NMR: 7.49 (d, 4H), 7.05 (d, 4H), 5.44 (t, 2H), 3.6-3.9 (m, 4H), 1.5-2.1 (m, 13H), 0.31 (s, 12H), 0.08 (s, 6n H). ^{13}C NMR (CDCl$_3$): 158.4 (Ar), 134.4 (Ar), 132.2 (Ar), 116.0 (Ar), 96.4 (THP), 62.0 (THP), 30.5 (THP), 25.3 (THP), 18.9 (THP), 1.14 (SiMe), 0.82 (SiMe).

Synthesis of α,ω-(p-tetrahydropyranyloxyphenyl)-oligo-(dimethyl-co-diphenylsiloxane) (8): A calculated amount of **5**, calculated amount of D_4 and D_4'', and 1% w of tetramethylammonium hydroxide and a stirring bar were placed in a flame dried, three-neck round bottomed flask equipped with an addition funnel and condenser. The mixture was maintained at 80 to 90 °C for two to three hours while purging with nitrogen. After equilibration, the mixture was cooled down to room temperature and pumped on a high vacuum line for 2 hours to remove residual D_4. IR (neat, cm^{-1}): 3068, 2956 (strong, CH), 1594 (strong, Ar), 1499 (SiAr), 1430, 1259 (strong, SiMe), 1180, 1121 (strong, SiO), 1040 (broad, Si-O-Si), 965, 919, 801 (doublet), 703 cm^{-1}. ^{13}C NMR (CDCl$_3$): 158.0 (Ar), 134.3 (Ar), 129.7 (Ar), 127.5 (Ar), 115.8 (Ar), 96.0 (THP), 62.0 (THP), 30.3 (THP), 25.2 (THP), 18.2 (THP), 0.8 (SiMe), 0.6 (SiMe).

Deprotection of THP-protecting group: The THP protected p-phenol terminated oligosiloxane was dissolved in CHCl$_3$. Dilute HCl (1 mL of 5% HCl in 25 mL of ethanol) was added. After stirring for 2 hours at room temperature, the solution was washed twice with water. The resulting solution was stripped of solvent by rotary evaporation and pumped on a high vacuum line for 4 hours. From the IR, a new broad peak appears at 3000-3500 cm^{-1}. For Bis[1,3-p-phenol]-1,1,3,3-tetramethyldisiloxane (**6**): IR(neat, cm^{-1}): 3349 (broad, OH), 3025, 2957 (strong, CH), 1595 (strong, Ar), 1501 (strong, Ar), 1417, 1361, 1256 (strong, SiAr), 1417, 1113 (strong, SiO), 1048 (strong, Si-O-Si), 835 (strong), 789 (strong), 675 cm^{-1}. ^{1}H NMR: 7.40 (d, 4H), 6.80 (d, 4H), 0.31 (s, 9H). ^{13}C NMR (CDCl$_3$): 157.0 (Ar), 134.6 (Ar), 131.0 (Ar), 115.0 (Ar), 0.9 (SiMe). For α, ω-(p-phenol) oligo (dimethylsiloxane) (**9**): IR (neat, cm^{-1}): 3386 (broad, OH), 2982 (strong, CH), 2904 (CH), 1597 (strong, Ar), 1503 (SiAr), 1412, 1359, 1280 (strong, SiMe), 1029 (broad, Si-O-Si), 799 (strong), 697 cm^{-1}. ^{1}H NMR: 7.40 (d, 4H), 6.80 (d, 4H), 0.31 (s, 9H), 0.07 (s, 60 H). ^{13}C NMR (CDCl$_3$): 156.7 (Ar), 134.8 (Ar), 130.9 (Ar), 115.0 (Ar), 1.1 (SiMe), 0.9 (SiMe). For α, ω-(p-phenol) oligo (dimethyl-co-diphenylsiloxane) (**10**): IR (neat, cm^{-1}): 3600-3000 (broad, OH), 2961 (strong, methyl), 1595 (strong, Ar), 1500 (SiAr), 1427, 1261 (strong, ArMe), 1100-1000 (strong, Si-O-Si), 801 (strong), 703 (strong).

Synthesis of bis[1,3-(4-cyanatophenyl)]-1,1,3,3-tetramethyl-disiloxane (11): Compound **6** (5.0 g, 0.016 mol) and cyanogen bromide (3.6 g, 0.034 mol) were dissolved in 100 mL of acetone in a 250 mL round bottomed flask. Triethylamine (3.2 g, 0.032 mol) was added dropwise to the above solution during 20 min while cooling with ice. After the addition was complete, the mixture was stirred for 20 min while cooling with ice and poured on to ice/water. The mixture was washed three times with water and the solvent was stripped by rotary evaporation. The product was pumped on a high vacuum line for several hours. IR (neat, cm^{-1}): 2959 (CH), 2268 (doublet, strong, CN), 1584 (Ar), 1493 (ArSi), 1258 (SiMe), 1197 (SiAr), 1101 (SiAr), 1114, 1058 (Si-O-Si), 831, 731 cm^{-1}. ^{1}H NMR (CDCl$_3$): 7.60 (d, 4H), 7.30 (d, 4H), 0.35 (s, 12H). ^{13}C NMR: 153.6 (Ar), 138.4 (Ar), 135.0 (Ar), 114.4 (Ar), 108.3 (-CN), 0.4-0.6 (-CH$_3$).

Results and Discussion

Utilizing McGrath's method[4], we were able to synthesize the THP-protected disiloxane **5** in high yield. The deprotection of the bisphenol under acid conditions can lead to the predominant cleavage of the phenyl-Si bond, but if done with care, the deprotection goes smoothly to produce the bis-phenol **6**. Reaction with cyanogen bromide with triethylamine catalysis leads to the 1,3-bis(4-cyanatophenyl)-1,1,3,3-tetramethyldisiloxane in high yields[5].

Equilibration of **5** with D_4 or D_4''/D_4 mixtures with tetramethylammonium hydroxide as catalyst leads to telechelic tetrahydropyranloxy siloxanes[5]. These were deprotected analogously to **5** and reacted with cyanogen bromide to yield reactive oligomers **12** and **13**. It should be noted that the oligomers showed less of a tendency to cleave at the phenyl-Si bond relative to the dimer during deprotection[6].

Hompolymerization of compound **11** by curing from 120-280°C leads to a glassy solid with T_g of 274°C. This compound exhibits a major mass process at 500°C and losses up to 70% of its mass at 700°C in either an air or nitrogen atmosphere. For compound **12**, the resultant material is a rubber with a T_g of -110°C and exhibiting a weak mass loss process occurring at 225°C and a major mass loss process starting at

504°C (with 90% of mass remaining) and resulting in a total 70% mass loss at 700°C in a nitrogen atmosphere and 50% in air. For compound **13** the T_g is 8°C and the mass loss process with onset at 335°C is more pronounced than observed in compound **12** and results in a 25% mass loss at 5010°C where a second loss process initiates. The total mass loss at 700°C is 60%. The higher temperature loss process is also observed in the commercial CER used in this study with onset at 482°C and a total mass loss of 72% at 700°C. The lower temperature process observed only in the siloxane systems is most likely due either to the scission of the silicon phenyl bond or loss of methyl units. The higher temperature is caused by the breakdown of the cyanate units which provide the cross-links in the system.

When compound **11** was blended with the commercial CER Primaset PT-30 (a novolac-based CER from Lonza, T_g = 275 °C) at 10 parts per hundred resin (phr), the resultant cured material exhibited a single glassy phase with T_g of 275 °C. However, for blends with compounds **12** or **13** at the same composition, a two-phase material is formed. In these case the oligomers were not soluble and agitation during the initial stages of cure was required. For the blend with the oligodimethylsiloxane network chain units (**12**), the morphology (**Figure 1**)

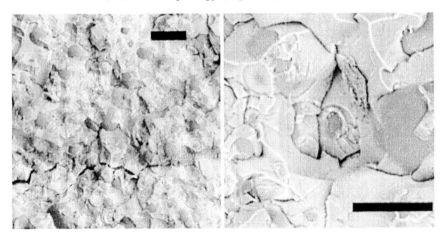

Figure 1 SEM photograph of the fracture surface of 10 phr telechelic siloxane **12** in PT-30 after curing. Left image is at 500x. (inset bar is 200μm), right image at 1,500 X (inset bar 120 μm).

of the blend is a continuous siloxane matrix (lighter materials in **Figure 1**) with large droplets (average diameter 100 μm) of the novolac CER (as determined by EDXA, darker material in **Figure 1**). For the blend with the diphenylsiloxane containing additive (**13**), the typical domain size of the rubbery phase is 20-30 μm (**Figure 2**) and the PT-30 rich phase is continuous. Note that, using EDXA, the cavities left behind after fracturing still have a thin layer of siloxane in place, indicating good adhesion between the two phases. The arrows indicate siloxane rubber particles which fractured and remained attached to the continuous novolac based phase. While this date is consistent with a high degree of adhesion between the two phases, mechanical studies need to be carried out to determine the quality of this interface.

Thermogravimetric analysis of the three blends are quite similar. The blend with **11** behaves identically to the pure PT-30. The blend with compound **12** shows a loss onset at 460°C and a total mass loss of 65%. Similar results are observed for the blend with compound **13**.

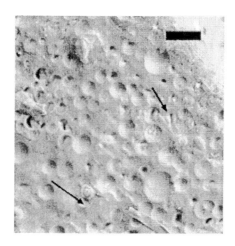

Figure 2. SEM photograph (500 x) of fracture surface of 10 phr telechelic siloxane **1 3** in PT-30 after curing. (inset bar is 100μm)

Conclusions

We have demonstrated the synthesis of a number of telechelic CERs which can be compounded with commercial CERs to form either single or two phase blends. In the case of the two-phase blends, the rubbery phase domain size is in the range necessary for improving the fracture toughness of the commercial resin. We are currently exploring the effect of oligomer composition and molecular weight on the mechanical and flammability properties of these new materials. We are also examining the thermal and mechanical properties of cured blends of the disiloxane CERs with oligosiloxane CERs.

Acknowledgments

We thank the NIST Building and Fire Research Laboratory (Grant Number 60NANB6D0016) for support of this research. We also wish to thank Dr. Jeffrey Gilman of NIST BFRL for his help in this effort. We also thank Mr. Steve Davis of R.J. Lee Instruments for providing the electron micrographs.

References

(1) Hamerton, I., *Chemistry and Technology of Cyanate Ester Resins,* Chapman & Hall, 1994.
(2) Arnold, C.; Mackenzie, P.; Malhotra, V.; Pearson, D.; Chow, N.; Hearn, M.; Robinson, G, *37th Int. SAMPE Symp. Exhib.* 1992, **37**, 128.
(3) Liao, Z.K.; Wang, C.S. United States Patent, 5 260 398, 1993.
(4) Knauss, D.; Yamamoto, T.; McGrath, J.E. *ACS Polymer Preprints* 1992, **33(1)**, 988.
(5) Fu, Z.D. and Pollack, S.K. *ACS Polymer Preprints*, 1998, **39(1)**, 577.
(6) Fu, Z.D., *Ph.D. Thesis, Howard University,* 1997.

Chapter 10

UV Curable Silicones from Acryloxymethyldimethylacryloxysilane

H. K. Chu

Loctite Corporation, 1001 Trout Brook Crossing, Rocky Hill, CT 06067

A facile preparation of acrylate terminated silicones by the condensation reaction between the readily available silanol terminated silicone fluids and acryloxymethyldimethylacryloxysilane is described. The simplicity of the reaction provides a practical route for the preparation of UV curable silicones. The surprising ease of the reaction between the silane and silanol is attributed to the possible hypervalent silicon transition state even though ^{29}Si-NMR evidence suggests the silane is tetracovalent.

The constant quest for faster curing silicones for various industrial applications has led to many important developments for radiation curable silicones in recent years (1,2). Notable examples among these developments include cationic curing silicones (3-7), free-radical curing thiol-ene (8-10), acrylate or acrylamide (11-14) functional silicones, and photohydrosilylation curing silicones (15-17). Each of these developments offers certain advantages but at the same time imposes limitations in their applications. Cationic curing silicones using iodonium or sulfonium salts as the photoinitiators for the curing of epoxy and vinyl ether functional silicones allows fast cure without the tacky surface that is often associated with the oxygen inhibition of the acrylate cure. However, the high cost of the photoinitiators and the compatibility of the photoinitiator with silicones are problems. Also, acid generated from the decomposition of the onium salt during UV irradiation is responsible for the cationic cure. However, the acid thus generated remains in the cured silicone network. The presence of the strong acid in a siloxane network clearly begs the question of whether the cured silicone will be durable enough for the intended applications. Thiol-ene crosslingking of mercaptosiloxanes and vinylsiloxanes in the presence of free radical initiator is also well known. The curing is fast and also free of oxygen inhibition. However, the shelf stability of the uncured silicones and the thermal stability of the cured silicones are both lacking. Photohydrosilylation using photolabile platinum catalysts such as platinum acetylacetonate offers an interesting mode of curing. The cured silicones should have all the advantages associated with the conventional two part thermal curing liquid silicone rubbers but in a convenient one part package. However, the stability appears to be lacking and the curing is slow. Therefore, in spite of its problems of inhibition of surface cure by oxygen, free radical curing of acrylate and acrylamide silicones is still by far the most widely practiced method for curing silicones photolytically. A variety of synthetic routes have been developed for the preparation of acrylate and acrylamide functional silicones. Many of these developments are carried out in industrial research labs and have been transformed into commercial products. However, most, if not all, of these developments still require tedious processes for the derivatization of the silicones that often result in inconsistent and inferior commercial products. Thus, a simple method for the preparation of acrylate or acrylamide polysiloxanes that gives consistently high quality UV curable silicones is still needed.

UV Curable Acrylated Silicones

Literature on the preparations of acrylated (and acrylamidated) silicones up to early 1990 has been expertly reviewed (*1*). Many further refinements have been reported since. It is outside the scope of this chapter to review all the developments. However, two prominent industrial developments from the influential silicone manufacturers, Dow Corning and Shin-Etsu, clearly warrant further discussions.

Intensive efforts have been carried out at Dow Corning with the objective of producing acrylamide substituted polydimethylsiloxanes (PDMS). Thus, various processes have been developed to react acryloyl chloride with aminosiloxanes or aminosilanes as shown below (*13, 18, 19*):

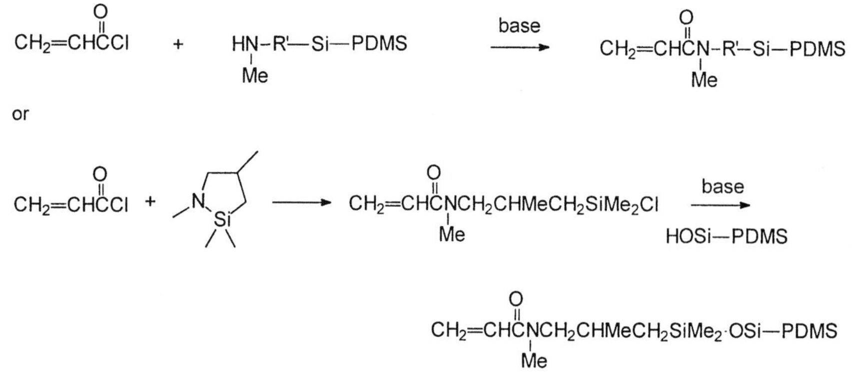

$$CH_2=CHCCl \overset{O}{\underset{\|}{}} + HN-R'-Si-PDMS \overset{|}{\underset{Me}{}} \xrightarrow{base} CH_2=CHCN-R'-Si-PDMS \overset{O}{\underset{\|}{}} \overset{|}{\underset{Me}{}}$$

or

$$CH_2=CHCCl \overset{O}{\underset{\|}{}} + \underset{\diagdown}{\overset{\diagup}{N-Si}}\diagdown \longrightarrow CH_2=CHCNCH_2CHMeCH_2SiMe_2Cl \overset{O}{\underset{\|}{}} \overset{|}{\underset{Me}{}} \xrightarrow[HOSi-PDMS]{base}$$

$$CH_2=CHCNCH_2CHMeCH_2SiMe_2\cdot OSi-PDMS \overset{O}{\underset{\|}{}} \overset{|}{\underset{Me}{}}$$

Scheme 1. Dow Corning's preparation of acrylamide terminated PDMS

In both reactions, hydrochloric acid liberated during reactions necessitated the addition of either an amine or a metal alkoxide for neutralization. The resulting salts thus formed are either filtered or dissolved into an aqueous solution. Removal of salts by filtration from the viscous silicone fluids is a tedious process whereas dissolution of salts into aqueous solution often results in emulsion formation. Furthermore, the liberation of the hydrochloric acid and the presence of strong bases such as metal alkoxide often cause siloxane cleavage resulting in polymers with lower stability and consequently inconsistency of the product.

Researchers at Shin-Etsu have concentrated their efforts on the synthesis of a novel acrylate functional silanol compound, acryloxymethyldimethylsilanol (20, 21). The silanol is used to endcap chlorine terminated PDMS which is in turn prepared from hydrosilylation of vinyl substituted PDMS with chlorosilanes as shown below:

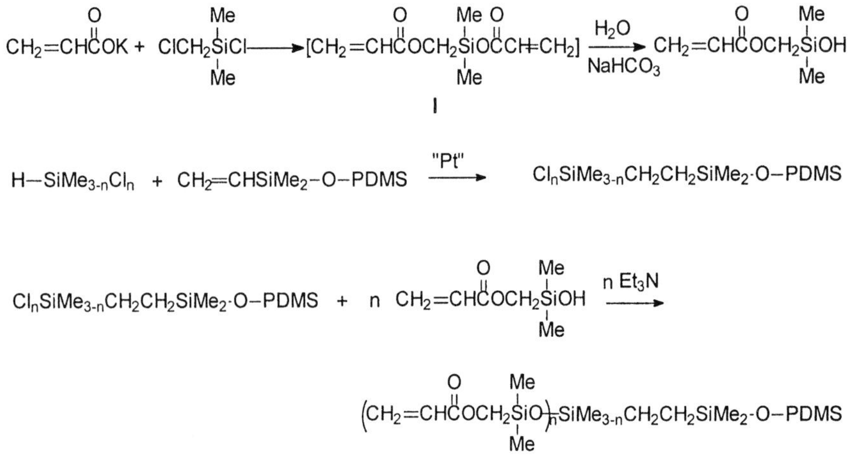

$$CH_2=CHCOK + ClCH_2SiCl(Me)_2 \longrightarrow [CH_2=CHCOCH_2SiOCCH=CH_2] \xrightarrow{H_2O, NaHCO_3} CH_2=CHCOCH_2SiOH(Me)_2$$

$$H-SiMe_{3-n}Cl_n + CH_2=CHSiMe_2-O-PDMS \xrightarrow{"Pt"} Cl_nSiMe_{3-n}CH_2CH_2SiMe_2-O-PDMS$$

$$Cl_nSiMe_{3-n}CH_2CH_2SiMe_2-O-PDMS + n\ CH_2=CHCOCH_2SiOH(Me)_2 \xrightarrow{n\ Et_3N}$$

$$\left(CH_2=CHCOCH_2SiO(Me)_2\right)_n SiMe_{3-n}CH_2CH_2SiMe_2-O-PDMS$$

Scheme 2. Shin-Etsu's preparation of UV curable silicones with multiple acrylate terminal groups

In the above reactions, a plurality of acrylate functional groups is placed on PDMS, presumably to increase the efficiency of UV cure. However, in order to achieve this goal, relatively inaccessible chlorine substituted PDMS prepared from a fairly expensive hydrosilylation reaction between chlorosilanes and vinyl substituted PDMS is used. Also, the formation of triethylamine hydrochloride necessitates the removal of solid from the viscous fluid by filtration. Finally, low molecular weight silanol such as acryloxymethyldimethylsilanol is notoriously unstable (22). Long term storage of the silanol may be troublesome.

UV Curable Silicones from Acryloxymethyldimethylacryloxysilane

Acryloxymethyldimethylacryloxysilane (**I**), the intermediate silane generated *in-situ* in Shin-Etsu's preparation of the silanol shown above, was first prepared by Gol'din et al. (23). The silane underwent hydrolysis and condensation reactions following its preparation to form the stable disiloxane. The ease of the disiloxane formation gives an indication of the instability of the silanol.

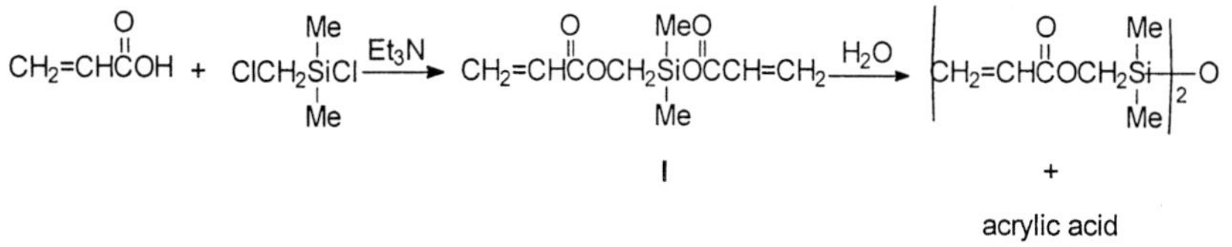

$$CH_2=CHCOH + ClCH_2SiCl \xrightarrow{Et_3N} CH_2=CHCOCH_2SiOCCH=CH_2 \xrightarrow{H_2O} [CH_2=CHCOCH_2Si-O]_2$$

with Me and MeO substituents shown; + acrylic acid

Scheme 3. Gol'din's synthesis of acrylate functional disiloxane

In our search for a simple preparation of acrylated PDMS, the condensation reaction between silane **I** and a model compound, nonamethyltetrasiloxane-1-ol (**II**), (24) was studied. The model compound was chosen since its reactions can be easily monitored by gas chromatography. The silanol of the model compound is configurationally similar to the silanol of the readily available silanol terminated PDMS. The reaction of silane **I** with the model compound therefore should be an excellent barometer for reactions with silanol terminated PDMS. To further simulate the kinetic effect of the model reaction, octamethyltrisiloxane (MDM) was used as the solvent/diluent as well as the GC internal standard for the progres of the reaction. Surprisingly, the condensation reaction proceeded instantaneously upon mixing to form the desired 1-acryloxymethyl-undecamethylpentasiloxane with the concomitant liberation of one equivalent of acrylic acid as shown below:

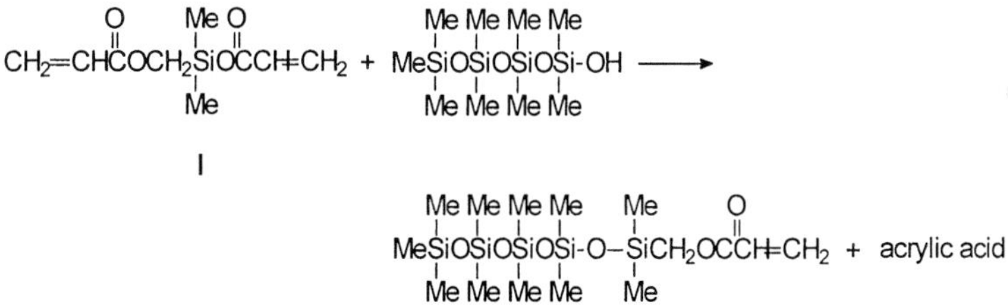

$$CH_2=CHCOCH_2SiOCCH=CH_2 + MeSiOSiOSiOSi-OH \longrightarrow$$
(**I**)

$$MeSiOSiOSiOSi-O-SiCH_2OCCH=CH_2 + \text{acrylic acid}$$

Scheme 4. Model silanol reaction with acryloxymethylacryloxydimethylsilane

The ease of the reaction was unexpected. In the condensation reactions of acetoxysilanes with silanol, the reactivity of acetoxysilanes is directly proportional to the number of acetoxy groups attached to silicon (25), presumably due to the electron withdrawing capability of the acetoxy groups that renders the silanes with more acetoxy groups more amenable for nucleophilic substitution. Thus, condensation with silanol takes place instantaneously with tetra- or triacetoxysilanes, but is orders of magnitude slower for diacetoxy- or monoacetoxysilanes. Silane **I**, with only one acryloxy group directly attached to silicon, was therefore expected to react with silanol very slowly.

Further studies with the silanol terminated PDMS of various molecular weights confirmed our findings. Silanol fluids with viscosity as high as 50,000 cps (~Mn =

60,000) can be readily endcapped with silane **I** to give the acryloxymethyldimethyl terminated PDMS by simply mixing silane **I** with silanol fluids. The reaction is generally complete within seconds after mixing at room temperature as evidenced by the transformation of the clear silanol fluid into a cloudy mixture due to the low solubility of the liberated acrylic acid in silicone. Removal of acrylic acid by vacuum stripping, if needed, yields the clear acrylate endcapped PDMS. Addition of a benzophenone type of photoinitiator, fillers and other type of additives common to RTV silicones then affords the UV curable silicones.

Is Silane I Pentacoordinate? (^{29}Si-NMR Studies). The surprising reactivity of silane **I** toward silanol beg explanation. Several similar silanes with carbonyl group γ- to silicon have been shown to possess pentacoordinate silicon structures (*26-31*) with an intramolecular coordinate Si ← O=C bond. Extraordinary reactivities of many hypervalent silicon compounds have been observed (*32, 33*) and are attributed to the hypervalency of these silanes. The high reactivity of silane **I** can be easily explained if we invoke the hypervalency of silane **I**. ^{29}Si-NMR chemical shifts are sensitive to the electronegativity of the silanes and have been used to distinguish the pentacoordinate silicon from the similar but tetracoordinate silanes. Thus, the tetracoordinate chloromethyltrifluorosilane (**II**) exhibits a chemical shift at -71.3 ppm whereas the pentacoordinate benzoylmethyltrifluorosilane (**III**) is at -94.8 ppm (*28*). The pentacoordinate silicon is nearly 25 ppm upfield from that of the tetracoordinate silane. On the other hand, chemical shifts of the two tetracoordinate silanes, chloromethyltriethoxysilane (**IV**; -59.9 ppm) and acetoxymethyltriethoxysilane (**V**; -58.2 ppm) differ by less than 2 ppm (*28*). In our current studies, ^{29}Si-NMR of silane **I** (+15.5 ppm) and the intermediate silane (**VI**) (prepared from reaction of chloromethyldimethylchlorosilane with one equivalent each of acrylic acid and triethylamine) differs by less than 1 ppm and clearly indicates that both silanes are tetracoordinate. This tetracoordinancy of silane **I** is further supported by the reported pentacoordinate silane N-methyl-N-chlrodimethylsilylmethyl-acetamide (**VII**) having a chemical shift of -37.6 ppm (*31*).

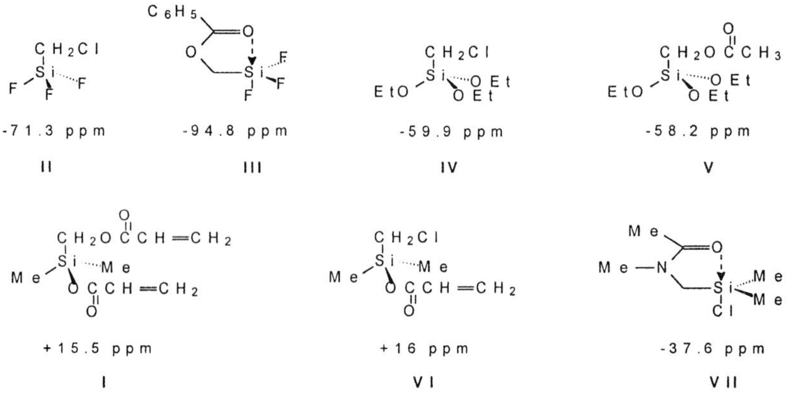

Figure 1. ^{29}Si-NMR chemical shifts of tetra- and penta-cordinate silanes

Reactions of Silane I and Related Compounds. Silane I was prepared by reacting two equivalents each of acrylic acid and triethylamine for each equivalent of chloromethyldimethylchlorosilane. When only one equivalent each of the acid and amine was used, an intermediate silane was isolated. Although chloromethyldimethylacryloxysilane (**VI**) is the expected product, neither ^{1}H- nor ^{13}C-NMR is capable of distinguishing it from the isomeric (and possibly more thermodynamically favorable) acryloxymethyldimethylchlorosilane (**VIII**). Nucleophilic substitution reactions on chloromethyldimethylchlorosilane can take place on both silicon and methylene carbon centers. Silicon chlorine bond is much more reactive than the carbon chlorine bond and is expected to undergo substitution reaction preferentially. However, precedence exists that substitution reaction actually takes place on carbon and leaves silicon center ostensibly unreacted. For example, trans-silylation of bis(trimethylsilyl)acetamide with chloromethyldimethylchlorosilane yielded a solid which was at first thought to be N,N-bis(chloromethyldimethylsilyl)acetamide but was later shown to be a pentacoordinate silane instead (*34,35*).

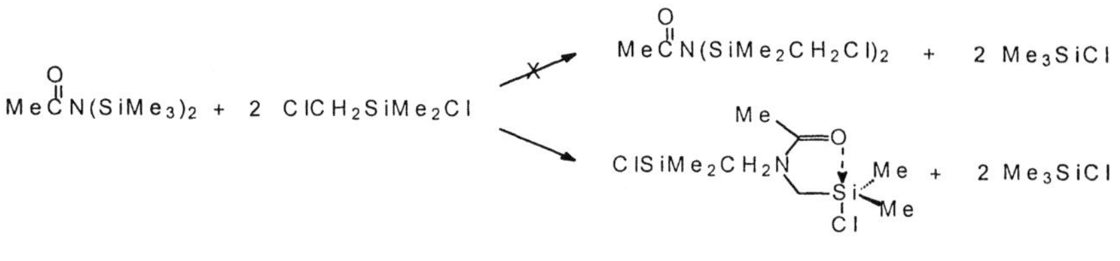

Scheme 5. Tran-silylation of bis(trimethylsilyl)acetamide

Further evidence supports the structure of **VI** comes from the hydrolysis of the intermediate yielding 1,3-bis(chloromethyl)-tetramethyldisiloxane and acrylic acid. The isomeric acryloxymethyldimethylchlorosilane would have been expected to form 1,3-bis(acryloxymethyl)-tetramethyldisiloxane instead as shown in Scheme 6.

In an attempt to prepare either the acryloxymethyldimethylacetoxysilane or the acetoxymethyldimethylacryloxysilane, the intermediate silane **VI** was further reacted with one equivalent each of acetic acid and triethylamine. The reaction resulted in the scrambling of acryloxy and acetoxy groups as shown in Scheme 7. GC monitoring of the reaction showed a total of three peaks formed in a GC area ratio of 1:2:1. Presumably the isomeric acryloxymethyldimethylacetoxysilane and acetoxymethyldimethylacryloxysilane co-elute and cannot be resolved.

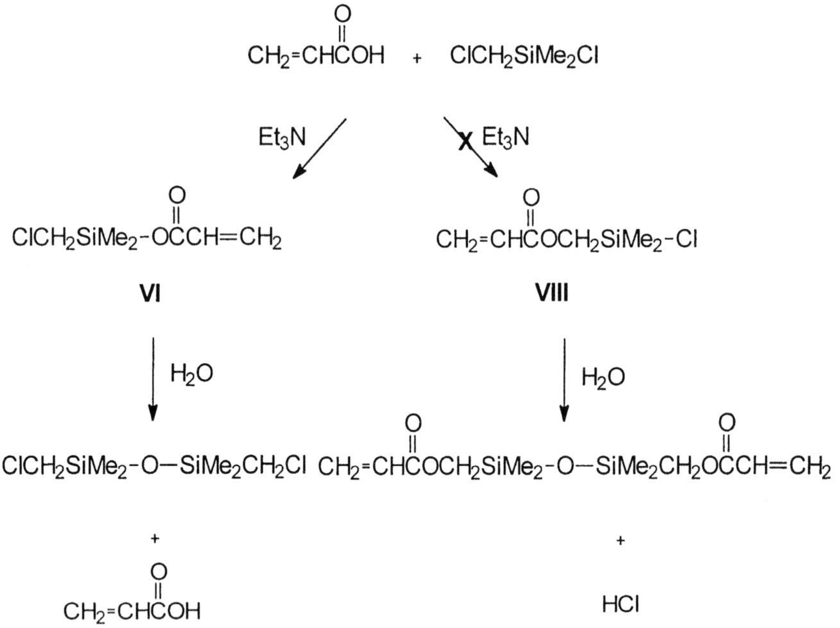

$$CH_2=CHCOOH + ClCH_2SiMe_2Cl$$

Et$_3$N ↙ ✗ ↘ Et$_3$N

$$ClCH_2SiMe_2\text{-}OCCH=CH_2 \quad\quad CH_2=CHCOCH_2SiMe_2\text{-}Cl$$
VI **VIII**

│ H$_2$O │ H$_2$O

$$ClCH_2SiMe_2\text{-}O\text{-}SiMe_2CH_2Cl \quad CH_2=CHCOCH_2SiMe_2\text{-}O\text{-}SiMe_2CH_2OCCH=CH_2$$

+ +

$$CH_2=CHCOOH \quad\quad\quad HCl$$

Scheme 6. Hydrolysis of the first intermediate silane established its identity of chloromethylacryloxydimethylsilane

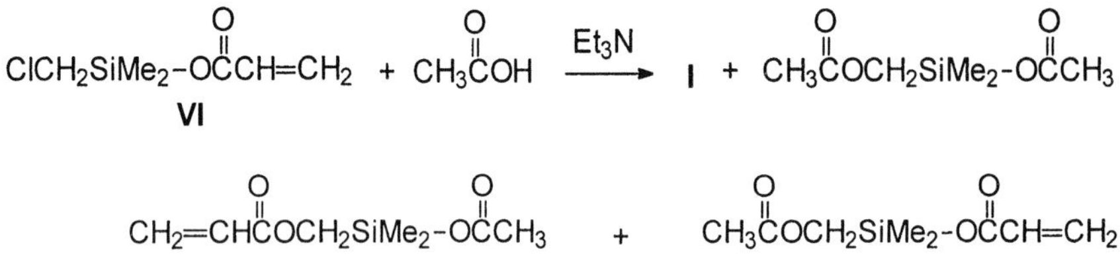

$$ClCH_2SiMe_2\text{-}OCCH=CH_2 + CH_3COOH \xrightarrow{Et_3N} \mathbf{I} + CH_3COCH_2SiMe_2\text{-}OCCH_3$$
VI

$$CH_2=CHCOCH_2SiMe_2\text{-}OCCH_3 \quad + \quad CH_3COCH_2SiMe_2\text{-}OCCH=CH_2$$

Scheme 7. Scrambling of acryloxy and acetoxy groups during preparation of mixed silanes

The intermediate silane **VI** was further found to react with the model compound, nonamethyltetrasiloxane-1-ol, very slowly, as expected. GC monitoring of the reaction indicates hours of reaction time was needed for any noticeable reaction taking place.

Reacting silane **I** in excess methanol resulted in the formation of acryloxymethyldimethylmethoxysilane. The resulting methoxysilane, however, was unreactive toward silanol presumably due to poorer ability for methoxy to leave during the nucleophilic substitution of the silanol.

Mechanism of the Formation of Silane I and Its Reaction with Silanol. Direct substitution of chlorine on methylene carbon by an acrylate group has been reported in the past. Thus, Merker prepared bis(methacryloxymethyl)-tetramethyldisiloxane from the corresponding bis(chloromethyl)-tetramethyldisiloxane using either methacrylic acid/triethylamine or potassium methacrylate under refluxing xylene or dimethylformamide (*36, 37*). However, we found the reaction proceeded very slowly in refluxing heptane, i.e. the reaction condition we used for the preparation of silane **I** (*38*). On the other hand, nucleophilic displacement at silicon of chloromethyl substituted silanes is known to undergo rearrangement reactions of organic groups from silicon to carbon. The rearrangement is induced by various nucleophiles and the pathway of the reaction is markedly affected by the reaction medium (*39-44*). Pentacordinate transition states are often invoked to account for the products formation. Therefore, the formation of silane **I** very likely follows the mechanism similar to the "*relay substitution reactions*" described by Eisch (*43,44*).

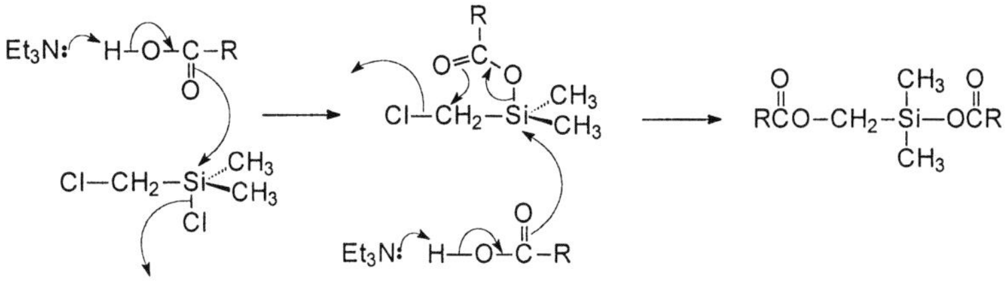

Scheme 8. Relay substitution of chloromethyldimethylchlorosilane

Similarly, the surprising ease of the reaction between silane **I** and silanol can be attributed to the anchimeric assistance of the acryloxy group on α-carbon during endcapping. A hexacoordinate transition state shown below is thus proposed. The transition state has a distinct similarity with the proposed but controversial transition state of the group transfer polymerization of methacrylate also shown below. Our proposed transition state shows the polymeric silanol nucleophilically attacks the silicon with the concomitant departure of the monomeric acrylic acid. In contrast, in the group transfer polymerization the monomeric methacrylate is the nucleophile and the polymer is the leaving group.

Conclusions

A simple preparation of acrylate terminated PDMS using the readily available silanol terminated PDMS is discovered. The ease of the reaction and the usage of the inexpensive raw materials made many industrial applications of UV curing silicones

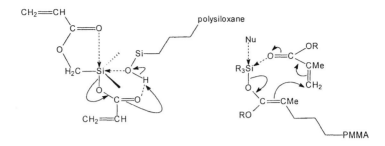

Figure 2. Transition states of silanol substitution and group transfer polymerization

practical. The surprising ease of the endcapping reaction is attributed to a hypervalent silicon transition state although the silane itself was shown to be tetracoordinate.

Acknowledgements

The author thanks Loctite management for the support of this work. The author further gratefully acknowledges Robert P. Cross for carrying out many of the experimental works.

References

1. Jacobine, A.F.; Nakos, S.T. In *Radiation Curing: Science and Technology*, ed. Pappas, S.P. Plenum Press, New York, 1992, pp 181-240.
2. Abdellah, L.; Boutevin, B.; Youssef, B. *Prog. Org. Coatings*, **1992**, *23*, 201.
3. Crivello, J.V.; Lee, J.L. *Polym. Mater. Sci. Eng. Prepr.* **1989**, *60*, 217.
4. Eckberg, R.P.; Riding, K.D. *Polym. Mater. Sci. Eng. Prepr.* **1989**, *60*, 222.
5. Crivello, J.V.; Eckberg, R.P. *US Patent 4 617 238*, 1986.
6. Pertz, S.V.; Glover, S.O.; Dill, T.J. *EP 0 625 533 A1*, 1994
7. Boutevin, B.J.; Abdellah, L.; Caporiccio, G. *US Patent 5 486 422*, (1996)
8. Legrow, G.E. *US Patent 3 655 713*, 1972.
9. Jacobine, A.F.; Glaser, D.M.; Nakos, S.T. *Polym. Mater. Sci. Eng. Prepr.* **1989**, *60*, 211.
10. Lee, C.L.; Lutz, M.A. *US Patent 5 169 879*, 1992.
11. Muller, U.; Jockusch, S.; Timpe, H. *J. Polym. Sci.: Part A Polym. Chem.* **1992**, *30*, 2755.
12. Inoue, Y.; Arai, M.; Fujioka, K.; Kimura, T. *US Patent 5 357 023*, 1994.
13. Kampling, M.J.; Lutz, M.A.; Scheibert, K.A. *US Patent 5 101 056*, 1992.
14. Wright, A.P. *US Patent 5 374 483*, 1994.
15. Oxman, J.D.; *US Patent 5 145 886*, 1992.
16. Boardman, L.D. presented at the *XXIII Organosilicon Symposium*, Midland, Michigan, April 21, 1990.
17. Cavezzan, J.; Dumas, S.; Prud'Homme, C.; Schue, F. *US Patent 4 939 06*, 1990.
18. Varaprath, P.J. *US Patent 4 608 270*, 1986.
19. Wright, A.P.; Varaprath, P.J. *US Patent 5 082 958*, 1992.
20. Satoh, S.; Arai, M.; Fujioka, K. presented at the *XXI Organosilicon Synposium*, Montreal, Canada, June 3-4, 1988.
21. Arai, M.; Kimura, T. *US Patent 4 845 259*, 1989.

22. Eaborn, C. *Organosilicon Compounds,* Butterworths, London1960; pp 246-250.
23. Gol'din, G.S.; Averbakh, K.O.; Yusim, M.A.; Fedotov, N.S.; Luk'yanova, I.A. *J. General Chem. (USSR) (English Translation),* **1973,** *43,* 781.
24. Chu, H.K.; Cross, R.P.; Crossan, D.I. *J. Organomet. Chem.* **1992,** *425,* 9.
25. Wake, W.C. *Crit. Rep. Appl. Chem.* **1987,** *16,* 89.
26. Voronkov, M.G.; Frolov, Yu.L.; D'yakov, V.M.; Chipanina, N.N.; Gubanova, L.I.; Gavrilova, G.A.; Klyba, L.V.; Aksamentova, T.N. *J. Organomet. Chem.* **1980,** *201,* 165.
27. Yoder, C.H.; Ryan, C.M.; Martin, G.F.; Ho, P.S. *J. Organomet. Chem.* **1980,** *190,* 1.
28. Albanov, A.I.; Gubanova, L.I.; Larin, M.F.; Pestunovich, V.A.; Voronkov, M.G. *J. Organomet. Chem.***1983,** *244,* 5.
29. Macharashvili, A.A.; Shklover, V.E.; Struchkov, Yu.T.; Baukov, Yu.I.; Kramarova, E.P.; Oleneva, G.I. *J. Organomet. Chem.* **1987,** *327,* 167.
30. Sidorkin, V.F.; Vladimirov, V.V.; Voronkov, M.G.; Pestunovich, V.A. *J. Molecular Structure (Theochem),* **1991,** *228,* 1.
31. Yoder, C.H.; Smith, W.D.; Buckwalter, B.L.; Schaeffer, Jr., C.D.; Sullivan, K.J.; Lehman, M.F. *J. Organomet. Chem.* **1995,** *492,* 129.
32. Corriu, R.J.P.; Young, J.C. In *The Chemistry of Organic Silicon Compounds*; Patai, S; Rappoport, Z. Ed.; John Wiley & Sons: Chichestr, UK, 1989, Part 2; pp 1241-1288.
33. Chuit, C.; Corriu, R.J.P.; Reye, C.; Young, J.C. *Chem. Rev.* **1993,** *93,* 1371.
34. Kowalski, J.; Lasocki, Z. *J. Organomet. Chem.* **1976,** *116,* 75.
35. Onan, K.D.; McPhail, A.T.; Yoder, Y.H.; Hillyard, R.W. *J. Chem. Soc. Chem. Comm.* **1978,** 209.
36. Merker, R.L.; Noll, J.E. *J. Org. Chem.* **1956,** *21,* 1537.
37. Merker, R.L.; Scott, M.J. *J. Org. Chem.* **1961,** *26,* 5180.
38. Chu, H.K.; Cross, R.P.; Crossan, D.I. *US Patent 5 179 134,* 1993.
39. Corey, J.Y.; Corey, E.R.; Chang, V.H.T.; Hauser, M.A.; Leiber, M.A.; Reinsel, T.E.; Riva, M.E. *Organometallics***1984,** *3,* 1051.
40. Damrauer, R.; Yost, V.E.; Danahey, S.E.; O'Connel, B.K. *Organometallics* **1985,** *4,* 1779.
41. Kreeger, R.L.; Menard, P.R.; Sans, E.A.; Shechter, H. *Tetrahedron Lett.* **1985,** *26,* 1115.
42. Hudrlik, P.F.; Abdallah, Y.M.; Hudrlik, A.M. *Tetrahedron Lett.* **1992,** *33,* 6743.
43. Eisch, J.J.; Chiu, C.S. *J. Organomet. Chem.* **1988,** *C1-C5,* 358.
44. Eisch, J.J.; Chiu, C.S. *Heteroatom Chem.* **1994,** *5,* 265.

Chapter 11

Advances in Non-Toxic Silicone Biofouling Release Coatings

Tim Burnell[1], John Carpenter[1], Kathryn Truby[1], Judy Serth-Guzzo[1,3], Judith Stein[1], and Deborah Wiebe[2]

[1]GE Corporate Research and Development, Building K1, Room 4A49, One Research Circle, Niskayuna, NY 12309
[2]Bridger Scientific, Inc., Sandwich, MA 02532

In this paper, we report two methods to control oil depletion from silicone foul release coatings: ablative networks and tethered incompatible oils. The synthesis of ablative and tethered diphenyldimethylsiloxane oils, the incorporation of such oils into the silicone room temperature vulcanized (RTV) network and the foul release properties of RTV coatings containing the ablative and tethered oils are discussed. The residence time of radiolabeled diphenyldimethylsiloxane oils in silicone RTV topcoats is also addressed. Synthesis of the radiolabeled diphenyldimethylsiloxane oil and incorporation of the radiolabeled oil into the silicone network are discussed. In addition, the environmental partitioning of the radiolabeled oils in both freshwater and marine systems is presented with the material balance.

Marine biofouling is a significant problem for ships and other structures submerged in a marine environment *(1)*. Both calcareous and non-calcareous fouling types present problems. Calcareous organisms, or "hard" foulers, are found in both marine and freshwater environments. Those found in marine water include barnacles, blue mussels, and encrusting bryazoans; those found in freshwater include zebra muscles. Examples of non-calcareous organisms, or "soft" foulers, are algae, slime, hydroids and tunicates. Barnacles are the biofouling organism of interest for this paper.

For ship owners, the fouling of the ship hull has many detrimental effects. Both soft and hard fouling leads to increased drag on the ship which decreases both the speed and fuel efficiency of the vessel, and consequently leads to an increase in operating expenses *(1)*. It has been reported that for oceangoing freighters, a 20% increase in fuel costs, or ~ $1 MM per year could be anticipated *(2)*.

[3]Corresponding author.

Traditional antifouling coatings, such as paints containing heavy metals, organic biocides and tin or copper ablative coatings, are highly effective at preventing biofouling. Such coatings release toxic substances into the water adjacent to the coating surface, thereby killing the biofouling before strong attachment can occur *(3)*. Consequently, antifouling coatings provide an environmental risk to marine organisms since they release toxins into the environment. Not surprisingly, the use of cupric oxide in paints is expected to be limited in the near future due to environmental concerns and triorganotin species, also very effective, are prohibited for use by the U.S. Navy.

For these reasons, much interest has evolved in nontoxic foul-release coatings, such as silicones *(4)*. These coatings inhibit the strong attachment of marine organisms via an "easy release" mechanism. Several desirable properties of silicone coatings minimize the adhesive strength of biofouling attachment and once fouling does occur, it can be removed easily by physical processes such as water pressure washing or gentle scrubbing. The easy release properties of silicones have been demonstrated empirically to be related to the glass transition temperature (Tg) and surface energy of the silicone coatings. The low Tg (-127°C) is attributed to the flexible siloxane backbone, which gives it its very high molecular mobility, even as a high molecular weight elastomer. Thus, coatings based on polymers with high Tg's tend to have poor fouling release, even at very low surface energy values. For example, Teflon (DuPont), $-(CF_2CF_2)_n-$ or poly(tetrafluouroethylene), has a Tg of 130°C due to its rigid backbone and exhibits poor foul release properties. Silicones also have a critical surface tension that coincides with the minimum of a plot of relative attachment versus surface free energy *(5)*. GE's foul release coatings, described below, exhibit both low Tg and low surface free energy.

GE foul release coatings are comprised of a silicone topcoat and a silicone oil additive, typically at 10 or 20 weight percent. The silicone topcoat, RTV11 (GE Silicones) is a room temperature condensation moisture cure system, which contains a silanol terminated polydimethylsiloxane (PDMS), $CaCO_3$ filler, tetraethoxy-orthosilicate (TEOS) crosslinker and dibutyltin dilaurate, a Sn(IV) catalyst. The chemistry of this system is shown below in Figure 1.

Barnacle adhesion testing is a technique used to measure the foul release performance of foul release coatings and is performed at static exposure sites such as the Indian River Lagoon in Melbourne, Florida. Barnacle adhesion testing is performed using a force gauge according to ASTM D5618-94. This technique measures the shear force required to remove barnacles adhered to the surface of a coating. Using barnacle adhesion data, it has been shown empirically that improvements in foul release are observed in coatings containing oils. First demonstrated by International Paint in the 1970's *(6)*, the Navy and then GE began incorporating oils into their foul release topcoats in the late 1980's and early 1990's, respectively. As shown below in Figure 2, RTV11 exhibits superior barnacle adhesion relative to an epoxy control; however, RTV11 containing 10% free diphenyldimethyl siloxane oil performs superior to both the epoxy control and RTV11. Since free oils in the silicone coating demonstrate improved foul release and since it is desirable to maximize the lifetime of the oil in the topcoat for maximum foul release performance, the following questions need to be addressed. Does oil need to be at the surface? Can oil diffusion from the matrix be controlled by attaching either ablative or tethered oils into the silicone topcoat?

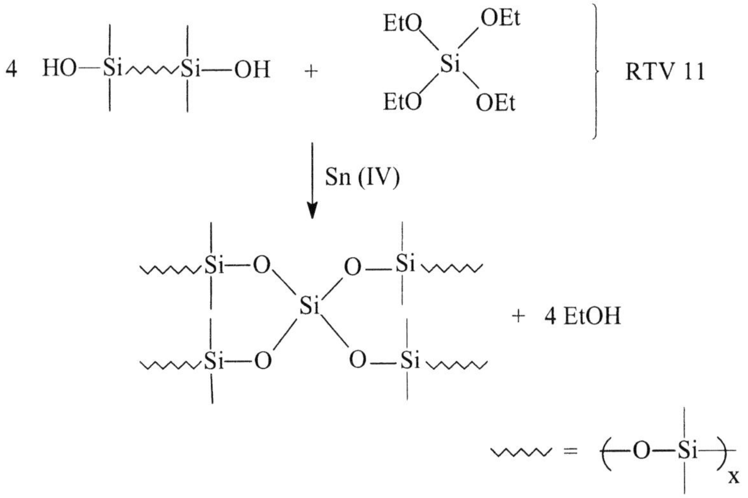

Figure 1. RTV11 Chemistry

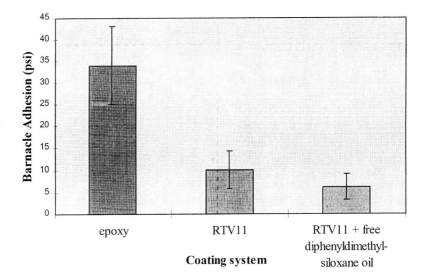

Figure 2. Barnacle Adhesion Data for Various Coatings

Mechanisms for Oil Retention in RTV11 Topcoat: Ablative Networks and Tethered Incompatible Oils

The incorporation of ablative and tethered oils into the silicone topcoat of fouling release coatings is a desirable mechanism for slow, controlled release of the silicone oil from the RTV topcoat. Once incorporated into the silicone network, the hydrolytically unstable Si-O-C bond in the ablative oil (Figure 3) should slowly degrade in water. Conversely, the tethered oil is chemically bonded into the silicone network and one end (the non-miscible portion) should phase separate to the surface of the PDMS. Both ablative and tethered oils contain diphenyldimethylsiloxane functionality, based on previous studies of the free oil. The approach was to synthesize both ablative and tethered diphenyldimethylsiloxane copolymers, incorporate the copolymers into the RTV topcoat and then measure the foul release performance of the coatings. Both oils are shown below in Figure 3.

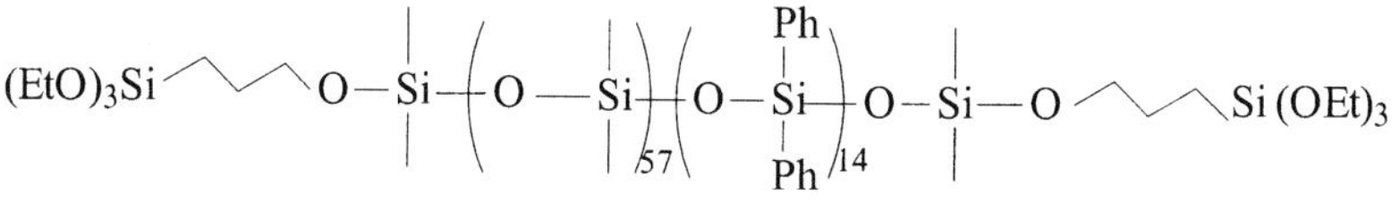

$(EtO)_3Si \sim\sim O-Si-(O-Si)_{57}-(O-Si(Ph)(Ph))_{14}-O-Si-O \sim\sim Si(OEt)_3$ **1**

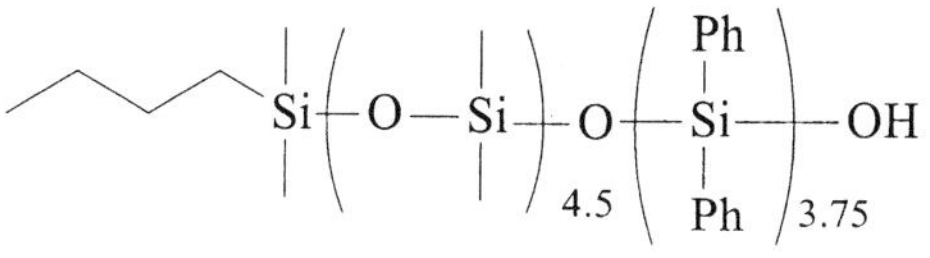

$\sim\sim Si-(O-Si)_{4.5}-(O-Si(Ph))_{3.75}-OH$ **2**

Figure 3. Summary of Ablative (1) and Tethered (2) Diphenyldimethylsiloxane Oils

The synthesis of the ablative diphenyldimethylsiloxane oil involved three steps, shown below in Figure 4. A silanol terminated diphenyldimethylsiloxane copolymer **3** containing 16 mole % diphenylsiloxane was reacted with dimethyldichorosilane at -5°C in the presence of triethylamine to give the bis-chlorosilane terminated derivative **4**. The chlorosilane derivative was subsequently reacted with allyl alcohol and triethylamine to yield the bis-allyl terminated diphenylmethylsiloxane **5**. Hydrosilyation with triethoxysilane and Karstedt's catalyst gave the product, bis-triethoxy terminated diphenyldimethylsiloxane **1**. Note some chain extension was observed in **4**, where the molecular weight doubled from approximately 3,300 to 7,000.

The tethered diphenyldimethylsiloxane oil was prepared by a kinetically controlled anionic ring opening polymerization of hexamethylcyclotrisiloxane (D_3) and hexaphenylcyclotrisiloxane (D_3^{Ph}) in the presence of n-BuLi (Figure 5) *(7)*. Once the lithium salt of D_3 and D_3^{Ph} **6** was formed in a two step process, it was then quenched with water to give a silanol terminated diphenyldimethylsiloxane product (**2**).

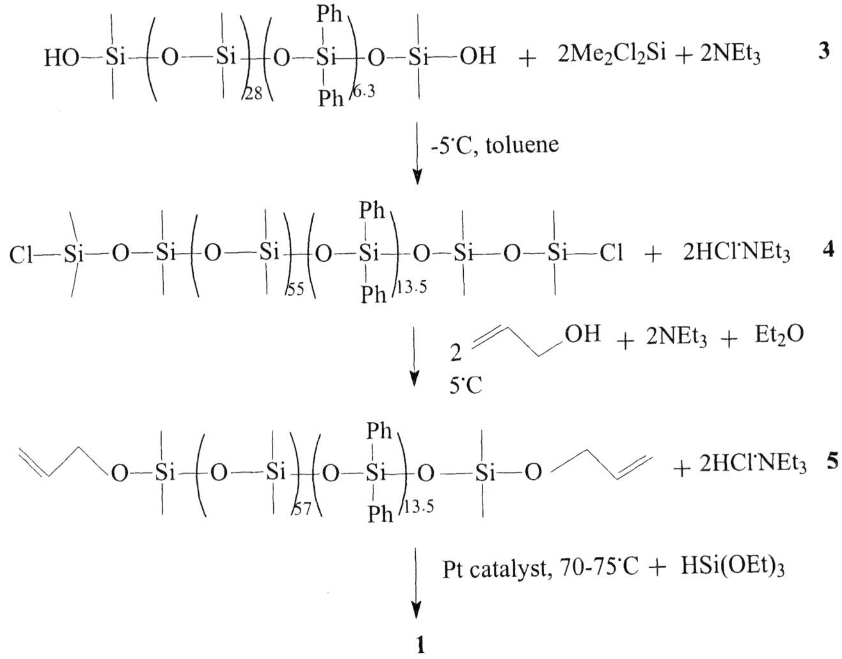

Figure 4. Synthesis of Ablative Diphenyldimethylsiloxane Oil

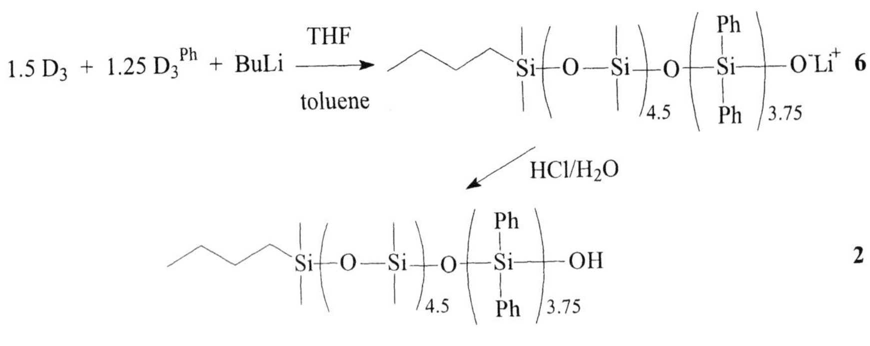

Figure 5. Synthesis of Tethered Diphenyldimethylsiloxane Oil

Incorporation of Ablative and Tethered Oils into Silicone Network

The ablative diphenyldimethylsiloxane copolymer was incorporated at 10 wt % into the PDMS network (Figure 6). The bis-triethoxysiloxane end groups of **1** condense with the silanol terminated PDMS in the presence of Sn(IV) to give a crosslinked network containing the hydrolytically unstable Si-O-C moiety.

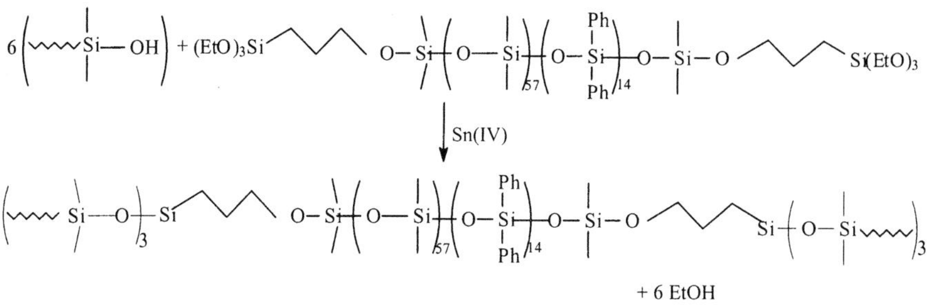

Figure 6. Incorporation of Ablative Diphenyldimethylsiloxane Oil into RTV Network

Likewise, the tethered diphenyldimethylsiloxane oil was also incorporated at 10 wt % into the PDMS network (Figure 7). The silanol endgroups of **2** condense with TEOS in the presence of Sn(IV) to give the triethoxy-terminated copolymer, which subsequently condense with the silanol-terminated RTV11 to form a crosslinked network. The incompatible butyl-terminated diphenyldimethyl fragment should phase separate to the surface of the PDMS network.

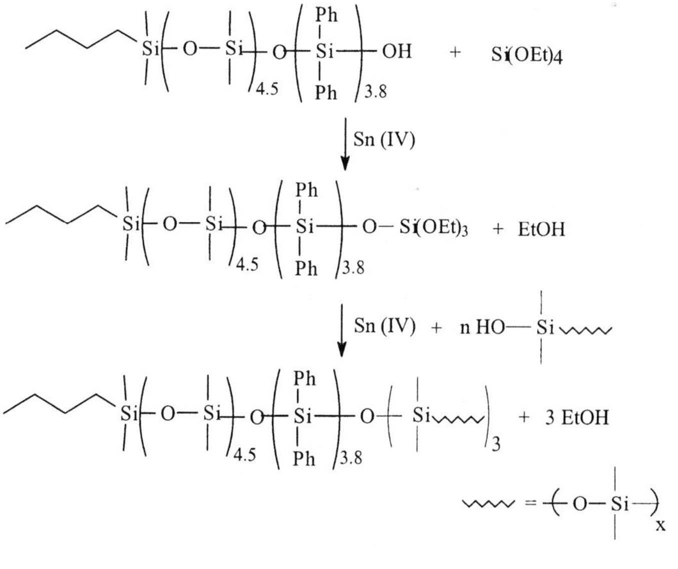

Figure 7. Incorporation of Tethered Diphenyldimethylsiloxane Oil into RTV Network

Foul Release Performance of Ablative and Tethered Oils

RTV topcoats containing either the ablative or tethered diphenyldimethylsiloxane copolymers were applied to steel panels previously coated with the epoxy and a tie layer developed at NRL *(8)*. After the panels were allowed to cure for one week, they were deployed at both northeast and southeast static test sites for 9 months. Controls of RTV11 and RTV11 containing 10% free diphenyldimethyl siloxane oil were also immersed in these marine environments. The overall fouling coverage was recorded for the northeast site and barnacle adhesion values were measured for the southeast site. Results are shown below in Figures 8 and 9.

RTV11

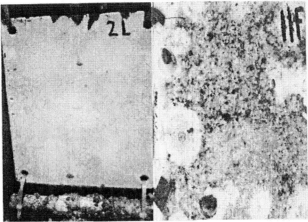

Free Oil   Ablative Oil

Figure 8.

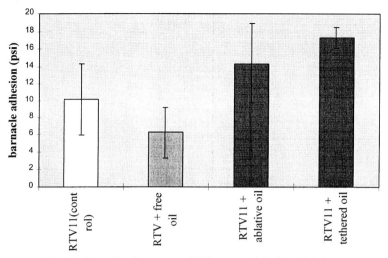

Figure 9. Foul Release Performance of Ablative and Tethered Oils

From Figure 8, we observe that the ablative oil, which is chemically bound to the PDMS topcoat, provides inferior antifouling performance compared to both the free diphenyldimethylsiloxane oil in RTV11 and the RTV11 control. Likewise, the barnacle adhesion data in Figure 9 suggests the coatings containing the tethered and the ablative oils performed poorly relative to the controls (RTV11 and RTV11 + free oil). From these results, we can conclude that free oil is necessary in the silicone coating for optimal foul release performance. However, it is desirable to also understand the rate of oil depletion of silicone foul release coatings containing free oil since depletion of the oil is anticipated to adversely affect biofouling release performance. Thus, does the free oil deplete from the matrix and decrease foul release performance of these coatings?

^{14}C Radiolabeled Oil Study

The leach rate of silicone oil additives from the silicone topcoat was readily determined using ^{14}C radiolabeled oils in both fresh and marine water systems *(9)*. Use of radiolabeled oils simplifies the study of their environmental partitioning, since each component of the matrix can be easily analyzed and quantified using radiometric detection. The approach was to synthesize ^{14}C radiolabeled polydiphenyldimethylsiloxane oil which is similar in composition to that used in determination of barnacle adhesion measurements (see Figure 2). Next, the ^{14}C oil was added to RTV11, catalyzed and applied to metal coupons which were subsequently suspended in fresh and salt water fish tanks. Coupons, soil and sediment were analyzed monthly for one year for mass balance determination.

The radiolabeled ^{14}C polydiphenyldimethylsiloxane oil was synthesized as shown in Figure 10. The precursor, ^{14}C labeled octamethylcyclotetrasiloxane (D_4), was first prepared by reaction of ^{14}C-labeled methyl Grignard reagent with tetrachlorotetramethylcyclotetrasiloxane. The precursor, ^{14}C-D$_4$, was then equilibrated with dodecamethylpentasiloxane in the presence of potassium trimethylsilanolate catalyst. Analysis of the resultant oil by ^{29}Si NMR revealed a dimethylsiloxy:diphenylsiloxy ratio of 46.7:53.3 (wt:wt) with 10.5 wt% trimethoxysiloxy endroups.

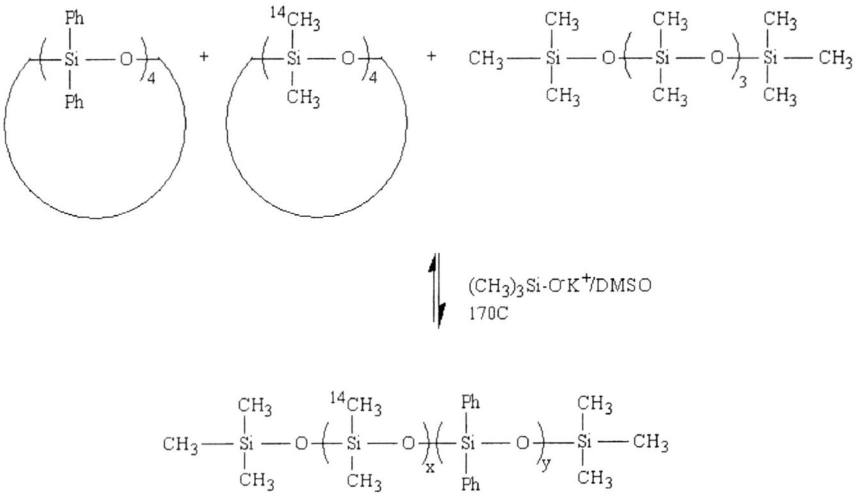

Figure 10. Synthesis of ^{14}C-labeled Polydiphenyldimethylsiloxane

Results of Radiolabeled Studies

After 8 months in the fresh water system, <0.4% and <0.07% of the total ^{14}C-labeled oil were detected in the water and sediment, respectively. Similarly, after 8 months in the marine system, <0.06% and <0.04% of the total ^{14}C-labeled oil were detected in the water and sediment, respectively. Mass balance data (Table I and II) reveal that, on average, >99% of the theoretical amount of ^{14}C-labeled oil remained in the silicone topcoat. Thus, levels of ^{14}C in the water are only slightly higher than background levels. Also, note the total % of ^{14}C-labeled oil does not consistently sum to 100% each month. This is probably because the oil was not uniformly dispersed in the PDMS network due to its immiscibility.

Conclusions

From the foul release results of the ablative and tethered oils, we can conclude that free oil is necessary in the silicone coating for optimal foul release performance. In

Table I. Material Balance for Fresh Water

time (months)	% of total ^{14}C				
	in Sn rinse	in water	in sediment	in RTV	total
1	0.4	0.11	0.02	103.49	104.02
2	0.4	0.15	0.02	94.84	95.41
3	0.4	0.19	0.03	102.52	103.14
4	0.4	0.34	0.05	94.11	94.9
5	0.4	0.08	0.05	89.7	90.23
6	0.4	0.06	0.05	100.26	100.77
7	0.4	0.64	0.06	98.37	99.47
8	0.4	0.33	0.06	99.69	100.48
Mean		0.238	0.043	97.87	98.55
std dev				4.67	4.67

Table II. Material Balance for Marine Water System

time (months)	% of total ^{14}C				
	in Sn rinse	in water	in sediment	in RTV	total
1	0.4	0.09	0.04	103.98	104.51
2	0.4	0.08	0.03	92.36	92.87
3	0.4	0.11	0.03	102.94	103.48
4	0.4	0.11	0.02	105.02	105.55
5	0.4	0.07	0.05	101.97	102.49
6	0.4	0.2	0.05	104.55	105.2
7	0.4	0.09	0.06	93.21	93.76
8	0.4	0.05	0.03	107.65	108.13
Mean		0.1	0.039	101.46	102
std dev				5.61	5.61

addition, the radiolabeled studies indicate that the free diphenyldimethysiloxane oil leaches very slowly from the RTV11. Thus, if loss of biofouling release performance was to be observed over time in these silicone paint systems, it could not be explained by loss of silicone oil from the topcoat. Retention of oils by the silicone topcoat also suggests that silicone paint systems of this type should not result in significant accumulation of silicone oils in marine and fresh water environments.

Experimental

All reagents were purified and/or dried over sieves prior to use. Triethylamine was distilled from CaH$_2$ and stored over 3Å molecular sieves (55 ppm H$_2$O). Toluene was distilled from sodium/benzophenone. Dimethyldichlorosilane was distilled under nitrogen and stored over 3Å molecular sieves. Diethylether was stored over molecular sieves for several days (16 ppm H$_2$O). PDS-1615 was obtained from Gelest and heated to 60°C under high vacuum with stirring for 1 hour to remove residual water. Allyl alcohol was dried over crushed CaSO$_4$ and 3Å molecular sieves (140 ppm H$_2$O). Standard schlenk techniques and a nitrogen atmosphere were used.

Synthesis of Bis-Chlorosilane-Terminated Diphenylsiloxanedimethylsiloxane 4

A 500 mL 3-neck round bottom flask was dried overnight and fitted with an overhead stirrer and a 250 mL pressure equalizing addition funnel. MeSi$_2$Cl$_2$ (29.0 g, 225 mmol) was added to the flask via cannula. Upon addition of NEt$_3$ (12.54 mL, 90 mmol) via syringe to the system, a cloudy white gas and a white precipitate formed. An ice bath (5°C) was immediately placed under the round bottom flask and a 50 wt % solution of PDS-1615 (silanol-terminated diphenylsiloxane-dimethylsiloxane copolymer) in toluene was added drop wise from the pressure equalizing addition funnel. The temperature of the ice bath was dropped to -5°C. The silanol-terminated solution was added over 120 minutes and a cloudy white precipitate formed (HCl·NEt3). The reaction mixture was filtered two times and residual MeSi$_2$Cl$_2$ and toluene were removed by distillation. A clear viscous liquid (118.7 g, 35 mmol) was obtained and characterized by ^{1}H and ^{29}Si NMR. The ^{1}H NMR was identical to the starting material, silanol terminated diphenyldimethylsiloxane copolymer, except the disappearance of the -OH proton was observed and appearance of methyl protons adjacent to the chlorosilane were observed. ^{1}H NMR: Si-Cl: 0.3 ppm, 12 H, s; Aryl H (Ph): average δ 7.6 ppm, m, 54H; Aryl H (Ph): average δ 7.2 ppm, m, 81 H; Si-Me$_2$ (D groups): multitude of singlets at 0-0.5 ppm (342 H).

Synthesis of Bis-Allyl-Terminated Diphenyldimethylsiloxane 5

A 1 L 3-neck round bottom flask was fitted with an overhead stirrer and a 250 mL pressure equalizing addition funnel containing the chlorosilane-terminated diphenylsiloxane dimethylsiloxane copolymer (118.0 g, 35 mmol). Allyl alcohol (7.14 mL, 105 mmol) and triethylamine (9.76 mL, 35 mmol) were added via syringe to the 1 L flask followed by 500 mL of diethyl ether. The chlorosilane was added drop wise to the ether solution at 5°C over 90 minutes. The solution was allowed to stir an additional 15 min. and was then filtered to remove HCl·NEt3. The organic layer washed with acid, water, dried over MgSO$_4$ followed by distillation of the ether at 30°C. A clear viscous liquid (116 g) resulted with a molecular weight of 7,000. By ^{1}H NMR, disappearance of the methyl protons adjacent to the chlorosilane and appearance of the allyl protons were observed. ^{1}H NMR: terminal CH$_2$ of allyl: 5.1 ppm, d, 4H; center CH of allyl: 5.9 ppm, ddt, 2H; CH$_2$-O: 4.2 ppm, m, 4 H. Aryl H

(Ph): average δ 7.6 ppm, m, 54H; Aryl H (Ph): average δ 7.2 ppm, m, 81 H; Si-Me$_2$ (D groups): multitude of singlets at 0-0.5 ppm (354 H).

Synthesis of Bis-Triethoxy-Terminated Diphenyldimethylsiloxane 1

To a 250 mL 3-neck round bottom flask fitted with an overhead stirrer, 5 wt% Karstedt's catalyst in xylenes (0.28 mL, 100 ppm H$_2$O) and triethoxysilane, which had been distilled under nitrogen (6.12 mL, 33.14 mmol), was added. The solution turned brown upon which ~ 5 ml of the bis-allyl copolymer was added directly with stirring. The remaining bis-allyl copolymer was added drop wise at 70-75°C over 60 minutes. ^{1}H NMR revealed disappearance of the allyl protons and appearance of the propyl and the ethoxy protons. ^{1}H NMR: propyl CH$_2$-Si 0.6 ppm, t, 4H; propyl -CH$_2$-: 1.6 ppm, m, 4H; CH$_2$-O: ~1.8 ppm, t, 4H; Aryl H (Ph): average δ 7.6 ppm, m, 54H; Aryl H (Ph): average δ 7.2 ppm, m, 81 H; Si-Me$_2$ (D groups): multitude of singlets at 0-0.5 ppm (354 H).

Synthesis of Tethered Diphenyldimethylsiloxane Oil 2

To a 250 mL round bottom flask, 13.6g of D$_3$Ph, 30g of D$_3$, 30 mL of THF and 100 mL of toluene was allowed to stir overnight. The reaction mixture was cooled to 0°C in and 27 mL of n-BuLi (1.6M) was added. The mixture was then heated to 60°C for 6 hrs and follow by GC until no n-BuLi remained. Ether was added and then the lithium salt was quenched with water/HCl until neutral. The aqueous fraction was extracted 3 times with ether and then washed with water/NaCl. The organic fraction was then dried over Na$_2$SO$_4$ and then stripped at 150°C and 2 mmHg to remove low boiling volatiles. ^{1}HNMR: CH$_3$-CH$_2$- 2.7 ppm, t, 3 H; -CH$_2$-CH$_2$-: 1.2 ppm, m, 4H; CH$_2$-Si: 0.5 ppm, t, 2 H; OH: 2.5 ppm, broad s, 1H; Aryl H (Ph): average δ 7.6 ppm, m, 15 H; Aryl H (Ph): average δ 7.2 ppm, m, 22.5 H; Si-Me$_2$ (D groups): multitude of singlets at 0-0.5 ppm, 27 H.

Preparation of ^{14}C Polydiphenyldimethylsiloxane Oil

To a 50mL one-neck round-bottom flask equipped with a magnetic stir bar and a condenser with a nitrogen inlet, was added 0.03 grams (5 mCi) of ^{14}C D$_4$, 1.97 grams unlabeled D$_4$, 3.80 grams dodecamethylpentasiloxane (MD$_3$M), 4.20 grams octamethylcyclotetrasiloxane(D$_4^{Ph}$) and 100 μ of a solution of potassium trimethylsilanolate in methyl sulfoxide. The flask was heated to 170°C for 6 hours while stirring under nitrogen. After 6 hrs, the mixture was allowed to cool to room temperature and 30.2 mg of a solution of phosphoric acid in silicone fluid (silyl phospate) was added and the mixture was stirred for a minimum of 30 minutes at room temperature. The resulting oil was then vacuum distilled at 250°C/0.03 mm Hg for 3 hrs to remove volatiles. The flask was cooled to room temperature and a clear fluid (a small amount of flocculent white precipitate is usually evident) was obtained (70% yield). Analysis of the resultant oil by ^{29}Si NMR revealed a dimethylsiloxy:diphenylsiloxy ratio of 46.7:53.3 (wt:wt) with 10.5 wt% trimethoxysiloxy endgroups.

Preparation of Silicone Coatings Containing ^{14}C Polydiphenyldimethylsiloxane Oil

Aluminum strips coated with epoxy were weighed and then coated with RTV11 containing 0.5 wt% dibutyltindilaurate catalyst and 10 wt% ^{14}C-labeled silicone oil. The strips were cured for 1 week and then weighed to determine the total weight of silicone topcoat (RTV + oil) per strip. The amount of ^{14}C-radiolabled oil was then calculated, assuming a uniform distribution of ^{14}C-oil throughout the network. The tin from the catalyst was then removed from the coupons by soaking them in water for several days. The tin rinse water was subsequently analyzed for total ^{14}C using liquid scintillation analysis to account for residual radiolabeled oil in the water.

Analysis of Radiolabeled Samples Containing ^{14}C Polydiphenyldimethylsiloxane Oil

One aluminum strip was removed each month and the amount of ^{14}C-labled oil in the RTV11 was measured by thermal oxidation, based on 3-5 samples. This process is described in a report by J. Carpenter *(9)*. The total amount of oil in each coupon was calculated based on the total weight of the silicone topcoat for a specified aluminum coupon. In addition, the amount of ^{14}C in the water and sediment was determined by liquid scintillation analysis and thermal oxidation/ liquid scintillation analysis, respectively.

Experimental Design for ^{14}C Radiolabeled Study

The coated aluminum strips were suspended in fresh and salt water tanks containing fresh and marine water sediments, respectively (Figure 11). Fresh water sediment was obtained from the Scioto River (Columbus, OH) and marine sediment was obtained from Tampa Bay (FL) and the pH of the sediments was 7.5 and 7.9, respectively. Each three gallon fish tank contained 6 liters of distilled water and 305 grams of the appropriate sediment. To the marine tank, sea salt (0.5 cup/gallon of water) was added. The tanks were stored in the dark and the water was stirred to simulate movement of a boat in water.

Acknowledgments: The authors would like to thank Goeffrey Swain, Mike Schultz and Chris Kavanagh from the Florida Institute of Technology in Melbourne, Florida, for kindly performing the barnacle adhesion measurements. We would also like to thank Jean Montemarano and Karen Poole from NSWC, Hal Guard from ONR and Steve Wax from DARPA. This research was supported by US Government Contract N00014-96-C0145.

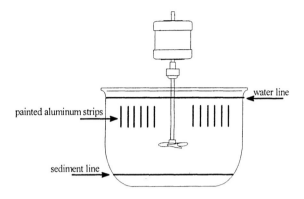

Figure 11. Fish Tank Study Design

References

(1) Bausch, G. G.; Tonge, J. S. *Silicone Technology for Fouling Release Coating Systems*, presented at the Waterborne, High-Solids, and Powder Coatings Symposium, February 14-16, 1996, pp 340-353.

(2) Marine Coatings by H.R. Bleile and S. Rodgers Federation Series on Coatings Technology, March 1989.

(3) Kannan, K.; Senthilkumar, K.; Loganathan, B. G.; Takahashi, S.; Odell, D. K. and Tanabe, S. *Environ. Sci. Technol.*, 1997, Vol. 31, pp 296-301.

(4) Edwards, D. P.; Nevell, T. G.; Plunkett, B. A.; and Ochiltree, B. C., International Biodeterioration & Biodegradation, 1994, pp 349-359.

(5) Baier, R. E. and Meyer, A. E. *Surface Analysis of Fouling-Resistant Marine Coatings*, *Biofouling*, Vol. 5, pp 165-180.

(6) Milne, A., U.S. Patent 4025693.

(7) Yang, M-H; Chao, C.; and Lin, C-H *Journal of Polymer Research*, July 1995, Vol. 2, No. 3, pp 197-201.

(8) Griffith, J. R., U.S. Patent 7847401, 1995.

(9) Carpenter, J.; Burnell, T.; Carroll, K.; Serth-Guzzo, J.; Stein, J.; Truby, K.; "Advances in Nontoxic Silicone Biofouling Release Coatings," Silicones in Coatings II, 24-26, Florida, March 1998.

Chapter 12

Poly(dimethylsiloxane) Gelation Studies

R. F. T. Stepto[1], D. J. R. Taylor[1], T. Partchuk[2], and M. Gottlieb[2]

[1]Polymer Science and Technology Group, Manchester Materials Science Centre, UMIST and University of Manchester, Grosvenor Street, Manchester M1 7HS, United Kingdom
[2]Ben Gurion University, Chemical Engineering Department, 84105 Beer Sheva, Israel

Recent experimental gelation studies[1] involving the endlinking of poly(dimethyl siloxane) (PDMS) chains of various lengths at various reactive-group concentrations (in a linear PDMS diluent) are interpreted using Ahmad-Rolfes-Stepto (A-R-S) theory of gelation[2]. These experimental systems have larger network-chain lengths, and cover wider ranges of initial dilutions of reactive-groups compared with polyester- and polyurethane-forming polymerisations analysed previously[3]. The PDMS data also include the effects of off-stoichiometry, a previously untested feature of A-R-S theory.

The experimentally-measured delays in the gel-point, relative to the predictions of Flory-Stockmayer (F-S) theory[4], as functions of the initial dilutions of reactive groups, follow the trends predicted by A-R-S theory. Quantitative interpretation of the PDMS gel-points using A-R-S theory yields values for the effective bond length, b (a measure of chain stiffness), which are in the range expected from the known unperturbed dimensions of linear PDMS chains.

Experimental

A series of PDMS networks was prepared[1] at 298K, using various initial dilutions of reactants, from the reaction of linear, vinyl-terminated PDMS chains (denoted R'B$_2$) with an f-functional, hydrogen-terminated dimethyl siloxane endlinker (RA$_f$). The diluent was inert, linear PDMS. The molar masses of the linear PDMS components used are listed in Table 1, along with their corresponding numbers of skeletal bonds, n. For a given reaction system, at each dilution, the reactive-group ratio was systematically adjusted, by increasing the initial concentration of the minority -H group (A) on the f-functional endlinker, until gelation was eventually observed. The critical reactive-group ratio, r_c, required for gelation at complete reaction of A groups, was therefore determined in each case.

The series of model networks may be denoted RA$_3$+R'B$_2$ and RA$_4$+R'B$_2$, where A represents an H atom on the tri- or tetra-functional endlinker species, and B a vinyl group (–CH=CH$_2$) at either end of the linear PDMS pre-polymer. The structures of the reactants are shown in Figure 1, and the six reaction systems are listed in Table 2, where the linear (R'B$_2$) components are coded according to their approximate molar masses. For example, 5K denotes the linear PDMS chain with M = 4900 ($\approx$ 5000 or "5K") g mol^{-1}, listed in Table 1.

Theory of non-linear polymerisations and analysis of experimental data

According to the classical, non-linear polymerisation theory of Flory[5] and Stockmayer[6] (F-S), α_c, the product of the critical extents of reaction at gelation, p_{ac} and p_{bc}, of A and B reactive groups, respectively is related to the functionalities f_a and f_b of the A and B units, respectively, as follows:

$$\alpha_c = p_{ac} p_{bc} = \frac{1}{(f_a - 1)(f_b - 1)} \quad (1)$$

For non-stoichiometric polymerisation mixtures, α_c can be rewritten in terms of the reactant ratio, r (= $c_{a0}/c_{b0} = p_b/p_a$, where c_{a0} and c_{b0} are the initial concentrations of A and B groups, respectively) as follows:

$$\alpha_c = (p_a p_b)_c = r p_{ac}^2 \quad (2)$$

The gel points for the polymerisation reactions discussed in this paper were recorded as the critical reactant ratios, r_c, for which gelation occurs at complete reaction of the minority (A) component. The value of p_{ac} in Equation 2 is therefore equal to 1, and, hence, $\alpha_c \equiv r_c$. Equation 1 can therefore be rewritten

$$r_c(f_a - 1)(f_b - 1) = 1 \quad (3)$$

Table 1: molar mass (M) of the linear ($R'B_2$) PDMS chains used, and the corresponding numbers of skeletal bonds, n, that they contain

notation	M /g mol^{-1}	number of Si-O repeat units	n
5K	4990	65.76	136
7K	6900	91.52	187
11K	11475	153.19	310
13K	12900	172.41	349

RA_3
$$H-Si(CH_3)_2-O-Si(CH_3)-O-Si(CH_3)_2-H$$
$$|$$
$$O-Si(CH_3)_2-H$$

RA_4
$$O-Si(CH_3)_2-H$$
$$|$$
$$H-Si(CH_3)_2-O-Si-O-Si(CH_3)_2-H$$
$$|$$
$$O-Si(CH_3)_2-H$$

$R'B_2$ $\quad CH_2=CH-Si(CH_3)_2-[O-Si(CH_3)_2-]_xCH=CH_2$

Figure 1. chemical structures of the RA_f and $R'B_2$ reactants.

Table 2: numbers of skeletal bonds, ν, in the linear subchain forming the smallest loop structures, for PDMS-forming non-linear polymerisations; "(short)" denotes the use of a low molar-mass, linear PDMS diluent

System reference	ν bonds
1. RA_3 + 5K	141
2. RA_4 + 5K	141
3. RA_4 + 5K(short)	141
4. RA_4 + 7K	192
5. RA_4 + 11K	315
6. RA_4 + 13K	354

The key F-S assumptions state that reactive groups react randomly, and that no pre-gel intramolecular reaction occurs between A and B groups. If these assumptions were satisfied for the series of PDMS polymerisations used here, the values of r_c would be as follows:

$$r_c = 0.500 \qquad (f_a = 3)$$

$$r_c = 0.333 \qquad (f_a = 4)$$

It is well known[7] that the neglect of pre-gel intramolecular reaction is rarely justified in the case of non-linear polymerisations. The rapid increases in the numbers of reactive groups attached to the complex molecular species prior to the gel point present large numbers of opportunities for intramolecular reaction between A and B groups. For example, the experimentally measured gel points, r_c, for the reaction systems of Table 2 are shown as functions of the reactive-group dilution, $1/c_{b0}$, in Figure 2. Due to the increased incidence of pre-gel intramolecular reaction, an increase in dilution results in a greater delay in gel point relative to the predictions of F-S theory ($r_c = 0.5$ and 0.333). However, in the limit of an infinite concentration of reactive groups ($1/c_{b0} \rightarrow 0$), all pre-gel reactions are intermolecular and extrapolation of both experimental plots reproduces the F-S gel points.

For a particular A group, the competition between intermolecular and intramolecular reaction can be quantified[7] by defining "internal" and "external" concentrations of B groups around the chosen A group, as shown in Figure 3. The external concentration is determined by the reactive-group concentration, c_b, and thus an increase in c_b will tend to promote intermolecular reactions at the expense of intramolecular ones. Figure 3 also indicates that the internal concentration is dependent upon the lengths of the linear subchains (the shortest has v skeletal bonds) between A and B groups on the same molecule, the stiffness (b) of the subchains, and the numbers of reactive-group pairs. If it is assumed that the linear subchain of v bonds obeys Gaussian statistics, the mutual concentration of A and B groups, P_{ab}, which can react intramolecularly to form the smallest possible loop structure, is given by[7]

$$P_{ab} = \frac{P(0)}{N_{Av}} = \left(\frac{3}{2\pi v b^2}\right)^{3/2} \frac{1}{N_{Av}} \qquad (4)$$

...where N_{av} is Avogadro's number. As defined, $P(0)$ is the probability of a zero end-to-end vector between the A and B groups, and b is the effective bond length, a chain-stiffness parameter[7], relating the mean-square end-to-end distance, $<r^2>$, and the number of skeletal bonds in the subchain:

$$<r^2> = v b^2 \qquad (5)$$

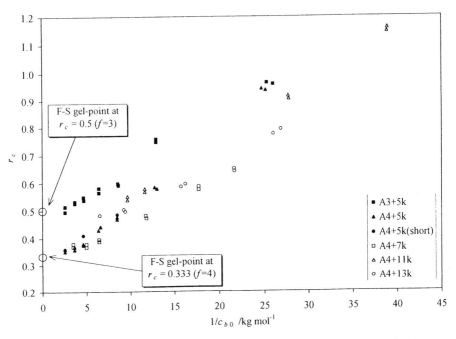

Figure 2. PDMS network formation; the critical reactive-group ratio for gelation, r_c, versus reactive-group dilution, $1/c_{b0}$, for the reaction systems of Table 2.

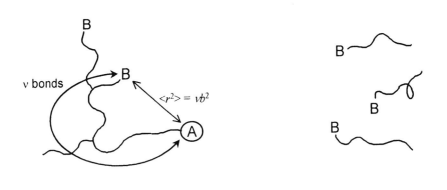

"internal" concentration due to a **B** group on the same molecule

"external" concentration of **B** groups from *other* molecules

Figure 3. diagram to illustrate the concepts of "internal" and "external" concentrations of reactive (B) groups around a chosen A group, and their influence on the proportions of intramolecular and intermolecular reactions between pairs of A and B groups.

The dimensions of the linear subchain corresponding to a particular experimental polymerisation system can be determined via knowledge of the reactant structures. The values of ν for the series of PDMS-forming polymerisations were calculated from the number of bonds (n) in the linear PDMS chains (Table 1) plus the number (=5) from the endlinker residue. The resulting values of ν are listed in Table 2.

The competition between intramolecular and intermolecular reaction of the chosen A group in Figure 2 is quantified by the ring-forming parameter, λ_{b0}, defined[4,7] as follows:

$$\lambda_{b0} = \frac{P_{ab}}{c_{b0}} \qquad (6)$$

Having characterised the intramolecular component of the polymerisation reaction, the critical reactant ratio including the effects of intramolecular reaction can be calculated via A-R-S theory[2]. The A-R-S gelation criterion (analogous to the F-S expression in Equation 3) for the condition $p_{ac} = 1$ and $p_{bc} = r_c$ becomes:

$$r_c(f_a - 1)(f_b - 1) = \\ \{1 + (f_a - 2)[\phi(1,\tfrac{3}{2})]\lambda_{b0} / r_c + (f_b - 2)(f_a - 1)[\phi(1,\tfrac{3}{2})]\lambda_{b0}\} . \\ \{1 + (f_b - 2)[\phi(1,\tfrac{3}{2})]\lambda_{b0} + (f_a - 2)(f_b - 1)\phi(1,\tfrac{3}{2})]\lambda_{b0}\} \qquad (7)$$

The term $\phi(1,3/2) = \sum i^{-3/2} = 2.612$ is the Truesdell function[4], and represents the sum of ring-forming opportunities from the smallest ring ($i = 1$) up to rings of infinite size. The A-R-S expression in Equation 7 is quadratic in both r_c and λ_{b0}, and is therefore soluble analytically for either quantity. Knowledge of λ_{b0} for specific reactant structures can be used to estimate r_c, or, alternatively, the experimental gel points can be used to calculate λ_{b0} and hence characterise the reactant structures. In addition, it can be seen that in the limit of an infinite concentration of reactive groups (*i.e.* $1/c_{b0} \to 0$), λ_{b0} is equal to zero (Equation 6), and intramolecular reaction is effectively disallowed. In the limit $\lambda_{b0} \to 0$, therefore, the A-R-S expression reduces to the F-S gelation criterion (Equation 3).

A-R-S Analysis of the PDMS Gelation Data

The experimental gel points, r_c, for the PDMS network-forming reactions were used to determine the corresponding values for λ_{b0}, by solving Equation 7:

$$[\phi(1,\tfrac{3}{2})]\lambda_{b0} = \frac{-B + \sqrt{B^2 - 4AC}}{2A} \qquad (8)$$

where

$$A = \left[\frac{(f_a - 2)}{r_c} + (f_b - 2)(f_a - 1)\right] \cdot \left[(f_b - 2) + (f_a - 2)(f_b - 1)\right]$$

$$B = \frac{(f_a - 2)}{r_c} + (f_b - 2)(f_a - 1) + (f_b - 2) + (f_a - 2)(f_b - 1)$$

$$C = 1 - r_c(f_a - 1)(f_b - 1)$$

In Figure 4 the results of the analysis of PDMS data are shown with plots of $v^{3/2}\lambda_{b0}$ against c_{b0}^{-1}. They are also compared with the results for polyester and polyurethane polymerisations published previously[2,3], with plots of $v^{3/2}\lambda_{b0}$ against c_{b0}^{-1}. The factor $v^{3/2}$ in the variable allows the effects of chain stiffness on intramolecular reaction to be considered independently of effects due to chain length. Equation 4 and Equation 6 show that the slopes of the plots are equal to $(3/2\pi b^2)^{3/2}/N_{Av}$.

It is immediately obvious that the PDMS polymerisations cover a much wider range of dilutions relative to the polyester- and polyurethane-forming systems and, therefore, represent a more severe test of A-R-S theory. The PDMS experimental data appear to fall into two distinct categories, for systems with a low molar-mass R'B$_2$ component (up to 7000 g mol^{-1}), and those with R'B$_2$ molar masses greater than 7000 g mol^{-1}. The experimental points for the polyester- and polyurethane-forming systems are clustered very close to the origin, but they appear to give a similar slope to the PDMS-forming reactions.

The initial, linear portions of the experimental plots in Figure 2 were used to determine the values of b for each PDMS-forming system, and the results are listed in Table 3. The values of b for the polyurethane- and polyester-forming systems[2,3] are also shown in Table 3. It can be seen that, in spite of the very large ranges of $v^{3/2}$ covered (148 - 6520), the slopes of the lines and the calculated values of b for the three types of system are of similar magnitudes.

The value of b calculated[9,10], via conformational analyses of a series of linear PDMS chains, using the rotational-isomeric-state (R-I-S) model of Flory, Crescenzi and Mark[11] and a temperature of 298K, is also shown in Table 3. It was found to be relatively insensitive to the number of skeletal bonds for chain lengths corresponding to the range of values of v listed in Table 2. The conformational calculations are also in good agreement with the value[12] of $b = 0.395$ nm determined by small-angle neutron scattering of linear PDMS in bulk at 298K. A complex series of chain conformational analyses[13] were performed on the polyurethane- and polyester-type subchains, yielding the value of b ≈ 0.32 nm (at the polymerisation temperatures of 333 - 353K). Again, the calculated chain stiffness is found to be relatively insensitive to chain length. In all cases it can be seen that the values of b suggested by the chain-conformational analyses lie in the ranges of values based on the A-R-S interpretation of experimental gel points.

The individual values of b for the PDMS data in Table 3 are plotted against $1/v$ in Figure 5. The A-R-S values of b display a much greater dependence upon v than is suggested by the linear chain-conformational analysis. Values for the PDMS-forming systems with the shorter subchains are greater than the calculated value for unperturbed linear PDMS chains. Increasing the subchain lengths to ca. 300 bonds (systems RA$_4$ +

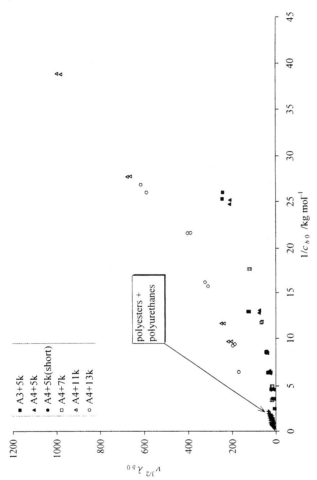

Figure 4. $v^{3/2}\lambda_{b0}$ versus $1/c_{b0}$ for the six PDMS-forming systems listed in Table 2; the corresponding data for the polyurethane- and polyester-forming systems analysed previously[3] are also plotted, for comparison (+).

Table 3: ranges of values of b based on the experimentally observed gel-point versus dilution relationships, calculated via A-R-S theory, and values of b calculated via chain-conformational analyses

Polymer type	v (range)	$v^{3/2}$ (range)	b /nm (range)
PDMS (5K, 7K)	141 – 192	1674 - 2660	0.48 – 0.61
PDMS (11K, 13K)	315 - 354	5591 – 6660	0.28 – 0.30
PDMS	conformational calcs.		*ca.* 0.41
polyurethane (PU)	28 – 65	148 – 524	0.23 – 0.27
polyester (PES)	36 - 135	216 - 1569	0.32 – 0.39
PU/PES	conformational calcs.		*ca.* 0.32

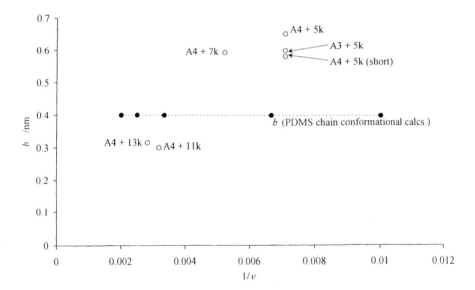

Figure 5. a plot of b versus $1/v$ for the PDMS-forming systems listed in Table 2. Individual data points were calculated via the A-R-S interpretation of the experimental gel points; the plot for the chain-conformational analysis of a series of unperturbed, linear PDMS chains is also shown.

11K and $RA_4 + 13K$) results in values of b smaller than the corresponding values derived from the chain-conformational calculations. The values are in approximate agreement with the PU and PES data generated previously (see Table 3).

The reasons for the greater sensitivity of b from gel points to chain length compared with values of b derived from conformational analyses are not yet understood. It is also evident in the PU and PES systems, with the values of b from PU gel points being less than expected from chain statistics and those for PES systems being slightly greater. The reasons could be linked with the changes in chain conformational statistics as highly branched species are formed.

Conclusions

This paper has presented the first gel-point data from PDMS-forming polymerisations. It is shown that A-R-S theory holds reasonably satisfactorily over wide ranges of r, dilution and molar masses of reactants.

The paper has established the occurrence of intramolecular reaction in polymerisations which are typical of those which have been used to form model PDMS networks[14]. Hence, like model polyester and polyurethane networks[7], model PDMS networks will not have perfect structures. The loop defects caused by intramolecular reaction will result in moduli less than those expected from the reactant molar masses.

References

1. Partchuk, T., Gottlieb, M., to be published
2. Rolfes, H., Stepto, R.F.T., *Makromol. Chem., Macromol. Symp.*, **1990**, *76*, 1
3. see Stepto, R.F.T., Taylor, D.J.R., *Polymer Gels and Networks*, **1996**, *4*, 405; Dutton, S., Stepto, R.F.T., Taylor, D.J.R., *Macromol. Symp.*, **1997**, *118*, 199
4. Stepto, R.F.T., *Comprehensive Polymer Science*, First Supplement, eds. S.L. Aggarwal and S. Russo, Pergamon Press, **1992**, Chapter 10
5. Flory, P.J., *J. Amer. Chem. Soc.*, **1941**, *63*, 3083, 3091, 3097; *Chem. Revs.*, **1946**, *39*, 137
6. Stockmayer, W.H., *J. Chem. Phys.*, **1943**, *11*, 45; **1944**, *12*, 125
7. Stepto, R.F.T., "*Non-linear polymerization, gelation and network formation, structure and properties*", in *Polymer Networks*, ed. R.F.T. Stepto, Blackie Academic & Professional, London **1998**; chap. 2
8. Z. Ahmad, Z., Stepto, R.F.T., *Colloid Polym. Sci.*, **1980**, *258*, 663
9. Stepto, R.F.T., Taylor, D.J.R., *J. Chem. Soc., Faraday Trans.*, **1995**, *91*(16), 2639
10. Stepto, R.F.T., Taylor, D.J.R., in preparation
11. Flory, P.J., Crescenzi, V., Mark, J.E., *J. Am. Chem. Soc.*, **1964**, *86*, 146
12. Mark, J.E. (ed.), *Physical Properties of Polymers Handbook*, American Institute of Physics Press, New York, **1996**; chap. V(24)
13. Stepto, R.F.T., Taylor, D.J.R., data to be published
14. Mark, J.E., Erman, B., "*Molecular aspects of rubber elasticity*", in *Polymer Networks*, ed. R.F.T. Stepto, Blackie Academic & Professional, London **1998**; chap. 7

Chapter 13

NMR Spin–Spin Relaxation Studies of Reinforced Poly(dimethylsiloxane) Melts

Terence Cosgrove[1], Michael J. Turner[1], Ian Weatherhead[1], Claire Roberts[1], Tania Garasanin[1], Randall G. Schmidt[2], Glenn V. Gordon[2], and Jonathan P. Hannington[2]

[1]School of Chemistry, University of Bristol, Cantock's Close, Bristol BS8 1TS, United Kingdom
[2]Dow Corning Corporation, Midland, MI 48686–0994

NMR relaxation analysis has been used to investigate the dispersion of silicate type materials in polydimethylsiloxane melts. The results show that several different regions of mobility are found, within the bulk melt, which depend on the particle size and chemical properties of the interface.

The dispersion of particles in polymer melts is used in many formulations to improve their mechanical strength, rheological behaviour and to reduce weight and cost. The segmental mobility in polymer melts has been widely discussed in the literature, both from a theoretical (1) and experimental (2) point of view. Of the molecular methods that can be used to investigate chain mobility, NMR relaxation data has proven particularly useful: NMR can be used to make in-situ measurements and the results are sensitive to chemical structure. In a recent paper (3) the two state model proposed by Brereton (4) has been shown to describe the segmental mobility of bulk polydimethylsiloxane [PDMS] melts rather successfully, giving a semi-logarithmic dependence of the spin-spin relaxation data [T_2] on the weight average molecular weight [M_w]. The polymers in that earlier study were narrow fractions but even so a slight non-exponential character was found.

On adsorption at an interface polymer segments can no longer orient isotropically and as a result the spin-spin relaxation times will be reduced. In the extreme narrowing regime corresponding to $\omega^2 \tau_c^2 < 1$, where ω is the Larmor frequency and τ_c is a correlation time describing a particular motion [e.g. rotation or translation], T_2 decreases with decreasing mobility. However, beyond a certain mobility limit T_2 becomes independent of τ_c. For polymers adsorbed as trains, i.e. in direct contact with the surface very short values of T_2 can be found (5). However, for segments that retain some mobility the value of T_2 will depend on the local monomer concentration and the segmental mobility. In a previous study of terminally attached polymer (6) it was found that the relaxation decay could be fitted by means of a distribution of relaxation times dependent on the volume fraction profile of the attached chains. In a more recent paper (7) T_2 values for PDMS/silica mixtures were found which showed that the presence of the filler particles dramatically changed the mobility of the PDMS chains. With solids concentration greater than 25% virtually all of the polymer segments in the dispersion experienced some degree of restricted mobility. In the present study NMR spin-spin relaxation measurements have been used to investigate the mobility of polydimethylsiloxane polymers when reinforced by silicate based particles; a wide range of sizes being investigated. Different molecular weight PDMS samples, above and below the critical entanglement molecular weight, were studied as a function of silicate loading, size and surface modification.

Experimental

NMR. The NMR results were obtained using a JEOL FX200 instrument modified with a SMIS vector processor and a digital RF console. All results were obtained using the Carr-Purcell-Meiboom-Gill (CPMG) pulse sequence (8) with a 180° pulse spacing between 0.2 and 5 ms. Up to 8K data points were collected by sampling the echo maxima. The resultant decays were fitted to multiple exponential decays using a non-linear least-squares analysis program and a novel application of the DISCRETE algorithm. (7).

System. PDMS samples and silicates were supplied by Dow Corning and Petrarch. Larger silica particles were prepared using the Stober method (8) A nanosized trimethylsilyl-treated polysilicate material was supplied by Dow Corning and was used as received. Surface modified silicas were prepared using the method described by Edwards (10) to coat the silica particles with terminally grafted PDMS and the method of Heldon et al (11) to graft C_{18} alkyl chains. The samples were blended together mechanically and also by first forming a dispersion of both particles and polymer in toluene followed by subsequent solvent evaporation. Measurements were made approximately one month after preparation to enable the samples to reach equilibrium. The polymer samples used are described in more detail in Table 1 and the particles in Table 2

Table 1. PDMS samples

Sample	Source	M_W /kg mol^{-1}	M_W/M_N
1.8K	Petrarch	1.8	1.2
32K	Petrarch	32	1.65
10K	Dow Corning	9.7	1.05
38K	Bristol U.	38	1.10
110K	Bristol U	110	1.68

Table 2. Silicate particles

Sample	Surface area/ $m^2 g^{-1}$	Size /nm
Aerosil A200	211	-
Silica	-	198
Polysilicate R3	-	2-3

Results and Discussion

Aerosil Silica Particles Dispersed in Polymer Melts. The first set of experiments focuses on the dispersion of unmodified silica particles in low molecular weight PDMS melts. In Figure 1 we show the experimental transverse magnetisation decays from aerosil silica dispersed in a 32K polymer. The two decay curves have been normalised to unit intensity to make it possible to compare their relative shapes. It is clear that the decay rate increases as the particle concentration increase from 2.1% to 10.1% solids. The change in T_2 is due to two mechanisms; the space filling of the particles means that the translational diffusion of the polymer will be limited and the change in the segmental correlation times brought about by the presence of the substrate making the resultant motion anisotropic. It is likely that the latter effect will be more important.

With mechanical mixing there must always be some concern about the reproducibility of the dispersions. To assess this we have made samples using mechanical mixing and solvent evaporation as described above. Figure 2 shows the relative relaxation times for two series of samples using the 32K polymer with Aerosil A200. The lower concentration samples were made with mechanical mixing but the ones that extend over a wider range of concentrations were made by solvent evaporation. Mechanical mixing becomes very difficult above 12% and there is the possibility that the shear required may desorb, degrade or chemically graft the chains to the surface. The relaxation data have been fitted to a single exponential, for simplicity, and have been normalised by dividing by the T_2 of the pure melt. As can be seen the trends in the two sets of data are very similar. As the loading of the silica increases the average relaxation time of the polymer can be seen to diminish. However, the decays are composites of several different relaxation processes and Figure 3 shows how the data for the 10.1% sample can be fitted rather well to two exponentials. For this sample two distinct populations can be found with relaxation times of 100 ms and 34 ms and intensities of 0.32 and 0.68 respectively. The bulk melt has a relaxation time of 100 ms indicating that 68% of the polymer segments are influenced by the inclusion of the silica powder. The radius of gyration of the 32K polymer is approximately 37Å and assuming that the particles are spherical and uniformly dispersed and that the adsorbed layer stretches to ~ 2Rg , it is expected

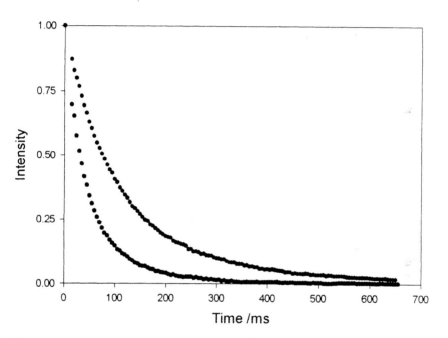

Figure 1. CPMG decays for PDMS 32K melt dispersed in Aerosil A200 at two loadings. Upper curve 2.1% w/w A200 and lower curve 10.1%

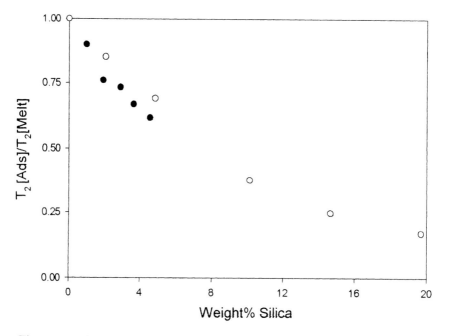

Figure 2. Relative single component T_2 relaxation times for samples of Aerosil dispersed in a 32K PDMS melt. The full circles correspond to samples that have been mixed mechanically and the open circles to those that have been prepared by dispersion in toluene.

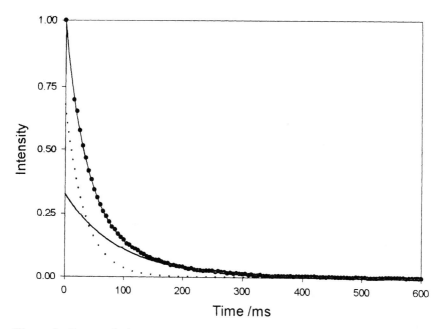

Figure 3. Deconvolution of the relaxation decay for PDMS 32K dispersed in 10.1 % w/w Aerosil.. The two lower curves represent the two different components found. The lower solid line represents free polymer and the dotted line perturbed polymer.

that a large change in relaxation time will occur at first overlap i.e. at ~ 11% particle concentration.

Polysilicate Particles Dispersed in Polymer Melts. The second set of experiments concerns PDMS melts with incorporated trimethylsilyl-treated polysilicate particles. The NMR spectrum of the polysilicate itself shows a very fast relaxation decay that is not observable in the present CPMG experiments. Thus the whole of the signal observed is due to PDMS in different dynamic environments. By using the DISCRETE algorithm (7) the decays can be deconvoluted into the optimum number of components consistent with the quality of the data. For the polysilicate-reinforced samples longer relaxation decays were found than with Aerosil inclusion and improvements in the instrumentation made it possible to obtain up to 8K data points in each relaxation experiment. Figure 4 gives the values of the various T_2s and Figure 5 the relative amounts of each component. These two sets of data must be taken together. The first and rather striking observation is that there is one component which has a short T_2 which is constant across the composition range with a value of 5 ±1 ms. [see Figure 4]. In Figure 5 however, it can be seen that the percentage of this component rises monotonically. We attribute this component to chain segments very near to the interface whose mobility is restricted and anisotropic. The two long relaxation times can be attributed to free polymer and perturbed polymer. Above 70% polysilicate the population of the free component [longest T_2] is effectively zero. The 'perturbed' chains are in a similar environment to those contributing to the shorter relaxation in Figure 3. The proportion of perturbed chains rises with the addition of the polysilicate and then decreases as all the polymer becomes bound.

Terminally Attached and Alkylated Silicas in Polymer Melts. In this part two different types of particles have been used, one with grafted C12 chains and one with terminally grafted PDMS with different molecular weight polymers viz 35K and 65K.

Figure 6 shows the effects of the incorporation of three different surface modified silicas on the bulk PDMS relaxation times for the 1.8K melt. The effect of the added silica is two-fold. Firstly, the addition of particulates reinforces the melt by their physical presence and secondly this primary effect can be enhanced by interactions of the melt with the surface. The C12 chains are very short and thus entanglement/penetration of the bulk melt will be minimal. Consequently the effect of the particulates upon the observed intensity is primarily due to the physical presence of the particles and not to the interactions of the melt with the grafted C12 chains. This may be useful in differentiating these two populations in future studies. The terminally grafted PDMS chains, however, give a much stronger effect, which increases with the grafted chain molecular weight. Similar effects are seen with different bulk polymer molecular weights. In these samples the surface is screened from the bulk polymer and the effects seen are due to entanglement of the surface chains with the bulk. Further experiments using deuterated grafted PDMS will

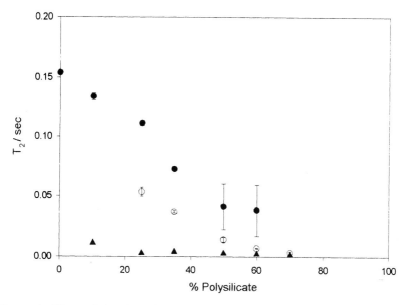

Figure 4. The variation in T_2 for blends of 10K PDMS with the polysilicate R3. Free polymer ●, influenced polymer O and bound polymer ▲.

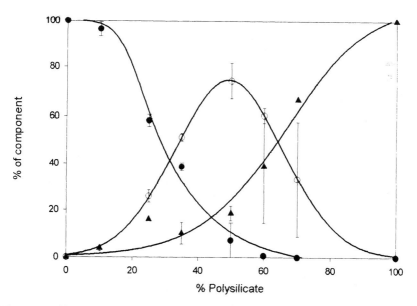

Figure 5. The variation in the proportions of the components of T_2 for blends of 10K PDMS with the polysilicate R3. Free polymer ●, influenced polymer O and bound polymer ▲

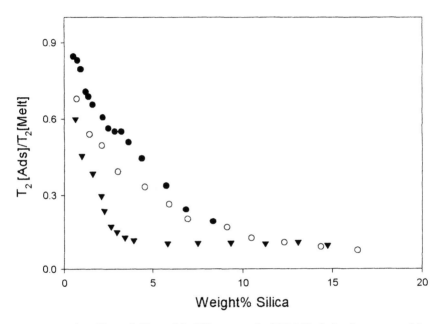

Figure 6. The effect of silica with different grafted PDMS chains incorporated in 1.8K PDMS melts on the relative spin-spin relaxation times. ● C-18 grafted C12 chains: ○ 35K grafted PDMS on silica and ▼ 65K PDMS grafted on silica.

enable us to differentiate signals from attached and bulk chains. The results show that the incorporation of filler particles significantly reduces the bulk chain mobility. The results can be compared with rheology measurements and theoretical calculations using the Scheutjens-Fleer (12) model and these data will be presented in a future publication.

Conclusions

NMR relaxation has been used to estimate the molecular basis behind the reinforcement of polymer melts by surface modified particles. The results show that both chain molecular structure and filler particle size have effects on the properties of the melt and enable an estimate of the proportion of the bound, perturbed and free segments to be obtained. Chemically modified surfaces can be use to modify the impact of the filler on the dynamics of the host polymer.

Acknowledgements

The authors are grateful to Dow Corning in Barry for supporting studentships [IW, TG, CR and MJT] and the EPSRC for a CASE award [CR].

References

1) Edwards, S. F.; Doi, M. *The Theory of Polymer Dynamics*; OUP: Oxford, **1986**.
2) Cosgrove, T.; Griffiths, P. C. *Advances in Colloid and Interface Science* **1992**, *42*, 175-204.
3) Cosgrove, T.; Turner, M. J.; Griffiths, P. C.; Hollingshurst, J.; Shenton, M. J.; Semlyen, J. A. *Polymer* **1996**, *37*, 1535-1540.
4) Brereton, M. G. *Macromolecules* **1991**, *24*, 2068.
5) Cosgrove, T. *Macromolecules* **1982**, *15*, 1290-1293.
6) Cosgrove, T.; Ryan, K. *Langmuir* **1990**, *6*, 136-142.
8) Meiboom, S.; Gill, G. *Rev. Sci. Instrum.* **1958**, *23*, 68.
9) Stober, W.; Fink, A.; Bohn, E. *J. Colloid and Interface Sci.* **1968**, *26*, 62.
10) Edwards, J. ;PhD Thesis Bristol, **1984**.
11) van Heldon, A. K.; Vrij, A. J. *J. Colloid and Interface Sci.* **1981**, *81*, 354.
12) Cosgrove, T.; Heath, T.; Vanlent, B.; Leermakers, F.; Scheutjens, J. *Macromolecules* **1987**, *20*, 1692-1696.
7) Cosgrove, T.; Weatherhead, I.; Roberts, C; Garasanin, T; Duval, D; Schmidt, R; and Gordon, G; *Langmuir* to be submitted

Chapter 14

Synthesis of Organosilicon Linear Polymer with Octaorganooctasilsesquioxanes Structural Units in Main Chain

N. A. Tebeneva, E. A. Rebrov, and A. M. Muzafarov

Institute of Synthetic Polymeric Materials, Russian Academy of Sciences, 117393 Profsoyuznaya Ul., 70, Moscow, Russia

A macromonomer T_8, 1,4-divinylhexamethyloctasilsesquioxane (3), was prepared by hydrolytic condensation of vinyltris(diethoxymethylsiloxy)silane (2a). Hydrosilylation of (3) by dimethylchlorosilane, methylphenylisopropoxysilane and phenyldimethylsilane leads to functionalized derivatives 1,4-bis(2-chlorodimethylsilylethyl)-hexamethyloctasilsesquioxane (4) and 1,4-bis(2-isopropoxymethylphenylsilylethyl)hexamethyloctasilsesquioxane (6) and a nonfunctionalized derivative 1,4-bis(2-phenyldimethylsilylethyl)hexamethyloctasilsesquioxane (5). Treating 6 with thionyl chloride yields 1,4-bis(2-chloromethylphenylsilylethyl)-hexamethyloctasilsesquioxane (7). Two approaches to the synthesis of polymers with fragments T_8 in the main chain have been proposed: (i) hydrolytic polycondensation of 7 and (ii) direct hydrosilylation of the starting macromonomer 3 by 1,3-dimethyl-1,3-diphenyldisiloxane. Data obtained by GPC, DSC and ^1H NMR spectroscopy have shown the latter approach to be preferable over the former. Thermogravimetric analysis of polymers with fragments T_8 in the main chain has provided evidence for their high thermal stability.

Polyhedral organosiloxanes are unique entities in the chemistry of organosilicon compounds. They represent a structural variety of polyorganosilsesquioxanes along with other species with a cross-linked, ladder-type and dendritic structure. To date, a range of individual compounds of this type of the general formula $(RSiO_{1.5})_n$ have been prepared, where R is an organic radical or a functional group, and n = 6, 8, 12, 14, 16 (1-7) (Fig. 1). In this series, the cubic molecular structure of octaorganooctasilsesquioxanes (T_8) made up of six fused eight-membered organosiloxane rings was found to be the most favorable thermodynamically. For this reason, the polyhedra (T_8) (whose molecules may be imagined as representing a unit cell of quartz) are produced in highest yield by either hydrolytic condensation of trifunctionalized organosilanes or thermal (*8*) and thermocatalyzed depolymerization of polyorganosilsesquioxanes in the presence of sodium or potassium hydroxides (*7*). A structural study of polymeric organosilsesquioxanes has shown that, irrespective of the reaction conditions, their molecules are composed of incompletely fused fragments (**T$_8$**) only linked to each other in a random manner (*9*). Despite the apparent imperfection of their molecular structure, polyorganosiloxanes in question rank among the most thermally stable polymers. Staying within the framework of a classical method of their synthesis, viz., hydrolytic polycondensation of trifunctionalized organosilanes, one is rather limited in the choice of means that would enable one to control effectively structure and, consequently, properties of this class of compounds. For this reason, the development of methods for synthesis of linear polymers with a regularly distributed fragments (**T$_8$**) in the main chain of macromolecules is a challenging and, simultaneously, promising problem in this field of research.

Within the framework of such an approach, soluble thermally stable polymers can be prepared and their properties controlled at a molecular level by varying the length and nature of the linear bridges between fragments (**T$_8$**) and placing various radicals at silicon atoms at the cube vertices. At present, a number of purely organosilicon and hybrid polymers of this type have been obtained whose molecules contain incompletely condensed fragments (**T$_8$**) only in the main chain (*10*). Completely condensed moieties (**T$_8$**) could be inserted only in the side chain of polymer molecules (*10-11*). On the other hand, a study of synthesized polymers with incompletely condensed fragments (**T$_8$**) has lent support to the usefulness of the aforementioned approach. It was shown that insertion of (**T$_8$**) in a linear polymer chain brings about a significant improvement in the basic characteristics of end polymer products: thermal and thermal oxidative stability and gas permeability were observed to increase, and specific weight, combustibility and heat conductance, to decrease. Viewed in this aspect, synthesis of polymers with completely condensed fragments (**T$_8$**) seems to hold much promise. However, for realization of this idea, a number of basic problems are to be resolved, in particular, those concerned with preparation of starting macromonomers whose molecules would contain two functional groups only at silicon atoms in a specified fixed position the structure (**T$_8$**). Following the traditional method of synthesis of polyorganosilsesquioxane polyhedrals – hydrolytic polycondensation of trifunctionalized organosilanes – either nonfunctionalized or fully functionalized (**T$_8$**) only can be prepared in reasonably acceptable yield. Attempts to synthesize the desired macromonomers by co-hydrolysis of a mixture of two differently trifunctionalized organosilanes, for example, n-C$_3$H$_7$SiCl$_3$ and 3-ClC$_3$H$_6$SiCl$_3$, have led, not unexpectedly, to a fairly complex mixture of structural isomers from which, using a preparative HPLC method, only three target isomers (**T$_8$**) whose molecules contained two 3-chloropropyl groups at silicon atoms in different positions of the structure (**T$_8$**) could be separated and characterized (*13*) (Fig. 2).

With all respect due to the experimental skill of the authors cited, it must be confessed, in all candor, that synthesis of the starting macromonomer by that procedure was not quite simple and sufficiently effective. In the present study, we

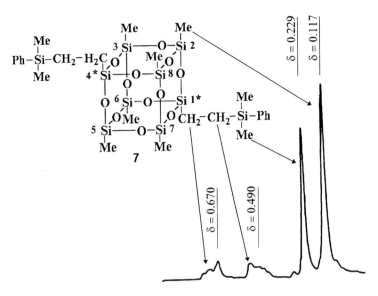

Figure 1. The first members of polyhedral organosilsesquioxane homologous.

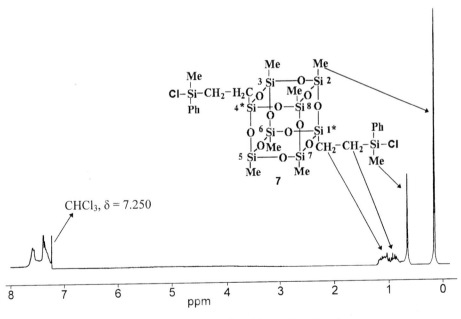

Figure 2. Structural isomers T₈ with two functional groups.

propose an alternative approach to the synthesis of (T$_8$) which makes use of individual hexafunctionalized organotetrasiloxanes with branched structure rather than monomeric organosilanes, at the step of hydrolytic polycondensation :

$$\begin{array}{c} R^2-Si{<}^X_X \\ O \\ R^1-Si-O-Si{-}R^2 \\ O \quad \quad X \\ R^2-Si{<}^X_X \end{array}$$

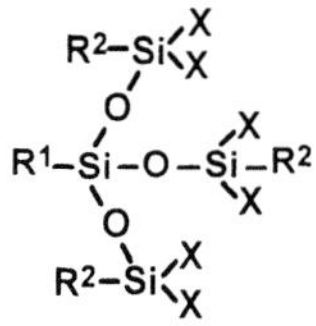

R^1, R^2 being organic radicals, and $X = OR, Cl$

The molecular organization of the starting compounds (whose structure features one completed vertex of the cube and provides for the required positioning of substituents at silicon atoms) is expected to be best suited to tailoring molecules (T$_8$) with desired properties.

It is pertinent to note that targeted synthesis of the starting organosilanes with branched structure was the subject of a separate study whose guiding principle was the use of new high-quality reagents (1)-sodiumoxyorganoalkokxysilanes (*14-15*).

Scheme 1

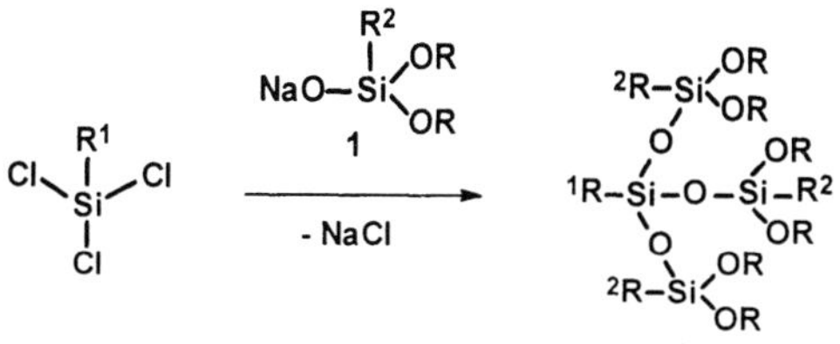

$$Cl{-}\underset{Cl}{\underset{|}{Si}}{-}Cl \;+\; NaO{-}\underset{R^2}{\underset{|}{Si}}{<}^{OR}_{OR} \quad \xrightarrow{-\,NaCl} \quad \begin{array}{c} ^2R-Si{<}^{OR}_{OR} \\ O \\ ^1R-Si-O-Si{-}R^2 \\ O \quad \quad OR \\ ^2R-Si{<}^{OR}_{OR} \end{array}$$

1 → 2

Development of reliable methods for preparing these reagents has allowed, in the final analysis, successful implementation of the goals outlined in the present work.

Discussion of results

A study of the reaction of hydrolytic polycondensation of hexafunctionalized organosiloxanes (2) has shown that this reaction is predominantly intermolecular even in dilute (~ 3%) solutions in the presence of catalytic amounts of HCl or AcOH. This fact is favorable to the formation of cubic molecules via interaction of two molecules of the starting oligomers 2. Thus, in the course of hydrolytic polycondensation of vinyl-tris-(diethoxymethylsiloxy)silane (2a) whose molecules bear a vinyl radical at the central silicon atom, a target macromonomer, 1.4-divinylhexamethyloctasilsesquioxane (3), was obtained.

Two reactive vinyl groups in the molecule of macromonomer 3 are attached to the silicon atoms that are located in one of the major corporal diagonals of the cube. Formation of the structure (T$_8$) during the course of hysrolytic polycondensation is not a process sufficiently explicit in all particulars. Therefore Scheme 2 presents, in some detail, one of the possible routes of chemical transformations leading to the occurrence of macromonomer **3** in the reaction system. The yield of target product was 23%. This value is sufficiently high for the cascade of condensation processes involved if one takes into account that the end molecules (**T8**) can form only from all-cis isomers of cyclic hydroxyl-containing intermediates (Scheme 2).

Scheme 2

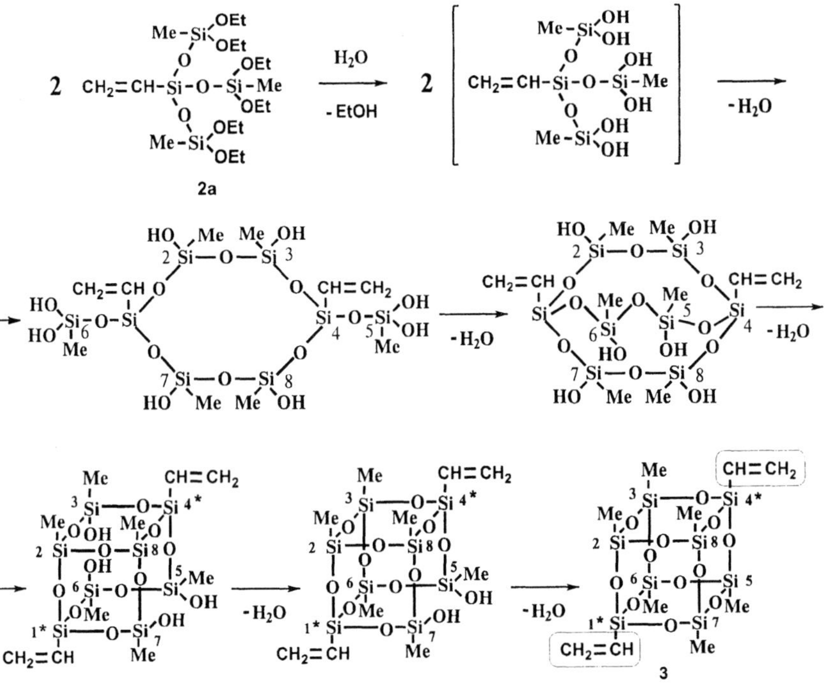

The structure of macromonomer **3** was established by X-ray diffraction analysis and ^{1}H NMR spectroscopy of its derivatives (*16*). Using a hydrosilylation reaction (Scheme 3), both functionalized (**4, 6, 7**) and nonfunctionalized (**5**) derivatives of the starting macromonomer **3** were obtained in good yield (~ 95%). Thus, at the end of the first step of our study, macromonomers required for obtaining target polymers have been synthesized.

Structure of the compounds synthesized was established by ^{1}H NMR and FTIR spectroscopic methods and further supported by elemental analysis. As a example, the ^{1}H NMR spectra both of non-functional **5** and di-functional **7** are shown in Fig. 3 and 4 correspondingly. A singlet proton signal of the methyl radicals located at silicon atoms of fragment **T8** is characteristic of 1,4 (2-p) isomer only. The singlet signal from protons of the methyl radicals of side branches provides evidence for stereospecificity of the hydrosilylation reactions of the starting compound **3** by phenylmethyl- and iso - propoxymethylphenylsilanes (Scheme 3); this reactions proceeds exclusively as β -addition. The ratio of integrated intensities of proton signals is in agreement with theory. (Fig. 3 and 4)

At the second step of our study using the synthesized macromonomers, two approaches to obtaining the target linear polymers containing fragments **T8** in the

Scheme 3

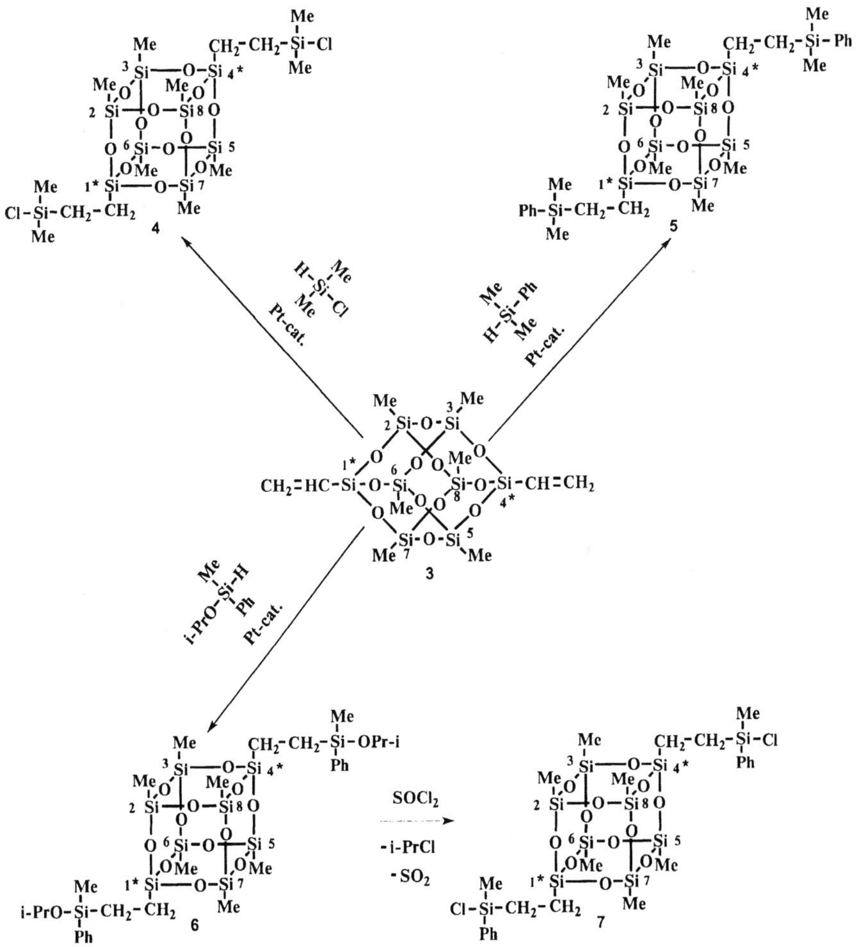

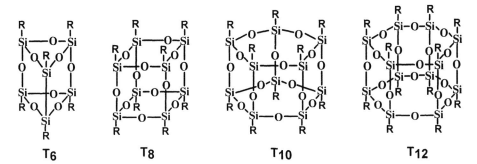

Figure 3. ^{1}H NMR spectra of 1,4-bis(2-phenyldimethylsilylethyl)hexamethyloctasilsesquioxane.

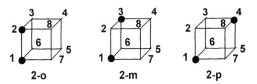

Figure 4. ^{1}H NMR spectra of 1,4-bis(2-chloromethylphenylsilylethyl)hexamethyloctasilsesquioxane.

main chain were attempted. The first approach involved the reaction of hydrolytic polycondensation of the chlorinated macromonomer **7**, and the second approach, a direct hydrosilylation of the starting macromonomer **3** by 1,3-dimethyl-1,3-diphenylsiloxane:

Scheme 4

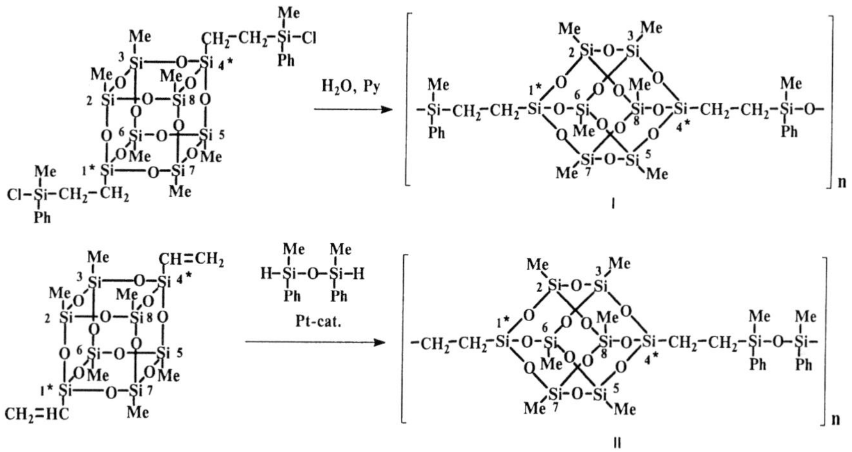

In principle, these two approaches are expected to lead to end polymers **I** and **II** of identical composition and structure. The synthesized polymers were examined using GPC, DSC and ^{1}H NMR-spectroscopy methods. A comparative analysis of the data obtained has shown the second approach to be preferable over the first for the reason that it allows preparation of higher molecular polymers with a rather narrow molecular-weight distribution.

A high-molecular fraction with Mw = 253000 g/m was separated by fractionation from polymer **II** (Fig. 5). The glass transition temperature as determined by DSC was Tg = 57.9 °C (Fig. 6). No sign of a crystallization process was revealed.

The TGA data lend support to the assumption as to the high thermal stabillity of polymers with fragments T_8 in the main chain (Fig. 7). The incipient decomposition temperature in a gel medium was ~ 460 °C, and the wieght loss at 1000 °C was ~ 14%. Simultaneously, thermal oxidative destruction in air causes a rapid structural degradation of the polymer starting at 300 °C. Here the weight loss was several times that observed in thermal destruction and could be as high as 60%.

Conclusion

The synthetic scheme proposed has allowed preparation of a linear polymer with a regular arrangement of structural silsesquioxane fragments in the main chain. Undoubtedly, physicomechanical and thermal properties of the polymer deserve a more detailed investigation. Thus, the infeience about the lack of crystallization in the polymer should be regarded as merely tentative, since no attempts have been made to gain deeper insight into crystallization at temperatures higher than the glass

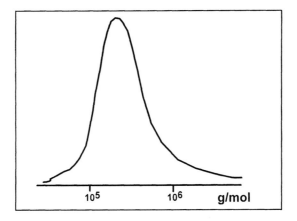

Figure 5. SEC data for high molecular fraction of polymer I.
$Mn=1.308 \times 10^5$, $Mw=2.534 \times 10^5$, $d=1.937$

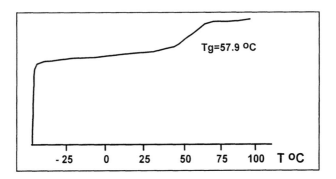

Figure 6. DSC curve for high molecular fraction of polymer I.

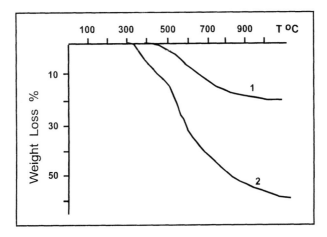

Figure 7. TGA-data for high molecular fraction of polymer I, (1)-He, (2)-air.

transition temperature. On the other hand, the presence or absence of the tendency in such polymers to ordering is a matter of special concern in view of the well-known fact that the ordering pattern of polymer chains exerts effect on mechanical properties of the polymer. By analogy with the conventional cyclic linear polymers (*17*), the degree of ordering can depend on the length and flexibility of linear portions of the polymer structure. Viewed in this light, the occurrence of ethylene bridges between fragments T_8 in the main chain is not obviously an optimum variant. The differing results of thermal and thermal oxidative destructions of the polymer in question argue for this. A comparison of the corresponding curves implies that the ethylene bridges are, in all likelihood, elements involved in the degradation of the polymer structure to low-molecular fragments.

The obtained results may be regarded as an important intermediate step forward to a synthesis of fully siloxane polymers with a controllable distance between the constitutive organosilsesquioxane fragments of their molecules.

Experimental

^{1}H NMR spectra were recorded on a Bruker WP-250 spectrometer in CDCl$_3$. IR spectra were recorded on a Bruker IFS-110 spectrophotometer. SEC analyses were carried out using Waters m-Styragel columns with a porosity of 10^5, 10^4, 10^3 and 10^6 Å and Lichrosphere Si 100 (7.5-µm particles) columns. Toluene was used as the eluent. A Waters 410 or RIDK 102 differential refractometer was used as the detector. Molecular masses were determined using polystyrene calibration. Synthesis of initial O_8-macromonomer - 1,4-divinylhexamethyloctasilsesquioxane (**3**) and its non-functional derivative - 1,4-bis(2-phenyldimethylsilylethyl)hexamethyloctasilsesquioxane (**5**) was described elsewhere (*16*).

Synthesis of 1,4-bis(2-iso-propoxymethylphenylsilylethyl)hexamethyloctasil-sesquioxane (6). To suspense of **3** (1.5 g, 2.6×10^{-3} mol) and iso-propoxymethylphenylsilane (1.06 g, 5.87×10^{-3} mol) in 1.5 ml of dry toluene was added solution of Pt-catalyst (7.7×10^{-5} g Pt). Reaction mixture was stirred at 80 °C during 80 h. The completeness of reaction was controlled by disappearance of Si-CH=CH$_2$ proton multiplet signals in ^{1}H NMR spectra at 6 ppm. After precipitation by dry ethanol 2.21 g **6** (90 %) was isolated. ^{1}H NMR (CDCl$_3$, 250 MHz) : δ 0.117 (s, CH$_3$SiO$_{1.5}$), δ 0.315 (s, CH$_3$SiOC$_3$H$_7$-i), δ 0.513 (m, -CH$_2$SiO$_{1.5}$), δ 0.772 (m, -CH$_2$SiOC$_3$H$_7$-i), δ 1.102 (d, (CH$_3$)$_2$CHO-), δ 3.844 (sept., (CH$_3$)$_2$CHO-), δ 7.042 (m, C$_6$H$_5$Si).

Synthesis of 1,4-bis(2-chloromethylphenylsilylethyl)hexamethyloctasil-sesquioxane (7). To solution of of **6** (2.0 g, 2.17×10^{-3} mol) in thionyl chloride (0.78 g, 6.55×10^{-3} mol) was added 0.01 ml of DMFA as catalyst. Reaction mixture was stirred at 40 → 60 °C with removing of SO$_2$ and i-C$_3$H$_7$Cl. The completeness of reaction was controlled by disappearance of Si-OC$_3$H$_7$-i proton multiplet signals in ^{1}H NMR spectra at 1.102 and 3.844 ppm. The rest after removing of SO$_2$, i-

C_3H_7Cl and excess of $SOCl_2$ *in vacuo* (1 Torr) was dissolved in 5 ml of dry n-hexane. Precipitated $Cl_2CHN(CH_3)$ was separated by filtration. Filtrate was evaporated *in vacuo* (1 Torr). As result 1.8 g (95 %) nearly pure **7** was obtained.. Found (%): Cl 8.073. Calculated (%) Cl 8.108. ^1H NMR ($CDCl_3$, 250 MHz) : δ 0.134 (s, $CH_3SiO_{1.5}$), δ 0.656 (s, CH_3SiCl), δ 0.924 (m, $-CH_2SiO_{1.5}$), δ 1.069 (m, -CH_2SiCl), δ 7.307 (m, C_6H_5Si).

Synthesis of polymer I. The solution of **7** (1.5 g, 1.71×10^{-3} mol) in 100 ml of dry toluene stirred at room temperature and added dropwise mixture of water (0.62 g, 3.44×10^{-2}), pyridine (0.27 g, 3.42×10^{-3} mol) and 20 ml toluene. Excess of water was removed as azeotrope with toluene. Precipitated Py·HCl was separated by filtration. After evaporation of toluene 1.26 g (90 %) of polymer was obtained. SEC analysis of obtained polymer shown wide molecular-mass distribution and average Mw=3000.

Synthesis of polymer II. The solution of Pt-catalyst (8.4×10^{-5} g Pt) was added in argon to suspense of **3** (1.15 g, 2.05×10^{-3} mol) and 1,3-methyl,3-diphenyldisiloxane (0.53 g, 2.05×10^{-3} mol) in dry toluene (0.5 ml) and stirring the mixer at temperature 60°C. Time of the reaction 200 h. The end of reaction was determined by disappearance of band 2100 cm^{-1} assigned to Si-H group. The final reaction mixture was nearly homogeneous. Obtained polymer was fractionated by precipitation (toluene-ethanol). SEC: monomodal distribution, Mn=130800 $gmol^{-1}$, Mw=253400 $gmol^{-1}$.

References

1. Sprung, M.M.; Gunter F.O. *J. Am. Chem. Soc.* **1955**, *77*, 3990.
2. Voronkov, M.G.; Lavrent'yev, V.I. *Topics Curr. Chem.* **1982**, *102*, 199.
3. Vasil'eva, T.V.; Kireev, V.V.; Chukova, V.M.; Chernov, A.I.; Derevyanko, S.L. *Vysokomolek. Soedinen.* **1988**, *30A*, 487 [*Polym. Sci.* **1988**, *30A* (Engl. Transl.)].
4. Agaskar, P.A.; Day, V.W.; Klemperer, W.G. *J. Am. Chem. Soc.* **1987**, *109*, 5554.
5. Agaskar, P.A.; *Synth. React. Inorg. Met. Org. Chem.* **1990**, *20*, 483.
6. Harrison, P.G. *J. Organomet. Chem.* **1997**, *542*, 141.
7. Barry, A.J.; Daudt, W.H.; Domicone, J.J.; Gilkey, J.W. *J. Am. Chem. Soc.* **1955**, *77*, 4248.
8. Papkov, V.S.; Il'ina, M.N.; Makarova, N.N.; Zdanov, A.A.; Slonimskii, G.L.; Andrianov, K.A. *Vysokomolek. Soedinen.* **1975**, *17A*, 2700 [*Polym. Sci.* **1975**, *17A* (Engl. Transl.)].
9. Andrianov, K.A.; Spektor, V.N.; Kamaritskii, B.A.; Nedorosol, V.D.; Ivanov, P.V. *Zh. Prikl. Khim.* **1976**, *49*, 2295 [*J. Appl. Chem. USSR.* **1976**, *49* (Engl. Transl.)].
10. Lichtenhan, J.D. *Comments Inorg. Chem.* **1995**, *17*, 115.
11. Haddad, T.S.; Oviatt, H.W.; Schwab, J.J.; Mather, P.T.; Chaffee, K.P.; Lichtenhan, J.D. *Polym. Prepr.* **1998**, *39*, 611.
12. Lichtenhan, J.D.; Haddad, T.S.; Schwab, J.J.; Carr, M.J.; Chaffee, K.P.; Mather, P.T. *Polym. Prepr.* **1998**, *39*, 489.

13. Hendan, B.J.; Marsmann, H.C. *J. Organomet. Chem.* **1994**, *483*, 33.
14. Rebrov, E.A.; Muzafarov, A.M.; Zdanov, A.A. *Dokl. Akad. Nauk SSSR.* **1988**, *302*, 346 [*Dokl. Chem.* **1988**, *302* (Engl. Transl.)].
15. Rebrov, E.A.; Muzafarov, A.M.; Papkov, V.S.; Zdanov, A.A. *Dokl. Akad. Nauk SSSR.* **1989**, *309*, 376 [*Dokl. Chem.* **1989**, *309* (Engl. Transl.)].
16. Rebrov, E.A.; Tebeneva, N.A.; Muzafarov, A.M.; Ovchinnikov, Yu. E.; Struchkov, Yu.T.; Strelkova, T.V. *Russ. Chem. Bull.* **1995**, *44*, 1286.
17. Makarova, N.N.; Petrova, I.M.; Matukhina, E.V.; Godovskii, Yu.K.; Lavrukhin, B.D. *Vysokomolek. Soedinen.* **1997**, *39A*, 1616 [*Polym. Sci.* **1997**, *39A*, 1078 (Engl. Transl.)].

Chapter 15

Halogenated Siloxane-Containing Polymers via Hydrosilylation Polymerization

J. Hu and D. Y. Son[1]

Department of Chemistry, P.O. Box 750314,
Southern Methodist University, Dallas, TX 75275-0314

The synthesis and characterization of new halogenated poly(carbosiloxane)s are described. The polymers are prepared via hydrosilylation polymerization using platinum-divinyldisiloxane complex as catalyst (Karstedt's catalyst). The monomers are either commercially available or can be prepared via a simple literature preparation. The polymers can be modified by either performing reactions on the pre-made polymer, or by modifying the monomers prior to polymerization. Insertion of cyclic tetrasiloxanes into the polymers is possible, and in certain cases the resulting polymers undergo spontaneous crosslinking reactions to form insoluble gels. A mechanism for this crosslinking process is proposed.

Organosilicon polymers have attracted much research interest due to their potentially useful properties. For example, the preparation, characterization, and applications of poly(siloxanes) (1), poly(carbosilanes) (2), poly(carbosiloxanes) (3-8), and poly(silanes) (9) have been extensively documented in the literature. Recent research has focused on developing methods for modifying these polymers in order to prepare a variety of substituted derivatives with altered properties. Different approaches used to modify organosilicon polymers have included free-radical hydrosilylation of polysilanes (10), nucleophilic substitution in poly(chlorosilylenemethylenes) (11), hydrosilylation of poly(silylenemethylenes) (12), and deprotonation of poly(carbosilanes) (13). The development of such types of tunable polymers is a major goal in current research.

Hydrosilylation polymerization has proven to be an extremely effective method for the synthesis of many types of organosilicon polymers. For example, hydrosilylation polymerization has been utilized extensively in the synthesis of both poly(carbosilanes) (14-23) and poly(carbosiloxanes) (3-8). These polymerizations

[1]Corresponding author.

usually proceed cleanly with a minimum of side-reactions. Since a typical polymerization involves reaction between A_2 and B_2 monomers, near exact stoichiometry is required to obtain polymers of reasonably high molecular weight. In this report we outline the synthesis and characterization of new halogenated poly(carbosiloxanes), as well as results indicating that these unique polymers can be successfully and easily modified. An additional benefit of this system is that the chemical substitutions can be brought about by either modifying the preformed polymer, or by polymerizing the substituted monomers. Certain cyclic siloxanes can also be inserted into the polymers, whereupon a rapid crosslinking reaction takes place after isolation of the products.

Results and Discussion

Polymer Synthesis and Characterization. As part of our investigation into the synthesis of modifiable organosilicon polymers, we recently reported a general synthetic route to a variety of *bis*(trialkylsilyl)dihalomethane compounds (*24*), of which compounds **1** and **2** seemed to be ideal monomers for hydrosilylation polymerization (Scheme 1). Unfortunately, hydrosilylation polymerizations utilizing various combinations of **1** and **2** were not successful in the presence of platinum catalyst. For example, reaction between **1-Cl$_2$** and **2-Cl$_2$** did not occur. Use of elevated temperatures induced reactions to occur in some instances, but the products were impure and highly colored. It was at this point that we decided to include commercially available reactive monomers in our experiments. As a result, we discovered that hydrosilylation polymerization occurred between **1-Cl$_2$** and 1,1,3,3-tetramethyldisiloxane or between **1-Cl$_2$** and 1,1,3,3,5,5-hexamethyltrisiloxane to yield polymers **3-Cl$_2$** and **4-Cl$_2$**, respectively, in quantitative yield (Scheme 2). The reaction period was rather lengthy but necessary for complete reaction to occur. The reaction progress was monitored by noting the disappearance of the Si-H absorbance peak in the IR spectrum. Immediately after polymerization, the polymers were viscous, transparent oils. Polymer **3-Cl$_2$**, however, slowly crystallized on standing.

Polymers **3-Cl$_2$** and **4-Cl$_2$** were characterized using ^{1}H and ^{13}C NMR spectroscopy, IR spectroscopy, elemental analysis, and gel permeation chromatography. The ^{1}H NMR spectra show a lack of endgroup resonances and also indicate that the polymerizations proceeded cleanly with a minimum of side reactions. An expanded view of the ^{1}H NMR spectrum for polymer **3-Cl$_2$** is shown in Figure 1(a). GPC molecular weight data are listed in Table I (entries 1 and 2). On repeating these polymerization reactions, we found the GPC data to be consistent from batch to batch as long as we utilized a starting monomer molar ratio as close to 1:1 as possible. These molecular weight data in addition to the lack of visible endgroups in the ^{1}H NMR spectra suggest that cyclization during polymerization cannot be ruled out. Indeed, during the preparation of polymer **4-Cl$_2$**, small amounts of a crystalline material were isolated and determined by X-ray diffraction studies to be an intramolecular cyclization product (Figure 2) (*25*).

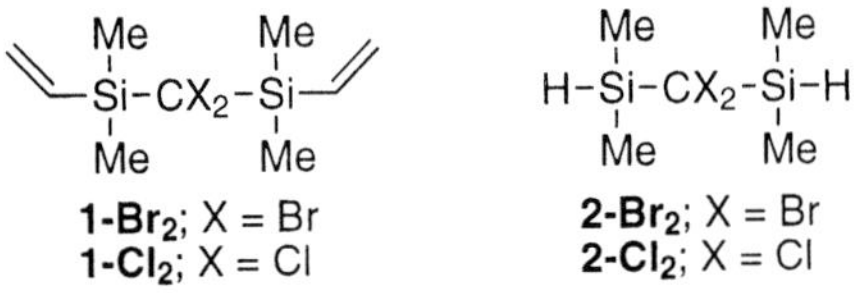

$$\begin{array}{cc} \text{Me} \quad \text{Me} & \text{Me} \quad \text{Me} \\ \diagdown\text{Si-CX}_2\text{-Si}\diagup & \text{H-Si-CX}_2\text{-Si-H} \\ \text{Me} \quad \text{Me} & \text{Me} \quad \text{Me} \\ \textbf{1-Br}_2; X = Br & \textbf{2-Br}_2; X = Br \\ \textbf{1-Cl}_2; X = Cl & \textbf{2-Cl}_2; X = Cl \end{array}$$

Scheme 1. Potential monomers for hydrosilylation polymerization.

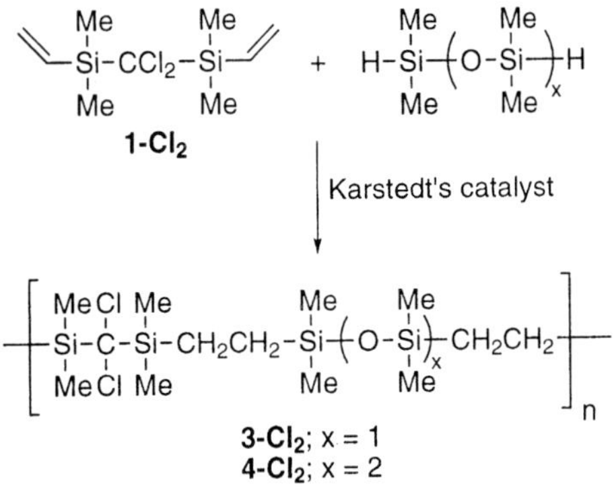

Scheme 2. Hydrosilylation polymerization.

$$\begin{array}{c} \text{Me} \quad \text{Me} \qquad\qquad \text{Me} \quad \text{Me} \\ \diagdown\text{Si-CCl}_2\text{-Si}\diagup \;+\; \text{H-Si}\text{-}(\text{O-Si})_x\text{-H} \\ \text{Me} \quad \text{Me} \qquad\qquad \text{Me} \quad \text{Me} \\ \textbf{1-Cl}_2 \end{array}$$

$$\Big\downarrow \text{Karstedt's catalyst}$$

$$\left[\begin{array}{c} \text{MeCl Me} \quad\quad \text{Me} \quad \text{Me} \\ \text{-Si-C-Si-CH}_2\text{CH}_2\text{-Si-(O-Si)}_x\text{-CH}_2\text{CH}_2\text{-} \\ \text{MeCl Me} \quad\quad \text{Me} \quad \text{Me} \end{array}\right]_n$$

3-Cl$_2$; x = 1
4-Cl$_2$; x = 2

Table I. GPC molecular weight data (referenced to polystyrene standards).

entry	Polymer	M_n	M_w
1	**3-Cl$_2$**	8600	14500
2	**4-Cl$_2$**	16500	26700
3	**3-Cl(Me)**[a]	4800	7700
4	**3-Cl(Me)**[b]	3200	5600
5	**3-Cl(SiMe$_2$H)**	4400	6700
6	**3-Cl(Me)**[c]	7800	13100

[a] from reaction of **3-Cl$_2$** with one equivalent of methyllithium
[b] from reaction of **3-Cl$_2$** with 1.5 equivalents of methyllithium
[c] prepared from monomer modification route

Polymer Modification Reactions. As part of our investigation we wished to demonstrate that the polymers we obtained could indeed be modified in a relatively facile manner. Previous work has shown the -SiCCl$_2$Si- moiety in small molecules to be very reactive towards a variety of reagents, particularly organolithium compounds (26,27). To demonstrate that this chemistry is applicable to the same moiety in polymers, we performed two basic reactions on polymer **3-Cl$_2$** outlined in Scheme 3. In both cases, reaction did proceed as expected. The ^{1}H NMR spectrum of polymer **3-Cl(Me)** obtained from the reaction of **3-Cl$_2$** with one equivalent of methyllithium exhibited a new peak at δ 1.53 due to the added methyl group (Figure 1(c)). However, the integration indicated that approximately 25% of the CCl$_2$ groups remained unreacted. Utilizing an excess of methyllithium decreased the amount of unreacted CCl$_2$ groups in **3-Cl(Me)**. The second reaction depicted in Scheme 3 is simply a lithium-halogen exchange followed by nucleophilic attachment of a dimethylsilyl group. Unlike the first reaction, the ^{1}H NMR spectrum of **3-Cl(SiMe$_2$H)** indicated that nearly complete substitution occurred. However, unidentified side reactions may also have occurred during this reaction, evidenced by the presence of several minor new peaks in the spectrum (around δ 0.0). It is also apparent that some cleavage of the polymer backbones took place in both of these reactions, as evidenced by a decrease in the GPC molecular weight values (Table I, entries 3-5).

When modifying polymers, it is often possible to modify the monomer prior to polymerization. The advantage of this method is that the resulting polymer would possess essentially 100% substitution and would be free of contaminants from polymer modification reactions. To illustrate the feasibility of this approach, we investigated the possibility of modifying the monomer, **1-Cl$_2$**, prior to polymerization. We first prepared (CH$_2$=CHSiMe$_2$)$_2$CClMe in 54% isolated yield from the reaction of **1-Cl$_2$** with methyllithium. Subsequent hydrosilylation polymerization utilizing (CH$_2$=CHSiMe$_2$)$_2$CClMe and 1,1,3,3-tetramethyldisiloxane as monomers proceeded smoothly in the manner used previously, yielding polymer **3-Cl(Me)** in quantitative yield as a viscous oil. The ^{1}H NMR spectrum of **3-Cl(Me)** (Figure 1(b)) prepared

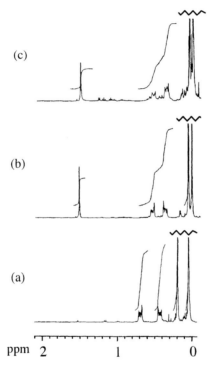

Figure 1. ^{1}H NMR spectra of (a) polymer **3-Cl$_2$**, (b) polymer **3-Cl(Me)** prepared via modification of monomer, and (c) polymer **3-Cl(Me)** prepared via modification of polymer (1.0 equiv MeLi).

231

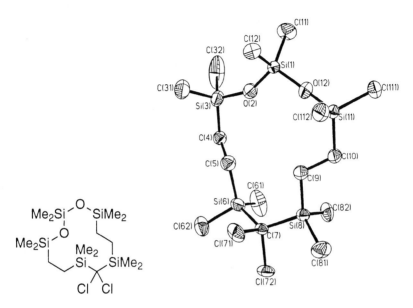

Figure 2. Structure of intramolecular cyclization product isolated during polymerization.

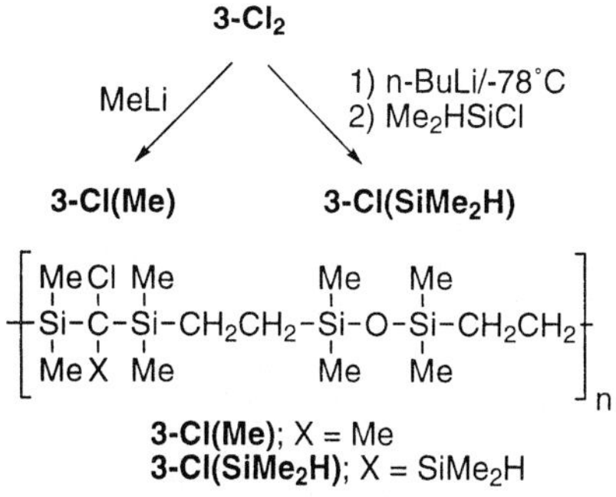

Scheme 3. Modification of polymer **3-Cl$_2$**.

using this approach indicated that this product was considerably cleaner than polymer **3-Cl(Me)** prepared by the polymer modification route. The molecular weight data is also similar to that obtained for polymer **3-Cl$_2$** (Table I, entry 6). The limitation of this method is that the choice of substitution on the monomer is restricted to groups that would not be reactive under the hydrosilylation polymerization conditions. For example, polymer **3-Cl(SiMe$_2$H)** would be difficult to obtain via this route since the -SiMe$_2$H group in (CH$_2$=CHSiMe$_2$)$_2$CCl(SiMe$_2$H) would be reactive during the polymerization.

Polymer Insertion Reactions. It was of interest to us to increase the siloxane content of the synthesized polymers and thus change their properties. Since polymer **3-Cl$_2$** contains an -Si-O-Si- linkage, it should be possible to insert cyclic siloxanes into the polymer backbone via the common ring-opening processes used to make polymeric silicones (Scheme 4) (28,29). This method would be an alternative to using long-chain siloxanes in the original polymer synthesis (Scheme 2, x>>2).

An initial experiment utilizing octamethylcyclotetrasiloxane (D$_4$) as the insertion monomer proved to be successful. In a typical experiment, a sample of **3-Cl$_2$** and two equivalents of D$_4$ were heated in the presence of trifluoroacetic acid for 7 days. The resulting thick polymer gave a single peak in the GPC chromatogram and indicated a higher molecular weight than that of **3-Cl$_2$** (M$_n$=13507, M$_w$=18935). The ^{1}H NMR spectrum of this product is shown in Figure 3, which indicates a relatively clean product.

Reaction of polymer **3-Cl$_2$** with tetramethylcyclotetrasiloxane (D'$_4$) was next attempted, as this would provide additional reactive sites in the polymer in the form of Si-H groups. These reactions were carried out in chloroform using trifluoromethanesulfonic acid as catalyst (trifluoroacetic acid can be used but the reaction is slower). As expected, a flowable colorless viscous oil was obtained after workup and removal of volatiles. However, the oil gradually turned into an insoluble gel-like solid on standing. This process occurred slowly in nitrogen but rapidly in air. In addition, we observed the formation of gaseous acid during the crosslinking by noting a color change in a piece of wet litmus paper held over the mixture. It is also noteworthy that gelation did not occur if the polymer was kept in chloroform solution.

A similar crosslinking occurred with the product from the reaction of polymer **3-Cl$_2$** with tetramethyl-tetravinylcyclotetrasiloxane (D$^{Vi}_4$). After removal of chloroform solvent, the viscous oil slowly crosslinked and became insoluble. Like the D'$_4$ crosslinking, this crosslinking occurred slowly in nitrogen but rapidly in air. Overall however, the D$^{Vi}_4$ product crosslinking occurs at a much slower rate than the D'$_4$ product crosslinking.

To further probe the mechanism of these crosslinking reactions, a reaction between polymer **3-Cl$_2$** and 1,1,3,3,5,5-hexamethyltrisiloxane was prepared (Scheme 5). After mixing the reagents thoroughly, the flask was exposed to regular room lighting conditions. After one month, the mixture had become an insoluble gum-like solid. In a separate experiment, the mixture was exposed to a weak ultraviolet light source (a thin-layer chromatography plate visualizer). In this case, the crosslinking was more rapid with complete gelation occurring after one week. During this experiment, the reaction progress was monitored by IR spectroscopy. The decreasing intensity of the Si-H absorbance was readily apparent during the course of the reaction, with the peak disappearing completely after one week (Figure 4).

Based on the results and observations described above, we believe the crosslinking reactions proceed via a radical mechanism. One possible mechanism is

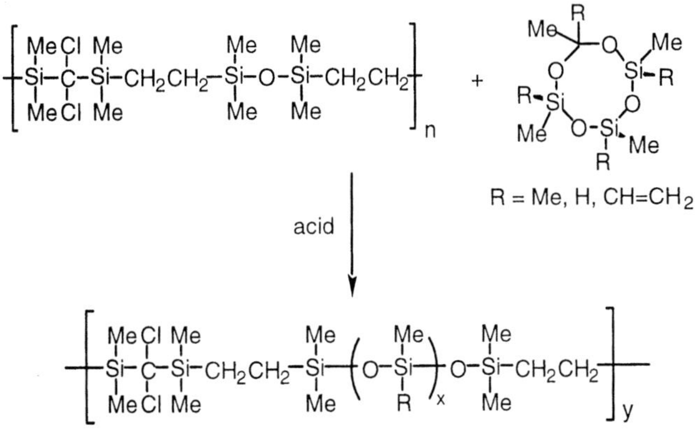

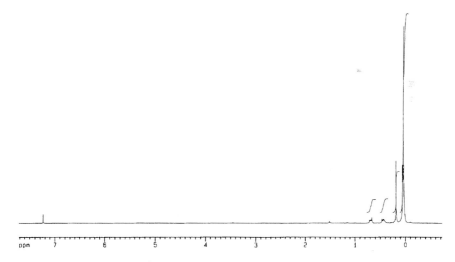

Scheme 4. Insertion of siloxanes into polymer **3-Cl$_2$**.

Figure 3. ^{1}H NMR of product from **3-Cl$_2$/D$_4$** reaction.

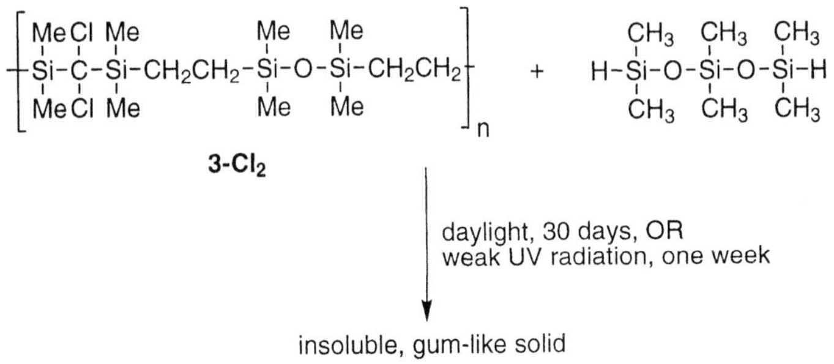

$$\left[\begin{array}{cccc} \text{Me} & \text{Cl} & \text{Me} & \text{Me} & \text{Me} \\ | & | & | & | & | \\ -\text{Si}-\text{C}-\text{Si}-\text{CH}_2\text{CH}_2-\text{Si}-\text{O}-\text{Si}-\text{CH}_2\text{CH}_2- \\ | & | & | & | & | \\ \text{Me} & \text{Cl} & \text{Me} & \text{Me} & \text{Me} \end{array} \right]_n \quad + \quad \begin{array}{ccc} \text{CH}_3 & \text{CH}_3 & \text{CH}_3 \\ | & | & | \\ \text{H}-\text{Si}-\text{O}-\text{Si}-\text{O}-\text{Si}-\text{H} \\ | & | & | \\ \text{CH}_3 & \text{CH}_3 & \text{CH}_3 \end{array}$$

3-Cl$_2$

↓ daylight, 30 days, OR weak UV radiation, one week

insoluble, gum-like solid

Scheme 5. Reaction between polymer **3-Cl$_2$** and hexamethyltrisiloxane.

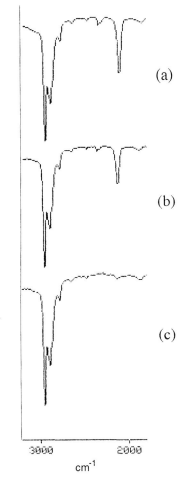

Figure 4. IR spectrum of crosslinking reaction after (a) 2 days, (b) 3 days, and (c) one week.

shown in Scheme 6. A similar type of mechanism has been proposed previously for the crosslinking of polyhydridosilanes with alkyl halides (30). The formation of the radical shown in equation 1 of Scheme 6 is certainly a reasonable step; it is well known that silicon atoms stabilize radicals in the α-position (31). The crosslinking with the vinyl siloxane-substituted polymers likely proceeds via a similar type of mechanism, with radical addition to a double bond being the crosslinking step. The presence of trace amounts of platinum catalyst remaining from the original polymer synthesis may also play a role in facilitating these crosslinking reactions.

Conclusion

New halogenated siloxane-containing polymers have been successfully prepared via a hydrosilylation polymerization procedure. These polymers are stable to air and moisture and can be easily modified by either performing reactions on the polymer, or by functionalizing the monomer prior to polymerization. Cyclotetrasiloxane insertion reactions are also possible, resulting in new polymers with added siloxane content. Insertion reactions with vinyl- or hydrido-substituted cyclotetrasiloxanes lead to polymers that spontaneously crosslink to form insoluble materials. This crosslinking likely proceeds via a radical mechanism. These "self-crosslinking" polymers could be used in adhesives applications or as a component in the synthesis of interpenetrating network materials.

Experimental Section

Materials and general information. Unless otherwise noted, all reactions were carried out in a nitrogen atmosphere. Compound **1-Cl$_2$**, $(CH_2=CHSiMe_2)_2CCl_2$, was prepared in 91% yield via the deprotonation of dichloromethane with LDA in the presence of vinyldimethylchlorosilane (24). $(CH_2=CHSiMe_2)_2CClMe$ was prepared from methyllithium and **1-Cl$_2$**, according to a literature procedure used for the preparation of $(Me_3Si)_2CClMe$ (32). 1,1,3,3-Tetramethyldisiloxane, 1,1,3,3,5,5-hexamethyltrisiloxane and Karstedt's catalyst in xylene (platinum-divinyltetramethyldisiloxane complex, Pt-DVTMDSO) were obtained from Gelest, Inc. and used as received. n-Butyllithium was obtained from Aldrich as a 2.5 M solution in hexane. Methyllithium was obtained from Acros/Fisher as a 1.6 M solution in ethyl ether. Elemental analyses were obtained from E+R Microanalytical Laboratory, Corona, NY. All NMR spectra were obtained on a Bruker AVANCE 400-MHz spectrometer using $CDCl_3$ as solvent. Gel permeation chromatography measurements were performed on a Waters Associates GPC II instrument using 500, 10^4, 10^5 and 10^6 Å μ-Styragel columns and UV or refractive index detectors. The operating conditions consisted of a flow rate of 1.5 mL/min of unstabilized HPLC-grade THF containing 0.1% tetra-n-butylammonium bromide [(n-Bu)$_4$NBr], a column temperature of 30 °C, and sample injection volume of 0.05 to 0.1 mL of a 0.1% solution. The system was calibrated with a series of narrow molecular weight polystyrene standards in the molecular weight range of ca. 10^3 to 10^6 g/mol.

Preparation of polymer 3-Cl$_2$. A 100-mL Schlenk flask containing a magnetic stir bar was charged with **1-Cl$_2$** (3.0 g, 0.0118 mol) followed by 2 drops of Pt-DVTMDSO and 1,1,3,3-tetramethyldisiloxane (1.59 g, 0.0118 mol). The mixture was freeze-thaw-degassed two times. The reaction mixture was then vigorously stirred while heated in an oil bath at 45•C. The reaction progress was measured by

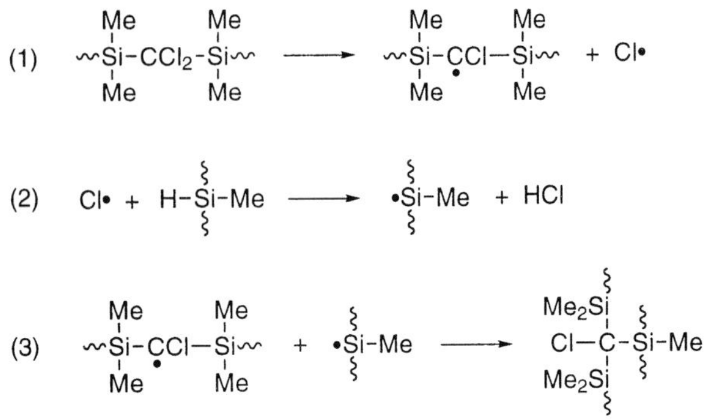

(1) $\underset{\underset{\text{Me}}{|}}{\overset{\overset{\text{Me}}{|}}{\sim\!\!\text{Si}}}\text{-CCl}_2\text{-}\underset{\underset{\text{Me}}{|}}{\overset{\overset{\text{Me}}{|}}{\text{Si}}}\!\!\sim \longrightarrow \underset{\underset{\text{Me}}{|}}{\overset{\overset{\text{Me}}{|}}{\sim\!\!\text{Si}}}\text{-}\overset{\bullet}{\text{C}}\text{Cl}\text{-}\underset{\underset{\text{Me}}{|}}{\overset{\overset{\text{Me}}{|}}{\text{Si}}}\!\!\sim + \text{Cl}\bullet$

(2) $\text{Cl}\bullet + \text{H-Si-Me} \longrightarrow \bullet\text{Si-Me} + \text{HCl}$

(3) ~Si-CCl–Si~ + •Si-Me ⟶ Cl–C(SiMe₂)(SiMe₂)-Si-Me

Scheme 6. Possible crosslinking mechanism.

monitoring the disappearance of the Si-H absorbance in the IR spectrum. The reaction was complete after three days, yielding polymer **3-Cl$_2$** as a viscous, transparent oil in quantitative yield. M$_n$ = 8641, M$_w$ = 14508, polydispersity = 1.68. Repeat polymerizations yielded polymers with similar molecular weights. ^{1}H NMR (400 MHz), δ: 0.045 (s, 12H), 0.19 (s, 12H), 0.43 (m, 4H), 0.68 (m, 4H). ^{13}C NMR (100 MHz), δ: -4.83 (SiCH$_3$), -0.600 (SiCH$_3$), 5.02 (CH$_2$), 10.1 (CH$_2$), 74.9 (CCl$_2$). IR (NaCl, neat, cm^{-1}): 2956 (s), 2922 (m), 1406 (m), 1253 (s), 1136 (m), 1055 (s), 836 (s), 790 (s). Elemental analysis: calcd, C 40.28, H 8.32, Cl 18.29; found, C 40.43; H 8.54, Cl 18.15.

Preparation of polymer 4-Cl$_2$. This polymerization was carried out using the **3-Cl$_2$** preparative procedure. The following reagents were used: **1-Cl$_2$** (3.0 g, 11.85 mmol), HSiMe$_2$OSiMe$_2$OSiMe$_2$H (2.47 g, 11.85 mmol), and Pt-DVTMDSO (2 drops). Polymer **4-Cl$_2$** was obtained as a viscous colorless oil in quantitative yield. M$_n$ = 16500, M$_w$ = 26700, polydispersity = 1.62. Repeat polymerizations yielded polymers with similar molecular weights. ^{1}H NMR (400 MHz), δ: 0.015 (s, 6H), 0.067 (s, 12H), 0.20 (s, 12H), 0.45 (m, 4H), 0.69 (m, 4H). ^{13}C NMR (100 MHz), δ: -4.84 (SiCH$_3$), -0.752 (SiCH$_3$), 1.05 (SiCH$_3$), 4.98 (CH$_2$), 9.97 (CH$_2$), 74.8 (CCl$_2$). Elemental analysis: calcd, C 39.01, H 8.29, Cl 15.35; found, C 39.22; H 8.03, Cl 15.29.

Preparation of polymer **3-Cl(SiMe$_2$H) from polymer 3-Cl$_2$.** A 100-mL Schlenk flask containing a magnetic stir bar was charged with polymer **3-Cl$_2$** (2.73 g, 7.04 mmol) and 50 mL of THF. The flask was cooled in a dry ice/acetone bath and then n-BuLi (2.9 mL, 7.25 mmol) was added via syringe. The reaction mixture was stirred for an additional three hours in the cold bath, at which point chlorodimethylsilane (0.68 g, 7.19 mmol) was added. The reaction mixture was stirred in the cold bath for 30 minutes, and then stirred at room temperature overnight. After an aqueous workup, evaporation of the organic layer gave a brown viscous oil. The oil was redissolved in ether and decolorizing carbon was added. The mixture was filtered and all volatiles were removed by heating under vacuum at 70•C overnight. Polymer **3-Cl(SiMe$_2$H)** was obtained as a viscous oil in 92% yield. M$_n$ = 4400, M$_w$ = 6700, polydispersity = 1.52. The ^{1}H NMR spectrum for this product was slightly messy with many small peaks in the region δ 0.0 to 0.5 in addition to single peaks at δ 0.031, 0.14, and a doublet at δ 0.24. The Si-H resonance appears at δ 4.13 as a multiplet.

Preparation of polymer 3-Cl(Me) from polymer 3-Cl$_2$ (one equivalent of MeLi). A 100-mL Schlenk flask containing polymer **3-Cl$_2$** (1.43 g, 3.68 mmol), 25 mL of Et$_2$O, and a magnetic stir bar was cooled in an ice bath. Methyllithium (2.3 mL, 3.68 mmol) was added via syringe. The reaction mixture was allowed to warm to room temperature and stirred overnight. After an aqueous workup, evaporation of the organic layer followed by heating under vacuum at 70•C afforded polymer **3-Cl(Me)** as a pale yellow, viscous oil (1.01 g, 75%). M$_n$ = 4800, M$_w$ = 7700, polydispersity = 1.59. The reaction of polymer **3-Cl$_2$** with 1.5 equivalents of MeLi is carried out in a similar manner. See **Figure 1(c)** for an expanded ^{1}H NMR spectrum of this product.

Preparation of polymer 3-Cl(Me) via hydrosilylation polymerization. This polymerization was carried out using the **3-Cl$_2$** preparative procedure. The following reagents were used: $(CH_2=CHSiMe_2)_2CClMe$ (1.0 g, 4.30 mmol), $(HSiMe_2)_2O$ (0.58 g, 4.30 mmol), and Pt-DVTMDSO (2 drops). Polymer **3-Cl(Me)** was obtained as a viscous, pale gray oil in quantitative yield. M_n = 7800, M_w = 13100, polydispersity = 1.68. ^{1}H NMR (400 MHz), δ: 0.037 (s, 12H), 0.076 (d, 12H), 0.39 (m, 4H), 0.53 (m, 4H), 1.5 (s, 3H). ^{13}C NMR (100 MHz), δ: -4.99 (SiCH$_3$), -4.91 (SiCH$_3$), -0.602 (SiCH$_3$), 5.04 (CH$_2$), 10.3 (CH$_2$), 22.7 (CH$_3$C), 48.8 (CCl). Elemental analysis: calcd, C 45.79, H 9.61, Cl 9.65; found, C 45.90, H 9.70, Cl 9.93.

Insertion reactions of [SiMe$_2$O]$_4$ (D$_4$) into polymer 3-Cl$_2$. A 100-mL Schlenk flask containing a magnetic stir bar was charged with **3-Cl$_2$** (0.5507 g, 2.29 mmol) and 2 equivalents of [SiMe$_2$O]$_4$ (1.3580 g, 4.58 mmol). The flask was flushed with N$_2$, and F$_3$CCOOH (0.202g 0.458 mmol) was added. The reaction mixture was freeze-thaw degassed three times. The mixture was then heated in a 40°C oil bath for 7 days. During this period, the mixture became more viscous. The volatiles were removed in vacuo, and the resulting double insertion product was purified by precipitation (methanol/toluene (8:1)) several times, which gave a colorless viscous oil (1.40 g, 73%). GPC exhibited a single peak; M_n = 13507, M_w = 18935, polydispersity = 1.40. When one equivalent of [SiMe$_2$O]$_4$ was employed under similar reaction conditions, the corresponding product was obtained in 78% yield.

Insertion reaction of [SiMeHO]$_4$ (D'$_4$) into polymer 3-Cl$_2$ and a subsequent crosslinking reaction. A mixture of polymer **3-Cl$_2$** (0.58 g, 1.25 mmol), [SiMeHO]$_4$ (0.32 g, 1.25 mmol), and 20 mL of CHCl$_3$ was freeze-thaw degassed 3 times. Two drops of CF$_3$SO$_3$H were added. The mixture was then stirred overnight at room temperature. The resulting solution was shaken with a small amount of powdered NaHCO$_3$ until a pH of 6 was obtained. Filtration followed by removal of CHCl$_3$ in vacuo afforded a flowable colorless viscous oil (0.8 g). Using CF$_3$COOH as catalyst under the same reaction conditions required a longer reaction time. The viscous oil product gradually turned into an insoluble gel-like solid under a nitrogen atmosphere. If the oil was exposed to room air, it almost immediately turned into an insoluble solid. In a separate experiment, performing the insertion reaction without solvent at 80•C resulted in the direct formation of an insoluble solid.

Insertion reaction of [SiMe(Vi)O]$_4$ (D$^{Vi}_4$) into polymer 3-Cl$_2$ and a subsequent crosslinking reaction. A mixture of polymer **3-Cl$_2$** (0.76 g, 1.64 mmol), [SiMe(Vi)O]$_4$ (0.57 g, 1.64 mmol), and 10 mL of CHCl$_3$ was freeze-thaw degassed 3 times. Two drops of CF$_3$SO$_3$H were added. The mixture was then stirred for 24 hours at room temperature. The resulting solution was shaken with a small amount of powdered NaHCO$_3$ until a pH of 6 was obtained. Filtration and removal of CHCl$_3$ in vacuo afforded a flowable colorless viscous oil (1.1 g). The oil gradually turned into an insoluble gel-like solid under a nitrogen atmosphere. If the oil was exposed to room air, it almost immediately turned into an insoluble solid.

Crosslinking reaction between SiMe$_2$(OSiMe$_2$H)$_2$ and polymer 3-Cl$_2$ with UV radiation. Polymer **3-Cl$_2$** (0.495 g, 1.072 mmol) and SiMe$_2$(OSiMe$_2$H)$_2$ (0.223 g, 1.072 mmol) were mixed thoroughly and placed under UV light (Model UVGL-25 Mineralight Lamp Multiband UV-254/366 NM 115 VOLTS 60 HZ 0.16 AMPS) until the v(Si-H) band in the IR spectrum completely disappeared (one week). An insoluble gum-like solid was obtained in quantitative yield. Alternatively, the same product can be obtained by prolonged (up to one month) stirring under normal daylight conditions.

Acknowledgments

The authors thank the following sources for financial support of this research: Southern Methodist University, American Chemical Society/Petroleum Research Fund, the Welch Foundation, the National Research Council, and the Office of Naval Research (Grant No. N000149610210). The authors are also grateful to Professor Patty Wisian-Neilson for providing the use of the GPC facilities, and to Dr. Cuiping Zhang for performing the GPC analyses.

Literature Cited

1. Mark, J. E.; Allcock, H. R.; West, R. *Inorganic Polymers*; Prentice Hall: Englewood Cliffs, NJ, 1992; p 141.
2. Birot, M.; Pillot, J-P.; Dunogues, J. *Chem. Rev.* **1995**, *95*, 1443.
3. Dvornic, P. R.; Gerov, V. V. *Macromolecules* **1994**, *27*, 1068.
4. Dvornic, P. R.; Gerov, V. V.; Govedarica, M. N. *Macromolecules* **1994**, *27*, 7575.
5. Mathias, L. J.; Lewis, C. M. *Macromolecules* **1993**, *26*, 4070.
6. Jallouli, A.; Lestel, L.; Tronc, F.; Boileau, S. *Macromol. Symp.* **1997**, *122*, 223.
7. Narayan-Sarathy, S.; Neilson, R. H.; Smith, Jr., D. W. *Polym. Prepr.* **1998**, *39(1)*, 609.
8. Kaganove, S. N.; Grate, J. W. *Polym. Prepr.* **1998**, *39(1)*, 556.
9. Miller, R. D.; Michl, J. *Chem. Rev.* **1989**, *89*, 1359.
10. Hsiao, Y-L.; Waymouth, R. M. *J. Am. Chem. Soc.* **1994**, *116*, 9779.
11. Rushkin, I. L.; Interrante, L. V. *Macromolecules* **1996**, *29*, 3123.
12. Rushkin, I. L.; Interrante, L. V. *Macromolecules* **1996**, *29*, 5784.
13. Seyferth, D.; Lang, H. *Organometallics* **1991**, *10*, 551.
14. Tsumura, M.; Iwahara, T.; Hirose, T. *Polym. J.* **1995**, *27*, 1048.
15. Tsumura, M.; Iwahara, T.; Hirose, T. *J. Polym. Sci. Pt. A, Polym. Chem.* **1996**, *34*, 3155.
16. Son, D. Y.; Bucca, D.; Keller, T. M. *Tetrahedron Lett.* **1996**, *37*, 1579.
17. Itsuno, S.; Chao, D.; Ito, K. *J. Polym. Sci. Pt. A: Polym. Chem.* **1993**, *31*, 287.
18. Jallouli, A.; Lestel, L.; Tronc, F.; Boileau, S. *Macromol. Symp.* **1997**, *122*, 223.
19. Boury, B.; Corriu, R. J. P.; Douglas, W. E. *Chem. Mater.* **1991**, *3*, 487.
20. Pang, Y.; Ijadi-Maghsoodi, S.; Barton, T. J. *Macromolecules* **1993**, *26*, 5671.
21. Boury, B.; Corriu, R. J. P.; Leclercq, D.; Mutin, P. H.; Planeix, J-M.; Vioux, A. *Organometallics* **1991**, *10*, 1457.
22. Kuhnen, T.; Ruffolo, R.; Stradiotto, M.; Ulbrich, D.; McGlinchey, M. J.; Brook, M. A. *Organometallics* **1997**, *16*, 5042.
23. Chen, R-M.; Deng, C. Z. B.; Sun, G.; Lee, S-T.; Luh, T-Y. *Polym. Prepr.* **1998**, *39(1)*, 89.
24. Yoon, K.; Son, D. Y. *J. Organomet. Chem.* **1997**, *545-6*, 185.
25. We thank Dr. Hongming Zhang for performing the X-ray analysis.

26. van Eikema Hommes, N. J. R.; Bickelhaupt, F.; Klumpp, G. W. *Tetrahedron Lett.* **1988**, *29*, 5237.
27. Ayoko, G. A.; Eaborn, C. *J. Chem. Soc. Perkin Trans.* **1987**, 381.
28. Chujo, Y.; McGrath, J. E. *J. Macromol. Sci. Pure Appl. Chem.* **1995**, *A32*, 29.
29. Kunzler, J.; Ozark, R. *J. Appl. Polym. Sci.* **1997**, *65*, 1081.
30. Shieh, Y-T.; Sawan, S. P. *Eur. Polym. J.* **1996**, *32*, 625.
31. Barton, T. J.; Boudjouk, P. In *Silicon-Based Polymer Science: A Comprehensive Resource*; Zeigler, J. M., Fearon, F. W. G., Eds.; American Chemical Society: Washington, D.C., 1990; p. 3.
32. Al-Hashimi, S.; Smith, J. D. *J. Organomet. Chem.* **1978**, *153*, 253.

Chapter 16

Radially Layered Poly(amidoamine-organosilicon) Copolymeric Dendrimers and Their Networks Containing Controlled Hydrophilic and Hydrophobic Nanoscopic Domains

Petar R. Dvornic[1], Agnes M. de Leuze-Jallouli[1], Michael J. Owen[2], and Susan V. Perz[2]

[1]Michigan Molecular Institute, 1910 West St. Andrews Road, Midland, MI 48640
[2]Dow Corning Corporation, Mail # C041D1, P.O. Box 994, Midland, MI 48686-0994

A new family of rare, radially layered copolymeric dendrimers is described. These dendrimers are comprised of hydrophilic polyamidoamine (PAMAM) interiors and hydrophobic (oleophilic) organosilicon (OS) exteriors and may be viewed as globular, amphiphilic, "covalently bonded inverted micelles". While derivatives of these dendrimers with trimethylsilyl- terminal groups show quite unusual surface properties, the methoxysilyl- substituted homologues are excellent precursors for the first tractable dendrimer-based networks that can be prepared as elastomers, plastomers or coatings on various substrates. Because of their unique composition comprising interconnected nanoscopic hydrophilic and hydrophobic domains, these dendrimer-based networks show highly unusual properties which include permselectivity and the ability to encapsulate various inorganic or organic electrophiles. These properties lead to exciting application possibilities, such as "molecular sponges", nanoscopic "reactors" or inorganic-organic nano-composites.

The importance of the copolymer concept to polymer science and engineering can not be overemphasized. Essentially, it was the realization of various methods to incorporate two or more compositionally different repeating units into the large molecules that provided practically unlimited versatility to macromolecular synthesis and the ability to tailor-make a vast variety of new polymer structures for specifically desired applications from only a limited number of polymerizable monomers available. By the widely accepted general understanding of traditional polymers, such as linear, randomly branched or cross-linked ones, a copolymer is a macromolecule comprising of two or more compositionally different main building blocks (i.e., repeat units) which may be organized within its molecular structure in different relative amounts and/or orders of placement. Consequently, many well known step-growth polymers, such as polyamides, polyesters, and alike, do not qualify as copolymers, because although they are produced from two different monomers, they comprise entirely of identical repeat units formed by linking together those monomers involved.

In recent years, a new architectural class of macromolecules (*1*): the dendritic polymers, has attracted outstanding scientific attention (*2*). In fact, these polymers in general, and among them, the structurally most regular members of the family: dendrimers in particular, have become one of the fastest growing research areas in polymer science. However, with one notable exception (*3*), most of the activities in dendrimer research have focused exclusively on "homopolymeric" dendrimers, and their "copolymeric" derivatives have been seriously overlooked. On the other hand, from the well documented experience of polymer science, the copolymer approach would expectedly bring to this still relatively young field the same potentials and versatility that it has so generously delivered to the conventional polymer architectures. Additionally, we anticipate that synergism of different compositions of basic building blocks and unique macromolecular architecture, will also broaden the scope of the resulting properties of copolymeric dendrimers and provide new application potentials that are not available either from classical compositional copolymers, or from homopolymeric dendrimers alone.

In this chapter we describe the first family of radially layered poly(amidoamine-organosilicon) (PAMAMOS) copolymeric dendrimers, which contain concentric layers of hydrophilic polyamidoamine (PAMAM) branch cells in their interiors and hydrophobic organosilicon (OS) branch cells in the exteriors (*4,5*). In the first part of the chapter, we present a brief overview of some general characteristics of the dendrimer molecular architecture and their main architecturally driven properties. Following this, we describe the synthesis of PAMAMOS dendrimers and some of their unusual surface properties that result from amphiphilic character of these "covalently bonded inverted micelles". In the third part, we report how these dendrimers can be utilized for preparation of cross-linked dendrimer-based networks, which, depending on the composition of the PAMAMOS precursor(s) used, may be tailor-made into elastomers, plastomers or coatings that contain very well defined (i.e., in size and shape) hydrophilic and hydrophobic nanoscopic domains. It will be clearly seen from the presented data that unprecedented properties indeed arise from the synergism of combined PAMAM and OS compositions with characteristic features of dendrimer molecular architecture.

Dendrimers

Dendrimers are globular, nano-scaled macromolecules which comprise of two or more tree-like dendrons emanating from a single central atom or atomic group, called the core (*6*). Ideally, a generalized structure of a tetradendron dendrimer for example, can be represented as shown in Figure 1. This figure shows that dendrons are constructed of branch cells, which are the main building blocks of dendritic structure and which may be considered as three-dimensional branched analogues of repeat units in classical linear polymers. Each branch cell contains a branch juncture and they are organized in mathematically precise architectural arrangement that gives rise to a series of regular, radially concentric layers, called generations, around the core (7-9). As illustrated in Table I for the well-known polyamidoamine (PAMAM) dendrimers, depending on the functionality of the branch junctures, the number of dendrimer end-groups (i.e., Z of Figure 1) increases in a geometrically progressive manner with generations, reaching at higher generations some of the highest values for the molecular density of functionality known to organic and/or polymer chemistry.

As a consequence of high degree of synthetic control, dendrimers can be prepared with very precise molecular organization and structural monodispersity (10) even at very high molecular weights, which may range into several hundreds of thousands to a million (see Table I). For example, it has been shown by electron microscopy (7), small angle X-ray (10) and neutron scattering studies (11), that they have very well defined shapes and sizes, and (if grown from a point-like core, such as single atom, or symmetrical atomic group) at higher generations tend to become almost spherical (12,13). Recent rheological investigations (14,15) and various "host-guest" interaction experiments (7,16-22) suggest that dendrimers consist of relatively "soft and spongy" interiors and "dense" exteriors, which enable these macromolecules to exhibit unique "encapsulation" properties. Their diameters span the lower nanoscopic size range from about 1 to about 15 nm (see Table I), which is just above the size range of molecules of classical organic chemistry, but below that typically found for macromolecules of polymer science (23,24). It is indeed provocative to note that many important natural macromolecules, including for example hemoglobin, mioglobin, the T-cells, etc., fit exactly into this nanoscopic size range, so that in a sense dendrimers may be considered as synthetic size-analogues of such natural molecules.

Although most of the dendrimer research to date has been focused on their chemistry (i.e., synthesis and chemical behavior) and structural characteristics, information about dendrimer physical properties is slowly beginning to emerge. Recently, the first comparative studies of selected structure-property relationships of two best known families of dendrimers: polyamidoamines (6,25) and polybenzylethers (26), and their corresponding linear homologues of identical chemical composition and similar molecular weights have been described (27,28). These results clearly showed that substantial architecturally driven property differences exist between different classes of macromolecular architectures (1), and some of these differences may be summarized as shown in Table II (21). In fact, some dendrimer properties are so architecturally specific that they are not found in any other type of conventional macromolecular topologies, including linear, cross-linked and randomly branched ones. In this sense, dendritic polymers in general, and among them dendrimers in particular, may be considered as the fourth main class of macromolecular architecture (1,23,24). However, it is also intrinsically clear that potential utilization of these unique architecturally driven properties will depend on dendrimer chemical composition and that for any particularly desired application, specific chemical

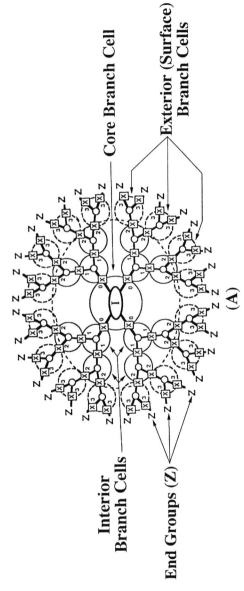

Figure 1: Schematic representation of homopolymeric (A) and two types or ordered copolymeric dendrimers: radially layered (B) and segmented (C) ones. Small circles denote the branch junctures, larger circles represent the branch cells, while numerals indicate dendrimer generation layers and each Z is an end-group. Note that copolymeric dendrimers are composed of at least two different compositional types of branch cells (1 and 2).

245

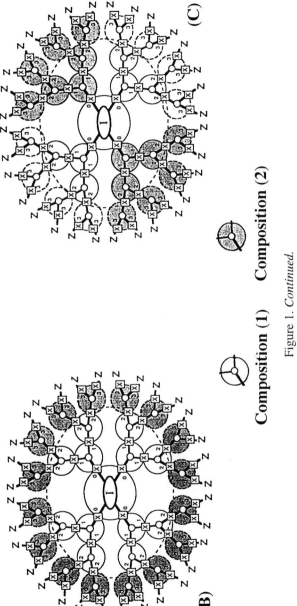

Composition (1)   Composition (2)

Figure 1. *Continued.*

Table I: Selected Molecular Characteristics of Ethylenediamine Core Poly(amidoamine) (PAMAM) Dendrimers

Generation	Number of surface groups	Molecular weight[a]	SEC[c]	Hydrodynamic radius Å[b] DSV	SANS[d]
0	4	517	7.6	-	-
1	8	1430	10.8	10.1	-
2	16	3256	14.3	14.4	-
3	32	6909	17.8	17.5	16.5
4	64	14215	22.4	25.0	19.7
5	128	28826	27.2	32.9	24.3
6	256	58048	33.7	-	30.3
7	512	116493	40.5	-	35.8
8	1024	233383	48.5	-	-
9	2048	467162	57.0	-	49.2
10	4096	934720	67.5	-	-

(a) Nominal values calculated for ideal dendrimer growth.
(b) 25°C; 0.1 molar citric acid in water.
(c) Relative to linear PEO standards (Source: R. Yin, unpublished data).
DSV: Dilute solution viscometry. Calculated values from intrinsic viscosity data (Source: S. Uppuluri, Ph. D. thesis, Michigan Technological University, 1997).
(d) Source: E. Amis et al, 29th ACS Central Regional Meeting, Midland, MI, May 28-30, 1997.

Table II: Architecturally Driven Property Differences Between Dendrimers and Corresponding Linear Polymers

LINEAR POLYMERS	DENDRIMERS
Medium dependent molecular configuration	Persistent molecular shape and size
Specific and generally broader molecular weight distribution	Narrow molecular weight distribution
Configurationally controlled exposure of end-groups	Exo-presented end-groups High density of functionality
Lower solubility for a given chemical composition and similar molecular weight	Generally high solubility
Low molecular density	High molecular density
Larger elution volumes	Smaller elution volume
Power-law shear-thinning flow	Newtonian flow
Well defined break in the zero-shear *vs.* molecular weight relationship	No break in the zero-shear viscosity *vs.* molecular weight relationship
*Inter*molecular interactions through entanglement coupling	No interpenetration to form transient networks
Mobility by reptation	Entire molecule is kinetic flow unit
High viscosity (solution and/or bulk) at high molecular weights	Substantially lower viscosity for comparable molecular weights and/or solution concentrations
Semicrystalline with well defined melting temperature range	Do not crystallize (no melting)
Higher glass temperatures due to *inter*segmental interactions	Lower glass temperatures
Linear dependence of intrinsic viscosity on molecular weight	Maximum in the intrinsic viscosity vs. molecular weight relationship
	Pronounced molecular encapsulation potential

composition will be required to accentuate these unique architectural features under any given set of specific operating conditions.

Copolymeric Dendrimers

Until present, over 50 different compositional families of dendrimers, with over 200 types of different end-group modifications (i.e., Z of Figure 1), have been reported (2). These predominantly include "fully organic" compositions, based entirely on carbon, hydrogen, nitrogen and oxygen, and considerably fewer representatives of silicon-, phosphorus- and metal-containing derivatives (2,7-9). The first silicon-containing dendrimers (29) were polysiloxanes, reported in 1989 by Rebrov and his co-workers (30), following which different polycarbosilane (31-33), polysilane (34-36) and most recently silicon/germanium-containing analogues (37) were also described. Of these, a variety of different surface-modified derivatives (particularly from polycarbosilane dendrimers) were prepared utilizing various surface group modification methods. It has been clearly shown that such modifications may result in dramatic changes of dendrimer properties, particularly those regarding their interactions with the surrounding environment and including solubility, surface characteristics, reactivity and chemical behavior, as well as in completely new features which the starting dendrimers did not posess, such as electrical conductivity, catalytic ability, luminescence, biological activity, etc. (2,7-9).

Consequently, diversification of the assortment of available dendrimer compositions is and will remain an important task for dendrimer synthetic chemists. Based on the experience with conventional polymer architectures, one of the most versatile approaches to successfully accomplish this task seems to be to combine compositionally different basic building blocks of dendritic architecture (i.e., the branch cells) in different relative amounts and/or order of placement in new types of *copolymeric* dendrimers. By extension of widely accepted concepts of conventional copolymers, truly copolymeric dendrimers would be those containing at least two types of compositionally different branch cells, organized either randomly or, as pointed out by Fréchet and Hawker (38), in ordered structures, including radially layered (B of Figure 1) and segmented derivatives (C of Figure 1) (i.e., those where main dendrons are of different chemical composition). Clearly, successful development of practical strategies for preparation of such copolymeric dendrimers would not only provide the ability to greatly expand the variety of different dendrimer compositions, but it would also enable their properties to be tailor-made for specifically targeted applications. However, this approach also involves the challenge of finding "compatible" combinations of different reiterative branching chemistries and appropriate cross-over reaction(s) to switch between the compositions, so that it is not surprising that until present, with a notable exception of poly(benzylether-benzylester) dendrimers (3), this approach has not been utilized to any more significant extent.

Preparation of Radially Layered Poly(amidoamine-organosilicon) (PAMAMOS) Copolymeric Dendrimers

The synthesis of radially layered poly(amidoamine-organosilicon) (PAMAMOS) copolymeric dendrimers starts from amine terminated PAMAM dendrimers, which, in turn, are obtained by a well-known "excess-reagent" divergent growth method that involves a reiterative sequence of (a) Michael addition reactions of methyl acrylate (MA) to primary amines, and (b) amidation of the resulting methyl ester intermediates with ethylenediamine (EDA), as shown in Reaction Scheme 1 (39-41). These PAMAM dendrimers are commercially obtained from Dendritech Inc., (Midland, MI) and they can be used for PAMAMOS preparation without any further purification. The synthesis then involves another Michael addition reaction, this time of a silylated acryl ester, such as (3-acryloxypropyl)dimethoxymethylsilane, as shown in Reaction Scheme 2 (4).

This reaction can be very effectively monitored by NMR because the H displacement of the methylene protons adjacent to the -COO- groups is conveniently different for the ester reagent and for the resulting dendrimer product (4). Thus, the degree of conversion can be easily and precisely determined at any stage of the reaction occurrence by following the relative ratio of the disappearing signal characteristic for the former and the appearing signal characteristic for the latter. Similarly, in the case of PAMAMOS dendrimers having two or more OS branch cell layers around the PAMAM interior, each generation (i.e., layer) of the respective silicons can be easily identified by ^{29}Si NMR.

It was found that temperature has little effect on the rate of these Michael addition reactions and that methanol is a very good solvent. However, it was also observed that the presence of small amounts of water is beneficial, which seems to indicate that water may act at least as a co-catalyst (see Figure 2) (4,21). As shown in Figure 3, the order of reactivity of PAMAM dendrimers in these reactions was: Generation 2 = Generation 3 > Generation 4, which is the same as for alkylation with methyl acrylate in the synthesis of homopolymeric PAMAM dendrimers (see Reaction Scheme 1). It can be also seen from Figure 3 that after a fast start, the reaction rate considerably decreases in the later stages, which may indicate increasing steric hindrance at higher degrees of dendrimer surface substitution with rather large and bulky silyl-ester groups.

Using this synthetic strategy, a number of different PAMAMOS dendrimers, having reactive and non-reactive end-groups have been prepared, and some representative examples of the obtained products are shown in Figure 4. As shown in Reaction Scheme 2 and Figure 4, they all contained hydrophilic PAMAM interior and hydrophobic (oleophilic) organosilicon (OS) exterior in covalently bonded dendrimer architectural organization and were, therefore, expected to exhibit properties arising from the synergism of both of these chemical compositions and dendrimer molecular architecture (5,22,42). More details about this and other synthetic methods for preparation of these copolymeric dendrimers, as well as their full structural characterization will be described elsewhere (43).

Reaction Scheme 1

$$\text{G}_x\text{-(NH}_2\text{)}_z + 2.4z\ CH_2=CH\text{-}COO\text{-}(CH_2)_3\text{-}SiR_1R_2R_3$$

$$\Big\downarrow \text{Methanol, r.t., } N_2$$

$$\text{G}_x\text{-}\left[\underset{\mid}{\overset{H_{(2-y)}}{N}}\text{-}\big(CH_2\text{-}CH_2\text{-}COO\text{-}(CH_2)_3\text{-}SiR_1R_2R_3\big)_y\right]_z$$

(A): $R_1 = \text{-}CH_3$; $R_2, R_3 = \text{-}OCH_3$

(B): $R_1, R_2, R_3 = \text{-}O\text{-}Si(CH_3)_3$

(C): $R_1 = \text{-}CH_3$; $R_2, R_3 = \text{-}O\text{-}Si(CH_3)_2CH=CH_2$

Reaction Scheme 2

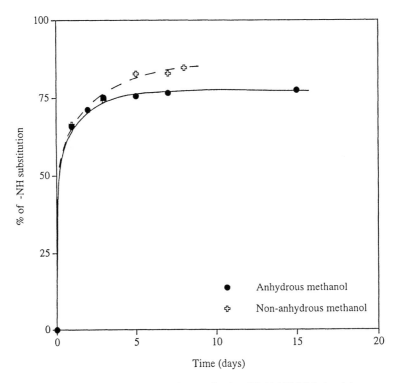

Figure 2: Effect of solvent on the synthesis of PAMAMOS dendrimers

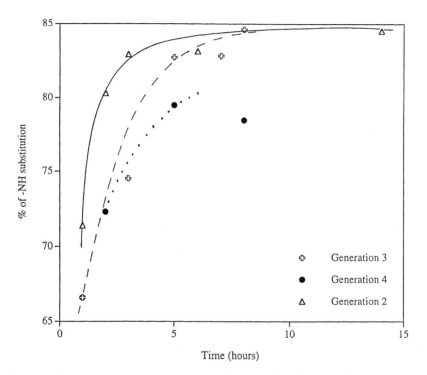

Figure 3: Effect of the generation of PAMAM dendrimer reagent on the rate of the synthesis and composition of the resulting PAMAMOS dendrimer product

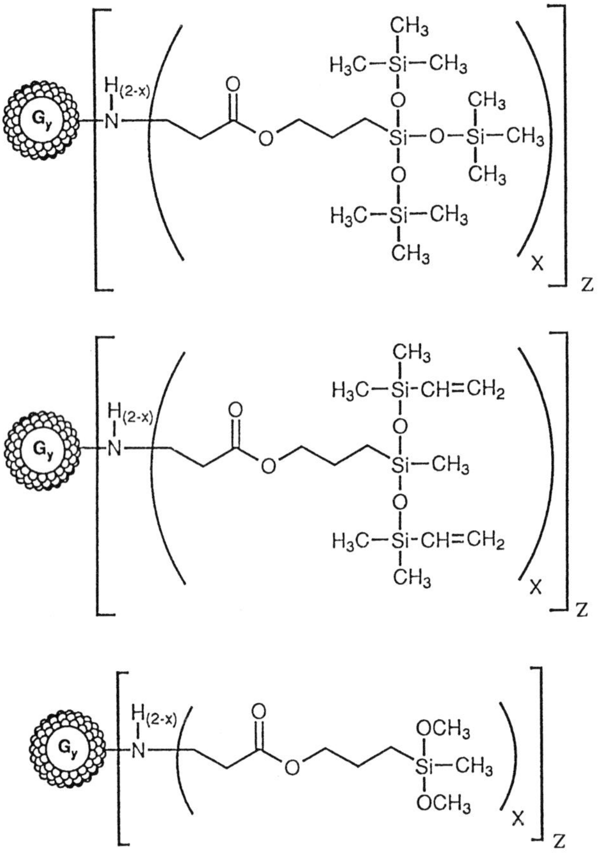

Figure 4: Selected examples of PAMAMOS dendrimers

Surface Properties of PAMAMOS Dendrimers

As the degree of PAMAM dendrimer surface substitution by OS branch cells increases, the resulting PAMAMOS dendrimers cease to be water soluble. For example, this transition occurs at *circa* 50 % -NH substitution for the first OS branch cell layer around a generation 3 PAMAM interior.

Surface activity of these PAMAMOS dendrimers was determined by the Wilhelmy plate method for the water soluble dendrimers and by the Langmuir trough technique for the insoluble ones. Some preliminary data on water soluble PAMAMOS have already been published (5,42). It was shown that the best of these materials lower the surface tension of water to just below 30 mN/m at 5 wt. %, with no break to a constant surface tension that would indicate micelle formation. Thus, these PAMAMOS behave more like considerably surface active water soluble polymers than surfactants. However, it is probable that their homologues with longer siloxane dendrons than the trimethylsilyl- groups studied so far will have considerably more surfactant-like behavior.

The water insoluble PAMAMOS dendrimers form spread monolayers on water with chloroform as the spreading solvent. Figure 5 shows a typical surface pressure/area isotherm obtained in the Langmuir trough experiments. It can be seen from this figure that these dendrimers typically attained surface pressures in excess of 40 mN/m. This offers a good example of the different behavior resulting from the presentation of organosilyl- groups in very different polymer architectures, since very flexible linear polydimethylsiloxane (PDMS) chains, for example, are only capable of attaining surface pressure of *circa* 10 mN/m (44). With respect to this, the only linear siloxane polymers that show a steady increase to significantly higher surface pressures on compression are those that contain both methyl- and very hydrophilic groups such as -CH_2NH_2. For PAMAMOS dendrimers, molecular modeling studies suggest that even at high degrees of surface derivatization of -NH groups, there is still ready access to the nitrogen atoms from the PAMAM interior. This, we assume, is the source of the anchoring of these dendrimers to the water surface that enables the high surface pressures to be sustained.

These Langmuir trough isotherms are liquid film type with no steps or abrupt slope changes at low surface pressure, indicative of significant reorientation of the film as is normally seen with PDMS Langmuir films. Similarly shaped surface pressure-area isotherms have been reported for other amphiphilic dendritic polymers, for example, quaternized polypropyleneimine dendrimers (45). Our measurements were made by first compressing the film and then expanding it. The expansion curve did not usually follow the compression curve indicating hysteresis effects. Moreover, differences were seen in the expansion curves which varied with barrier speed and number of compression/expansion cycles. These hysteresis effects were more marked with the generation 2 modified materials and with those having only one OS layer. These are the dendrimers that are the least densely packed in the outermost regions and possible explanations include partial solubility in the water or water-induced reorientation or association of dendrimer molecules. They may also be accounted for by interpenetration or formation of multilayers. In the example in Figure 5, the hysteresis loop at high surface pressure is suggestive of formation of a second dendrimer layer.

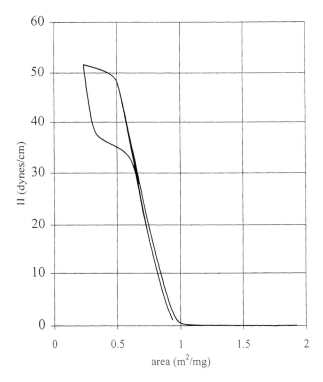

Figure 5: Surface pressure measurements results for a PAMAMOS dendrimer comprising a generation 3 PAMAM interior and two exterior layers of OS branch cells

Surface areas per dendrimer molecule can be obtained by extrapolating to zero pressure the steepest portion of the curves, giving one value for the compression and a smaller value for the expansion curve. Expectedly, these areas increase with size (i.e., generation) of the parent PAMAM dendrimer and with the degree of substitution of its -NH groups. The compression curve values are in good accord with the PAMAM dendrimer dimensions shown in Table I. For example, the PAMAMOS dendrimer that showed behavior presented in Figure 5 (i.e., generation 3 PAMAM with two layer/generation of external OS; MW = 28,598) has a compression curve area of 39 nm^2/mol, which correlates well to the unmodified generation 5 PAMAM dendrimer (MW = 28,826) for which the area of 34 nm^2/mol was obtained from the dilute solution viscometry data.

PAMAMOS Dendrimer-Based Networks Containing Hydrophilic and Hydrophobic Nanoscopic Domains

Functionalized PAMAMOS dendrimers of Figure 4 are extremely useful precursors for preparation of unique dendrimer-based networks (21,22) which contain well defined, hydrophilic PAMAM and hydrophobic (oleophilic) OS nanoscopic domains. For example, from methoxysilyl- end-functionalized PAMAMOS derivatives, such networks can be obtained as shown in Reaction Scheme 3. The process consists of two steps: (a) a water hydrolysis of methoxysilyl- dendrimer end-groups into the corresponding silanols, and (b) subsequent condensation of these silanol intermediates into siloxane *inter*dendrimer bridges. The latter is self-catalyzed by the basic PAMAM interior of the PAMAMOS dendrimer precursor used. In practice, this cross-linking is easily accomplished either by simple, direct exposure of methoxysilyl- functionalized PAMAMOS dendrimers to atmospheric moisture, or by controlled addition of water, either in the form of vapor (for example in a humidity chamber), or as a liquid into appropriate dendrimer solution. Of course, since the process is in effect a chain-reaction in which water used up in the methoxysilane hydrolysis is liberated in the silanol condensation (see Reaction Scheme 3), less than stochiometric amounts of water are needed to trigger its onset. On the other hand, as long as the methoxysilyl- functionalized PAMAMOS dendrimers are kept in dilute solution (preferably in methanol) their shelf-life is at least six months at room temperature.

A typical procedure for the preparation of these networks may be described as follows. If a methanol solution of a methoxysilyl- terminated PAMAMOS dendrimer is poured into an aluminum pan, covered with aluminum foil in order to prevent dust contamination, and left exposed to air for a period of time, slow curing will soon became evident by gradual "densification" of the original liquid. Furthermore, the sample will show a time-dependent weight loss which results from evaporation of methanol that is present as a solvent but that is also being formed as a by-product of the cross-linking reaction. This change of weight will eventually became undetectable (i.e., after about 5 days under these conditions), but the curing will continue long after that (i.e., it may extend over several months) as evidenced by a slow increase of the glass temperature (T_g) of the resulting network product, as shown in Figure 6. This behavior obviously results from an increase in viscosity (and eventual solidification) of the cross-linking reaction mixture, but it also seems to indicate that reactivity of the condensing methoxysilyl- dendrimer end-groups is dependent on steric effects at the

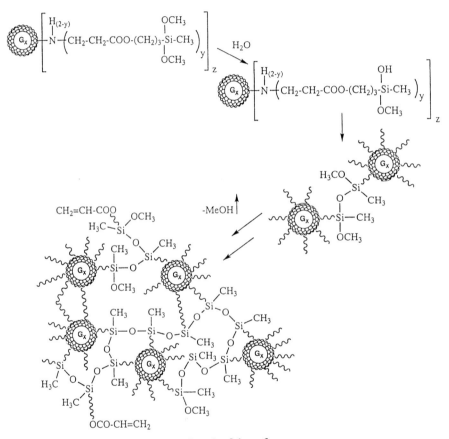

Reaction Scheme 3

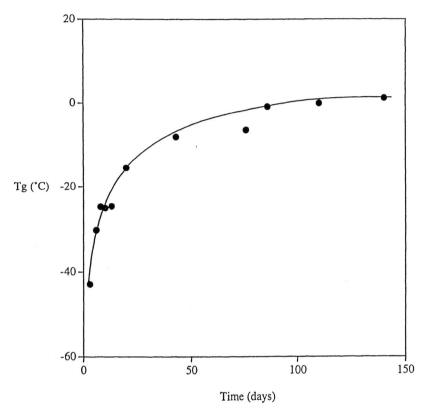

Figure 6: Effect of the time of curing at room temperature on the glass temperature (T_g) of the resulting PAMAMOS dendrimer-based network

outer dendrimer surface, and that not all of the end-groups are accessible to the cross-linking reaction. If so, this would represent another architecturally driven specific property of dendrimers, which in this particular case creates a situation favorable for *intra*dendrimer silanol condensation, leading to the formation of closed siloxane loops at the outer surfaces of individual dendrimer building blocks.

These PAMAMOS dendrimer-based networks are clear, transparent and colorless films which have unusually smooth surfaces (see Figure 7), and are insoluble in solvents, such as methanol, methylene chloride, THF, acetone or water. Depending on the dendrimer precursor used, and on its density of functionality, these films may be either elastomeric or plastomeric. The rate of cure can be accelerated by heating, which, of course, must not exceed the thermo-oxidative stability limit of the PAMAMOS dendrimer precursor. Efficient curing can be achieved at temperatures between about 60 and 100°C, with stable T_g products being obtained within several hours of curing time. In addition to this, if cross-linking is performed with as-obtained PAMAMOS dendrimers (i.e., without isolation or purification from the reaction mixture), the excess organosilicon acrylate reagent (see Reaction Scheme 2) that may have remained unreacted at the end of the PAMAMOS dendrimer synthesis will also incorporate into the network by condensation through its own methoxysilyl- groups. This yields acryl-functionalized networks (see Reaction Scheme 3), and opens up new synthetic possibilities for their further chemical modification. It is also interesting to note that regardless of the degree of cross-linking, the cure of PAMAMOS dendrimers is very efficient. For example, the weight % of extractables from a typical network, as determined by soxhlet extraction in methanol for three days, was only 6.5% of the weight of the network product.

Properties of the PAMAMOS Dendrimer-Based Networks

Thermal Properties. As expected, glass temperatures (T_g) of the PAMAMOS dendrimer-based networks depend significantly on the selected dendrimer precursor and the composition of the curing system. In general, PAMAMOS dendrimers containing only one layer of the organosilicon branch cells in their exteriors, show T_gs which are not very much different from those of the PAMAM dendrimers from which they have been derived (6,25). The T_gs decrease, however, with introduction of the second layer of organosilicon branch cells, reflecting an increase in the relative content of flexibilizing -Si-C- and/or -Si-O-Si- units. Thus, the thicker the OS exterior of the PAMAMOS dendrimer precursor, the lower its T_g and the more elastomeric the resulting network. In addition to this, the flexibility of these networks could be also affected by the composition of the curing system. For example, addition of $Si(OEt)_4$ into the curing system led to an increase of T_g from about 10-14°C for an unmodified network to about 37°C for the network obtained from a mixture containing 60 mole % of $Si(OEt)_4$ relative to the amount of methoxysilyl- end-groups present. Conversely, addition of α,ω-telechelic linear oligopolydimethylsiloxane having M_w of about 600, decreased the glass temperature of the resulting networks to -25°C.

Thermal and thermo-oxidative stability of the PAMAMOS dendrimer-based networks is predetermined by the stability of their less stable compositional parts, i.e., the PAMAM dendrimer interior. In nitrogen, thermal degradation occurs in two steps.

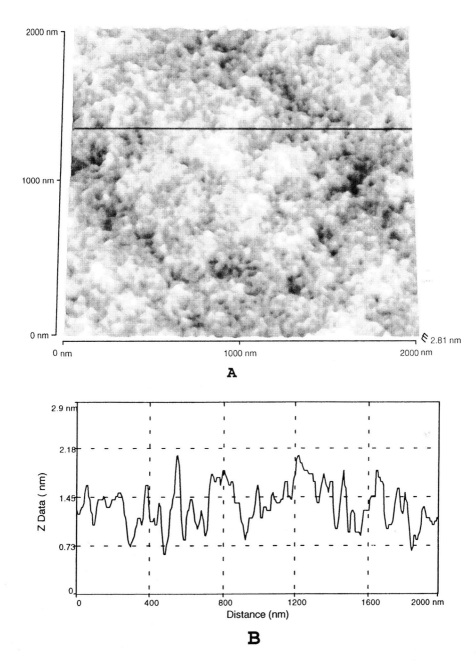

Figure 7: Tapping mode atomic force microscopy image of a PAMAMOS dendrimer-based network cast on mica surface (A), and the height profile associated with the line drawn on the image (B)

The first step typically starts at about 200°C, shows maximum rate of weight loss at about 300°C and results in the total loss of about 30 % of the original sample weight at about 350°C. The second step follows from about 350°C, shows maximum rate of weight loss at about 390°C and ends at about 600°C with the total weight loss of about 55 % of the original weight, i.e., the total remaining residue of about 15 % of the original sample weight.

Surface Properties. Quasi-equilibrium advancing contact angles for one of the PAMAMOS network samples showed for water, methylene iodide and n-hexadecane, the values of 109, 74 and 31 deg, respectively. These translate to a solid surface energy of 23.1 mN/m using the geometric mean Owens and Wendt approach (46), which is close to the value obtained for an all-methyl surface such as that provided by conventional polydimethylsiloxane (22.8 mN/m) or paraffin wax (25.4 mN/m). This indicates that the surface consisted of methyl groups on silicon from the PAMAMOS end-groups, and also suggests that any nano-roughness deriving from the dendritic structure is of too small a scale to affect the measured contact angles. Atomic force microscopy (AFM) further confirmed this conclusion, showing maximum nano-roughness of less than 2 nm for a sample cast on the mica support (see Figure 7).

The same elastomeric dendrimer film was also characterized by scanning electron microscopy (SEM), energy dispersive spectroscopy (EDS) and electron spectroscopy for chemical analysis (ESCA). SEM (see Figure 8) showed fairly featureless, uniform cross-sectional organization across the network sample, as expected for nanoscopic sizes of the domains involved. EDS (see Figure 9), as well as ESCA confirmed the presence of carbon, oxygen and silicon in the surface region. The C 1s spectrum (see Figure 10) showed five different states, corresponding to the following peaks: 285 eV for CH_3 (from Si-CH_3 end-groups in OS exterior layers) and CH_2 (from both PAMAM and OS); 286 eV for C-N and 289 eV for N(H)-C=O from PAMAM; as well as 287 eV for C-O and 290 eV for C(O)O from OS. Thus, carbons from both the organosilicon part and the PAMAM dendrimer interior were contributing to this spectrum.

Permeability. Permeability of the PAMAMOS dendrimer-based networks was examined for pure water, and water solutions of NaCl and methylene blue. To pure water, the networks were found quite permeable, with the permeate flow flux decreasing with increasing degree of organosilicon substitution (i.e., OS coverage of the PAMAM interior) in the PAMAMOS dendrimer precursor used for film preparation. This behavior probably reflects an increase in the number of organosilicon end-groups or *intra*dendrimer siloxane loops, at the PAMAMOS outer surface (due to the branching functionality of silicon junctures equal to 2), as well as an increase in the degree of network cross-linking through the formation of *inter*dendrimer -Si-O-Si- bridges.

However, when tested with NaCl and methylene blue water solutions, these PAMAMOS dendrimer-based networks behaved completely differently. For example, while permeability was excellent for sodium chloride dissolved in water, the films turned out practically impenetrable for methylene blue. The latter showed strong tendency towards complexation and encapsulation into the hydrophilic PAMAM domains, as manifested by coloration of the resulting samples.

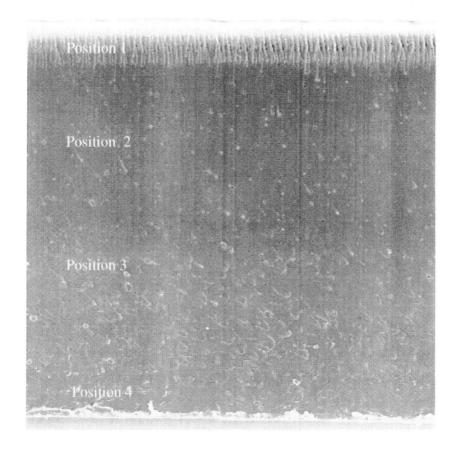

Figure 8: Scanning electron microscopy image of a cross-cut of a film of PAMAMOS dendrimer-based network. Magnification 190 X (10 kV)

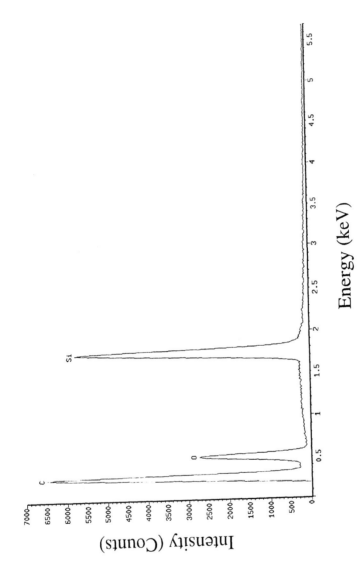

Figure 9: SEM/EDS spectrum of a PAMAMOS dendrimer-based network surface

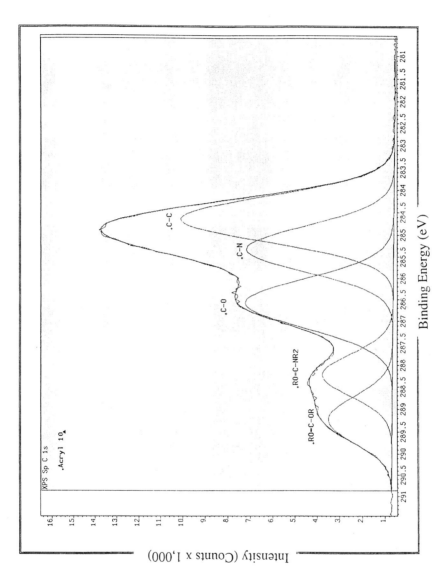

Figure 10: C 1s high resolution ESCA spectrum of a PAMAMOS dendrimer-based network surface

Dendrimer-Based "Molecular Sponges", Nanoscopic "Reactors" and Inorganic-Organic Composites

Observation that methylene blue was easily complexed and entrapped into the PAMAM domains of the PAMAMOS dendrimer-based networks, prompted a series of similar experiments with a variety of different electrophiles, including inorganic salts and water soluble organic dyes. The former included various salts of Ag^+, Cu^+, Cu^{2+}, Ni^{2+}, Cd^{2+}, Fe^{2+}, Fe^{3+}, Au^{3+}, Rh^{3+}, etc., while the latter involved methyl red and green ink. It was found that in all cases examined, these PAMAMOS dendrimer-based networks easily (i.e., by simple immersion into water solutions of these solutes) absorbed and retained (when dried) electrophilic solute species, most probably by ligating them to strongly nucleophilic nitrogen centers of the PAMAM domains (7,16-22) like unique "nanoscopic molecular sponges".

Furthermore, the absorbed cations could be chemically transformed while entrapped within the encapsulating PAMAM domains of these dendrimer-based networks, which served as "confined nanoscopic reactors". For example, reactions of complexed Cu^{2+} and Ni^{2+} with H_2S led to the formation of corresponding metal sulfides, while reduction of Ag^+ and Au^{3+} yielded elemental metals encapsulated inside the dendrimer-network domains.

Conclusions

Radially layered copolymeric PAMAMOS dendrimers and their networks described in this chapter represent truly unique new materials which can be prepared in a variety of different chemical compositions. They consist of hydrophilic PAMAM and hydrophobic (i.e., oleophilic) OS nanoscopic domains, the sizes of which can be precisely controlled by the selection of the reagents used for the dendrimer synthesis and by the chemistry involved. Hence, this synthetic strategy is highly versatile and it enables close control of the resulting structure(s), permiting precise tailor-making of these unique new materials.

The amphiphilic PAMAMOS copolymeric dendrimers may be viewed as globular, covalently bonded "inverted micelles" which show highly unusual surface properties. Among others, these properties include the following. They form liquid-like films capable of sustaining much higher surface pressures than linear polydimethylsiloxane. Values of over 40 mN/m are readily attained, providing a marked example of the impact of macromolecular architecture on surface properties of organosilicon-containing polymers. Surface areas obtained from Langmuir trough studies are in good agreement with expected PAMAMOS dendrimer sizes, with increases in area occurring both with increasing degree of organosilicon substitution of the parent PAMAM dendrimer surface groups and with increasing PAMAM generation number.

The PAMAMOS dendrimer-based networks can be prepared in a variety of different chemical compositions as elastomers, plastomers or coatings, depending on the particularly desired end-application(s). They also contain well defined (in size and shape) hydrophilic PAMAM and oleophilic (hydrophobic) OS nanoscopic domains which are covalently interconnected in a three-dimensional network continuum. These tractable dendrimer-based materials have unusually smooth surfaces, that are hydrophobic when in contact with air, but permeable when exposed to liquid water. The presence of organosilicon domains imparts useful mechanical properties to these dendrimer-based networks, while the presence of hydrophilic and nucleophilic PAMAM domains provides unique ability to attract and encapsulate various electrophiles, such as metal ions or water soluble organic and organometallic molecules. The confined nanoscopic PAMAM domains can encapsulate electrophilic "guests" and serve as nano-scaled reactors for the chemical transformations of the latter, resulting in unique elastomeric or plastomeric inorganic- or organic-organosilicon nano-composites.

In conclusion, these unprecedented properties of PAMAMOS dendrimers and their networks, clearly result from the synergism of two distinctly different chemical compositions and unique dendrimer molecular architecture. These materials provide provoking new possibilities for applications in various fields, including optoelectronics, protective coatings, separation processes, etc.

Acknowledgments

Financial suport for this program was provided by Dow Corning Corporation and Dendritech Inc., both of Midland, MI. Contributions of Drs. Douglas Swanson and Lajos Balogh, both of Michigan Molecular Institute, and very helpful discussions with and suggestions by Dr. Ralph Spindler also of Michigan Molecular Institute, are gratefully acknowledged. We also thank Drs. Dale Meier and Jing Li of Michigan Molecular Institute for providing atomic force microscopy images, and Dr. Chris McMillan of Dow Corning Corporation for ESCA assistance.

Literature Cited

1. Dvornic, P. R.; Tomalia, D. A., *Sci. Spectra*, **1996**, 5, 36.
2. Dvornic, P. R.; Tomalia, D. A., *Curr. Opin. Coll. Interf. Sci.*, **1996**, *1*, 221.
3. Hawker, C. J.; Fréchet, J. M. J., *J. Am. Chem. Soc.*, **1992**, *114*, 8405.
4. de Leuze-Jallouli, A. M.; Swanson, D. R.; Perz, S. V.; Owen, M. J.; Dvornic, P. R., *Polym. Mater. Sci. Eng.*, **1997**, *77*, 67.
5. de Leuze-Jallouli, A. M.; Swanson, D. R.; Dvornic, P. R.; Perz, S. V.; Owen, M. J., *Polym. Mater. Sci. Eng.*, **1997**, *77*, 93.
6. Tomalia, D. A.; Dvornic, P. R., in *"Polymeric Materials Encyclopedia"*, Salamone, J. C., Ed., CRC Press, Boca Raton (FL), 1996, Vol. 3, pp. 1814-1830.
7. Tomalia, D. A.; Naylor, A. M.; Goddard III, W. A., *Angew. Chem. Int. Ed. Engl.*, **1990**, *29*, 138.
8. Tomalia, D. A.; Durst, H. D., *Topics Curr. Chem.*, **1993**, *165*, 193.
9. Newkome, G. R.; Moorefield, C. N.; Vögtle, F., *"Dendritic Molecules. Concepts, Synthesis and Perspectives"*, VCH Verlag, Weinheim, 1996.
10. Prosa, T.J.; Bauer, B. J.; Tomalia, D. A.; Scherrenberg, R., *J. Polym. Sci., Part B: Polym. Phys.*, **1997**, *35*, 2913,
11. Valachovic, D. E.; Bauer, B. J.; Amis, E. J.; Tomalia, D. A., *Polym. Mater. Sci. Eng.*, **1997**, *77*, 230.

12. Naylor, A. M.; Goddard, III, W. A.; Kiefer, G. E.; Tomalia, D. A., *J. Am. Chem. Soc.*, **1989**, *111*, 2339.
13. Mattice, W. L., *"Masses, Sizes and Shapes of Macromolecules from Multifunctional Monomers"*, Chapter 1 in *"Dendritic Molecules. Concepts, Synthesis, Perspectives"*, Newkome, G. R.; Moorefield, C. N.; Vögtle, F., Eds., VCH Verlag, Weinheim, 1996.
14. Dvornic, P. R.; Uppuluri, S.; Tomalia, D. A., *Polym. Mater. Sci. Eng.*, **1995**, *73*, 131.
15. Uppuluri, S.; Tomalia, D. A.; Dvornic, P. R., *Polym. Mater. Sci. Eng.*, **1997**, *77*, 116.
16. Ottaviani, M. F.; Bossmann, S.; Turro, N. J.; Tomalia, D. A., *J. Am. Chem. Soc.*, **1994**, *116*, 661.
17. Jansen, J. F. G. A.; de Brabander-van der Berg, E. M. M.; Meijer, E. W., *Science*, **1994**, *266*, 1226.
18. Newkome, G. R.; Moorefield, C. N.; Keith, J. M.; Baker, G. R.; Escamilla, G. H., *Angew. Chem. Int. Ed. Engl.*, **1994**, *33*, 666.
19. Balogh, L.; Swanson, D. R.; Spindler, R.; Tomalia, D. A., *Polym. Mater. Sci. Eng.*, **1997**, *77*, 118.
20. Beck Tan, N.; Balogh, L.; Trevino, S., *Polym. Mater. Sci. Eng.*, **1997**, *77*, 120.
21. Dvornic, P. R.; de Leuze-Jallouli, A. M.; Owen, M. J.; Perz, S. V. in *"Silicones in Coatings II"*, Paint Research Association, London, UK, 1998.
22. Dvornic, P. R.; de Leuze-Jallouli, A. M.; Owen, M. J.; Perz, S. V., *Polym. Preprints*, **1988**, *39(1)*, 473.
23. Dvornic, P. R.; Tomalia, D. A., *Macromol. Symp.*, **1994**, *88*, 123.
24. Dvornic, P. R.; Tomalia, D. A., *J. Serb. Chem. Soc.*, **1996**, *61*, 1039.
25. Uppuluri S., Ph.D. Thesis, Michigan Technological University, Houghton, 1997.
26. Fréchet, J. M. J.; Hawker, C. J.; Wooley, K. L., *J. Macromol. Sci., Pure Appl. Chem.*, **1994**, *A31(11)*, 1627.
27. Tomalia, D. A.; Dvornic, P. R.; Uppuluri, S.; Swanson, D. R.; Balogh, L., *Polym. Mater. Sci. Eng.,* **1997**, *77*, 95.
28. Hawker, C. J.; Malmström, E. E.; Frank, C. W.; Kampf, P. J.; Mio, C.; Prausnitz, J., *Polym. Mater. Sci. Eng.*, **1997**, *77*, 61.
29. Mathias, L. J.; Carothers, T. W., in *"Advances in Dendritic Macromolecules"*, Newkome, G. R., Ed.; JAI Press Inc., Greenwich (CT), 1995, Vol. 2, pp. 101-121.
30. Rebrov, E. A.; Muzafarov, A. M.; Papkov, V. S.; Zdanov, A. A.; *Dokl. Akad. Nauk. SSSR*, **1989**, *309*, 376.
31. van der Made, A. W.; van Leeuwen, P. W. N. M., *J. Chem. Soc., Chem. Commun.*, **1992**, 1400.
32. van der Made, A. W.; van Leeuwen, P. W. N. M.; de Wilde, J. C.; Brandes, R. A. C., *Adv. Mater.*, **1993**, *5*, 466.
33. Zhou, L.-L.; Roovers, J., *Macromolecules*, **1993**, *26*, 963.
34. Lambert, J. B.; Pflug, J. L.; Stern, C. L., *Angew. Chem. Int. Ed. Engl.*, **1995**, *34*, 98.
35. Sekiguchi, A.; Nanjo, M.; Kabuto, C.; Sakurai, H., *J. Am. Chem. Soc.*, **1995**, *117*, 4195.
36. Suzuki, H.; Kimata, Y.; Satoh, S.; Kuriyama, A., *Chem Lett.*, **1995**, 293.

37. Nanjo, M.; Sekiguchi, A., *Organometallics*, **1998**, *17*, 492.
38. Hawker, C. J.; Fréchet, J. M. J., *"Three-Dimensional Dendritic Macromolecules: Design, Sybthesis and Properties"*, Chapter 8 in *"New Methods of Polymer Synthesis"*, Ebdon, J. R.; Eastmond, G. C., Eds., Blackie Academic and Professional, London, UK, 1995, Vol. 2, pp. 290-330.
39. Tomalia, D. A.; Dewald, J. R.; Hall, M. J.; Martin, S. J.; Smith, P. B., *"Preprints 1st Soc. Polym. Sci. Japan Int. Polym. Conf."*, Kyoto, Japan, August 20-24, 1984.
40. Tomalia, D. A.; Baker, H.; Dewald, J. R.; Hall, M. J.; Kallos, G.; Martin, S. J.; Roeck, J.; Ryder, J.; Smith, P. B., *Polymer J. (Tokyo)*, **1985**, *17*, 117.
41. Tomalia, D. A,; Baker, H.; Dewald, J. R.; Hall, M. J.; Kallos, G.; Martin, S. J.; Roeck, J.; Ryder, J.; Smith, P. B., *Macromolecules*, **1986**, *19*, 2466.
42. de Leuze-Jallouli, A. M.; Dvornic, P. R.; Perz, S. V.; Owen, M. J., *Polym. Preprints*, **1998**, *39(1)*, 475.
43. Dvornic, P. R.; de Leuze-Jallouli, A. M.; Swanson, D.; Owen, M. J.; Perz, S. V., *U.S. Patent* 5,739,218, 1998.
44. Noll, W.; Steinbach, H.; Sucker, C., *Progr. Colloid Polym. Sci.*, **1971**, *55*, 131.
45. Elissen-Roman, C.; van Hest, J. C. M.; Baars, M. W. P. L.; van Genderen, M. H. P.; Meijer, E. W., *Polym. Mater. Sci. Eng.*, **1997**, *77*, 145.
46. Owens, D. K.; Wendt, R. C., *J. Appl. Polym. Sci.*, **1969**, *13*, 1741.

Chapter 17

Organic–Inorganic Hybrid Polymers from Atom Transfer Radical Polymerization and Poly(dimethylsiloxane)

Krzysztof Matyjaszewski, Peter J. Miller, Guido Kickelbick,
Yoshiki Nakagawa, Steven Diamanti, and Cristina Pacis

Department of Chemistry, Carnegie Mellon University,
4400 Fifth Avenue, Pittsburgh, PA 15213

> A summary of the synthesis of hybrid materials composed of inorganic siloxanes and polymers prepared by atom transfer radical polymerization (ATRP) is given. Hydrosilation of vinyl or hydrosilyl terminal and pendant poly(dimethylsiloxane) (PDMS) with an attachable initiator yielded macroinitiators for ATRP. Polymerization of styrene and (meth)acrylates resulted in triblock copolymers with increased molecular weights and reduced polydispersities. Low polydispersity living anionic PDMS was terminated with an attachable initiator containing a benzyl chloride moiety. ATRP of styrene from such a monofunctional macroinitiator resulted in formation of a block copolymer with well-defined segments. A tetrafunctional initiator was synthesized from a cyclotetrasiloxane core toward formation of star polymers by ATRP. Finally, a silsesquioxane initiator was used in styrene ATRP yielding a polymer with a bulky inorganic tail group.

As technology in a variety of fields improves, the need for specialized high performance materials becomes a necessity. One major goal is the combination of properties not accessible by one kind of material. For example, in microlithography and electronic applications, a combination of low dielectric properties and mechanical strength is required while in medicine, oxygen permeability and biocompatibility is paramount. The list of examples is ever increasing. To fulfill these requirements scientists and engineers are using copolymers of varying compositions and/or architectures to obtain the desired properties. One such set of copolymers that is gaining interest are the inorganic / organic hybrid materials.

Inorganic polymers generally have specific properties which organic analogues do not possess, making them desirable in hybrid materials. Polyphosphazenes – which have high thermal stability and biocompatibility – and polysilylenes – possessing photoconductive, photorefractive and nonlinear optical properties – are two common examples (*1*). However, the most widely studied inorganic polymers are the polysiloxanes. Poly(dimethylsiloxane) (PDMS) has a high oxygen permeability, chain flexibility and thermal conductance, making it attractive for a variety of applications ranging from biomedicine to thermal transfer fluid technologies (*1*). Unfortunately, the homopolymer of PDMS is unattractive from a mechanical standpoint due to its low dimensional stability. Therefore, the copolymerization of PDMS with tougher materials can lead to improved properties. In many cases, organic polymers have been found good candidates to fill this role. Typically, polysiloxanes are synthesized by ionic mechanisms leading to polymers with a high degree of terminal functionality (*1*).

Materials related to polysiloxanes are silsesquioxanes (Scheme 1) (*2*). These cubic siloxane molecules have been produced with a number of functional species such as styrene and methacrylate moieties protruding from one corner of the structure (*3, 4*). Classic free radical polymerization of these molecules has been shown to produce an inorganic / organic hybrid homopolymer. Furthermore, cubes have been synthesized with pendant silyl hydrido and silanolate moieties which can be used for further functionalization (*5*).

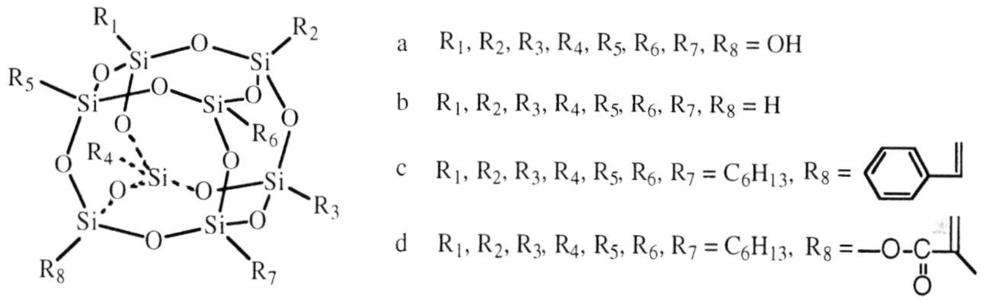

a $R_1, R_2, R_3, R_4, R_5, R_6, R_7, R_8 = OH$

b $R_1, R_2, R_3, R_4, R_5, R_6, R_7, R_8 = H$

c $R_1, R_2, R_3, R_4, R_5, R_6, R_7 = C_6H_{13}$, $R_8 =$ (styryl group)

d $R_1, R_2, R_3, R_4, R_5, R_6, R_7 = C_6H_{13}$, $R_8 = -O-C(=O)-$ (methacrylate)

Scheme 1

Organic polymers have been more widely studied than the inorganic analogues. In terms of block copolymers functionality is again paramount. Typically, this functionality is achieved through living ionic polymerizations. However, the underlying mechanisms rely on stringent reagent purity and are limited to a select number of monomers. Recently, controlled free radical polymerization has provided an alternative to the ionic techniques. In particular, atom transfer radical polymerization (ATRP) is most effective (*6, 7*). The method utilizes a dynamic equilibrium between active ($P_n\cdot$) and dormant (P_n–X) radical species facilitated by halogen atom (X) transfer mediated by a transition metal (Mt) species (Scheme 2) (*8*). For polymerizations that exhibit first order kinetic consumption of monomer and molecular weights predetermined by the ratio of consumed monomer to initially infused initiator, the value of the equilibrium constant must be sufficiently low such

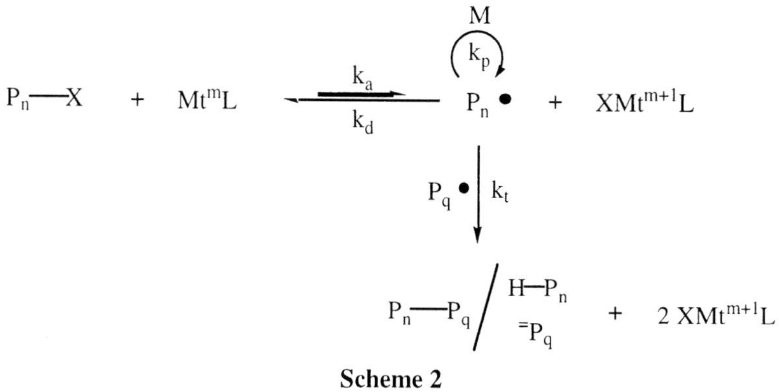

$$P_n\text{—}X + Mt^mL \underset{k_d}{\overset{k_a}{\rightleftarrows}} P_n\cdot \;(\overset{M}{\underset{k_p}{\circlearrowleft}}) + XMt^{m+1}L$$

$$P_q\cdot \;\Big|\; k_t$$

$$P_n\text{—}P_q \;/\; \begin{matrix} H\text{—}P_n \\ =P_q \end{matrix} + 2\,XMt^{m+1}L$$

Scheme 2

that the steady-state radical concentration is on the order of that observed in conventional free radical systems. This low radical concentration limits termination to negligible levels. Narrow molecular weight distributions ($M_w/M_n < 1.3$) are obtained when the rates of initiation and deactivation are greater than or equal to that of propagation. ATRP has been shown to be effective for polymerization of styrenes (*6, 8-10*), acrylates (*6*), methacrylates (*6, 11-16*), and acrylonitrile (*17*) using metals such as copper (*6*), iron (*12*), ruthenium (*11*) and nickel (*18*).

The combination of control over chain length and functionality has allowed for the synthesis of numerous (co)polymers exhibiting a variety of compositions and architectures. Figure 1 demonstrates the general form of said structures. Linear, diblock, triblock, graft, network, star and hyperbranched polymers can all be synthesized with relative ease (*19, 20*). Since all that is needed to initiate ATRP are activated alkyl halides (*21-23*), a number of research groups have synthesized segments by different mechanisms (ionic, conventional free radical, step growth, etc.) as macroinitiators for the ATRP of vinyl monomers in the synthesis of block (*24-34*) and graft (co)polymers (*34-39*). This combination of mechanisms with ATRP is not limited to strictly organic segments. Two primary examples resulting in inorganic / organic hybrid materials are the grafting of polystyrene from a poly(methylphenylsilylene) backbone (*36*) and growth of polyacrylamide from a modified silica surface (*40*).

This paper will summarize work performed in our laboratory on the synthesis of hybrid materials using cyclic oligo- and linear polysiloxanes in conjunction with polystyrenes, acrylates and methacrylates synthesized by ATRP. Architectural variation will also be discussed including linear, graft and star polymers.

Results and Discussion

ABA Triblock Copolymers. There has been considerable interest in the use of poly(dimethylsiloxane) in di- and triblock copolymers for applications such as thermoplastic elastomers and pressure sensitive adhesives (*41-44*). In terms of inorganic / organic hybrids, copolymers have most often consisted of polystyrene and PDMS. For example, a diblock copolymer of poly(styrene-*b*-dimethylsiloxane) was produced by polymerization of hexamethylcyclotrisiloxane (D3) from living polystyryl lithium (*45*). Here, ABA triblock copolymers could not be produced by subsequent addition of styrene monomer to the living lithium silanolate solution due to the inability of the active species to reinitiate styrene polymerization. However, in other contributions triblock copolymers were prepared by coupling reactions using hydrosilation techniques (*46*) or by condensation of silanolates (*47, 48*) with organic polymers containing reactive silyl chlorides (*49*). In these cases, block copolymers of high purity and yield were difficult to obtain due to the need for exact stoichiometry and reagent purity required in the coupling of two macromolecular species.

In terms of free radical chemistry, the best examples of inorganic / organic linear hybrid materials is the use of PDMS containing in-chain silylpinacolate moieties to initiate styrene and methacrylate polymerizations (*41, 42*). Due to the predominance of termination by radical coupling over disproportionation in styrene polymerization, segmental copolymers were produced. However, the polymerizations were not "living" and polydispersities were often greater than 3. Azo terminal PDMS macroinitiators have also been reported (*44*). Here, formation of a mixture of block and homopolymers resulted due to initiation from both sides of the homolytically cleaved initiating species (*44*).

In our laboratory the motivation for the synthesis of di- and triblock copolymers was three-fold: 1) synthesis of the organic blocks by a "living" polymerization technique such that segment length and copolymer composition could be controlled, 2) prevention of coupling of macromolecular species and 3) use of a technique that would allow for the synthesis of a plethora of copolymers based on variation of the organic monomer. Considering these qualifications, ATRP of vinyl monomers from mono- and difunctional PDMS macroinitiators was used. Scheme 3 illustrates the synthesis of ABA triblock copolymers from terminal functionalized PDMS. The method utilizes hydrosilation of attachable initiators to commercially available vinyl

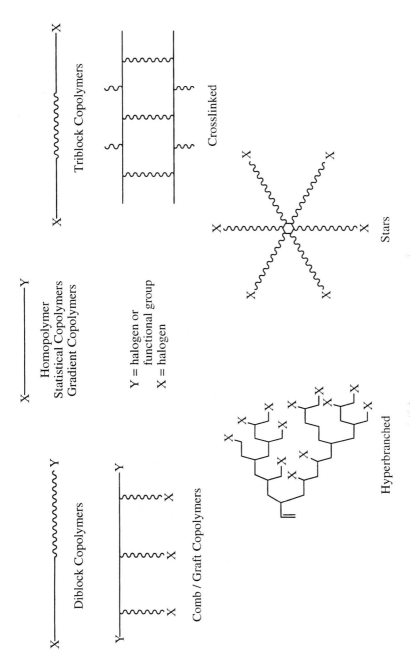

Figure 1. Summary of architectures available through ATRP.

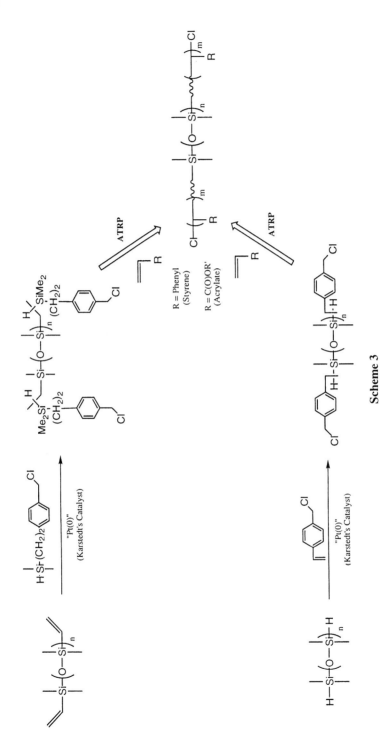

Scheme 3

or hydrosilyl terminal PDMS. The benzyl chloride moieties of the attachable initiators are unreactive toward the platinum catalyst and remain intact following attachment to the inorganic macromolecule. Quantitative coupling is confirmed by ^{1}H NMR where – depending on the functionality of the PDMS – either vinyl (5.5 - 7.0 ppm) or hydrosilyl (4.5 ppm) protons completely disappear concomitant with the appearance of the benzyl chloride methylene protons at 4.7 ppm.

ATRP of styrene was performed from the PDMS macroinitiator described above. The reaction was performed under homogeneous conditions using copper chloride complexed with di-(5-nonyl)bipyridyl (dNbpy) ligands at 130 °C in 50% *p*-xylenes. The kinetic and molecular weight plots are illustrated in Figures 2 and 3 respectively. The linear first order consumption of monomer as a function of time demonstrates conservation of active species throughout the reaction. The molecular weight plot (Figure 3) shows a monotonous increase in chain length with monomer conversion. The deviation in measured molecular weight from that predicted by the ratio of monomer to macroinitiator is currently under investigation but is most likely due to differences in hydrodynamic volume between the triblock copolymer and the SEC calibration curve constructed from linear polystyrene standards (THF mobile phase, differential refractive index detection). The purified poly(styrene-*b*-dimethylsiloxane-*b*-styrene) was isolated as a white powder.

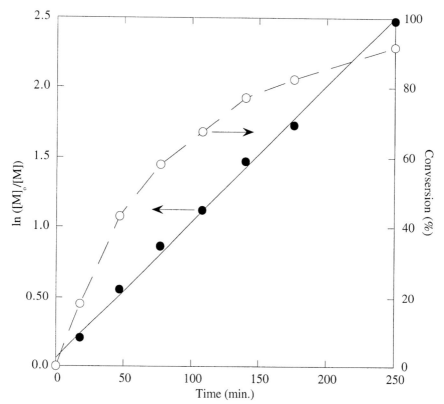

Figure 2. Kinetic plot for the ATRP of styrene from a difunctional PDMS macroinitiator. Conditions: $[M]_o/[I]_o/[CuCl(dNbpy)_2]_o$ = 29:1:0.5, 50% *p*-xylene, 130 °C.

Studies with other ATRP systems have indicated that while the benzyl chloride moiety will initiate polymerization of acrylates and methacrylates, it does not have the optimum ratio of k_p to k_i observed with other species such as 2-bromopropionates and 2-bromoisobutyrates. Therefore, an additional PDMS macroinitiator containing the 2-bromoisobutyryloxy moiety was synthesized by hydrosilation of hydrosilyl terminal silicone with allyl 2-bromoisobutyrate *(50)*. Again, ^{1}H NMR demonstrated quantitative conversion of the silyl hydride group at 4.5 ppm to the 2-bromoisobutyrate (ester methylene at 3.5 ppm, 2-bromoisobutyryl methyls at 1.9 ppm) moiety.

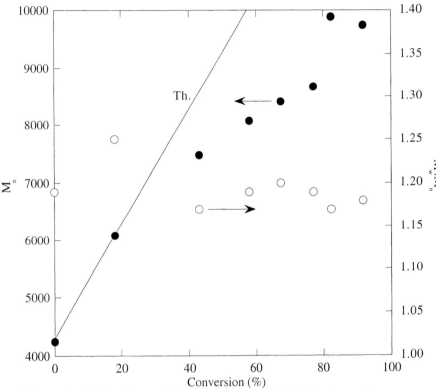

Figure 3. Molecular weight plot for the ATRP of styrene from a difunctional PDMS macroinitiator. Conditions: $[M]_o/[I]_o/[CuCl(dNbpy)_2]_o$ = 29:1:0.5, 50% *p*-xylene, 130 °C.

A series of polymerizations from the benzyl chloride and 2-bromoisobutyryloxy terminal PDMS macroinitiators (Scheme 4) was performed using a variety of monomers listed in Table 1. The table shows that with the benzyl chloride functionalized macroinitiators (**I** in Scheme 4) increased molecular weights were observed with ATRP of styrene, isobornyl acrylate and butyl acrylate. Furthermore, with the higher molecular weight macroinitiator significant decreases in polydispersity resulted indicating addition of segments to the copolymer under controlled conditions. 2-Bromoisobutyryloxy functionalized PDMS macroinitiators (**II** in Scheme 4) were also used to produce triblock copolymers. Again, increased molecular weights and decreased polydispersities were observed over that of the macroinitiators. The final entry, ATRP of HEMA-TMS, is interesting since, after deprotection of the alcohol, the product is an amphiphilic triblock copolymer composed of two biocompatible segments.

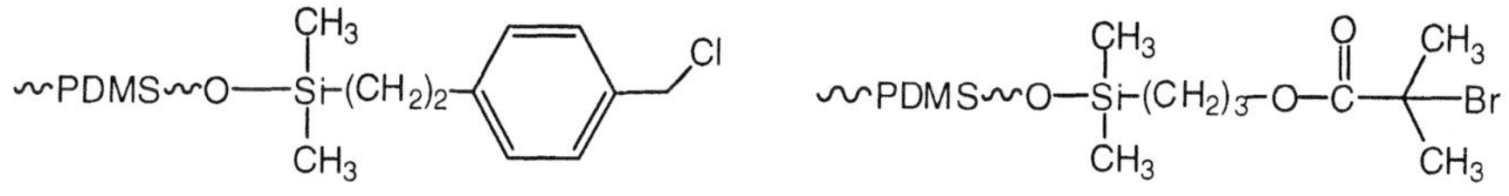

Scheme 4

Table I. ATRP of vinyl monomers from PDMS macroinitiators

PDMS[1]	M_n^2	PDI[3]	Monomer	Solvent	T (°C)	Conv.[4]	M_n^2	PDI[3]
I	4.5	1.2	Styrene	–	130	90	10.1	1.3
I	9.8	2.4	Styrene	Xylene	130	70	20.6	1.7
I	4.5	1.6	^{i}BnA[5]	–	90	53	12.5	1.6
I	9.8	2.4	^{n}BA[6]	DMB[7]	100	81	24.6	1.5
II	4.0	1.3	MMA[8]	Xylene	90	70	20.4	1.2
II	33.9	1.8	MMA	Xylene	90	34	89.6	1.3
II	33.9	1.8	HEMA-TMS[9]	DMB	80	11	62.8	1.6

[1]Macroinitiator depicted in Scheme 4, [2]x 10^{-3}, [3]M_w/M_n, [4]conversion (%), [5]isobornyl acrylate, [6]n-butyl acrylate, [7]1,4-dimethoxybenzene, [8]methyl methacrylate, [9]2-(trimethylsilyloxy)ethyl methacrylate

AB Diblock Copolymers. In an effort to quantify variables such as initiator functionality, ATRP initiation efficiency and copolymer structure-property relationships, it was necessary to synthesize PDMS macroinitiators of narrow molecular weight distribution regardless of the chain length of the silicone. This was achieved by anionic ring-opening polymerization (ROP) of D3. Shown schematically in Scheme 5, living anionic PDMS was terminated with an attachable initiator containing a silyl chloride moiety and a species (benzyl chloride or 2-bromoisobutyryloxy) capable of initiation in ATRP. ROP's of D3 followed by termination with either of the attachable initiators has been performed and ^{1}H NMR analysis of the purified products has demonstrated that the lithium silanolate reacts selectively with the silyl chloride over any other fragments of the molecules. Furthermore, in the case of benzyl chloride terminal PDMS, NMR also showed a good correlation between the benzyl methylene and butyl protons of the polymer head and tail groups respectively. ATRP of styrene from this macroinitiator (M_n = 3900, M_w/M_n = 1.14) resulted in a diblock copolymer with M_n = 7100, M_w/M_n = 1.19. Efforts are currently underway to examine ATRP from PDMS macroinitiators composed of either of the terminal "I" groups shown in Scheme 5 as a function of both silicone chain length and vinyl monomer incorporated into the copolymer.

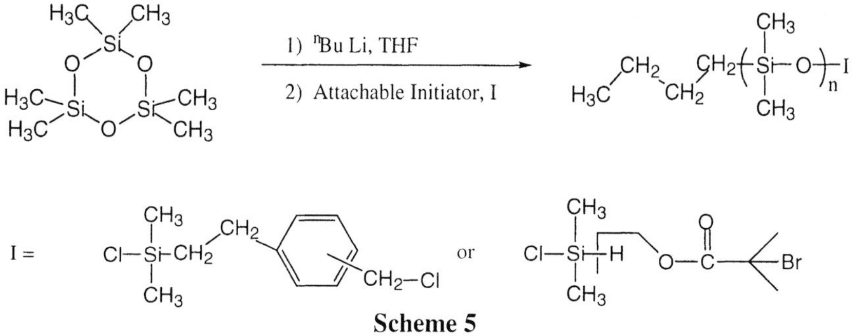

Scheme 5

Graft Copolymers. The hydrosilation technique described above for difunctional PDMS macroinitiators is also suitable for pendant functional polymers. For example, commercially available poly(dimethylsiloxane-*stat*-vinylmethylsiloxane) was reacted with 2-(4-chloromethylphenyl)ethyldimethylsilane in a hydrosilation analogous to the top reaction shown in Scheme 3 to yield PDMS containing pendant benzyl chloride moieties. ATRP of styrene proceeded with first order kinetic consumption of monomer. An increase in molecular weight was observed from M_n = 6600 to 14,800. Unlike the triblock copolymers, here the polydispersity increased from 1.8 to 2.1 due to an inconsistent number of initiating sites per PDMS chain. Again, the product was isolated as a soluble white powder after purification (*34*).

Star Polymers. Yet another example of the utility of hydrosilation reactions is the synthesis of multifunctional initiators from inorganic heterocycles (*51*). Shown in Scheme 6 is the conversion of 1,3,5,7-tetramethylcyclotetrasiloxane into an ATRP initiator containing four benzyl chloride moieties. ATRP of styrene at 130 °C from these initiators results in first order kinetics and linear growth of molecular weight with conversion. The polydispersity is low throughout the polymerization (<1.4) and the SEC curves illustrated for three of the samples withdrawn at 19, 44, and 72% conversion are symmetrical (Figure 4). Only after long reaction time is a low molecular weight shoulder observed in the SEC traces – most likely due to thermal self-initiation of styrene which then reacts with copper(II) chloride in the system to commence controlled growth of a monofunctional macromolecule.

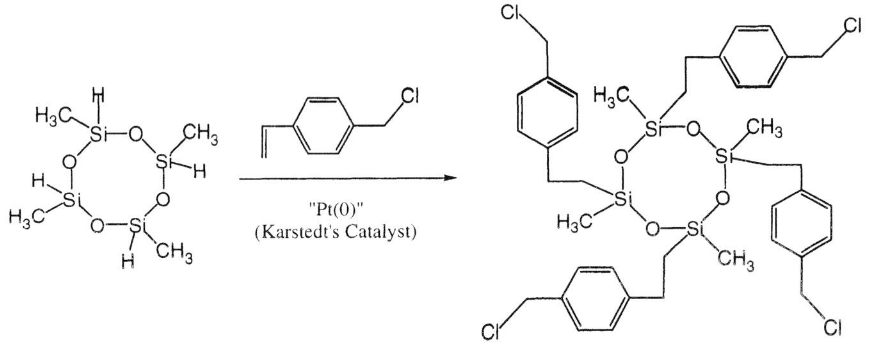

Scheme 6

ATRP from Silsesquioxane Initiators. In addition to silsesquioxanes with vinyl based monomeric units attached to corner silicon atoms, other species can be bonded to the cubes. One such example is the silsesquioxane shown in Scheme 7. Termed benzyl chloride functionalized polyhedral oligomeric silsesquioxane (benzyl chloride-POSS), the molecule contains seven cyclopentyl groups, which provide increased solubility in organic solvents, and one benzyl chloride moiety which can serve as an initiator for ATRP. Studies of such initiators may be of interest due to the bulky nature of the initiating species. Furthermore, the high temperature stability of the inorganic cubes may lend some interesting thermal properties to polymers initiated from it.

The polymerization of styrene from benzyl chloride-POSS was run in 50% *p*-dimethoxybenzene at 120 °C using a copper chloride / N,N,N',N'',N''-pentamethyldiethylenetriamine (PMDETA) catalyst system. The kinetic and molecular weight plots are illustrated in Figures 5 and 6 respectively. They show that the concentration of active species was conserved throughout the polymerization in the absence of transfer reactions, and that the condition of fast initiation is not hampered by the inorganic fragment.

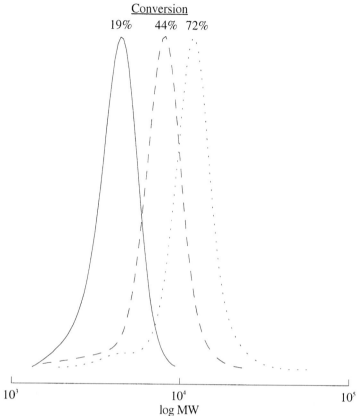

Figure 4. SEC traces of the ATRP of styrene initiated from a tetrafunctional cyclotetrasiloxane. Conditions: $[M]_o/[I]_o/[CuCl(dNbpy)_2]_o$ = 384:1:4, bulk, 130 °C.

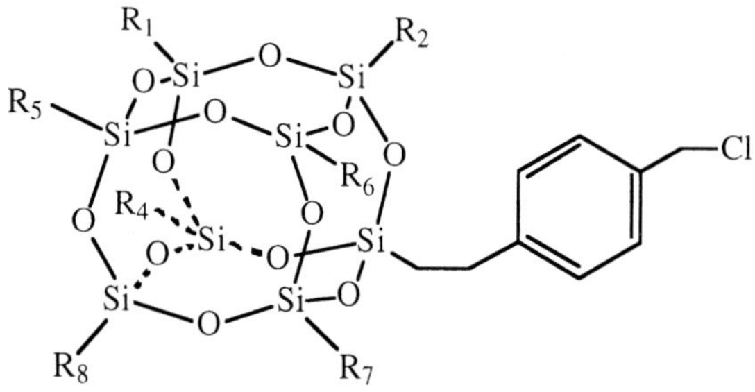

R = Cyclopentyl

Scheme 7

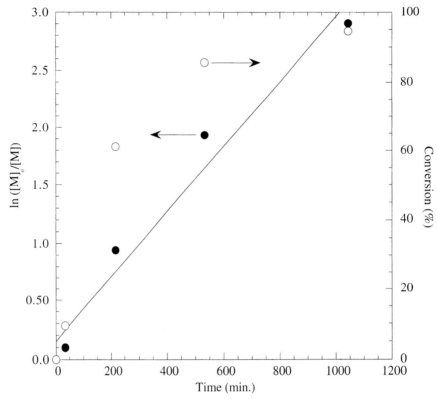

Figure 5. Kinetic plot for the ATRP of styrene initiated from benzyl chloride-POSS. Conditions: $[M]_o/[I]_o/[CuCl(PMDETA)]_o = 100:1:0.5$, 50% p-dimethoxybenzene, 80 °C.

Conclusions

Atom transfer radical polymerization from a variety of small molecule and polymeric siloxane bearing (macro)initiators can be used to create a plethora of inorganic / organic hybrid materials. Triblock copolymers have been synthesized from difunctional PDMS macroinitiators of predetermined composition with narrow polydispersities. In an analogous fashion, organic polymeric grafts were initiated from a silicone backbone. Similarly, linear and star polymers were synthesized from inorganic cubic and cyclic molecules respectively. All of these polymerizations demonstrate the versatility of both ATRP as a free radical polymerization technique and siloxanes as conjugates in inorganic / organic hybrid materials.

Acknowledgments The authors would like to thank Dr. S. Rubinstajn of GE Corporation for donation of several of the terminal / pendant vinyl and hydrosilyl functionalized PDMS samples and Dr. J. Schwab of Ratheon for donation of the benzyl chloride-POSS initiator. The following funding sources are acknowledged for their financial support: the industrial sponsors of the CMU ATRP Consortium, Kaneka Corporation for YN, Fonds zur Förderung der wissenschaftlichen Forschung, Austria (#J01423-CHE) for GK and NSF for PJM.

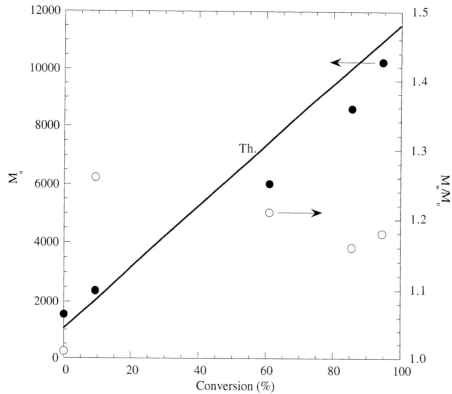

Figure 6. Molecular weight plot for the ATRP of styrene initiated from benzyl chloride-POSS. Conditions: $[M]_o/[I]_o/[CuCl(PMDETA)]_o$ = 100:1:0.5, 50% p-dimethoxybenzene, 80 °C.

References

1) Mark, J. E.; Allcock, H. R.; West, R. *Inorganic Polymers* Prentice Hall: Englewood Cliffs, 1992.
2) Harrison, P. G. *J. Organomet. Chem.* **1997**, *542*, 141.
3) Lichtenhan, J. D.; Otonari, Y. A.; Carr, M. J. *Macromolecules* **1995**, *28*, 8435.
4) Haddad, T. S.; Lichtenhan, J. D. *Macromolecules* **1996**, *29*, 7302.
5) Hoebbel, D. *Z. Anorg. Allg. Chem.* **1971**, *43*, 384.
6) Wang, J.-S.; Matyjaszewski, K. *J. Am. Chem. Soc.* **1995**, *117*, 5614.
7) Sawamoto, M.; Kamigaito, M. *Trends Polym. Sci.* **1996**, *4*, 371.
8) Patten, T. E.; Xia, J.; Abernathy, T.; Matyjaszewski, K. *Science* **1996**, *272*, 866.
9) Matyjaszewski, K.; Patten, T. E.; Xia, J. *J. Am. Chem. Soc.* **1997**, *119*, 674.
10) Qiu, J.; Matyjaszewski, K. *Macromolecules* **1997**, *30*, 5643.
11) Kato, M.; Kamigaito, M.; Sawamoto, M.; Higashimura, T. *Macromolecules* **1995**, *28*, 1721.
12) Matyjaszewski, K.; Wei, M.; Xia, J.; McDermott, N. E. *Macromolecules* **1997**, *30*, 8161.
13) Grimaud, T.; Matyjaszewski, K. *Macromolecules* **1997**, *30*, 2216.

14) Haddleton, D. M.; Jasieczek, C. J.; Hannon, M. J.; Shooter, A. J. *Macromolecules* **1997**, *30*, 2190.
15) Haddleton, D. M.; Crossman, M. C.; Hunt, K. H.; Topping, C.; Waterson, C.; Suddaby, K. G. *Macromolecules* **1997**, *30*, 3992.
16) Wang, J.-L.; Grimaud, T.; Matyjaszewski, K. *Macromolecules* **1997**, *30*, 6507.
17) Matyjaszewski, K.; Jo, S. M.; Paik, H.-J.; Gaynor, S. G. *Macromolecules* **1997**, *30*, 6398.
18) Granel, C.; Dubois, P.; Jerome, R.; Teyssie, P. *Macromolecules* **1996**, *29*, 8576.
19) Gaynor, S. G.; Matyjaszewski, K. in *Controlled Radical Polymerization* Matyjaszewski, K., Ed.; ACS Publishing: Washington, D. C., 1998; ACS Symp. Series Vol. 685, 411.
20) Kusakabe, M.; Kitano, K. **1997**, European Patent EP 0 789 036 A2.
21) Wang, J.-S.; Matyjaszewski, K. *Macromolecules* **1995**, *28*, 7901.
22) Ando, T.; Kamigaito, M.; Sawamoto, M. *Tetrahedron* **1997**, *53*, 15445.
23) Percec, V.; Kim, H. J.; Barboiu, B. *Macromolecules* **1997**, *30*, 8526.
24) Wang, J.-S.; Greszta, D.; Matyjaszewski, K. *Polym. Mater. Sci. Eng.* **1995**, *73*, 416.
25) Kotani, Y.; Kato, M.; Kamigaito, M.; Sawamoto, M. *Macromolecules* **1996**, *29*, 6979.
26) Coca, S.; Matyjaszewski, K. *Macromolecules* **1997**, *30*, 2808.
27) Coca, S.; Matyjaszewski, K. *J. Polym. Sci., Part A: Polym. Chem.* **1997**, *35*, 3595.
28) Chen, X.; Ivan, B.; Kops, J.; Batsberg, W. *Polym. Prepr.* **1997**, *38(1)*, 715.
29) Gaynor, S. G.; Matyjaszewski, K. *Macromolecules* **1997**, *30*, 4241.
30) Gao, B.; Chen, X.; Ivan, B.; Kops, J.; Batsberg, W. *Polym. Bull.* **1997**, *39*, 559.
31) Hawker, C. J.; Hedrick, J. L.; Malmström, E. E.; Trollsås, M.; Mecerreyes, D.; Moineau, G.; Dubois, P.; Jérôme, R. *Macromolecules* **1998**, *31*, 213.
32) Betts, D. E.; Johnson, T.; LeRoux, D.; DeSimone, J. M. in *Controlled Radical Polymerization* Matyjaszewski, K., Ed.; ACS Publications: Washington, D. C., 1998; Vol. 685, ACS Symp. Series, 418.
33) Ellzey, K. A.; Novak, B. M. *Macromolecules* **1998**, *31*, 2391.
34) Nakagawa, Y.; Miller, P. J.; Matyjaszewski, K. *Polymer* **1998**, *39*, 5163.
35) Grubbs, R. B.; Hawker, C. J.; Dao, J.; Frechet, J. M. J. *Angew. Chem., Int. Ed. Engl.* **1997**, *36*, 270.
36) Jones, R. G.; Holder, S. J. *Macromol. Chem. Phys.* **1997**, *198*, 3571.
37) Matyjaszewski, K.; Beers, K. L.; Kern, A.; Gaynor, S. G. *J. Polym. Sci. A. Polym. Chem.* **1998**, *36*, 823.
38) Paik, H.-J.; Gaynor, S. G.; Matyjaszewski, K. *Macromol. Rapid Commun.* **1998**, *19*, 47.
39) Prucker, O.; Rühe, J. *Macromolecules* **1998**, *31*, 592.
40) Huang, X.; Wirth, M. J. *Anal. Chem.* **1997**, *69*, 4577.
41) Crivello, J. V.; Lee, J. L.; Conlon, D. A. *J. Polym. Sci.: Part A: Polym. Chem.* **1986**, *24*, 1251.
42) Crivello, J. V.; Conlon, D. A.; Lee, J. L. *J. Polym. Sci.: Part A: Polym. Chem.* **1986**, *24*, 1197.
43) Mazurek, M. H. **1986**, US Patent US4693935 A 870915.
44) Noguchi, T.; Mise, T.; Yoshikawa, H.; Inoue, J.; Ueda, A. **1991**, Japanese Patent JP04372675 A2 921225.
45) Saam, J. C.; Gordon, D. J.; Lindsey, S. *Macromolecules* **1970**, *3*, 1.
46) Chaumont, P.; Beinert, G.; Herz, J.; Rempp, P. *Eur. Polym. J.* **1979**, *15*, 459.
47) Owen, M.; Kendrick, T. C. *Macromolecules* **1970**, *3*, 458.

48) Morton, M.; Kesten, Y.; Fetters, L. J. *Appl. Polym. Symp.* **1975**, *26*, 113.
49) Mason, J. P.; Hattori, T.; Hogen-Esch, T. E. *Polym. Prepr. (Am. Chem. Soc., Div. Polym. Chem.)* **1989**, *30(1)*, 259.
50) Allyl 2-bromopropionate was synthesized by the esterification of allyl alcohol with 2-bromopropionyl bromide in the presence of triethyl amine in THF at 0 °C.
51) Kickelbick, G.; Miller, P. J.; Matyjaszewski, K. *Polym. Prep. (Am. Chem. Soc., Polym. Div.)* **1998**, *39(1)*, 284.

Chapter 18

Synthesis and Photoinitiated Cationic Polymerization of Monomers Containing the Silsesquioxane Core

James V. Crivello[1] and Ranjit Malik[2]

[1]Department of Chemistry, Rensselaer Polytechnic Institute, Troy, NY 12180
[2]Adhesives Research, Glen Rock, PA 17327

A synthetic scheme was developed to prepare cationically polymerizable octafunctional monomers with cubic silsesquioxane (T_8) cores. Epoxy and 1-propenoxy functional groups were attached to the core by the hydrosilation of T_8^H with an appropriate precursor. The steric constraints and the requirements for the hydrosilation reaction are discussed. The monomers were fully characterized and then polymerized by exposure to ultraviolet irradiation in the presence of onium salt photoinitiators. The polymerization conditions for the monomers were optimized and compared with each other using real-time infrared spectroscopy. Thermal analysis was also performed on the resulting crosslinked polymers.

In recent years, there has been a great deal of interest in functionalizing silsequioxanes to prepare interesting monomers and polymers.[1,2] Bassindale and Gentle[3] hydrosilated allyl- and vinyl-functional molecules onto the hydrogen functional silsesquioxane (T_8^H) cage to obtain eight arm "octopus" molecules. Sellinger and Laine[4] employed a similar strategy to prepare crosslinkable vinyl-functional silsesquioxane monomers. Previously, we have reported on the synthesis of siloxane containing epoxy resins for coatings and composite applications[5]. These resins were prepared by the hydrosilation of vinyl epoxides onto siloxane substrates bearing Si-H groups and these resins displayed exceptional reactivity on exposure to ultraviolet or electron beam radiation in the presence of onium salt inititors[6].

The present work focuses on the synthesis of novel multifunctional epoxy and vinyl ether monomers bearing the T_8 silsesquioxane core and a study of their behaviour under photoinitiated cationic polymerization conditions.

Results and Discussion.

Synthesis and Characterization of Monomers. T_8^H was the key starting material for the preparation of a series of cationically photopolymerizable multifunctional monomers prepared during the course of this work. Previously, Bassindale and Gentle[3] had prepared this compound by the ferric chloride catalyzed hydrolysis of trichlorosilane in solution. A 12% yield was obtained. We have found that a higher yield (23%) can be obtained when this reaction is carried out under essentially high dilution conditions. This is achieved by passing a stream of nitrogen saturated with trichlorsilane vapor into a methanolic solution solution of ferric chloride.

T_8^H

Octasubstituted silsesquioxane **I** was prepared as a model compound by the hydrosilation of 1-octene with T_8^H (eq. 1).

T_8^H →(Karstedt's catalyst, 70°C)→ **I** eq. 1

The hydrosilation of 4-vinylcyclohexene oxide with T_8^H using Wilkinson's catalyst was carried out under a variety of conditions. Even in the presence of excess epoxide, only four of the eight Si-H bonds could be successfully hydrosilated onto the T_8^H core (eq. 2).

T_8^H →$(Ph_3P)_3RhCl$→ **II** eq. 2

The product, **II**, is presumed to be symmetrically substituted, although this has not been fully established. More aggressive hydrosilation conditions resulted in the opening of the epoxide groups with consequent gelation. We ascribe the inability to fully functionalize the remaining Si-H groups as due to steric hindrance. Indeed, computer modeling of the the target octasubstituted molecule revealed it to be highly hindered due to the presence of the bulky epoxycyclohexyl groups.

In contrast, the hydrosilation of T_8^H with the linear, open chain epoxide, 1,2-epoxy-5-hexene took place to give the fully octafunctional epoxide **III**.

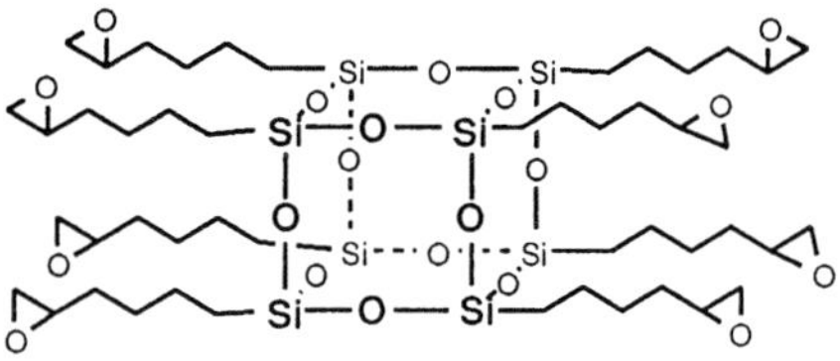

III

Since steric hindrance appeared to be a major factor in the hydrosilation of T_8^H, we prepared the novel vinyl epoxy compound **V** shown below in which the vinyl and a bulky cycloaliphatic epoxy groups are separated by a spacer group.

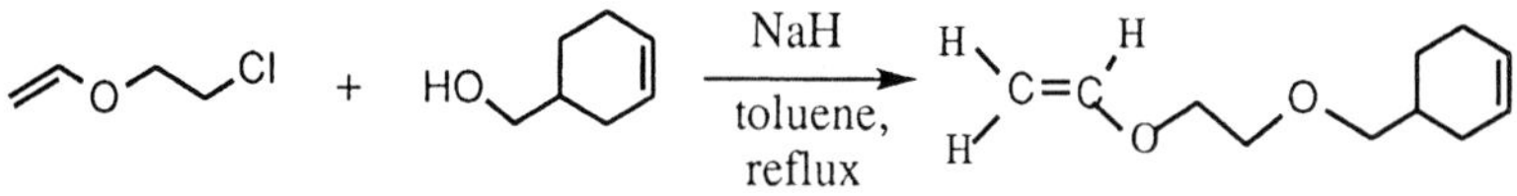

$$\text{CH}_2=\text{CHOCH}_2\text{CH}_2\text{Cl} + \text{HOCH}_2\text{-C}_6\text{H}_9 \xrightarrow[\text{reflux}]{\text{NaH, toluene}} \text{IV}$$

eq. 3

Hydrosilation of the above compound **IV** with T_8^H using the platinum containing Karstedt's catalyst followed by epoxidation with 3-chloroperoxybenzoic acid yielded octafunctional epoxy monomer **V** with a T_8 core.

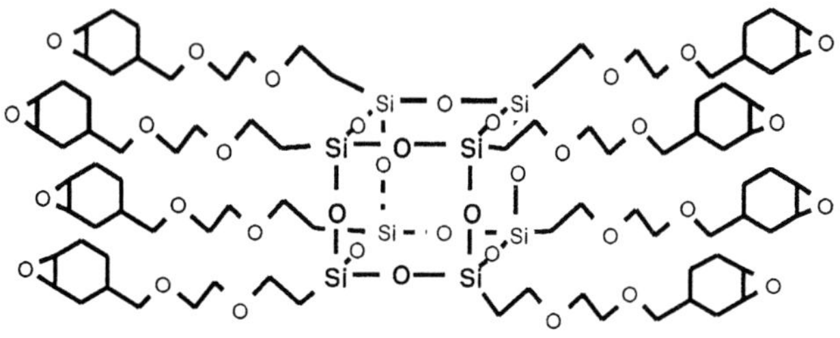

V

It was also observed that a similar strategy was effective for the synthesis of monomer **VI** shown below.

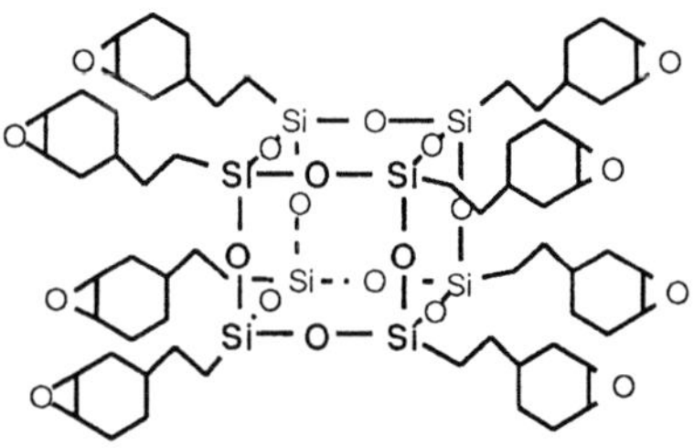

VI

Condensation of T_8^H with 4-vinylcyclohexene gave an intermediate silsesquioxane compound bearing eight ethylcyclohexene groups. Thus, reducing the steric bulk of 4-vinylcyclohexene oxide by removal of the epoxy group was effective in achieving a fully functionallized T_8 core. Subsequent epoxidation of the cyclohexenyl groups using Oxone® readily gave the desired monomer **VI** bearing eight epoxycyclohexyl groups.

In addition to epoxy functional silsesquioxane monomers, we wished to also examine analogous monomers bearing even more reactive functional groups. In recent years, we have found that 1-propenyl ether monomers display exceptional reactivitiy in photointiated cationic polymerization.[7,8] Moreover, these monomers are easily prepared by straightforward synthetic reactions. Accordingly, the precursor, 1-propenoxy-2-vinyloxyethane, **VIII**, was prepared using the methods shown in equations 3 and 4. The key step in this reaction sequence is the catalytic isomerization of the allyl ether **VII** using $(Ph_3P)_3RuCl_2$.

$$HO\text{-}CH_2CH_2\text{-}O\text{-}CH=CH_2 + Br\text{-}CH_2\text{-}CH=CH_2 \xrightarrow[benzene]{NaOH}$$

$$CH_2=CH\text{-}O\text{-}CH_2CH_2\text{-}O\text{-}CH_2\text{-}CH=CH_2$$
VII   eq. 4

$$\textbf{VII} \xrightarrow[154°C]{(Ph_3P)_3RuCl_2} CH_2=CH\text{-}O\text{-}CH_2CH_2\text{-}O\text{-}CH=CH\text{-}CH_3$$
VIII   eq. 5

The hydrosilation of T_8^H with 1-propenoxy-2-vinyloxyethane **VIII** takes place regeoselectively at the vinyl group.[9] All eight Si-H groups of T_8^H react to give an octafunctional monomer **IX** bearing cationically polymerizable 1-propenyl ether groups. This monomer is a mixture of isomers due to cis and trans isomerism at the terminal 1-propenyl ether groups.

IX

Characterization of Monomers. All the monomers and the model compound **I** bearing the T_8 silsesquioxane core were characterized by ^1H and ^{13}C NMR spectroscopy, by gel permeation chromatography as well as by elemental analysis. Typical NMR spectra are shown in Figure 1 for model compound **I** and Figure 2 for monomer **III**. Definitive assignments for each of the resonances in the ^1H NMR spectra can be made based on model compound **I**. The ^{13}C NMR spectrtum of **III** (Figure 2B) consists of two silicon resonances at 66.6 ppm and 66.0 ppm which correspond respectively, to silicon atoms which have undergone hydrosilylation at the α- and β-carbon atoms of the double bond. All the monomers, with the exception of monomer **VI** were liquids. The fact that these monomers are liquids despite their rather high molecular weights can be ascribed to the presence in these monomers of multiple chiral centers and geometrical isomers. Elemental analyses for all the monomers are within experimental error limits of the calculated values. The only exception to this is, again, monomer **VI** which is within 1-2% of the theoretical values for both carbon and silicon. Further insight into the structure of the monomers is provided by GPC analysis. The results of that study are shown in Figure 3. It may be noted that the GPC traces for all the monomers indicate that in addition to the

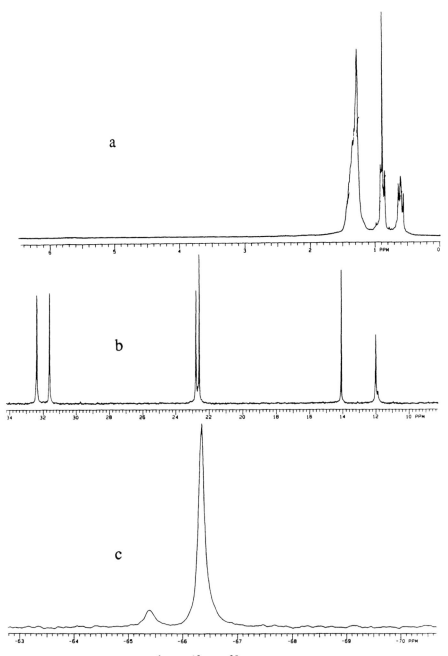

Figure 1. a)^{1}H, b) ^{13}C, c) ^{29}Si NMR spectra for **I**.

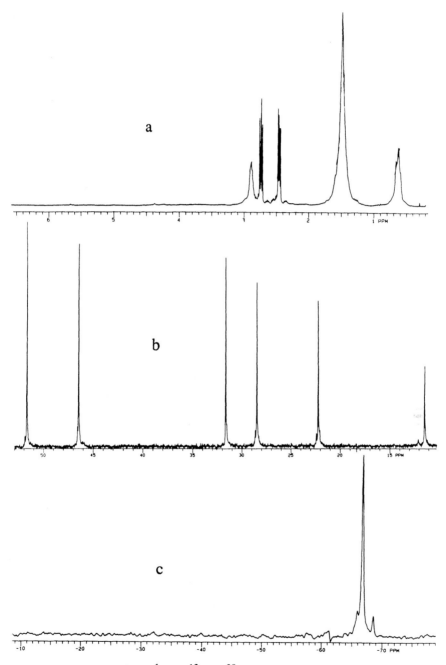

Figure 2. a) ^{1}H, b) ^{13}C, c) ^{29}Si NMR spectra for **III**.

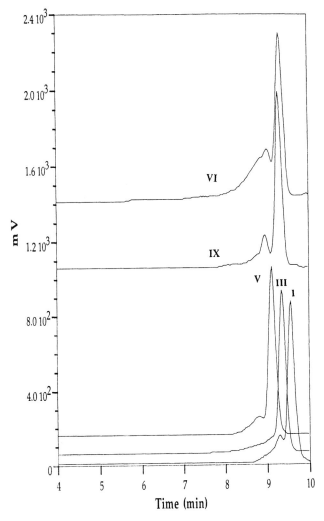

Figure 3. GPC traces of functionalized T_8 monomers.

expected molar mass which comprises the major component present, there is also a minor component at approximately twice the expected molar mass. This is indicative of the presence of dimers. In the case of monomer **VI**, the amount of this dimer is appreciable and there are further indication of the presence of higher molecular weight species as well. We speculate that these species arise by the platinum or rhodium reaction of the Si-H groups with water during the hydrosilation reaction to give silanols followed by the further condensation of the silanols to give siloxane bridged T_8 cores. With the exception of monomer **VI**, we have estimated that the other monomers contain less than 3% of such bridged species.

Photoinitiated Cationic Polymerization. The cationic photopolymerization of the monomers synthesized above was studied using real-time infrared spectroscopy (RTIR).[10] This technique involves monitoring the decrease of an IR absorption characteristic of the functional group undergoing polymerization. In these studies, 2 mol % (4-decyloxyphenyl)phenyliodonium SbF_6^- was used as the photoinitiator. Figure 4 gives individual plots of the percent conversion of the various T_8 monomers as a function of time at the optimum photoinitiator concentration for each of the monomers. The rate of photopolymerization of 1-propenyl ether functional monomer **IX** is the fastest followed by **III**, **V** and **VI**.

The limiting conversions achieved by epoxy and 1-propenyl ether functional T_8 monomers range from 78-90%. It is surprising to achieve conversions as high as these for octafunctional monomers. Typically, as higher and higher functional monomers are photopolymerized, the limiting conversions decrease markedly. This is generally attributed to the decrease in mobility of the reactive functional groups as the highly crosslinked network develops. One possible explanation for the present results is that a considerable amount of intramolecular reaction may be taking place. The resemblence of the silsequioxane monomers with polymerizable epoxy and 1-propenyl ether functional groups described in this communication to spherical dendrimeric materials is striking. In both cases, the concentration of the reactive functional groups at the surface of each monomer molecule is very high facilitating intramolecular reaction. In addition, since the functional groups are located at the same distance from the core, they are "preoriented" further favoring intramolecular reaction. At the same time, the extreme steric crowding present in these molecules together with the large steric bulk of each molecule prevents interpenetration of the arms containing the functional groups. This results in diminished intermolecular reaction.

It is also surprising to observe that monomer **III** containing open-chain epoxy groups are more reactive than those monomers (**V** and **VI**) which have epoxycyclohexyl groups. Again, the larger steric bulk of the arms containing the epoxycyclohexyl groups may inhibit the approach of one reactive functional group to another during the ring-opening polymerization reaction. It should be noted that a cationic epoxide ring-opening polymerization is a S_N2 reaction which requires highly specific backside approach of the incoming epoxide group on one of the carbons bearing the positively charged epoxide oxygen.

Thermal Analysis. The weight loss and Tg data for the polymers obtained by optimally photopolymerizing all four T_8 monomers are given in Table 1. If it is postulated that polymerization of the octafunctional monomers takes place by predominantly an intramolecular process, then the resulting polymers may not exhibit glass transitions because the rigidity of the resulting highly crosslinked structure may restrict chain mobility. This rationalé may explain why a T_g was not observed for photopolymerized samples of **V** and **VI** on heating from 40°C to their decomposition (>350°C). However, T_gs of 236°C and 228°C were observed for polymers obtained from **III** and **IX** respectively. This again supports the hypothesis that the polymers obtained from **III** and **IX** have relatively low crosslink densities even at conversions

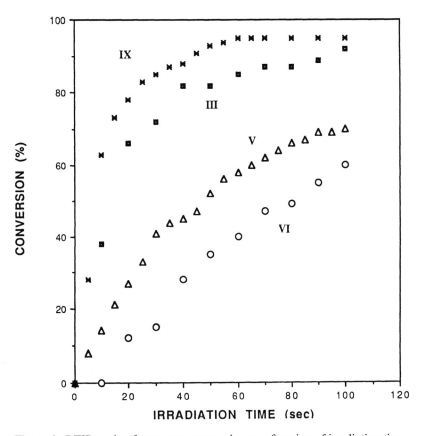

Figure 4. RTIR study of monomer conversion as a function of irradiation time.

of 90 and 94 % because the majority of the reaction takes place by an intramolecular process.

Thermal gravimetric studies of the polymers derived from photopolymerization of four T_8 monomers were carried out in nitrogen at a heating rate of 20°C/minute. The results are given in Table 1 and show that the polymer obtained from VI has the highest thermal stability. High weight retentions in these polymers can be attributed in part to the inherent stability of the polymers, but mainly to the low carbon content of the polymers. 5% Weight losses observed at 366°C and 339°C for polymers from VI and III respectively, can attributed to their high crosslink density as well as to a low hydrocarbon content. If, as has been suggested earlier, III reacts primarily through intramolecular mode and for this reason has a low crosslink density, the only factor contributing to a lower weight loss for this monomer is the low hydrocarbon content of the polymer. The higher weight loss seen for V appears due to fact that it has the highest hydrocarbon content among the four T_8 monomers. Similarly, the comparatively poor thermal stability of IX can be explained by the inherently poor thermal stability expected with polymers derived from 1-propenyl ethers.

Table 1

Results of the thermal analysis of polymers from T_8 monomers.

Monomer	TGA (N_2, 20°C/min)		DSC (20°C/min)
	5% wt. loss	10% wt. loss	T_g
III	339°C	382°C	236°C
IX	217°C	283°C	228°C
V	278°C	335°C	not observed
VI	366°C	392°C	not observed

Conclusions.

A series of octafunctional epoxy and 1-propenyl ether monomers bearing the T_8 silsesquioxane core have been prepared and characterized. Despite the high functionality and reactivity of these monomers, the efficiency with which the monomers undergo crosslinking is considerably less than expected. Instead, due to steric inhibition effects and the proximity of the functional groups to one another, photoinduced cationic polymerization proceeds mainly by an intramolecular process.

Experimental.

Preparation of T_8^H. T_8^H was prepared by a modification of the method described by Bassindale and Gentle[3]. A 3 L, three necked round bottom flask fitted with a mechanical stirrer, reflux condenser and a gas dispersing tube was charged with ferric chloride (75 g), methanol (98.6 g), 37% hydrochloric acid (74.4 g), toluene (300 g) and hexane (1000 g). While this mixture was stirred at high speed, a stream of nitrogen which was first passed through a reservoir containing trichlorosilane (83 g) was bubbled into the reaction mixture at such a rate that it required approximately 5 hours for all of the trichlorosilane to evaporate. The reaction mixture was filtered and dried first over potassium carbonate (46 g), then calcium chloride (31 g). The solvents were removed on a rotary evaporator until 50 mL of solution remained. On cooling, colorless crystals of T_8^H were collected, dried and subjected to purification by sublimation at 180-200 °C/15 mm Hg. A 23% yield (7.5 g) of T_8^H was obtained.

Synthesis of 1-Vinyloxy-2(2-propenoxy)ethane VIII. A 100 mL, 3 neck round bottomed flask fitted with a mechanical stirrer, addition funnel and a dry ice/acetone cooled condenser was charged with 4.4 g (50 mmol) of 2-hydroxyethyl vinyl ether, 5 mL of benzene and 3 g (75 mmol) of powdered sodium hydroxide. Allyl bromide (6.05 g, 50 mmol) was added dropwise to the flask and the flask was warmed slightly to initiate the reaction. The exothermic reaction that set in was controlled by cooling in a water bath. After the exotherm had subsided, the reaction was continued at reflux for 18 hours. The solid residue was filtered and the filtrate fractionally distilled to isolate the product, 1-vinyloxy-2(2-propenoxy)ethane (b.p. 79°C/40 mm Hg; yield 6 g, 97%).

Synthesis of 4(2-Vinyloxyethoxy)cyclohexene IV. A 1000 mL three neck round bottomed flask fitted with a nitrogen inlet, mechanical stirrer, addition funnel and a dry ice/acetone cooled condenser was charged with 100 mL of dry toluene, 4 g (0.17 mol) of sodium hydride and 0.5 g of 18-crown-6 ether.. Nitrogen was passed through the reaction vessel and 1,2,3,6-tetrahydrobenzylalcohol (18.7 g, 0.17 mol) was slowly added via the addition funnel until no further hydrogen evolution could be detected. The exotherm of the reaction was controlled by adjusting the addition rate. When the addition had been completed, the contents of the flask were heated to 80°C and then 17.7 g (0.17 mol) of 2-chloroethylvinyl ether was added. Heating was continued at 80°C for about 18 h. The reaction mixture was cooled, filtered, washed with water in a separatory funnel and dried over anhydrous magnesium sulfate. The toluene was removed using a rotary evaporator and the product was purified by first column chromatography (silica gel, 10:90 ethyl acetate:hexene) followed by fractional distillation (b.p. 77-78°C/0.25 mm Hg). The product, (yield: 15 g, 50%) was 99.3% pure as determined by gas chromatography.

The synthesis given below for the sythesis of monomer **V** is typical for the preparation of all the T_8 monomers in this communication.

Preparation of Monomer V. 4(2-Vinyloxyethoxy)cyclohexene (5 g, 0.027 mol) and 1.09 g (0.0026 mol) of T_8^H were placed in a vial together with a magnetic stirrer and sealed with a screw cap. Approximately 50 mL of Karstedt's platinum catalyst was added and the mixture heated to 80°C. The initially insoluble T_8^H went in solution as the reaction proceeded. The reaction was monitored by following the disappearance of the 2254 cm^{-1} infrared band assigned to the Si-H bond. After approximately 30 minutes, the reaction was complete. Excess 4(2-vinyloxyethoxy)cyclohexene was removed under vacuum leaving **V** as a colorless oil.

IR (NaCl); 3021, 1652 cm^{-1} (cycloaliphatic double bond).

^{1}H-NMR (CDCl$_3$, 200MHz); δ (ppm) 5.65 (s, H$_{9,10}$); 3.55 (br., H$_{2-4}$); 3.35 (d, H$_5$); 2.2-1.6 (m, H$_{7,8,11}$); 1.35-1.15 (m, H$_6$); 1.15-1.05 (t, H$_1$).

^{13}C-NMR (CDCl$_3$, 50MHz); δ(ppm) 126.5 (C$_9$); 125.5 (C$_{10}$); 76 (C$_5$); 70 (C$_4$); 69.5 (C$_3$); 66 (C$_2$); 33 (C$_6$); 28 (C$_{11}$); 25 (C$_8$); 24 (C$_7$); 14 (C$_1$).

^{29}Si-NMR (CDCl$_3$, 40 MHz); δ(ppm) -68.5.

Elemental Anal. Calc. for $C_{88}H_{152}O_{28}Si_8$: C, 56.14%; H, 8.14%. Found: C, 56.29%; H, 8.20%..

Cationic Photopolymerization. Real-Time infrared spectroscopy was used to monitor the kinetics and conversions of the polymerizations of the functionalized T8 monomers prepared in this investigation. The techniques employed were originally described by Decker and Mousa.[7] A thin film of the monomer containing 2 mol % (4-decyloxyphenyl)phenyliodonium SbF_6^- as photoinitiator was drawn onto a sodium chloride salt plate placed in the sample holder. Photopolymerizations were carried out at room temperature and at 100°C using a light intensity of 16 mW/cm². During polymerization the epoxide band at 850 cm^{-1} and the 1-propenyl ether band at 1660 cm^{-1} were monitored.

Thermal Analysis. Samples for thermal gravimetric analyses and differential scanning calorimetry were prepared in the following way. Optimized amounts of a photoinitiator as determined by a previous RTIR study were dissolved in the four T8 monomers (0.5 mole % IOC10/functional group for **IX**, 1.0 mole % (4-thiophenoxyphenyl)diphenylsulfonium SbF_6^- (SS)/functionality for **VI**, 2.0 mole % IOC10/functional group for both **III** and **V**. These compositions were cast as thin (~25mm) films onto glass plates and polymerized by exposure to an unfiltered 200 W medium pressure Hg arc lamp for 4 minutes. The distance between the UV source and the sample was adjusted to give a radiation intensity of 16 mW/cm² at the position of the sample as measured by a radiometer sensitive to 363 nm. The polymerized films were removed from the glass plates and ground to a fine powder which was used in the thermal analysis studies. Glass transition temperatures and thermal stabilities were obtained using a Perkin-Elmer DSC-7 Thermal Analysis System equipped with DSC-7 and TGA-7 modules.

References.

1. Feher, F. J.; Budzichowski, T. A., *J. Organometallic Chemistry*, **1989**, 379, 33-40.
2. Dittmar, U, Hendan, B. J., Florke, U., Marsmann, *J. Organometallic Chemistry*, **1995**, 489, 185-194.
3. Bassindale, A. R.; Gentle, T. E., *J. Mater. Chem.*, **1993**, 3(12), 1319-1325.
4. Sellinger, A.; Laine, R. M., *Polymer Preprints*, **1994**, 665-666.
5. Crivello, J. V.; Lee, J. L., *J. Polym. Sci.: Part A: Polym. Chem. Ed.*, **1990**, 28, 479-503.
6. Crivello, J. V.; Fan, M., Bi, D., *J. Applied Polym. Sci.*, **1992**, 44, 9-16.
7. Crivello, J.V.; Jo, K. D. *J. Polym. Sci., Polym. Chem. Ed.*, **1993**, 31(6), 1473.
8. Crivello, J.V.; Jo, K. D. *J. Polym. Sci., Polym. Chem. Ed.*, **1993**, 31(6), 1483.
9. Crivello, J.V.; Yang, B.; Kim, W.-G., *J. Polym. Sci., Polym. Chem. Ed.*, **1995**, 33(14), 2145.
10. Decker, C; Moussa, K., *J. Polym. Sci.: Part A: Polym. Chem. Ed.*, **1990**, 28, 3429.

Chapter 19

Low Modulus Fluorosiloxane-Based Hydrogels for Contact Lens Application

J. Künzler and R. Ozark

Department of Polymer Development, Bausch and Lomb Inc., 1400 North Goodman Street, Rochester, NY 14692-0450

Novel methacrylate functionalized fluorinated-siloxy silanes were evaluated for potential use in hydrogels for extended wear contact lens application: methacryloyloxypropyl-tris(3-(2,2,3,3,4,4,5,5-octafluoropentoxy)propyldimethylsiloxy)silane (Tris(F)), methacryloyloxypropyl-di(3-(2,2,3,3,4,4,5,5-octafluoropentoxy)propyldimethylsiloxy)methylsilane (Di(F)), and 1-(methacryloyloxypropyl)-3-(3-(2,2,3,3,4,4,5,5-octafluoropentoxy)propyl)tetra-methyldisiloxane (Mono(F)). The methacrylate fluorinated-silanes were synthesized by the hydrosilation reaction of methacrylate capped hydrido-siloxy silanes with allyloxyoctafluoropentane. An alternate synthetic procedure for Mono(F) was developed. Radical bulk polymerization of the methacrylate functionalized fluorinated-siloxy silanes with hydrophilic monomers, such as dimethylacrylamide, resulted in transparent hydrogels possessing a wide range of water contents, high oxygen permeability, and a low modulus of elasticity.

To design a successful hydrogel for contact lens application, the candidate polymer must satisfy a number of material requirements (*1-3*). The material must be optically transparent, possess chemical and thermal stability and be biologically compatible with the ocular environment. The material must also possess a low modulus of elasticity for patient comfort and high tear strength for lens handling durability. In addition, it is important that the material can be bulk polymerized and processed utilizing current contact lens manufacturing techniques (*4*). Finally, the material must be permeable to oxygen. Due to a lack of blood vessels within the corneal framework, the cornea obtains oxygen from the atmosphere. Without an adequate supply of oxygen, corneal edema may occur resulting in a number of adverse physiological responses (*5-6*). The key intrinsic material property that is a measure of oxygen diffusion is oxygen permeability (Dk, where D is the diffusion coefficient and k is a proportionality coefficient called the Henry's law coefficient). There is currently no generally accepted level of Dk for extended wear application. Many practitioners believe, however, that for a 0.1 mm thick lens, a Dk of 100 barrers ((cm^3O_2(STP)cm)($sec^{-1}cm^{-2}mmHg^{-1}$)) is suitable for extended wear application (5-6).

There exist two basic methods for the development of hydrogels with high oxygen permeability. The first approach involves the development of high water content hydrogels. The high water content lens material increases the supply of oxygen to the cornea (the higher the water content-the higher the oxygen permeability of the hydrogel)(1). The second approach for the development of high oxygen permeable hydrogels involves the design of silicone based hydrogels. Polydimethylsiloxane (PDMS) due to its low modulus of elasticity, optical transparency and high oxygen permeability is an ideal candidate for use in contact lens materials. PDMS possesses an oxygen permeability that is about 50 times higher than the oxygen permeability of the hydrogel poly(HEMA) and 15 times higher than the high water content hydrogels (7).

There are, however, several limitations to overcome before designing hydrogels based on PDMS. The primary obstacle is that PDMS is hydrophobic and insoluble in hydrophilic monomers. Thus, when attempts are made to copolymerize methacrylate functionalized siloxanes with hydrophilic monomers, opaque, phase-separated materials are usually obtained. In many cases, a co-solvent such as hexanol or isopropanol can be used to solubilize the siloxane and hydrophilic monomer. In addition, the copolymerization of methacrylate functionalized silicones with hydrophilic monomers results in materials with a reduction in water content, loss of surface wettability and an increase in lipophilic character. Lipid uptake can lead to a loss in material wettability.

In previous work, we had shown that copolymers of methacrylate end-capped fluoro substituted siloxanes with varying concentrations of fluorinated methacrylates, resulted in transparent, oxygen permeable, low water (<1%) materials possessing a low affinity for lipids (8-9). The higher the concentration of fluoro side-chain and fluoro methacrylate in the copolymer formulation, resulted in a dramatic reduction in lipid uptake. Further, we showed that the copolymerization of methacrylate end-capped siloxanes containing fluorinated side chains possessing a terminal [-CF_2-H] functionality, with high concentrations of hydrophilic monomers, resulted in transparent hydrogels possessing high levels of oxygen permeability without the use of a solubilizing co-solvent (10). The copolymerization of a fully fluorinated side chain [-CF_2-F] methacrylate end-capped polydimethylsiloxanes or a non-functionalized polydimethylsiloxane with hydrophilic monomers, such as DMA and NVP, lead to phase separated materials. One limitation of the methacrylate capped PDMS based hydrogel copolymers, however, is their high modulus of elasticity. Contact lens materials possessing a high modulus of elasticity generally result in poor on-eye comfort.

This paper describes the continued development of oxygen permeable silicone hydrogels based on the [-$(CF_2)_x$-H] functionalized siloxanes. A series of methacrylate functionalized fluoro-siloxy silane based monomers were synthesized. These fluorinated monomers, when copolymerized with hydrophilic monomers, such as dimethylacrylamide, resulted in transparent, low modulus hydrogels possessing high levels of oxygen permeability.

Experimental

Materials. The ultraviolet initiator Darocur 1173 (2-hydroxy-2-methyl-1-phenyl-propan-1-one) was purchased from EM Science and was used as received. Dimethylacrylamide (DMA), methacryloyl chloride (MC), allyloxytrimethylsilane and (tris(triphenylphosphine)rhodium)chloride were purchased from Aldrich Chemical Co. DMA and MC were distilled under nitrogen prior to use. 1,3-Tetramethyldisiloxane, methacryloylpropyltrichlorosilane, and 1,3-tetramethyldisiloxane platinum complex (2 % platinum in xylenes) were purchased from Gelest. The fluorinated allylic ether, allyloxy octafluoropentane, was prepared by the phase transfer catalyzed reaction of allyl bromide with octafluoropentanol using tetrabutylammonium hydrogen sulfate, tetrahydrofuran and 50% (w/w) NaOH (*11*). The fluorinated side-chain methacrylate end-capped siloxane (FSi) was prepared according to a literature procedure. All other solvents and reagents were used as received.

General procedure for the synthesis of Tris(F), Di(F) and Mono(F) by direct hydrosilation. Synthesis of methacryloxypropyl tris(3-(2,2,3,3,4,4,5,5-octafluoropentoxy)propyldimethylsiloxy)silane (Tris(F)): Methacryloxypropyl tris(dimethylsiloxy)silane (Tris(H)). To a three neck round bottom flask equipped with a thermometer and magnetic stirrer is added methacryloxypropyltrichlorosilane (2.5 g, 0.01 mole), dimethylchlorosilane (6.53 g, 0.069 mole), triethylamine (7.69 g, 0.076 mole) and 25 mLs of anhydrous ether. The reaction mixture is cooled to −15 °C and distilled water (95 g, 5.3 mole) is slowly added. The reaction is allowed to come to room temperature and the reaction is stirred overnight. The resultant solution is washed three times with distilled water. The ether layer is collected, dried over magnesium sulfate, filtered and the diethylether is removed using a rotoevaporator. The resultant oil is vacuum distilled (83-93 °C/1 mm Hg) to give a 51.4% yield of 97.5% pure Tris(H) (as determined by GC). ^{1}H NMR (CDCl$_3$, TMS, ∂, ppm): 0.1 (s, 18H, Si-CH$_3$), 0.5 (t, 2H, Si-CH$_2$-), 1.65 (m, 2H, Si-CH$_2$-C$\underline{H}_2$-CH$_2$), 4.1 (t, 2H, CH$_2$-O-C(O)), 4.6 (m, 3H, Si-H)), 5.6 (s, 1H, =C-H), 6.1 (s, 1H, =C-H).

Methacryloxypropyl tris(3-(2,2,3,3,4,4,5,5-octafluoropentoxy)propyl)silane (Tris(F)). To a 200mL round bottom flask is added Tris(H) (5.0 g, 0.0132 mole), allyloxyoctafluoropentane (21.4 g, 0.079 mole), 0.005 mL of a platinum divinylcomplex and 50 mLs of THF. The solution is refluxed for 1 hr. The THF and unreacted allyloxyoctafluoropentane is removed using a rotoevaporator (50 °C/30 mm Hg) to give Tris(F). ^{1}H NMR (CDCl$_3$, TMS, ∂, ppm): 0.1 (s, 18H, Si-CH$_3$), 0.5 (t, 8H, Si-CH$_2$-), 1.65 (m, 8H, Si-CH$_2$-C$\underline{H}_2$-CH$_2$), 1.95 (s, 3H, =C-CH$_3$), 3.55 (t, 6H, -CH$_2$-O), 3.9 (t, 6H, -O-CH$_2$-CF$_2$-), 4.1 (t, 2H, -CH$_2$-O-C(O)), 5.6 (s, 1H, =C-H), 5.8 (t, 0.33H, -CF$_2$-H), 6.1 (m, 1.33H, -CF$_2$-H and =C-H), and 6.3 (t, 0.33H, -CF$_2$-H).

Procedure for the synthesis of 1-(methacryloxypropyl)-3-(3-(2,2,3,3,4,4,5,5-octafluoropentoxy)propyl)tetramethyldisiloxane (Mono(F)) (Scheme 2-modified synthesis): 1-(3-Trimethylsiloxypropyl)-1-(hydrido)tetramethyldisiloxane. To a 1L round bottom flask is added 1,3-tetramethyldisiloxane (100 g, 0.744 mole), allyloxytrimethylsilane (97.0 g, 0.745 mole), (Tris(triphenylphosphine)rhodium) chloride (0.008 g, 8.8 x 10^{-6} mole) and 400 mLs of anhydrous toluene). The solution is heated at 80 °C for 2 hr at which time the silicone hydride is reacted as shown by ^{1}H NMR spectroscopy. The toluene is removed using a rotoevaporator and the resultant oil is vacuum distilled (65 °C/1.5 mm Hg) to yield 127.5 g (64.8% yield) of trimethylsilyl protected hydroxy propyl tetramethyldisiloxane. ^{1}H NMR (CDCl$_3$, TMS, ∂, ppm): 0.1 (s, 21H, Si-CH$_3$), 0.5 (t, 2H, Si-CH$_2$-), 1.65 (m, 2H, Si-CH$_2$-C$\underline{H}_2$-CH$_2$), 3.45 (t, 2H, -CH$_2$-O), 4.6 (m, 1H, Si-H), High resolution GC-MS, M=264, C$_{10}$H$_{28}$O$_2$Si$_3$.

1-(3-Trimethylsilyloxypropyl)-3-(3-(2,2,3,3,4,4,5,5-octafluoropentoxy)-propyl)tetramethydisiloxane. To a 1L round bottom flask is added trimethylsilyl protected hydroxy propyl tetramethyldisiloxane (60 g, 0.227 mole), allyloxyoctafluoropentane (74.1 g, 0.272 mole), platinum divinyl tetramethyldisiloxane complex (113 ul, 0.002 mole/ul catalyst), 200 mL of THF and 200 mL of 1,4-dioxane. The solution is refluxed for 3 hr at which time the solvent is removed using a rotoevaporator. The resultant oil is passed through 50 g of silica gel using a 10/1 mixture of pentane and methylene chloride. The solvent is removed using a rotoevaporator and the resultant oil is vacuum distilled (120 °C/0.2 mm Hg) to yield 103 grams of a 97% pure (by GC) 1-(3-trimethylsilyloxypropyl)-3-(3-(2,2,3,3,4,4,5,5-octafluoropentoxy)propyl)tetramethyl-disiloxane. ^{1}H NMR (CDCl$_3$, TMS, ∂, ppm): 0.1 (s, 21H, Si-CH$_3$), 0.5 (t, 4H, Si-CH$_2$-), 1.65 (m, 4H, Si-CH$_2$-C$\underline{H}_2$-CH$_2$), 3.45 (t, 2H, -CH$_2$-O-Si), 3.55 (t, 2H, CH$_2$-C$\underline{H}_2$-O), 3.9 (t, -O-CH$_2$-CF$_2$-), 5.8 (t, 0.33H, -CF$_2$-H), 6.1 (t, 0.33H, -CF$_2$-H), and 6.3 (t, 0.33H, -CF$_2$-H). High resolution GC-MS, M=536, C$_{18}$H$_{36}$O$_3$F$_8$Si$_3$.

1-(3-Hydroxypropyl)-3-(3-(2,2,3,3,4,4,5,5-octafluoropentoxy)propyl)tetramethyldisiloxane. 1-(3-trimethylsilyloxypropyl)-3-(3-(2,2,3,3,4,4,5,5-octafluoropentoxy)propyl) tetramethyldisiloxane (53.7 g, 0.1 mole) is dissolved in 540 mL of methanol and to this solution is added 8.8 mL of 10% solution of acetic acid at room temperature. The mixture is stirred for 1 hr and the solvent is removed on a rotoevaporator at 40 °C resulting in a quantitative yield of 1-(3-hydroxypropyl)-3-(3-(2,2,3,3,4,4,5,5-octafluoropentoxy) propyl)tetramethyldisiloxane. The deprotected product is dissolved in 300 mL of hexane and washed four times with distilled water. The organic layer is collected, dried over magnesium sulfate and filtered. ^{1}H NMR (CDCl$_3$, TMS, ∂, ppm): 0.1 (s, 12H, Si-CH$_3$), 0.5 (t, 4H, Si-CH$_2$-), 1.65 (m, 4H, Si-CH$_2$-C$\underline{H}_2$-CH$_2$), 3.3 (t, 2H, -C$\underline{H}_2$-O-H), 3.55 (t, 2H, CH$_2$-C$\underline{H}_2$-O), 3.9 (t, -O-CH$_2$-CF$_2$-), 5.8 (t, 0.33H, -CF$_2$-H), 6.1 (t, 0.33H, -CF$_2$-H), and 6.3 (t, 0.33H, -CF$_2$-H). High resolution GC-MS, M= 464, C$_{15}$H$_{28}$O$_3$F$_8$Si$_2$.

1-(Methacryloxypropyl)-3-(3-(2,2,3,3,4,4,5,5-octafluoropentoxy)propyl) tetramethyldisiloxane (Mono(F)). The deprotected hydroxypropyltetramethyldisiloxane reaction product (46.3 g, 0.1 mole) and triethylamine (11.1 g, 0.11 mole) is added to a 1L round bottom flask. The solution is cooled to 0°C and methacryloyl chloride (11.5 g, 0.11 mole) is slowly added. Following the addition, the solution is brought to room temperature and allowed to stir overnight. The next day the resultant solution is extracted two times with 1 N HCl, two times with 2 N NaOH and two times with distilled water. The organic layer is collected and dried over magnesium sulfate. The solution is filtered and the solvent is removed. The resultant oil is passed through 50 g of silica gel using a 10/1 mixture of pentane and methylene chloride. The solvent is removed using a rotoevaporator and the resultant oil is vacuum distilled (120 °C/0.1 mm Hg) to yield 34.1 grams (64% yield) of a 95% pure (by GC) 1-(methacryloxypropyl)-3-(3-(2,2,3,3,4,4,5,5-octafluoropentoxy)propyl) tetramethyldisiloxane (Mono(F)). ^{1}H NMR (CDCl$_3$, TMS, ∂, ppm): 0.1 (s, 12H, Si-CH$_3$), 0.5 (t, 4H, Si-CH$_2$-), 1.65 (m, 4H, Si-CH$_2$-C$\underline{\text{H}}$$_2$-CH$_2$), 1.95 (s, 3H, =C-CH$_3$), 3.55 (t, 2H, -CH$_2$-O), 3.9 (t, 2H, -O-CH$_2$-CF$_2$-), 4.1 (t, 2H, -CH$_2$-O-C(O)), 5.6 (s, 1H, =C-H), 5.8 (t, 0.33H, -CF$_2$-H), 6.1 (m, 1.33H, -CF$_2$-H and =C-H), and 6.3 (t, 0.33H, -CF$_2$-H). High resolution GC-MS, M=532.17, C$_{19}$H$_{32}$O$_4$F$_8$Si$_2$.

Techniques. Monomer purity was determined on a Hewlett-Packard HP5890A GC using a 15 m X 0.53 mm I.D. X 1.2 µm column of Alltech EC-5 (SE-4). Gas chromatography-mass spectrometry (GC-MS) was completed using a JEOL JMS-AX505HA mass spectrometer through a tandem HP-5890 Series II chromatograph using ammonia chemical ionization. The monomer structure was confirmed by 200 MHz ^{1}H NMR spectroscopy using a Varian 200 spectrometer. Films were cast between silanized glass plates with a 0.3 mm Teflon spacer. The optimum cure conditions consisted of 1h UV at room temperature using a UV intensity of 3500 µW/cm² and 0.5% Darocur 1173 as the initiator. The resultant films were extracted 16 hours in 2-propanol and two hours in distilled water followed by a 16 hr hydration in phosphate-buffered saline (pH 7.3). The water content was determined using the following equation:

% H$_2$O = (hydrated weight - dry weight/hydrated weight) X100

The mechanical properties of films were determined on an Instron Model 4500 using ASTM methods 1708 and 1938. Oxygen permeability (Dk) was determined using the polarographic probe method (*12*).

Results and Discussion

The goal of this study was to design low modulus hydrogels for extended contact lens wear application based on methacrylate functionalized fluoro-substituted polydimethylsiloxanes. In our investigation into the design of low modulus fluorosiloxane based hydrogels, we synthesized and evaluated a series of methacrylate functionalized fluoro-siloxy silanes (Figure 1). This work was in an attempt to design silicone hydrogels possessing both low modulus and lipid resistance.

Synthesis. Figure 2 outlines the synthetic procedure used to prepare Tris(F). The synthesis consisted of two steps: preparation of a methacrylate functionalized tris hydride (TrisH) followed by the hydrosilation of allyloxyoctafluoropentane. The tris hydride was prepared by the co-hydrolysis of methacryloxypropyltrichlorosilane with chlorodimethylsilane. It was imperative to wash all glassware with a mild acid solution prior to distillation of the TrisH to avoid hydride decomposition. The platinum-catalyzed hydrosilation of TrisH was completed using a 3.3 molar excess of allyoxyoctafluoropentane (*13-16*). In most cases we were able to obtain a high purity Tris(F)(>90%) as demonstrated by ^{1}H NMR and GC analysis, however, the results were not reproducible and scale-up was unsuccessful. An unknown was observed in both the ^{1}H NMR and GC that increased significantly in concentration at scale-up. The ^{1}H NMR showed two singlets at ∂ 1.14 ppm and ∂ 1.16 ppm and a multiplet at ∂ 2.5ppm. We were unable to identify this impurity by GC MS due to its low volatility,

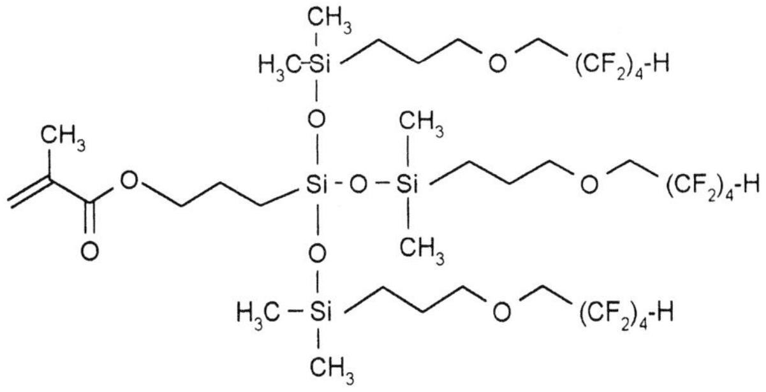

Figure 1. Molecular structure of Tris(F), Di(F), Mono(F) and FSi.

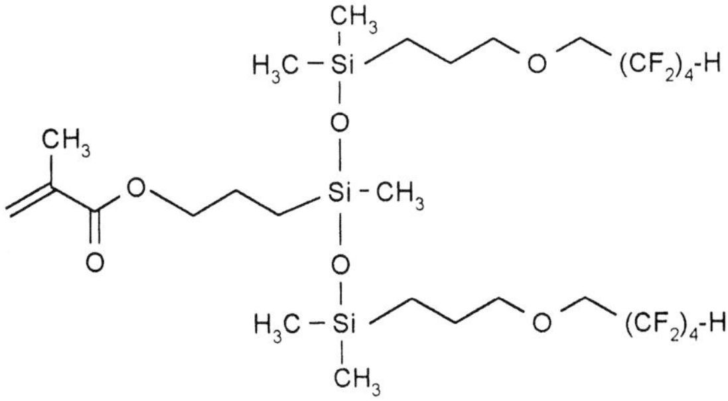

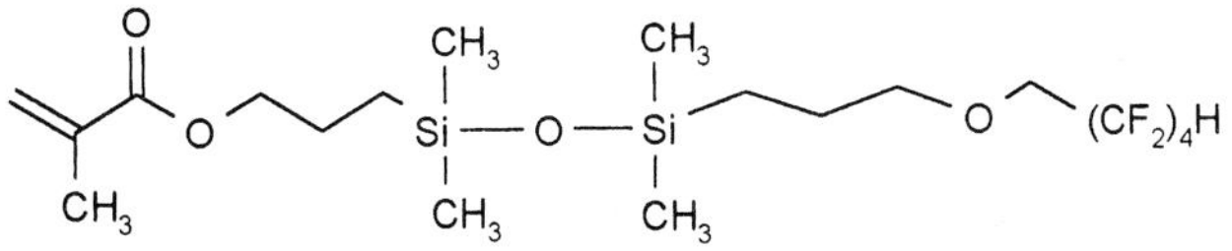

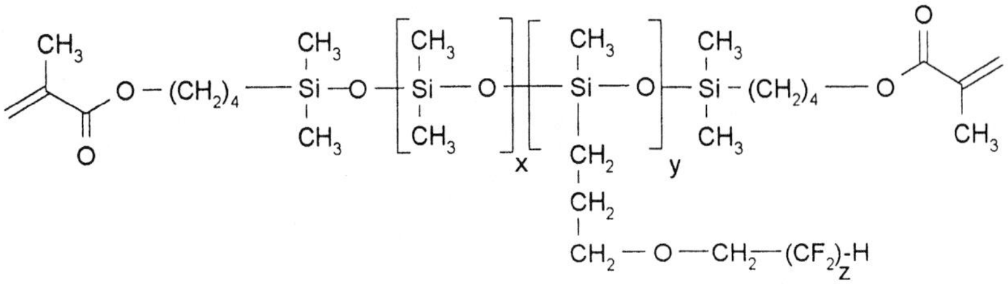

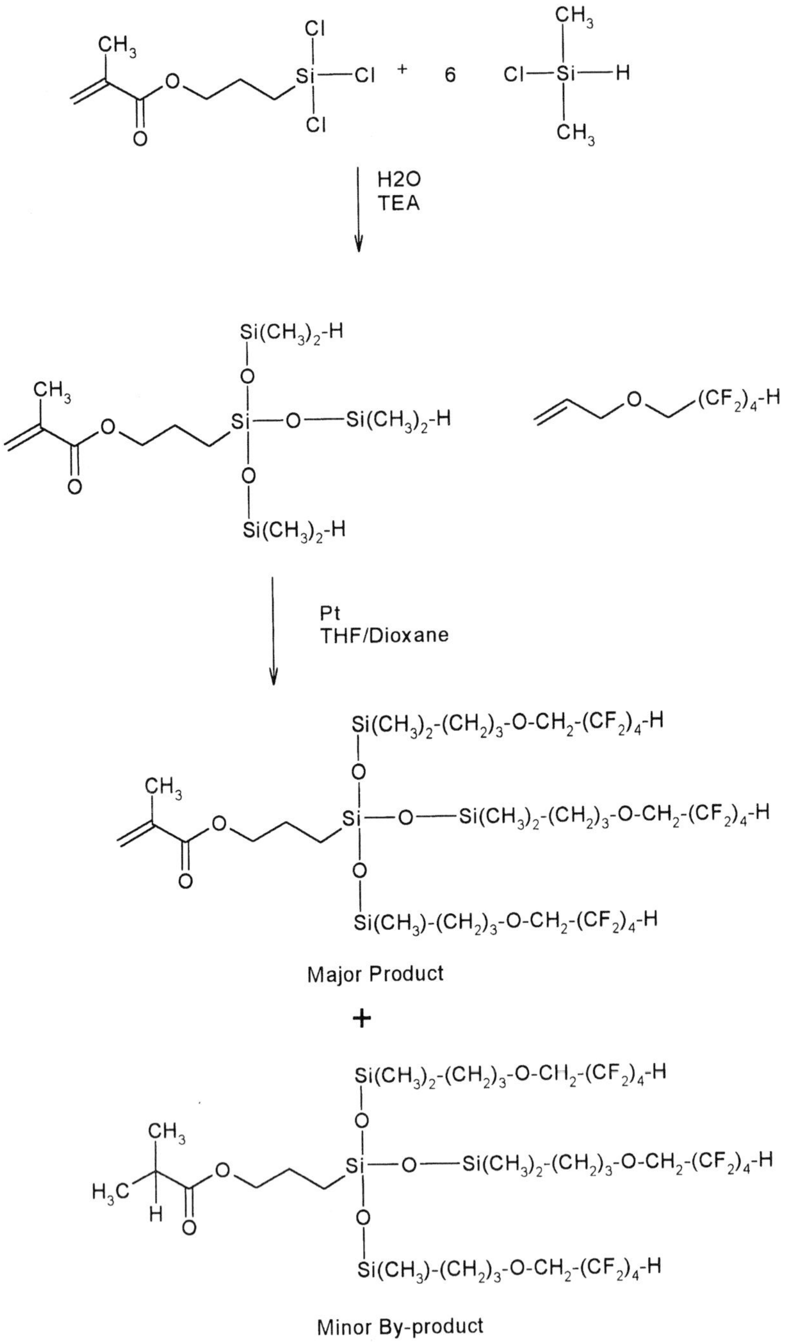

Figure 2. Synthetic procedure used to prepare the Tris(F) showing the reduced methacrylate by-product.

nor were we able to isolate the impurity by column chromatography. The synthesis of Di(F) was completed using the same synthetic procedure with identical results.

We then decided to explore the synthesis of Mono(F) using the same synthetic procedure. This monomer was selected primarily as a model reaction in an attempt to elucidate the unknowns that formed during the synthesis of Tris(F) and Di(F). Due to the higher volatility of Mono(F), GC-MS identification evaluation of the unknowns was considered possible. The synthesis consisted of first preparing the methacryloyloxypropyltetramethyldisiloxane. This was accomplished by the hydrolysis reaction of methacryloyloxypropyldimethylchlorosilane with a large excess of dimethylchlorosilane, followed by hydrosilation with allyloxyoctafluoropentane. Analysis of the resultant Mono(F) by ^{1}H NMR again shows the two identical singlets at ∂ 1.14 ppm and ∂ 1.16 ppm and a multiplet at ∂ 2.5 ppm. GC MS analysis shows one major impurity that was identified as a methacrylate reduction by-product. This impurity forms through the reaction of a silicone hydride with the methacrylate carbonyl. The reduced methacrylate was present at a 15-20% concentration, and as shown by NMR analysis, appears to be the primary impurity obtained in the synthesis of both Tris(F) and Di(F) (Figure 2). We presently believe that this impurity can be minimized through the use of a large excess of allylic fluoro ether during the hydrosilation reaction. Efforts to minimize this impurity by careful control of the reaction conditions (temperature, allylether concentration, solvent, catalyst concentration and type) are presently under evaluation.

We also explored an alternate synthesis of Mono(F). Figure 3 outlines the four step synthetic procedure used to prepare Mono(F). The first step consisted of the rhodium catalyzed hydrosilation reaction of tetramethyldisiloxane with one equivalent of allyloxytrimethylsilane (*17*). The next step consisted of the platinum catalyzed hydrosilation of the disiloxane silicone hydride intermediate with allyloxyoctafluoropentane. Both hydrosilation reactions were monitored for extent of reaction (loss of Si-H) by ^{1}H NMR spectroscopy. The third step in the reaction consisted of an acetic acid catalyzed deprotection of the trimethylsilyl group using a 10% solution of acetic acid in methanol. The deprotection was quantitative with no apparent degradation of the siloxane linkage. The final step consists of the reaction of the deprotected disiloxane (used as is) with methacryloyl chloride. The final purified product, as expected, is free of the methacrylate reduction by-products.

Hydrogel Formulation. The mechanical and physical properties that we hoped to achieve in this study included a Young's modulus between 30g/mm^2 and 100g/mm^2, a DK greater than 50 barrers, and water contents between 20 and 60%. These physical and mechanical property objectives were chosen on the bases of clinical experience from a variety of commercial and experimental lens materials (*18*).

Table I summarizes the mechanical and physical property results for films cast from Tris(F), Di(F) and Mono(F) with varying concentrations of dimethylacrylamide (DMA) using the photoinitiator Darocur 1173 (Novartis). All of the films were transparent without the use of a co-solvent. In contrast, films cast with varying concentration of DMA from a non-fluorinated TRIS, or Tris(F) possessing a fully

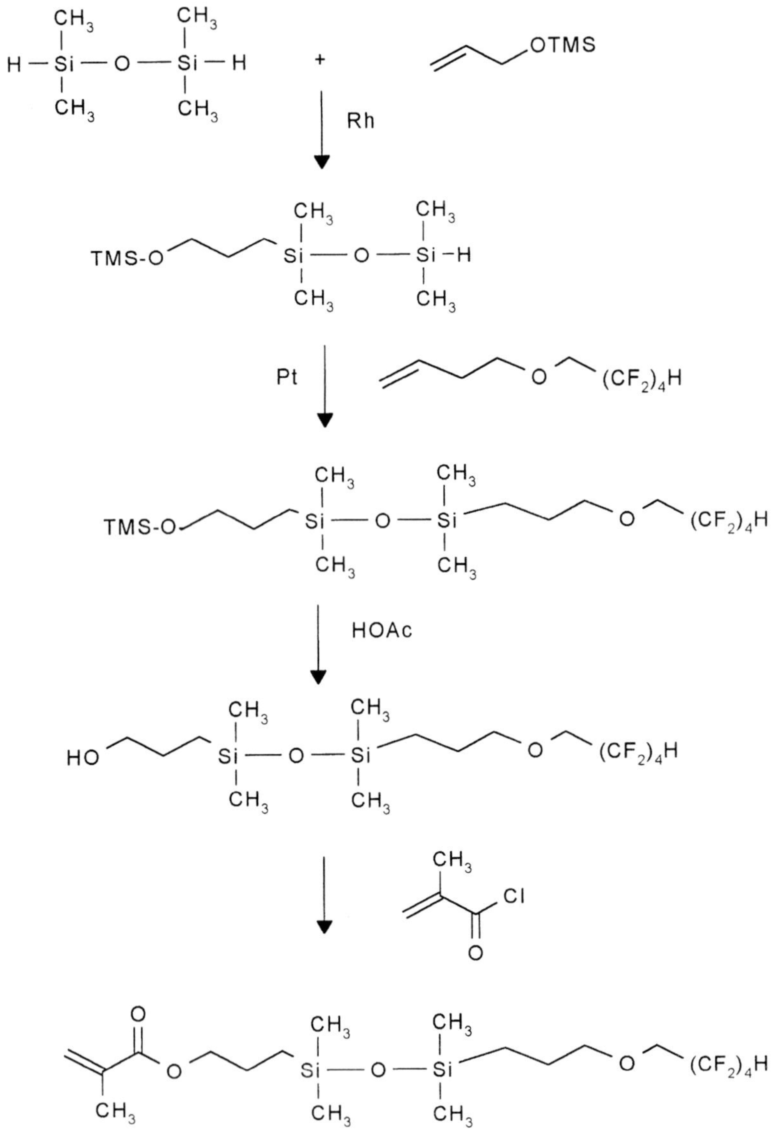

Figure 3 Modified synthetic procedure used to prepare the Mono(F).

Table I. Mechanical and physical property results for formulations based on the methacrylate functionalized fluoro-siloxy silanes (Tris(F), Di(F), Mono(F))

Composition (w/w)	% IPA Loss	% H_2O	Modulus (g/mm^2)	Tear (g/mm)	DK
Tris(F)/Dma					
80/20	12.4	15	113	3.3	40
70/30	11.5	31	89	2.7	36
60/40	11.1	42	113	2.0	36
Di(F)/Dma					
70/30	13	32	24	3.0	42
Mono(F)/FSi/Dma					
70/0/30	12	27	25	5.6	35
60/0/40	7.5	33	40	2.5	31
50/0/50	8.9	54	25	2.8	37
0/70/30	6.7	34	192	3.1	104
20/50/30	5.5	33	100	3.5	70
40/30/30	5.0	34	29	3.1	65

DK in units of $[(cm^3\ O_2(STP)\ cm)/(sec.\ cm^2\ mmHg)]10^{-11}$. All formulations contain 0.5% Darocur 1173 as UV initiator and 0.5% ethyleneglycol dimethacrylate as cross-linker.

fluorinated graft [-$(CF_2)_x$-F], resulted in phase separated opaque films. No morphological data is presently available on the fluoro-siloxy silane copolymers films, but the data strongly indicates that the incorporation of the terminal [-CF_2-H] functionality reduces or eliminates phase separation that occurs with conventional polydimethylsiloxanes. We attribute this to the hydrogel bond interactions between the terminal [-CF_2-H] and the amide linkage of DMA.

In this study the kinetics of polymerization was monitored by NIR spectroscopy. For all films cast, a complete loss of vinyl was shown to occur following one hour of irradiation (3,500 μW/cm²). The low level of isopropanol extractables also indicates a high incorporation of methacrylate as well as a high level of monomer purity.

Hydrogels possessing a wide range in water content were obtained. An increase in the concentration of DMA resulted in, as expected, a significant increase in water content for all fluoro-siloxy monomers tested. For example, the 80/20 (Tris(F)/DMA) copolymer resulted in a water content of 15% and the 60/40(Tris(F)/DMA) copolymer gave a water content of 42%. No statistical trend in tear strength was observed. Modulus values, however, were consistently higher for the Tris(F) based formulations. Surprisingly, the Dk for all three fluoro-siloxy monomers remained relatively constant. The 70/30(Tris(F)/DMA) and 70/30 (Mono(F)/DMA) gave respective Dk values of 35 and 36. This was a very significant result in that it directed our research efforts to focus on the Mono(F) copolymers, since not only was the modulus lower for these copolymers, but the Mono(F) alternate synthesis procedure yields a more reproducible and higher purity product. Although all of these films resulted in low modulus, low extractables, and acceptable levels of water content and transparency, the Dk for these materials is below that required for extended wear application.

In an attempt to design formulations with higher Dk and low modulus, we copolymerized a DP 100 methacrylate end capped 25 mol % fluoro side-chain siloxane (F-Si)(Figure 1) with DMA and varying concentrations of Mono(F). Table I summarizes the mechanical and physical property results for films prepared from F-Si, DMA and Mono(F). The mechanical data shows that with an increase in the concentration of Mono(F), a significant reduction in modulus occurred. Similar trends were also obtained with other silicone hydrogel formulations, i.e., reduction in modulus while maintaining moderately high levels of DK. The oxygen permeability for these formulations is in the generally acceptable region for extended wear application. We believe the Mono(F) acts as a polymerizable diluent in these systems, thus lowering the T_g of the final polymers.

Summary

Three novel methacrylate functionalized fluoro-siloxy silanes were evaluated for potential use in hydrogels for extended wear contact lens application: (Tris(F)), (Di(F)), and (Mono(F)). The methacrylate fluoro-siloxy silanes were synthesized by the hydrosilation reaction of methacrylate capped hydrido-siloxy silane with allyoxyoctafluoropentane. Poor lot to lot reproducibility was achieved using this synthetic scheme due primarily to a methacrylate reduction side-reaction. An alternate synthetic procedure was developed for Mono(F). Radical bulk polymerization of the methacrylate functionalized fluorinated-siloxy silanes with hydrophilic monomers, such as dimethylacrylamide, resulted in transparent hydrogels possessing a wide range of water contents, high oxygen permeability, and a low modulus of elasticity.

References

1. Künzler, J. In *Contact Lenses, Gas Permeable in Polymer Materials Encyclopedia;* Salamone, J.C., Ed.; CRC Press: Boca Raton, Fl, 1996, p. 1497.
2. Künzler, J.F.; McGee, J.M. *Chemistry and Industry* **1995**, *16*, 651.
3. Friends, G.D.; Künzler, J.F.; Ozark, R.M. *Macromol. Symp.* **1995**, *98*, 619
4. Ruscio, D.V. *Polym. Prep. Am. Chem. Soc. Div. Polym. Mat. Sci. and Eng.* **1993**, *69*, 221.
5. White, P. *Contact Lens Spectrum* **1990**, p46-63.
6. Holden, B.; Mertz, G.; McNally J. *Invest. Ophthalmol. Vis. Sci.* **1983**, *24*, 218.
7. *Polymer Handbook;* Brandrup, J.; Immergut, E.H., Eds.; 3rd. Ed.; Wiley-Interscience: New York, 1989, p. VI 435.
8. Künzler, J.F. *Trends in Polymer Science* **1996**, *4*, 52.
9. Friends, G.; Künzler, J.; Ozark, R.; Trokanski, M. In *ACS Symposium Series No. 540, Polymers of Biological and Biomedical Significance,* Shalaby, S.W.; Williams, J.; Ikada, Y.; Langer, R., Eds.; American Chemical Society, 1993.
10. Künzler, J.; Ozark, R. *J. of Applied Polymer Science* **1995**, *55*, 611.

11. Dehmlow, E.V.; Dehmlow, S.S. In *Phase Transfer Catalysis;* Ebel, H., Ed.; Verlag Chemie: Weinheim, 1983, p. 104.
12. Fatt, I.; Rasson, J.E.;Melpolder, J.B. *ICLC* **1984**, *14*, 38.
13. Britcher, L.J.; Kehoe, D.C.; Matisons, J.G.; Swincer, A.G. *Macromolecules* **1995**, *28*, 3110.
14. Lewis, L.N. *J. Am. Chem. Soc.* **1990**, *112*, 5998.
15. Speier, J. L. *Adv. Organomet. Chem.* **1979**, *17*, 407.
16. Marciniec, B.; Gulinski, J.; Urbaniak, W.; Kornetka, Z. In *Comprehensive Handbook on Hydrosilation;* Marciniec, B., Ed.; Pergamon Press: Oxford, 1992.
17. Crivello, J.V.; Bi, D. *J. Polym. Sci., Polym. Chem.* **1993**, *31*, 2729.
18. Bausch and Lomb in-house clinical data.

Chapter 20

Fluorosilicones Containing the Perfluorocyclobutane Aromatic Ether Linkage

Dennis W. Smith, Jr.[1], Junmin Ji[2], Sridevi Narayan-Sarathy[2,4], Robert H. Neilson[2], and David A. Babb[3]

[1]Department of Chemistry, Clemson University, Clemson, SC 29634
[2]Department of Chemistry, Texas Christian University, Fort Worth, TX 76129
[3]The Dow Chemical Company, Freeport, TX 77541

A novel class of fluorosilicones containing the perfluorocyclobutane (PFCB) aromatic ether linkage are described. Monomers and multi-functional resins are obtained by the direct delivery of a trifluorovinyl aromatic ether (TFVE) group to olefin containing substrates via Grignard or aryl lithium reagents prepared from p-BrC$_6$H$_4$OCF=CF$_2$. Mono-olefin intermediates undergo pendant hydrosilation and di-olefin intermediates undergo hydrosilation polymerization with silane (Si-H) functionalized precursors, respectively. When heated above 130°C, TFVE groups cyclodimerize to new chain extended or crosslinked PFCB fluorosilicones with low glass transition temperatures and good thermal stability.

Fluorosilicones combine excellent chemical and heat resistance with unique low temperature, surface, and electrical properties.[1-3] Traditional thermosetting, or network fluorosilicones, such as those containing pendant trifluoropropyl groups, are used commercially as high performance fluids, lubricants, surfactants, gels, coatings, adhesives, and sealants.[3] Current application of these products require additional components and catalysts for the crosslinking chemistry and the resulting network structures typically experience greater rates of thermomechanical degradation versus the parent fluorosilicone. Future demands for improved performance in extreme environments dictate that alternative non-traditional routes to fluorosilicone materials be considered.

[4]Current address: Discovery Group, Ashland Chemical Company, Columbus, OH 43216.

Our ongoing strategy to develop new fluorosilicone materials has been to combine the well known preparative chemistry of silicon containing compounds with the relatively new development of perfluorocyclobutane (PFCB) aromatic ether polymers. PFCB materials are prepared by the thermal cyclodimerization of trifluorovinyl ether (TFVE) monomers.[4-16] Recently we developed a synthetic strategy to deliver the *p*-(trifluorovinyloxy)phenyl group to silicon containing substrates which serve as precursors to fluorosilicones.[6,15-17] Here we describe our current working synthetic approach and initial property survey of these new materials.

Perfluorocyclobutane (PFCB) Chemistry

Practical application of crystalline perfluoropolymers such as Teflon™ is often limited due to prohibitive processing costs and / or degradation of thermal and mechanical properties with time and stress.[18-19] Hybrid fluoropolymers containing hydrogen or other atoms in their structure, however, typically exhibit improved strength and processability over perfluorinated systems. Our interest in polymers with partially fluorinated segments seeks to balance favorable fluoropolymer properties with enhanced processability, thermal resistance, and mechanical properties.

Perfluorocyclobutane (PFCB) polyaryl ethers are one such class of partially fluorinated polymers which combine the processability and durability of engineering thermoplastics with the optical, electrical, thermal, and chemical resistant properties of traditional fluoroplastics. Developed originally at The Dow Chemical Company[4] in Freeport, TX, PFCB polymers are prepared by the radical mediated thermal cyclopolymerization of trifluorovinyl ethers (Figure 1) and have, to date, provided a variety of thermoplastic and thermosetting materials possessing a tunable range of performance.[4-16]

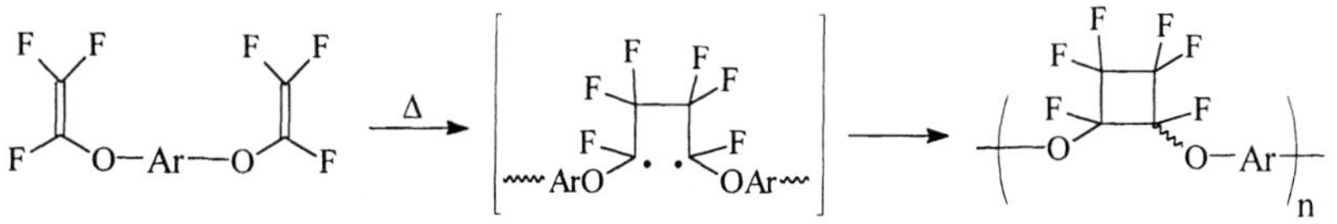

Figure 1. Thermal Polymerization of Trifluorovinyl Aryl Ethers.

The cyclodimerization of trifluorovinyl ethers does not require catalysts or initiators, yet proceeds thermally due to an increased double-bond strain, a lower C=C π-bond energy, and the strength of the resulting fluorinated C-C single bond adducts formed.[20] As generalized in Figure 1, step growth head-to-head cycloaddition generates the more stable diradical intermediate followed by rapid ring closure giving an equal mixture of *cis*- and *trans*-1,2-disubstituted perfluorocyclobutyl linkages.[4,6] In addition, cyclopolymerization results in well defined telechelic polymers containing known trifluorovinyl ether terminal groups.[6] PFCB polymers are easily processed from the melt or solution (e.g., spin coated) and afford high molecular weight amorphous thermoplastics or thermosets which exhibit high glass transition temperatures (T_g), good thermal stability, optical clarity, and isotropic dielectric constants around 2.3.[4]

Traditional PFCB Materials. Trifluorovinyl ether (TFVE) monomers are traditionally prepared in three steps from commercially available phenolic precursors such as, tris(hydroxyphenyl)ethane,[4,12] biphenol,[7] bishydroxyphenylfluorene,[7] and very recently bishydroxy-α-methylstilbene.[13] Deprotonation followed by fluoroalkylation with 1,2-dibromotetrafluoroethane in dimethylsulfoxide (<35°C) gives bromotetrafluoroethyl ether intermediates. Elimination of BrF with zinc metal in refluxing anhydrous acetonitrile provides trifluorovinyl aromatic ether monomers in good overall yield. For example, tris(hydroxyphenyl)ethane gives the trifunctional monomer (TVE) shown in Figure 2 and has been the primary focus for PFCB technology targeted for coatings and composites applications.

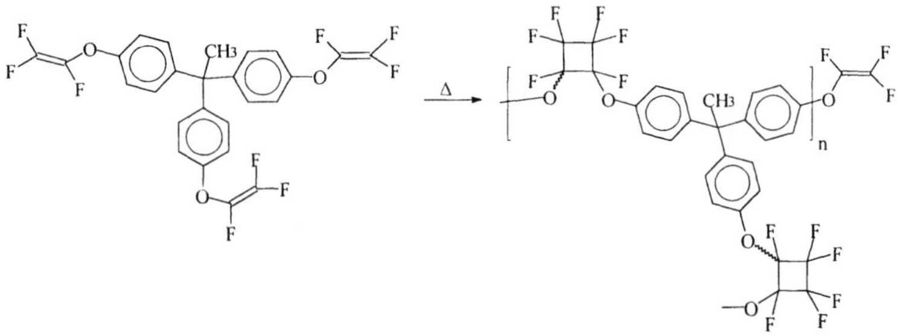

Figure 2. Polymerization of Tris-Vinyl Ether (TVE) Monomer.

TVE monomer can be solution advanced at 150°C in typical solvents to a precisely controlled viscosity, molecular weight, and polydispersity.[4] The pre-network branched oligomer solutions can be spin-coated and cured giving optically clear films which exhibit exceptional planarization and gap fill in microelectronics and flat panel display applications. Structures can also be molded from the melt and mechanical properties such as tensile and flexural moduli for the TVE thermoset (both near 2300 MPa) have been measured.[7]

Polymerization kinetics for TVE monomer has recently been studied in detail by Raman spectroscopy.[12] Second order rate constants for the neat disappearance of fluoroolefin gave half-lives which vary from 450 minutes at 130°C to less than 10 minutes at 210°C and an activation energy of 25 kcal/mol for the cyclopolymerization was determined.

The thermal and thermal oxidative stability of traditional PFCB thermosets, as well as the degradation mechanism and zeroth order kinetic analysis has been reported in detail.[5] The fully cured polymer shown in Figure 2 underwent catastrophic degradation at T_{onset} = 475°C (TGA 10°C/min in nitrogen) and gave isothermal weight loss rates of <0.05%/h in nitrogen and 0.7%/h in air at 350°C. Ultimately, all PFCB polymers degrade quickly above 400°C.[8]

Siloxane-Containing PFCB Materials

While traditional PFCB structures have provided high T_g polymers with good thermal stability and mechanical properties, we began a program to investigate structures which further complement existing PFCB material properties and extend low temperature use. Siloxane polymers are an obvious class of materials expected to show compatibility with PFCB systems in several useful areas. As mentioned previously, fluorinated silicones are currently used commercially in a variety of high performance technologies.[3] Modified silicone materials, in general, have enjoyed tremendous recent attention due to the unique combination of properties which siloxane incorporation imparts; including low temperature performance, low surface energy, thermal stability, bio-compatibility, optical clarity, and gas permeability.[1-3]

PFCB Silicones From Phenols. Initial efforts toward the synthesis of siloxane containing PFCB polymers focused on the preparation of bisphenol intermediates amenable to the traditional fluoroalkylation/elimination synthetic scheme (*vide supra*). However, this strategy was met with little success and the attempted preparation of 1,3-bis(hydroxyphenyl)-1,1,3,3,-tetramethyldisiloxane by published methods,[21-22] could not be accomplished in acceptable yields.

Focus was then shifted to well established hydrosilation techniques[23-29] using readily available olefin functionalized phenols.[25] The highly active Pt(0) complex, Pt$_2$[(CH$_2$=CHSiMe$_2$)$_2$O]$_3$ (Karstedt's catalyst),[26-27] was used under standard conditions and the siloxane bisphenol adduct of 2-allylphenol and tetramethyldisiloxane (Figure 3) was obtained as the pure β-adduct.[8] The corresponding bis-TFVE derivative was then easily prepared by our standard method. When heated at 210°C, this new monomer provided the first siloxane containing PFCB polymer, as well as the first *ortho*-substituted example. Moderate degrees of polymerization (M_n = 16000) and a T_g = -16°C were measured.

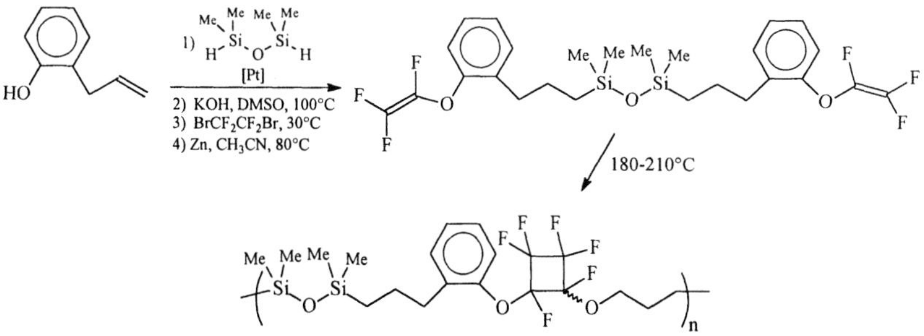

Figure 3. The First PFCB Siloxane Polymer (T_g = -16°C).

The hydrosilation method was further expanded briefly to siloxane thermosets containing pendant fluorovinyl ethers. Commercially available 2,2'-bis-allyl-bisphenol-A (Ciba-Geigy) was first converted to the bis-trifluorovinyl ether by standard procedures. Hydrosilation polymerization[25-29] of the resulting bis-allyl-bis-TFVE monomer with tetramethyldisiloxane gave hybrid carbosiloxane oligomers containing pendant fluorovinyl ethers. As expected, TFVE groups did not interfere with hydrosilation, and the reactive oligomers (M_w = 2600 and liquid at 25°C) underwent thermal crosslinking when heated and gave the first siloxane containing PFCB networks (T_g = 30°C).[8]

The Intermediate Strategy. Although the concept of preparing siloxane functionalized PFCB polymers was successfully demonstrated, early routes described briefly above were limited to expensive or not readily available olefin containing phenolic precursors. We therefore set out to establish a general method of delivering the trifluorovinyl aryl ether (TFVE) group "intact" to a variety of substrates. This "intermediate strategy" began with inexpensive and readily available derivatives of phenol, such as 4-bromophenol, which would provide a reactive TFVE intermediate and allow access to general methods of aryl halide chemistry.

Aryl Grignard and lithium reagents containing the trifluorovinyl ether group (Figure 4) were successfully pursued and have since proven to be quite general for a wide range of substrates including borate esters,[9] phosphine,[10] thiophene,[11] and phosphazene.[14] The initial development of siloxane containing PFCB polymers, featured in this chapter, have also been reported.[6,15-17]

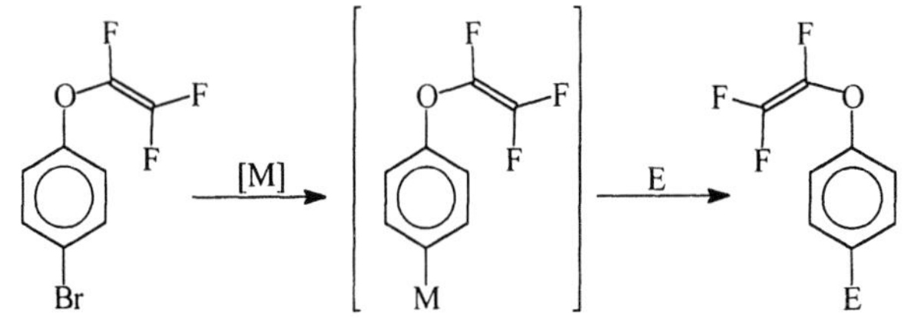

Figure 4. The Intermediate Strategy (E = electrophillic substrate).

Aryl bromide TFVE-Br is easily prepared on the multi-pound per batch scale and is isolated pure by vacuum distillation.[6] The corresponding Grignard reagent is generated by standard methods and is remarkably stable in THF solution, even at elevated temperatures. Recently discovered arly lithium reagents, on the other hand, are much more reactive and their formation is more specific (*vide infra*).[17] (***Caution!** When warmed above –20°C, solutions of TFVE-Li react very exothermically with subsequent decomposition.*) However, stable solutions are easily prepared and used at low temperature in diethyl ether.

The preparation and use of these nucleophilic organometallic reagents in the presence of electrophilic fluorinated olefins was a surprising result and prompted us to explore theoretical support for their stability. Upon calculating (at the *ab initio* level) deprotonation energies of trifluorovinyloxy substituted benzene and pyridine model compounds vs. the corresponding methyl ether, we found that perfluorovinyl ether stabilizes a *p*-carbanion by as much as 11 kcal/mol and destabilizes the protonated pyridinium analog by 10 kcal/mol.[17]

Linear PFCB Fluorosilicones. Application of the intermediate strategy was first demonstrated with Grignard chemistry. The room temperature reaction of TFVE-Br with magnesium proceeded smoothly in THF and gave Grignard reagent TFVE-MgBr.[6] Substitution with chlorodimethylsilane and subsequent self-condensation gave previously sought disiloxanyl monomer (n = 1) shown in Figure 5. Neat cyclopolymeriztion at 200°C afforded the first PFCB fluorosilicone thermoplastic prepared by the "intermediate strategy" (Table 1).

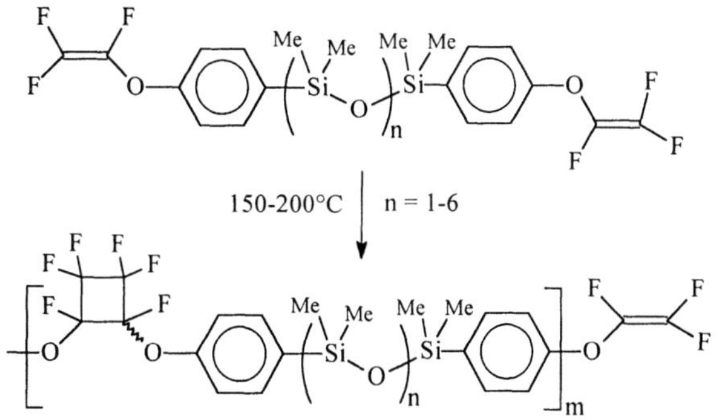

Figure 5. Linear Cyclopolymerization of Siloxanyl Monomers.

The disiloxanyl (n = 1) polymer also exhibits a somewhat elusive liquid-liquid thermal transition and may suggest some degree of liquid crystalline order. We are currently pursuing the synthesis and polymerization of a homologous series of polysiloxane monomers (Figure 5) from commercially available bis-chlorosiloxanes in order to systematically reveal the physical properties and potential liquid crystallinity of linear PFCB silicones.

To date, linear polymers for n = 2-3 have also been prepared. Polymerization of tri- and tetra-siloxane monomers (n = 2-3) gave lower GPC molecular weights and slightly narrower molecular weight distributions compared to that found for the disiloxanyl (n = 1) polymer. As expected, the T_g decreased dramatically with increasing siloxane incorporation. Table 1 contains selected properties for the linear siloxanyl polymers.

Table 1. Selected Properties of Linear PFCB Fluorosilicones from Figure 5.

n	GPC $M_n{}^a$	GPC $M_w/M_n{}^a$	DSC T_g (°C)[b]	TGA T_{onset}(°C)[c]	TGA Total loss (%)[c]
1	20000	3.4	18	434	100
2	12000	1.6	-19	432	86
3	12400	1.7	-34	430	86

[a]In THF vs. Polystyrene, [b]10°C/min, [c]10 °C/min in N_2 to 900°C.

Network PFCB Fluorosilicones. In addition to linear fluorosilicone materials, the intermediate strategy is currently used in combination with hydrosilation techniques to prepare silicones with pendant trifluorovinyl ether groups which crosslink when heated. As mentioned earlier, traditional fluoroelastomers – including fluorosilicones – typically experience a significant degradation in thermal and mechanical properties after crosslinking.[3] PFCB chemistry offers an alternative site-specific crosslinking mechanism and thus the possibility for improving fluorosilicone elastomer properties.

Mono-olefin Intermediates for Hydrosilation. Our entry into hybrid fluorosilicone polymers was led successfully by silylation of the Grignard reagent from TFVE-Br. However, further studies indicated that the Grignard reagent was insufficiently reactive toward many electrophiles, thus limiting its general synthetic utility. This prompted the investigation of the possible formation of the analogous organolithium derivative from TFVE-Br.

We found that aryl bromide (TFVE-Br) underwent smooth and quantitative metal-halogen exchange with *t*-butyllithium in ether at −78°C.[17] The aryllithium reagent reacted readily in ether solution at low temperature with a variety of inorganic and organic electrophiles. In combination with established Grignard chemistry, new TFVE derivatives **1-3** were obtained from readily available and inexpensive substrates (Figure 6).

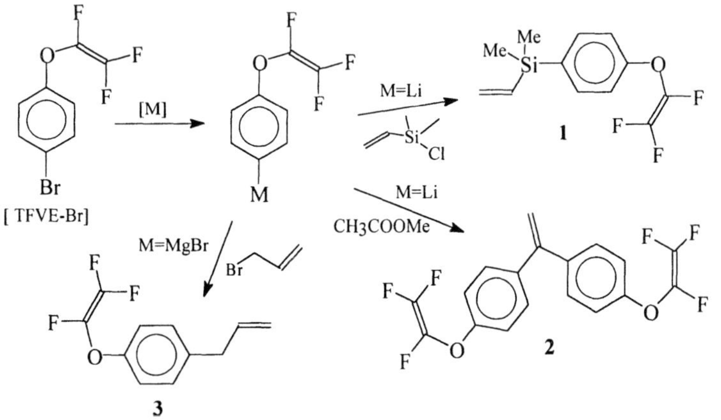

Figure 6. Mono-olefin TFVE Intermediates for Pendant Hydrosilation.

Grignard substitution on allylbromide gave allyl derivative **3** while chlorodimethylvinylsilane and methylacetate gave olefin intermediates **1-2** via substitution with the aryllithium derivative. Formation of bis-TFVE olefin **2** is illustrative of the pronounced reactivity of TFVE-Li toward organic electrophiles. Addition of methyl acetate to a solution of TFVE-Li, followed by protonation, yields the tertiary alcohol [*p*-F_2C=$CFOC_6H_4$]$_2$C(CH_3)OH, which undergoes rapid dehydration to afford the bis-TFVE olefin **2**. Isolated yields of **1-3** after short path distillation ranged from 42-78% and structures were confirmed by NMR spectroscopy and elemental analysis. The new mono-olefin containing compounds were then used to deliver the TFVE group via common hydrosilation techniques.

TFVE Functional Silicones 4-6. Mono-olefin derivatives were reacted with a variety of commercial siloxane copolymers containing Si-H functionality by Pt(0) catalyzed hydrosilation methods.[23-29] Polymer compositions containing varying amounts of TFVE were obtained based on the amount of Si-H linkages present in the copolymer (Figure 7).[15]

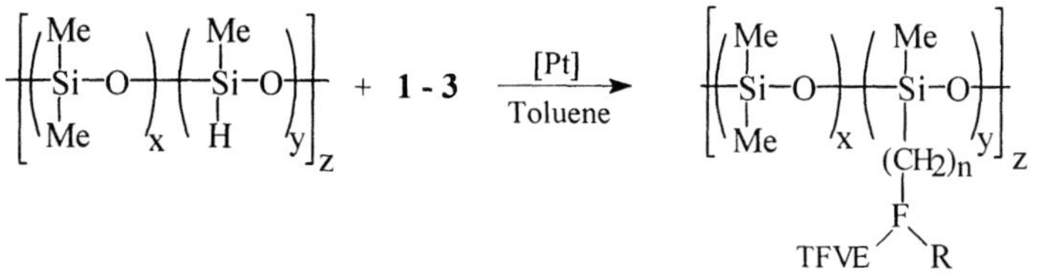

4: F = SiMe; n=2; R=Me
5: F = CH; n=1; R=TFVE
6: F = CH$_2$; n=2; R=H

Figure 7. TFVE Derived Silicone Copolymers via Hydrosilation of Mono-olefins.

Olefin derivatives **1-3** were stirred with methylhydrosiloxane-*co*-dimethylsiloxane copolymers in the presence of Pt$_2$[(CH$_2$=CHSiMe$_2$)$_2$O]$_3$ for 10-12 h at room temperature until the IR absorbance corresponding to Si-H at 2126 cm^{-1} was no longer detected by FTIR. Hydrosilated copolymers **4-6** were obtained as soluble transparent oils and were purified by precipitating in methanol and filtering (in chloroform) over activated charcoal. The presence of trifluorovinyl ether pendant or terminal groups were confirmed by FTIR and NMR spectroscopy.

Table 2. Selected Properties of Mono-olefin Hydrosilated Copolymers.

TFVE Oligomer	Mono-olefin Monomer	y (%) (Fig. 7)	DSC T_g (°C)[a]	GPC M_w[b]	GPC M_w/M_n[b]
4a	1	50-55	<-50	25,000	6.8
4b	1	32-35	<-50	-	-
4c	1	15-18	<-50	24,000	6.1
4d	1	7-9	<-50	-	-
4e	1	0-1	<-50	-	-
4f	1	H-terminal	<-50	41,000	1.9
5a	2	50-55	<-50	-	-
5b	2	15-18	<-50	-	-
6a	3	50-55	-35	8,200	2
6b	3	15-18	<-50	-	-

[a]From –50-200°C at 10°C/min. [b]In THF vs. Polystyrene

Table 2 contains some selected properties for hydrosilated copolymers **4-6**. In general, the copolymers retained good solubility, low glass transition temperatures, and molecular weights and molecular weight distributions similar to that of the parent Si-H functional precursor. In most cases, copolymer viscosity was only slightly (qualitatively) increased as a result of pendant functionalization and therefore should permit typical fluid processing in potential applications such as composites, sealants, and coatings where flow and void filling are required.

PFCB Fluorosilicone Networks 4'-6'. With the exception of TFVE terminated copolymer **4f**, upon heating TFVE functionalized copolymers **4-6** in aluminum pans at 165°C for 6-24 h, tack-free transparent free-standing films of fluorosilicones **4'-6'** were obtained (Figure 8).[15] TFVE terminated copolymer **4f** underwent substantial chain extension as characterized by a dramatic increase in viscosity yet remained completely soluble in common solvents.

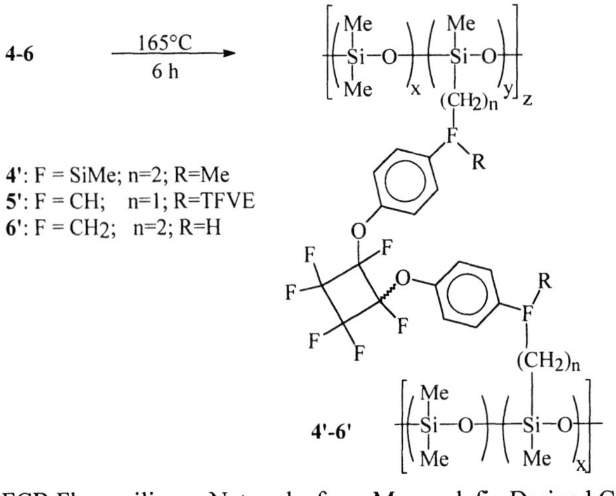

4': F = SiMe; n=2; R=Me
5': F = CH; n=1; R=TFVE
6': F = CH$_2$; n=2; R=H

Figure 8. PFCB Fluorosilicone Networks from Mono-olefin Derived Copolymers.

PFCB silicone networks are distinguished from typical commercial fluorosilicones since catalysts or initiators are not required and crosslinking occurs exclusively at the fluorocarbon sites. In most cases, the crosslinked polymer films could be easily peeled from the aluminum surface. The films exhibited considerable swelling in organic solvents and FTIR confirmed the distinct presence of the hexafluorocylclobutane linkage at 963 cm^{-1}. Some thermal properties are summarized in Table 3.

For the most part, PFCB fluorosilicones were generally stable below 400°C and glass transition temperatures remained remarkably low despite relatively high levels of crosslinking. A detailed thermal property comparison between the reactive precursor polymers and the corresponding networks is currently underway.

Table 3. Selected Thermal Properties of Fluorosilicone Networks.

PFCB Fluorosilicone	DSC T_g (°C)^a	TGA T_{onset}(°C)^b	TGA Total loss (%)^b
4a'	-22	429	64.4
4b'	<-50	440	70.9
4c'	<-50	423	66.3
4d'	<-50	429	76.6
4e'	<-50	371	70.7
4f'	<-50	441	90.3
5a'	<-50	342	61.2
5b'	<-50	334	68.9
6a'	-7.6	416	73.7
6b'	<-50	409	80.2

^{a}From -50-250°C at 10°C/min ^{b}10 °C/min in N_2 to 900°C.

Di-Olefin Intermediates for Hydrosilation. As similarly described for the preparation of mono-olefin TFVE derivatives by the aryl lithium intermediate strategy, di-olefin compounds **7-8** containing multiple TFVE groups were likewise obtained (Figure 9). Di-olefin TFVE derivatives provided monomers via a hydrosilation polymerization scheme with bis-silane (Si-H) functional monomers. Aryl lithium derivative TFVE-Li was generated in diethyl ether at –78°C as before and dichlorodivinylsilane or dichlorodimethyldivinylsiloxane were added and allowed to slowly warm to room temperature. Short-path distillation (0.02 mm Hg and 110°C) gave **7** in 58% yield and **8** in 44% yield (0.3mmHg and 120°C). The structures were confirmed by NMR spectroscopy and elemental analysis.[16]

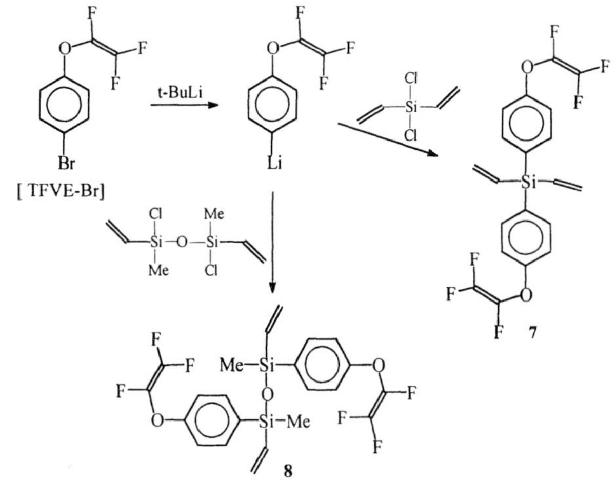

Figure 9. Di-Olefin TFVE Monomers for Hydrosilation Polymerization.

TFVE Functional Silicones 11-14. Hydrosilation polymerization is a well established method for the preparation of high molecular weight linear polycarbosilanes and polycarbosiloxanes.[24-29] Polymer can be obtained under mild conditions by the reaction of di-olefin monomers with dihydrido silicon compounds in the presence of Pt(0) catalyst. Polymerization reactions containing catalyst, di-olefins **7** or **8**, and diphenylsilane or tetramethyldisiloxane were stirred at room temperature for 10-12 h until the signal corresponding to Si-H at 2126 cm^{-1} was not detected by FTIR. Oligomers **11-14** were purified as before and gave transparent viscous oils in essentially quantitative yield (Figure 10).

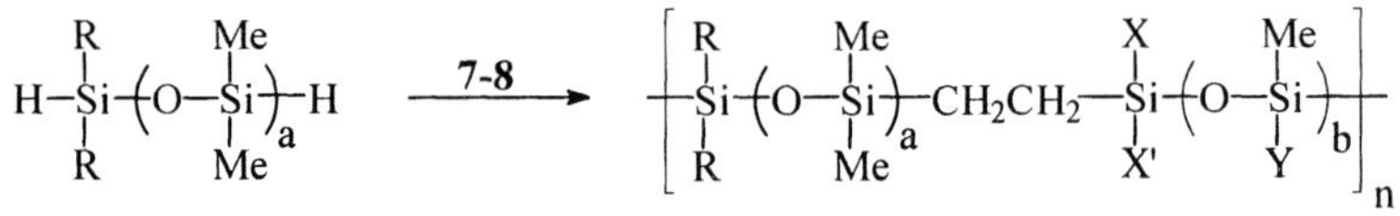

$$H-\underset{R}{\overset{R}{Si}}{+}O-\underset{Me}{\overset{Me}{Si}}{}_a H \xrightarrow{7-8} \left[{+}\underset{R}{\overset{R}{Si}}{+}O-\underset{Me}{\overset{Me}{Si}}{}_a CH_2CH_2-\underset{X'}{\overset{X}{Si}}{+}O-\underset{Y}{\overset{Me}{Si}}{}_b\right]_n$$

9: R = Ph, a = 0
10: R = Me, a = 1

11: R = Ph, a = 0, X = X' = TFVE, b = 0
12: R = Me, a = 1, X = X' = TFVE, b = 0
13: R = Ph, a = 0, X = Me, X' = Y = TFVE, b = 1
14: R = Me, a = 1, X = Me, X' = Y = TFVE, b = 1

Figure 10. Hydrosilation Polymerization of TFVE Di-olefin Monomers.

The hydrosilation molecular weights of oligomers **11-14** were somewhat lower than expected (Table 4). Previously reported high molecular weight hydrosilation polymers have been obtained by utilizing the "one monomer deficient" method which consisted of incremental portions of the dihydrido compound added to a solution containing an excess of the divinyl monomer.[28-29] In our case, efforts were not taken to improve the molecular weight at the time as the crosslinked products were the materials of interest. The reactive oligomer products ranged from highly viscous oils to tacky solids and melting transitions were not observed by DSC. As expected, phenyl substituted siloxane monomer **9** resulted in a substantially higher T_g oligomer (**13**) than did the methyl substituted analog (**10**).

Table 4. Selected Properties of Di-olefin Hydrosilation Oligomers.

TFVE Oligomer	Di-olefin & Siloxane	DSC T_g (°C)[a]	GPC M_w[b]	GPC M_w/M_n[b]
11	7 & 9	15	3,500	1.7
12	7 & 10	-	-	-
13	8 & 9	- 3	3,000	1.7
14	8 & 10	- 26	5,600	3.2

[a]10°C/min. [b]In THF vs. Polystyrene.

PFCB Fluorosilicone Networks 11'-14'. As similarly found for the thermal crosslinking of polymers **4-6**, reactive oligomers **11-14** underwent site-specific cylclodimerization crosslinking when heated at 165°C for several hours (Figure 11).[16] Although reactive oligomers **11-14** contain 1,3-pendant TFVE groups in close proximity, crosslinking is preferred over intramolecular cyclization. Preliminary model calculations at the semi-empirical level (AM1) indicate that the cyclized bicyclic, if formed, would force PFCB linked phenyl rings to an unreasonable restricted distance of just 3.2 Å apart. Fluorosilicone networks **11'-14'** were obtained as transparent films which, as similarly found for networks **4'-6'**, could be easily peeled from the aluminum surface.

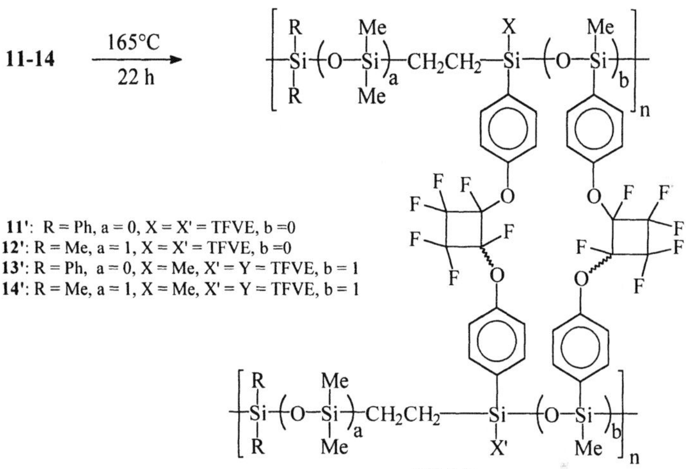

11-14 $\xrightarrow[22\,h]{165°C}$

11': R = Ph, a = 0, X = X' = TFVE, b = 0
12': R = Me, a = 1, X = X' = TFVE, b = 0
13': R = Ph, a = 0, X = Me, X' = Y = TFVE, b = 1
14': R = Me, a = 1, X = Me, X' = Y = TFVE, b = 1

11'-14'

Figure 11. PFCB Siloxane/Organosilane Networks

As mentioned earlier, exothermic polymerization of traditional trifluorovinyl ether monomers reach a measurable rate (DSC at 10°C/min) near 140°C and polymerizations are typically carried out at temperatures between 150-210°C.[12] In contrast, the exothermic polymerization event for the crosslinking of TFVE functionalized silicones was delayed somewhat as expected due to the concentration of TFVE groups vs. traditional neat di- and trifunctional monomers.[7] For example, DSC analysis (10°C/min) of reactive oligomer **11** exhibits an exothermic onset at 172°C and peak at 251°C. Selected thermal properties are summarized in Table 5.

Networks **11'-14'** show glass transition temperatures ranging from 15-47°C by DSC. The significantly higher T_g values for **11'-14'** compared to those measured for **4'-6'** is expected due to the much higher crosslink density, or lower average molecular weight between crosslinks. Film texture and appearance varied from brittle to rubbery as defined by the carbosiloxane precursors. Surprisingly, the T_g measured for network **13'** exceeded that found for **11'** and **12'** despite the expected opposite trend observed for their respective precursor oligomers. Although not soluble in organic solvents, the films showed considerable swelling in most cases and FTIR analyses confirmed the presence of the PFCB linkage at 963 cm^{-1}.

Table 5. Selected Thermal Properties of PFCB Fluorosilicone Networks.

PFCB Fluorosilicone	DSC T_g (°C)^a	TGA T_{onset}(°C)^b	TGA Total loss (%)^b
11'	47	443	73
12'	47	422	86
13'	54	435	80
14'	15	442	59

^{a}10°C/min ^{b}10 °C/min in N$_2$ to 900°C

Thermal stability in nitrogen exceeded 400°C in most network cases which is characteristic of both siloxane and PFCB containing materials. Not shown here is the thermal oxidative stability for thermosets **11'-14'**. In contrast to the TFVE reactive polymers prepared from mono-olefin TFVE derivatives, networks **11'-14'** prepared from TFVE functionalized di-olefin monomers do not contain oxidatively unstable benzylic carbons.

Conclusions. This work demonstrates the versatility of trifluorovinyl ether (TFVE) intermediates used to obtain novel fluorosilicone structures with tunable properties. Linear perfluorocyclobutane (PFCB) fluorosilicones were accessed by the cyclopolymerization of bis-TFVE–n-siloxanyl monomers prepared from TFVE Grignard reagents and chlorosilanes and siloxanes. TFVE containing mono- and di-olefin monomers were also prepared by reacting TFVE aryl Grignard or lithium reagents with readily available organic and inorganic substrates. Pendant functionalized fluorosilicones were synthesized by the hydrosilation functionalization of siloxane copolymers containing Si-H reactive sites or by hydrosilation polymerization of TFVE containing di-olefin monomers with commercial bis-silane monomers. Site-specific thermal crosslinking of the TFVE pendant groups resulted in perfluorocyclobutane (PFCB) fluorosilicone networks with low glass transition temperatures and good thermal stability.

Acknowledgments. The authors thank the Texas Higher Education Coordinating Board (Advanced Technology Program), Clemson University, The Robert A. Welch Foundation, and The Dow Chemical Company for support of the work.

References.
1. Clarson, S.J. and Mark, J.E. In *Polym. Mat. Encyclopedia*, Salamone, J.C., Ed.; CRC Press: Boca Raton, **1996**, Vol. 10, p. 7663.
2. Clarson, S.J.; Semlyen, J.A., Eds. *Siloxane Polymers*, Prentice Hall: London, **1993**.
3. Maxson, M.T.; Norris, A.W.; Owen, M.J. In *Modern Fluoropolymers*, Scheirs, J., Ed.; Wiley: New York, **1997**, p. 359.

4. Babb, D.A.; Ezzell, B.R.; Clement, K.S.; Richey, W.F.; Kennedy, A.P. *J. Polym. Sci., Part-A, Polym. Chem.* **1993**, *31*, 3465.
5. Kennedy, A.P.; Babb, D.A.; Bremmer, J.N.; Pasztor, A.J., Jr. *J. Polym. Sci., Part-A, Polym. Chem.* **1995**, *33*, 1859.
6. Smith, D.W., Jr.; Babb, D.A. *Macromolecules* **1996,** *29*, 852.
7. Babb, D.A.; Snelgrove, R.V.; Smith, D.W., Jr.; Mudrich, S. In *Step-Growth Polymers for High Performance Materials: New Synthetic Methods*, J. Hedrick, J. Labadie, Eds. *Am. Chem. Soc.* **1996**, *624*, p. 432.
8. Smith, D.W., Jr.; Boone, H.W.; Babb, D.A.; Snelgrove, R.V.; Latham, L.E. *Polym. Prepr. (Am. Chem. Soc., Div. Polym. Chem.)* **1997**, *38(2)*, 361.
9. Boone, H.W.; Smith, D.W., Jr.; Babb, D.A. *Polym. Prepr. (Am. Chem. Soc., Div. Polym. Chem.)* **1998**, *39(2)*, 812.
10. Babb, D.A.; Boone, H.W.; Smith, D.W., Jr.; Rudolf, P.W. *J. Appl. Polym. Sci.*, **1998**, *69*, 2005.
11. Xu, Yunlin; Loveday, D.C.; Ferraris, J.P.; Smith, D.W., Jr. *Polym. Prepr. (Am. Chem. Soc., Div. Polym. Chem.)* **1998**, *39(1)*, 143.
12. Cheatham, C.M.; Lee, S-N.; Laane, J.; Babb, D.A.; Smith, D.W., Jr. *Polymer International*, **1998**, *46*, 320.
13. Triphol, R.; Boone, H.W.; Perahia, D.; Ivey, K.; Ballard, E.; Hoeglund, A.B.; Babb, D.A.; Smith, D.W., Jr. *Polym. Mat. Sci. & Eng. (Am. Chem. Soc., Div. PMSE)*, **1999**, *80*, 197.
14. Junmin, J.; Narayan-Sarathy, S; Neilson, R.H; Smith, D.W., Jr. *Polym. Prepr. (Am. Chem. Soc., Div. Polym. Chem.)* **1998**, *39(1)*, 635.
15. Narayan-Sarathy, S; Junmin, J; Neilson, R.H; Smith, D.W., Jr. *Polym. Prepr. (Am. Chem. Soc., Div. Polym. Chem.)* **1998**, *39(1)*, 530.
16. Narayan-Sarathy, S; Neilson, R.H; Smith, D.W., Jr. *Polym. Prepr. (Am. Chem. Soc., Div. Polym. Chem.)* **1998**, *39(1)*, 609.
17. Junmin, J; Narayan-Sarathy, S; Neilson, R.H.; Oxley, J.D.; Babb, D.A; Rondan, N.G.; Smith, D.W., Jr *Organometallics* **1998**, *17(5)*, 783.
18. Scheirs, J. In *Modern Fluoropolymers*, Scheirs, J., Ed.; Wiley: New York, **1997**, 1.
19. Feiring, A.E. In *Organofluorine Chemistry Principles and Commercial Applications*; Banks, R.E.; Smart, B.E.; Tatlow, J.C., Eds.; Plenum Press: New York, **1994**, p. 339.
20. Smart, B.E. *Ibid*, p. 73.
21. Matsukawa, K.; Inoue, H. *Polymer* **1992**, *33*, 667.
22. Mironov, V.F.; Fedotov, N.S.; Kozlkov, V.L. U.S. Patent 3697569, **1972**.
23. Stark, F.O.; Falkender, J.R.; Wright, A.P. In *Comprehensive Organometallic Chemistry*; Wilkinson, G.; Stone, F.G.A.; Abel, E.W., Eds.; Pergamon Press, Ltd.: Oxford, **1982**, Vol. 2, p. 305.
24. Dvornic, P.R; Gerov, V.V. *Macromolecules* **1994**, *27*, 1068.
25. Mathias, L.J; Lewis, C.M.. *Macromolecules* **1993**, *26*, 4070.
26. Karstedt, B.D. U.S. Patent 3775452, Nov. 27, **1973**.
27. Hitchcock, P.B., Lappert, M.F.; Warhurst, N.J.W. *Angew. Chem., Int. Ed. Engl.* **1991**, *30*, 438.
28. Dvornic, P.R; Lenz, R.W. *J.Polym.Sci., Polym.Chem.Ed.* **1982**, *20*, 954.
29. Dvornic, P.R. *J.Appl.Polym.Sci.* **1983**, *28*, 2729.

Chapter 21

Surface Properties of Thin Film Poly(dimethylsiloxane)

H. She[1], M. K. Chaudhury[1], and Michael J. Owen[2]

[1]Department of Chemical Engineering and Polymer Interface Center,
Lehigh University, Bethlehem, PA 18015
[2]Dow Corning Corporation, Mail #C041D1, P.O. Box 994,
Midland, MI 48686-0994

Exposure of a silicon wafer to the vapor of undecenyltrichlorosilane, $Cl_3Si(CH_2)_9CH=CH_2$, results in the formation of a self-assembled monolayer film whose outer surface is composed of olefin groups. This surface can be further derivatized by reacting it with SiH-functional polydimethylsiloxane (PDMS) via platinum-catalyzed hydrosilylation reaction. For PDMS chains anchored at one end, the water contact angle hysteresis (difference between advancing and receding values) decreases as the molecular weight of the PDMS chains and the thickness of the layer they produce decreases. A sufficiently thin layer produced by the multiple attachments of the PDMS chain provides a useful low-hysteresis model PDMS substrate for contact angle and other surface studies. This model system is superior to the more usual systems based on PDMS fluids baked onto glass or metals, cross-linked coatings on paper or plastics, and PDMS elastomer surfaces.

The surface applications of silicone products, particularly those based on polydimethylsiloxane (PDMS), are many and varied (*1*). Familiar examples include release liners for pressure-sensitive adhesives (PSAs), antifoaming agents, and water-repellent treatments for a wide variety of substrates. This broad diversity of application is a direct consequence of the low surface energy of PDMS which is lower than most other polymers except for those based on aliphatic fluorocarbon moieties. Despite the commercial importance of this aspect of the properties of polydimethylsiloxane, there is no fully satisfactory contact angle characterization of PDMS yet available. There is no lack of potential candidates; part of the difficulty in identifying a definitive study lies in the breadth of these diverse wetting investigations of PDMS. However, even within apparently similar systems there is considerable range of reported data.

Previous Contact Angle Characterizations of PDMS

These contact angle studies of PDMS fall into three broad classes: (a) PDMS fluids baked or otherwise adsorbed onto solids such as glass or metals, (b) cross-linked PDMS coatings on flexible substrates such as paper and plastics, and (c) PDMS elastomer surfaces. Perhaps the most quoted values for PDMS are those of Zisman (2), an example of the first class. This was a critical surface tension of wetting study but Owens and Wendt (3) later used Zisman's data in their geometric mean solid surface tension approach. The Zisman quasi-equilibrium, advancing contact angles (θ_a) for the three most commonly reported contact angle test liquids are 101 deg for water, 70 deg for methylene iodide, and 36 deg for n-hexadecane. Typical examples of the second and third classes are Gordon and Colquhoun's (4) study of PDMS release liners for PSAs and Chaudhury and Whitesides'(5) characterization of elastomeric PDMS.

The range of the contact angle (θ_a) data reported in these and other studies is considerable; of the order of 20 deg for each of the most commonly used liquids, 95-113 deg for water, 66-89 deg for methylene iodide, and 27-49 deg for n-hexadecane. Some of this variability is due to neglect of the various pitfalls inherent in contact angle characterization, but in each of these three classes there are basic reasons specific to each class that explain why rigorous, definitive data are not readily available. PDMS is unique among polymers in maintaining its liquid nature to very high molecular weights. Essentially, these three classes are different strategies for obtaining a sufficiently immobile PDMS surface for contact angle study. Unfortunately, each is deficient in some fundamental manner.

Class One. When a PDMS film is adsorbed onto a rigid glass or metal substrate, the maximum hydrophobicity effect is not obtained. Evidently, the high chain flexibility and low intermolecular forces between polymer chains in PDMS must enable significant interaction to occur between polar water molecules and the hydroxylated surface. A thermal baking treatment is required to develop the familiar highly water-repellent character. Such increases of contact angle were first described by Hunter et al.(6) over fifty years ago. For example, when films of 500 cs PDMS were first formed on glass by dipping in benzene solution, water contact angles between 50 and 60 deg were initially observed, values greater than 100 were only obtained by heating to 200°C.

The causes of this behavior are still not fully understood. Obviously, conformational changes such as an increased number of polymer/surface adsorption sites due to the removal of adsorbed water are involved, but so too are possibilities of condensation of surface hydroxyls and residual silanols in the polymer, as well as decomposition, rearrangement and cross-linking catalyzed by the surface. This can happen at surprisingly low temperatures for certain substrates. For example, Willis (7) has reported significant changes in PDMS film structure at temperatures as low as 90-100°C on oxidized copper.

Class Two. The use of cross-linked PDMS coatings on paper and plastic substrates avoids this thermal variation difficulty. There is also the advantage that such materials are readily available as commercial products primarily to provide release liners for the facile delivery of PSA-coated labels, decals and tapes. Two types of well-understood cross-linking technology are utilized; tin-catalyzed systems based on the condensation of silanol and alkoxysilyl functionalities, and platinum-catalyzed systems based on hydrosilylation addition of SiH to vinyl functional siloxanes.

However, the practical experience in the surface characterization of these coatings is a large variation in contact angles similar to that experienced with the other classes of PDMS surfaces. The situation is yet more perplexing as x-ray photoelectron spectroscopy (XPS) surface composition studies (8) show similar atomic compositions fully consistent with pure PDMS outer surfaces across the range of contact angle variability. Inherent roughness effects, particularly on the most popular paper substrates, are part of the explanation. The type of coating also has a marked effect. They are typically available in three forms, solvent-based, aqueous emulsions, and as 100% neat, solventless materials. Wilson and Freeman (9) have demonstrated morphological differences by scanning electron microscopy (SEM) between these three types of PDMS coating. Other variables such as cross-link density also affect the surface behavior adding to the generally unsatisfactory nature of such coatings for rigorous contact angle characterization.

Class Three. Preformed elastomer surfaces would seem to avoid both the film baking and morphological difficulties of the other two systems if molded against an adequately smooth metal substrate. Unfortunately, the frequent use of release agents is a significant contamination danger with elastomer surfaces. More important is the propensity for PDMS elastomers to swell when contacted with many organic liquids. This results in dynamic contact angle variability that is hard to control. Swelling problems are particularly noticeable with the n-alkanes which are the preferred contact angle test liquids for Zisman critical surface tension of wetting determinations of low energy polymers.

Elastomers present other complications. They are usually filled systems and although the solid filler does not generally occupy the outermost layers according to the results of several X-ray photoelectron spectroscopy (XPS) studies, they tend to increase the contact angle hysteresis (difference between advancing and receding contact angles, $\theta_a - \theta_r$) which further reduces confidence in the data. A recent study (10) on the effect of saline exposure on the surface properties of medical grade silicone elastomers exemplifies the magnitude of this contact angle hysteresis. On both peroxide and hydrosilylation cured silica-filled PDMS elastomers, initial liquid drop advancing contact angles for water ranged from 110 to 115 deg, whereas initial liquid drop receding contact angles ranged from 48 to 64 deg.

Preparation of Silicone Surfaces

What appears to be needed is a thin film of PDMS attached by well-understood, low-temperature chemistry to a very smooth, rigid substrate, thereby avoiding the diverse drawbacks of the previous three classes of PDMS. Both freshly cleaved mica which is atomically smooth and cleaned silicon wafers, which typically have a rms roughness in the 0.1 to 0.3 nm range depending on cleaning procedure, would appear to be excellent candidates for the smooth, rigid substrate. Hydrosilylation addition offers the best-understood chemistry approach which can be carried out at temperatures as low as ambient. Greatest control with this chemistry is achieved by forming olefin-terminated, self-assembled monolayers of alkoxysiloxanes on the rigid supports and then forming thin films of PDMS by the reaction of SiH functional PDMS with these ordered terminal olefin groups.

Freshly cleaved muscovite mica was used without any other treatment. The silicon wafers (obtained from Silicon Quest International) were cleaned in hot piranha solution (H_2SO_4:H_2O_2 = 7:3, v:v - used with great care!), rinsed in distilled, de-ionized water, dried under nitrogen and plasma oxidized. This was done in a Harrick plasma Cleaner, Model PDC-32G, using oxygen as the plasma gas at a power of 100 W for 45 seconds. This treatment generates a thin layer of silica whose surface is converted to silanol groups (SiOH) by exposure to the water moisture in the air. A self-assembled monolayer of undecenyltrichlorosilane [$Cl_3Si(CH_2)_9CH=CH_2$] is then formed on these surfaces by exposing them at reduced pressure to the silane vapor from a solution of silane in paraffin oil, prepared and stored under nitrogen. The adsorbed, self-assembled vinyl-terminated alkylsiloxane monolayer (formed from the hydrolysis of the silane and reaction with itself and the surface silanols) was characterized by water and n-hexadecane contact angles and its thickness determined ellipsometrically. The data are shown in Table I. The data are in excellent agreement with a direct comparison already in the literature (11). The 1.6 nm thickness of the monolayer on the silicon wafer corresponds to the length of the molecule in the trans-extended configuration. Because of the close matching of the refractive index between mica and the monolayer, estimation of the thickness of the monolayer on mica by ellipsometry was not possible. XPS was also used to verify the presence of these monolayers.

The SiH functional PDMS polymers are grafted onto this surface by platinum catalyzed hydrosilylation, using DC platinum 4 catalyst:

$$SiH + H_2C=CH(CH_2)_9SiO_{3/2} \rightarrow Si(CH_2)_{11}SiO_{3/2}$$

Two types of SiH functional polymers were used; linear polymers of varying chain length that are SiH functional at one end only, the other end being trimethylsilyl, and a polymethylhydrogensiloxane/polydimethylsiloxane (PMHS/PDMS) copolymer. These materials were synthesized in-house. Table II shows the molecular weight, polydispersity, and layer thickness determined ellipsometrically of the linear monofunctional SiH polymers. The structure of the

Table I: Characterization of Undecenylsiloxane Monolayers

Substrate	θ_a H$_2$O (deg)	θ_a C$_{16}$H$_{34}$ (deg)	Thickness (nm)
Silicon Wafer (Si/SiO$_2$):			
This work	101	34	1.6
Wasserman et al.[11]	101	34	1.6
Mica	95	44	-

Table II: SiH Functional PDMS Polymers Grafted onto Silicon Wafer (Si/SiO$_2$)

Polymer	M$_n$ (GPC)	Polydispersity Index (GPC)	Thickness (nm)
PDMS-A	3882	1.101	5.14
PDMS-B	8520	1.222	7.78
PDMS-C	11080	1.166	10.16
PDMS-D	17670	1.319	11.98
PDMS-E	25840	1.223	15.16

PMHS/PDMS copolymer is $(H_3C)_3Si((OSiHCH_3)_{20}(OSi\{CH_3\}_2)_{145}OSi(CH_3)_3$. After 10 days of grafting time at room temperature, the samples were extracted to remove any ungrafted chains. Note that the highest molecular weight monofunctional polymer did not graft effectively at room temperature and the grafting process was aided by heating overnight at 75°C.

Contact Angle Measurement Technique

Contact angles were measured using a modification of the conventional method on a Rame-Hart instrument (12). The samples of PDMS films were seated on a moveable stage that could be driven laterally to the drop viewing axis. Small drops of water (1-2 µL) were formed from a microsyringe. The conventional method measures the advancing and receding contact angle by adding liquid to the drop or withdrawing it from the drop. In our modified method, the advancing and receding angles are measured simultaneously at the two edges of the triple-phase contact lines by moving the stage laterally against the static water drops (with the syringe needle still in the drop). The advancing contact angle is measured at the right hand side of the drop if the stage is driven from right to left where fresh surface is moving under the drop. The receding angle is observed at the left hand side of the drop where it is retreating from the sample and exposing solid surface to the air that had been previously covered by liquid. In the experiments reported here a lateral stage speed of 50 µm/s was used.

Most of our study has focused on the SiH functional polymers grafted to the monolayer on the silicon wafer because the monolayer on mica proved not to be hydrolytically stable. As shown in Table I the instantaneous advancing contact angle of water is *ca* 95 deg but it decreases with time. The receding contact angle of water is zero on this surface. The stability of the monolayer increases upon ageing either at room temperature or at elevated temperature. The monolayer formed on the silicon wafer is hydrolytically stable with the water contact angle of 101 deg being invariant with time.

Results and Discussion

Table III shows the variation of advancing and receding water contact angle with molecular weight of the PDMS chains grafted to the monolayer on the silicon wafer. The hysteresis increases as a function of molecular weight from 10 to 26 deg over the range studied. Clearly, considerable hysteresis is possible with long tethered chains even when they are not part of an elastomeric network. These results indicate that in our search for low hysteresis surfaces, the thinner the attached PDMS chain the better. However, for chains attached at only one end a sufficiently thin layer to have acceptably low contact angle hysteresis might have insufficient polymeric nature. Multiple attachments along the chain appeared to be a promising approach to the desired goal so for this reason we examined the PMHS/PDMS copolymer. The PDMS film formed in this manner is *ca* 1.5 nm thick from ellipsometric measurements. This is still significantly more than the 0.6 nm expected if the polymer chain were fully extended along the surface but much thinner than the films formed by the single end-attached PDMs polymers (Table 2).

Table III: Contact Angles of Water on PDMS Films End-Grafted onto Si/SiO$_2$

Polymer	θ_a (deg)	θ_r (deg)	H
PDMS-A	112.0	102.5	9.5
PDMS-B	115.3	99.5	15.8
PDMS-C	116.5	97.0	19.5
PDMS-D	117.5	94.0	23.5
PDMS-E	117.5	91.5	26.0

Table IV shows contact angle data for this PDMS film grafted onto silicon wafer. Note that these contact angles were measured in the conventional manner to allow a Zisman critical surface tension of wetting determination (σ_c) to be made from the advancing n-alkane and paraffin oil data. This is shown in Figure 1, a value of 22.7 mN/m being obtained. This surface exhibited small hysteresis and the σ_c value is gratifyingly close to the contact mechanics (Johnson, Kendall and Roberts approach) value of 22.6 mN/m (*13*). Note also that paraffin wax has a very similar value of 23 mN/m (*3*).

Table IV: Contact Angles on PHMS/PDMS Film Grafted onto Si/SiO$_2$

Liquid	σ (mN/m)	θ_a (deg)	θ_r (deg)	H
Water	72.8	108	105	3
Paraffin oil	32.4	56	48	8
n-hexadecane	27.7	41	37	4
n-tetradecane	26.6	35	32	3
n-dodecane	25.4	29	26	3
n-decane	23.9	17	16	1

σ is the surface tension of the contact angle test liquid, θ_a and θ_r are advancing and receding contact angles, respectively

The low hysteresis of the PDMS-grafted silicon wafer indicates no extensive swelling effects. There is no particular driving force for the PDMS molecules to pack into a tightly-packed structure; the grafted molecules should be in a liquid-like state. Interestingly, as the chain length of the hydrocarbon molecules increases so does the slight hysteresis. The overall observation of low contact angle hysteresis for this disordered liquid-like polymer film indicates that its surface properties are rather homogeneous - there are no significant defects of the type that pin the contact line at some solid/liquid/air interfaces.

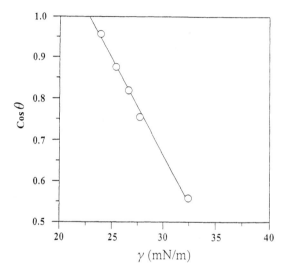

Figure 1. Critical surface tension of wetting plot of PMHS/PDMS copolymer grafted on Si wafer

The advancing (θ_a = 40 deg) and receding (θ_r = 37 deg) contact angles of *n*-hexadecane on the PMHS/PDMS copolymer film grafted onto mica are similar to those on the silicon wafer. However, this surface exhibits a high hysteresis in the water contact angle, advancing and receding angles are 106 deg and 53 deg respectively. We believe this high hysteresis to be due to the hydrolytic instability of the anchoring monolayer on mica discussed earlier, although it should be noted that grafting the copolymer to the olefin groups increases the hydrolytic stability compared to the unreacted monolayer. The hydrolytic stability increases either by aging the film at room temperature or by heating at higher temperature. These treatments result in a surface that exhibits lower hysteresis in water contact angles (θ_a = 103-107 deg, θ_r = 93-97 deg). The increased stability is most likely caused by further lateral crosslinking of the silanol groups of the alkylsiloxane monolayer as has been postulated by Kessel and Granick (*14*).

Alternative smooth substrates and coupling chemistries between the polymer and substrate may also prove suitable. In particular SiH functional PDMS of the type we have used are known to react directly with oxide surfaces such as silica (*15*) although any ordering benefits of the self-assembled monolayer would be lost and the hydrogen produced might also be disruptive. Plasma polymerization of hexamethyldisiloxane (HMDS) has been shown to provide an ultra-thin hydrophobic polysiloxane film on mica (*16*). This layer is homogeneous and water stable with no detectable contact angle hysteresis, but not resistant to peeling in water. The main drawback to this approach for our present purposes is that plasma polymers are usually highly cross-linked and the HMDS plasma polymer is unlikely to be a good model for the uniquely flexible PDMS chain.

Summary

Our expectation was that a smooth, rigid substrate coated with an ultra-thin PDMS layer anchored by a well-understood and controllable chemistry would avoid the complications associated with thermally baked and cross-linked elastomer surfaces and produce a useful, low hysteresis model substrate for contact angle and other surface studies. The data reported here confirm the validity of the self-assembling, reactive silane monolayer and SiH functional PDMS approach when silicon wafers are used as the substrate. Mica which is even smoother is not so suitable because the similar system on mica is not as hydrolytically stable on mica as it is on oxidized silicon. For PDMS chains anchored at one end, the hysteresis decreases as the molecular weight of the PDMS chains and the thickness of the layer they produce decreases. The thinnest layer is produced by multiple attachments of the PDMS chain using a PMHS/PDMS copolymer. This approach has produced the anticipated useful, low-hysteresis model PDMS substrate for contact angle and other surface studies. Examples of the use of this approach include an estimation of adhesion hysteresis using rolling contact mechanics (*12*) and an investigation of the effect of interfacial slippage in viscoelastic adhesion (*17*). Our contact angle values are a little higher than the generally accepted Zisman values (2) and our critical surface tension of wetting value correspondingly a little lower (22.7 mN/m compared to Zisman's 24 mN/m and to the JKR value of 22.6 mN/m obtained from direct work of adhesion measurements). We offer no explanation for this difference; Zisman baked his films of PDMS onto glass at 300°C which is precisely the temperature indicated by Hunter et al. (6) to achieve maximum hydrophobicity.

Literature Cited

1. Owen, M. J. In *Siloxane Polymers*; Clarson, S. J.; Semlyen, J. A., Eds.; PTR Prentice Hall: Englewood Cliffs, NJ, 1993; p. 309.
2. Zisman, W. A. In *Adhesion and Cohesion*; Weiss, P., Ed.; Elsevier Publishing Company: New York, NY, 1962; p.201.
3. Owens, D. K.; Wendt, R. C. *J. Appl. Polym. Sci.*, **1969**, *13*, 1741.
4. Gordon, D. J.; Colquhoun, J. A. *Adhesives Age*, **1976**, *19(6)*, 21.
5. Chaudhury, M. K.; Whitesides, G. M. *Langmuir*, **1991**, *7*, 1013.
6. Hunter, M. J.; Gordon, M. S.; Barry, A. J.; Hyde, J. F.; Heidenreich, R. D. *Ind. Eng. Chem.,* **1947**, *39,* 1389..
7. Willis, R. *Nature*, 1969, *221*, 1134.
8. Duel, L. A.; Owen, M. J. *J. Adhesion*, 1983, *16*, 49.
9. Wilson, J. E.; Freeman, H. A. *TAPPI*, **1981**, 64(2), 95.
10. Kennan, J. J.; Peters, Y. A.; Swarthout, D. E.; Owen, M. J.; Namkanisorn, A.; Chaudhury, M. K. *J. Biomed. Mater. Res.*, 1997, *36*, 487.
11. Wasserman, S. R.; Tao, Y-T.; Whitesides, G. M. *Langmuir*, **1989**, *5*, 1074.
12. She, H.; Chaudhury, M. K. submitted to Langmuir.
13. Chaudhury, M. K. *J. Adhesion Sci. Technol.*, **1993**, *7*, 669.
14. Kessel, C. R.; Granick, S. *Langmuir*, **1991**, *7*, 532.
15. Reihs, K.; Aguiar Colom, R.; Gleditzsch, S.; Deimel, M.; Hagenhoff, B.; Benninghoven, A. *Appl. Surf. Sci.*, 1995, *84*, 107.
16. Proust, J-E.; Perez, E.; Segui, Y.; Montalan, D. *J. Colloid Interface Sci.*, 1988, *126*, 629.
17. Newby, B-m. Z.; Chaudhury, M. K.; Brown, H. R. *Science*, 1995, *269*, 1407.

Chapter 22

Aggregation Structure and Surface Properties of 18-Nonadecenyltrichlorosilane Monolayer and Multilayer Films Prepared by the Langmuir Method

Ken Kojio, Atsushi Takahara, and Tisato Kajiyama[1]

Department of Materials Physics and Chemistry, Graduate School of Engineering, Kyushu University, 6–10–1 Hakozaki, Higashi-ku, Fukuoka 812–8581, Japan

The 18-nonadecyltrichlorosilane (NTS) monolayer with a vinyl end group was prepared by the Langmuir method. The monolayer with a hydrophilic carboxylated surface was subsequently obtained by oxidation of the vinyl end group. For the multilayer preparation, the NTS monolayer was immobilized onto the hydrophilic carboxylated NTS (NTS_{COOH}) monolayer. The NTS multilayer film was obtained by repetitions of depositions and oxidations. Aggregation structure of the NTS multilayer film was evaluated on the basis of X-ray reflectivity (XR), Brewster angle infrared (IR) spectroscopic measurements, electron diffraction (ED) and atomic force microscopic (AFM) observations. It was revealed that the NTS multilayer film with a well-ordered structure could be prepared by the Langmuir method. The local height control was performed by utilizing the phase-separated mixed monolayer being composed of the NTS and [2-(perfluorooctyl)ethyl]trichlorosilane (FOETS) molecule. The (NTS/FOETS) mixed monolayer showed phase-separated structure like sea-island, whose domain and matrix phases were composed of crystalline NTS and amorphous FOETS molecules, respectively. The (NTS_{COOH}/FOETS) mixed monolayer with the hydrophilic domain and the hydrophobic matrix was obtained by selective oxidization of the NTS phase in the (NTS/FOETS) mixed monolayer. Furthermore, the local height control on the NTS_{COOH} phase of the (NTS_{COOH}/FOETS) mixed monolayer was performed by chemisorption of NTS from the solution. The height increase in the NTS_{COOH} phase was confirmed by AFM and lateral force microscopy (LFM).

Immobilization of stable and defect-diminished hydrophobic monolayers on solid substrates is of significant importance in a wide variety of applications ranging from electronics to biotechnology.[1] Sagiv and co-workers first prepared organosilane monolayers on solid substrates by a chemisorption method.[2] However, it is difficult to control the molecular aggregation state of such a monolayer through chemisorption due to random adsorption process from an organosilane solution.

[1]Corresponding author.

Later, authors proposed a novel method for the preparation of the organotrichlorosilane monolayer by the Langmuir method.*(3-11)* It was shown that the film thickness and molecular orientation of the organostrichlorosilane monolayer prepared by this method is controllable at a molecular level.

In the case of the mixed organosilane monolayer prepared by Langmuir's method, it was shown that the (n-octadecyltrichlorosilane (OTS)/[2-(perfluorooctyl)ethyl] trichlorosilane (FOETS)) mixed monolayer was in a phase-separated state whose domain and matrix phases were composed of the OTS and the FOETS molecules, respectively.*(3-6)* In the cases of the (alkylsilane/fluoroalkylsilane) mixed monolayer with phase-separated structure, the method of precise structure control has been discussed.

It has been reported that the organosilane monolayers with hydroxylated and carboxylated surface were prepared by a hydrolysis of a surface ester group and a oxidation of a terminal double bond, respectively.*(12,13)* In the case of these monolayers, a multilayer film was constrcted by repetitions of the monolayer deposition-oxidation processes. In contrast, it is expected that the multilayer film prepared by the Langmuir method has highly-ordered structure in comparison with that prepared by the chemisorption method. The Y-type deposition in which the monolayers are deposited onto the substrate in upstroke and downstroke is the most common mode of the multilayer formation in the Langmuir-Blodgett technique. However, the interaction between hydrophobic-hydrophobic end groups in the Y-type multilayer film composed of the organosilane compound is not strong enough for thermal, mechanical and environmental attacks. In contrast, it is expected that the multilayer film prepared by repetition of the Langmuir organosilane monolayer with a hydrophilic end group has high stability owing to the strong interaction between hydrophilic-hydrophilic (covalent and hydrogen bonds) groups. Since the balance between hydrophilicity and hydrophobicity of the amphiphilic compound is an important factor for the formation of the Langmuir monolayer, the organosilane monolayer with hydroxyl or carboxyl end groups can not be directly prepared by the Langmuir method. Then, the monolayer with a hydrophilic carboxyl group could be prepared by oxidation of the hydrophobic vinyl end group of the amphiphile. The subsequent monolayer can be deposited on this hydrophilic surface. Therefore, it is easily expected that the stable multilayer film with highly-ordered structure can be constructed by utilizing an organosilane compound with a vinyl end group and Langmuir's method. Furthermore, if the phase with vinyl end groups (reactive part) can be formed in the phase-separated mixed monolayer, the local height control might be achieved in a two-dimensional state through localized multilayer deposition onto this reactive part.

In this study, the layered structure of the organotrichlorosilane monolayer with a vinyl end group was constructed by the Langmuir method and its structure was analyzed by X-ray reflectivity (XR), Brewster angle Fourier transform infrared (B-FT-IR) spectroscopic measurements *(14)*, electron diffraction (ED) and scanning force microscopic (SFM) observations. Furthermore, the domain height of the (NTS/FOETS) mixed monolayer surface with phase-separated structure was controlled by the combined method of Langmuir and chemisorption methods.

Experimental Section
Monolayer Preparation 18-nonadecyltrichlorosilane (NTS, $CH_2=CH(CH_2)_{17}SiCl_3$) and [2-(perfluorooctyl)ethyl]trichlorosilane (FOETS, $CF_3(CF_2)_7(CH_2)_2SiCl_3$) (Shin-Etsu Chemical Co., Ltd.) were used to prepare the monolayers. NTS and FOETS were purified by vacuum distillation. NTS and FOETS toluene solutions with a concentration of ca. 3×10^{-3} M were spread on the pure water surface at 293 K. Surface pressure-area ($\pi-A$) isotherms were measured with a computer-controlled home-made Langmuir-trough. In order to form the

polymeric monolayer, the monomer mononolayers were kept on the water subphase under a certain constant surface pressure for 15 minutes. After the formation of the polymeric monolayer on the water subphase, it was transferred onto the substrate by an upward drawing method and then, immobilized onto the substrate with Si-OH groups. The substrates of 2 cm wide x 5 cm long x 0.028 cm thick for Brewster angle IR measurement and 2 cm wide x 4 cm long x 0.2 cm thick for X-ray reflectivity measurement were cut from Si(111) wafers.

Oxidation of NTS Monolayer Oxidation of the vinyl end group of the NTS monolayer was carried out to prepare the monolayer surface with carboxyl groups.(15) Stock solutions of $KMnO_4$ (5 mM), $NaIO_4$ (195 mM) and K_2CO_3 (18 mM) in water were prepared and then, the oxidizing solution was obtained by mixing each 1 mL from these solutions and 7 mL of distilled water. The NTS monolayer was immersed in this oxidizing solution for a few minutes. Then, the samples were removed from the solution and rinsed in 20 mL each of $NaHSO_3$ (0.3 M), distilled water, 0.1 N HCl and ethanol. The NTS monolayer was then transferred onto the carboxylated NTS (NTS_{COOH}) monolayer, and these processes were repeated five times. Figure 1 shows the schematic representation of the formation process of the NTS multilayer film by the Langmuir method.

Characterization of Monolayer and Multilayer Film In order to study the molecular aggregation state of the monolayer, ED observation was carried out. The monolayer was transferred onto the collodion-covered electron microscope grids covered with an evaporated hydrophilic SiO layer by an upward drawing method.(16) The ED pattern was obtained with a transmission electron microscope (TEM) (Hitachi H-7000) which was operated at an acceleration voltage of 75 kV, a beam current of 0.5 µA and an electron beam spot size of 10 µm in diameter.

The surface morphology and mechanical behavior of the monolayer was investigated by atomic force microscopic (AFM) observation and lateral force microscopic (LFM) measurement. AFM and LFM images were obtained using an SPA 300 instrument with an SPI 3700 controller (Seiko Instruments Industry, Co., Ltd. Japan). These investigations were performed with constant force mode under air at room temperature, using a 20 µm x 20 µm scanner and a silicon nitride tip on a rectangular cantilever with a spring constant of 0.09 N m^{-1}. In order to evaluate the contribution of adhesion force to the lateral force due to a water capillary effect between tip and monolayer surface, LFM measurement was performed in vacuo with an SPA 300 HV instrument with an SPI 3800 controller (Seiko Instruments Industry, Co., Ltd. Japan).

X-ray photoelectron spectroscopic (XPS) measurement was carried out to characterize the oxidation of the vinyl end group at the NTS monolayer surface. The XPS spectra were obtained with PHI 5800 ESCA system (Physical Electronics, Co., Ltd.) at room temperature. XPS measurement was performed with a monochromatized AlKα source at 12 kV and 20 mA. The take-off angle of the photoelectron was 10° in order to enhance the signal from the carboxyl group at the monolayer surface. The main chamber of the XPS instrument was maintained at -10^{-7} Pa. All C_{1s} peaks were calibrated to a binding energy of 284.8 eV for hydrocarbon in order to correct the charging energy shifts.

In order to evaluate the surface free energy of the organosilane monolayers, static contact angle measurement was carried out. The contact angles of the water and the methylene iodide droplets on the monolayer surface were measured with a contact angle goniometer CA-D (Kyowakaimenkagaku, Co., Ltd.) at 293 K.

Brewster angle IR spectroscopic measurement (14) was performed to evaluate the formation of layered structure in the NTS multilayer films. The NTS monolayer was transferred onto double mirror polished silicon wafer substrate. The incident angle of p-polarized infrared beam was 73.7° (the Brewster angle of interface between silicon wafer and air) in order to eliminate the multiple reflection of the infrared beam within the silicon wafer substrate. The spectra were recorded at a resolution of 4 cm^{-1}

at 293 K with a Nicolet Magna 860 instrument which was equipped with a mercury-cadmium-telluride (MCT) detector. In order to obtain the spectra with high signal-to-noise ratios, 256 scans were collected.

XR measurement was carried out in order to evaluate the electron density distribution, interfacial roughness and the film thickness of the NTS monolayer and the NTS multilayer film. A monochromator (Multilayer synthetic crystal prepared by deposition of W and Si) was set next to the exit of the X-ray source (Mac Science, Co., Ltd. M18XHF). The CuKα X-ray beam was collimated by two slits with vertical width of 20 mm placed between sample and monochromator. The angular dependence of the specular reflectivity was measured by a series of θ-2θ scans at 293 K. The reflected beam was detected by a scintillation counter after passing through the detection slit. Wave vector was defined as q (=4π sin θ/λ), where θ and λ are the incident angle and the wavelength of X-ray (0.154 nm), respectively.

Results and Discussion

NTS Monolayer In order to investigate a molecular aggregation state of the NTS monolayer, ED studies and AFM observations were carried out. Figure 2 shows the π-A isotherm for the NTS monolayer on the water subphase at 293 K. The AFM image and the ED pattern were taken for the NTS monolayers transferred onto silicon wafer and hydrophilic SiO substrates, respectively, at the surface pressure of 15 mN m^{-1} at 293 K. The π-A isotherm showed a steep increase of surface pressure with a decrease in occupied area. The limiting occupied area estimated from the π-A isotherm was 0.22 nm^2 molecule^{-1}. This magnitude is smaller than that for the OTS monolayer on the water subphase at 293 K.*(3,7)* This indicates that the NTS monolayer formed more dense and closely packed monolayer in comparison with the OTS one. The ED pattern of the NTS monolayer exhibited hexagonal crystalline spots as shown in the inset of Figure 2, indicating the formation of the large-area crystalline monodomain of the NTS monolayer at 293 K. That is, this result indicates that the fairly large crystalline NTS domains in comparison with an electron beam diameter of 10 μm were formed by the surface pressure-induced sintering mechanism such as fusion or recrystallization at the domain interface during compression of the monolayer on the water subphase.*(17,18)* The (10) spacing of the NTS monolayer was evaluated to be ca. 0.42 nm in agreement with that of the OTS monolayer.*(3,7)* Also, AFM observation revealed that the NTS monolayer showed a fairly uniform, smooth and continuous morphology as shown in Figure 2. The mean square surface roughness was evaluated to be ca. 0.3 nm from AFM observation.

The molecular aggregation state in the NTS monolayer was investigated by XR measurement. Figure 3 shows the XR and the fitting curves for the NTS monolayer immobilized onto the silicon wafer substrate at a surface pressure of 15 mN m^{-1} at 293 K. The corresponding calculated electron density profile is shown in the inset. Fitting for the reflectivity curve was performed with MUREX118 software programmed by Dr. K. Sakurai.[19] The first minimum appeared due to a destructive interference between reflections from the top (monolayer surface) and the bottom (monolayer/substrate interface) of the NTS monolayer. This destructive interference was observed at q=1.24 nm^{-1} in the experimental XR curve. Then, the film thickness, L, was calculated by the following equation:

$$L = \frac{\pi}{q_{min}}$$

where q_{min} is the wave vector at the minimum of reflectivity. The film thickness of the NTS monolayer was experimentally calculated to be 2.53 nm. For the fitting analysis, the roughnesses at both the monolayer surface and the monolayer-silicon substrate interface were determined by AFM observations. Also, the thickness of the NTS monolayer was estimated to be 2.54 nm in the fitting analysis, which well corresponds to that magnitude calculated from the minimum of X-ray reflectivity. Since the bond length and the bond angle of C-C and C-C-C are 0.154 nm and 108°,

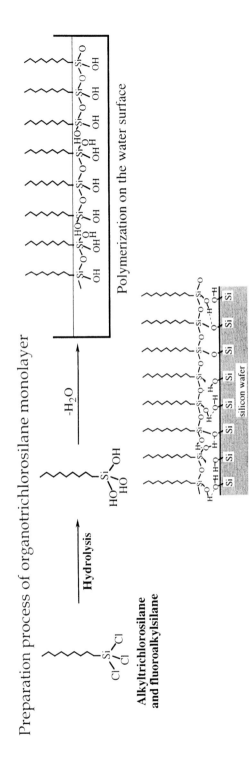

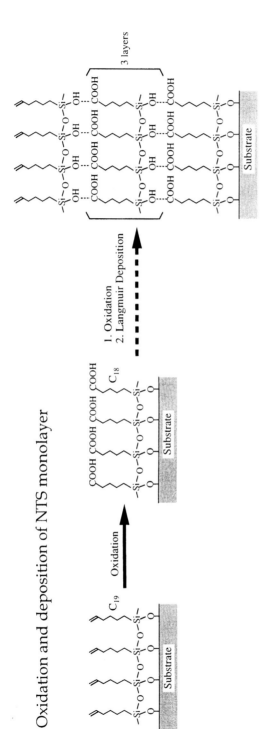

Figure 1 Schematic representation of the formation process of the NTS multilayer film prepared by the Langmuir method.

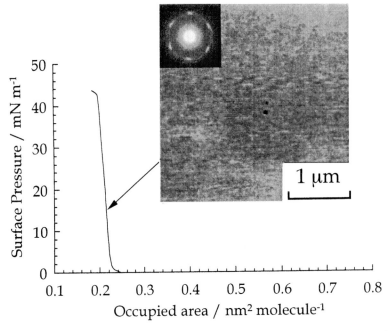

Figure 2   π-A isotherm for the NTS monolayer on the water subphase at 293 K, the AFM image and the ED pattern of the NTS monolayer transferred onto the silicon wafer substrate at a surface pressure of 15 mN m^{-1} at 293 K.

respectively, the length of alkyl chain composed of 17 carbons in *all trans* conformation can be estimated to be 2.15 nm. Therefore, it is reasonable to consider that the alkyl chain of a NTS molecule in the monolayer is in an *all trans* conformation with ca. 0.39 nm of a vinyl end group and, the NTS molecules are oriented perpendicular to the monolayer surface.

It has been reported that the organotrichlorosilane monolayer has a stable surface structure in an aqueous environment without the reorganization of molecules due to the immobilization onto the substrate with strong chemical interaction.*(2,3)* This indicates that a stable organotrichlorosilane monolayer with a hydrophilic surface can be obtained by oxidation of vinyl end groups on the NTS monolayer surface. Figure 4 shows the C_{1s} XPS spectra for the NTS and the carboxylated NTS (NTS_{COOH}) monolayers at the photoelectron take off angle of 10°. The peaks or the shoulders observed at 284.8 and 286.0 and 288.5 eV were assigned to hydrocarbon (C-C*-C), carbon next to carboxyl group (C*-COOH) and carbonyl carbon (-C*OOH), respectively. The C_{1s} XPS curve was separated into the peaks by curve fitting based on a nonlinear least-squares method. C*-COOH and -C*OOH were clearly observed only for the C_{1s} XPS spectra of the NTS_{COOH} monolayer. The integrated intensity ratio of C*-COOH : -C*OOH : C-C*-C was ca. 1:1:10. The only C-C*-C peak was observed for the NTS monolayer. The surface free energy was evaluated by static contact angle measurement of water and methylene iodide droplets on the monolayer surface. The surface free energy of the NTS monolayer was changed from 26.9 mN m^{-1} to 68.8 mN m^{-1} with an oxidation treatment. Therefore, it can be concluded from XPS and contact angle measurements that the higher surface free energy of NTS_{COOH} is attributed to the presence of stable carboxyl groups at the monolayer surface. The vinyl end group of NTS molecule can be easily changed to hydroxyl group, bromine group and so on.*(15)* This suggests that the surface free energy and reactivities of the NTS monolayer can be controlled by various procedures.

Construction of NTS Multilayer Film Since the hydrophobic vinyl end groups can be oxidized to hydrophilic carboxyl groups at the NTS monolayer surface, it is expected that the NTS monolayer can be immobilized onto the NTS_{COOH} monolayer with strong chemical bonds. Furthermore, the NTS multilayer film should be readily constructed by repetitions of the oxidation and monolayer deposition processes. Figure 5 (a) shows the B-FT-IR spectra for the 1- to 5-layers NTS films. The peaks observed at 2918 and 2850 cm^{-1} are assigned to the antisymmetric and the symmetric CH$_2$ stretching bands (v_a(CH$_2$) and v_s(CH$_2$)), respectively. The peak positions for the antisymmetric and the symmetric stretching modes of alkyl chain are reported typically to be in the ranges of 2915-2918 cm^{-1} and of 2846-2850 cm^{-1} for an *all-trans* extended chain *(20)* and at ~2928 cm^{-1} and ~2856 cm^{-1} for a liquid-like disordered chain.*(21)* Therefore, it seems reasonable to consider from B-FT-IR spectra that the methylene chain of the NTS molecule was in a *trans* conformation in the monolayer. This result agrees well with the conclusion of the XR measurement that the monolayer thickness corresponded to the NTS molecular length in *all trans* conformation of a methylene chain as discussed above. The plot of absorbances against the number of layers for v_a(CH$_2$) and v_s(CH$_2$) is shown in Figure 5 (b). The absorbances of v_a(CH$_2$) and v_s(CH$_2$) increased linearly with an increase in the deposition number of the NTS monolayer. Figure 5 (b) indicates that the NTS multilayer film could be formed by repetitions of the oxidation and deposition processes of the NTS monolayer. Since the peak position for v_a(CH$_2$) and v_s(CH$_2$) did not vary with an increase of the monolayer number, it is apparent that the NTS multilayer film was formed by a simple multi-transfer of the NTS monolayer without a change of molecular orientation or conformation.

In order to characterize the layered structure in the NTS multilayer film, XR measurement was carried out. Figure 6 shows the XR and the fitting curves for the 5-

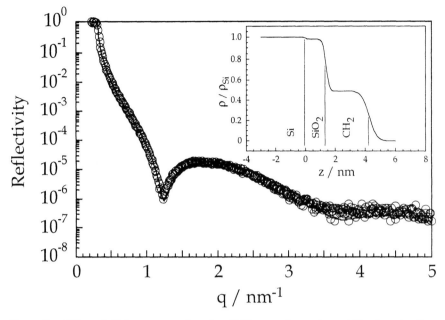

Figure 3   XR and fitting curves for the NTS monolayer immobilized onto the silicon wafer substrate at a surface pressure of 15 mN m^{-1} at 293 K. The corresponding calculated electron density profile is shown in the inset.

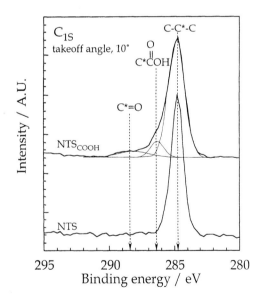

Figure 4   XPS C_{1s} core-level spectra for the NTS and NTS$_{COOH}$ monolayers. The take-off angle of photoelectrons was 10°.

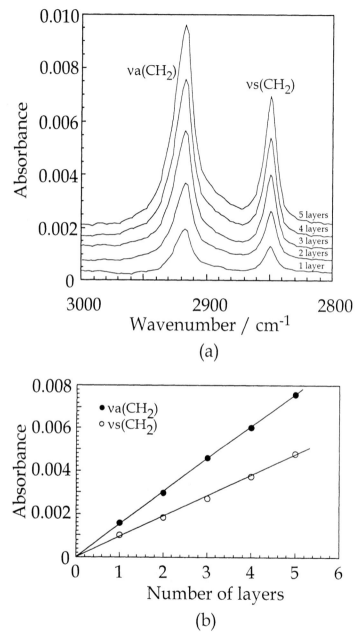

Figure 5 (a) Brewster angle IR spectra for the 1- to 5-layer NTS films produced by the Langmuir deposition. (b) The number of layer dependence of absorbance of the antisymmetric and the symmetric CH_2 stretching bands ($\nu_a(CH_2)$ and $\nu_s(CH_2)$) of the NTS multilayer.

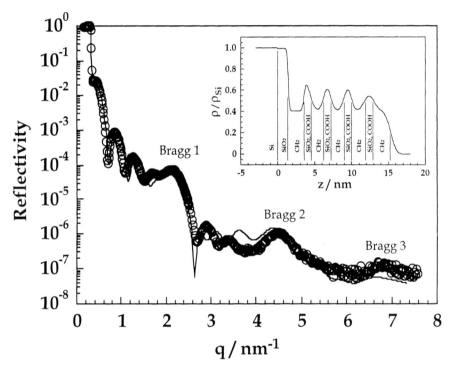

Figure 6 XR and fitting curves for the NTS multilayer film immobilized onto the silicon wafer substrate at a surface pressure of 15 mN m^{-1} at 293 K. The corresponding calculated electron density profile is shown in the inset.

layers NTS multilayer film transferred onto the silicon wafer substrate at a surface pressure of 15 mN m^{-1} at 293 K. The solid line in Figure 6 was the best fitting curve. The pronounced equidistant minima of q corresponded to the destructive interference of X-ray reflected at the top and the bottom of the NTS multilayer film. In this study, the total film thickness, L_{tot}, was calculated as follows:

$$L_{tot} = \frac{(2n-1)\pi}{q_{min}} \quad (n = 1, 2, ...)$$

where n is the index of the Kiessig fringes. The magnitude of L_{tot} was calculated to be ca. 14.13 nm. Furthermore, the several Bragg peaks were superimposed on the reflectivity curve, indicating that the NTS multilayer film is a well-ordered structure. The layer spacing estimated by Bragg peaks observed at q=2.17 (first), 4.52 (second) and 6.92 nm^{-1} (third) was ca. 2.70 nm. This layer spacing corresponded well to the magnitude of one fifth of the total film thickness.

Each layer in the NTS multilayer film can be separated simply into the two component parts, that is, the hydrophobic alkyl chain part and the hydrophilic part composed of COOH and Si-OH groups as shown in Figure 1. The thicknesses of the hydrophobic and the hydrophilic parts were estimated to be 2.15 nm and 0.70 nm based on the fitting analysis for the NTS multilayer film, respectively, as discussed below. These magnitudes correspond well to those calculated by the destructive interference and Bragg peaks. Since the bond length and the bond angle for C-C and C-C-C are 0.154 nm and 108°, respectively, the length of methylene chain composed of 17 carbons in *all trans* conformation was estimated to be 2.15 nm. This means that the extended NTS molecule oriented perpendicular to the surface of the NTS multilayer film. Since the bond lengths of Si-O and C-O bonds are 0.16 and 0.13 nm, respectively, and the bond angles of C-Si-O and C-C-O are 110° and 120°, respectively, the layer thickness occupied by hydrophilic Si-OH and COOH groups could be estimated to be ca. 0.40 nm. Therefore, it seems reasonable to consider that there is a water layer of ca. 0.30 nm thick between the COOH part and the Si-OH one. It was revealed from XR measurement that the NTS multilayer film consisted of a regularly aligned and oriented layer structure perpendicular to the film surface.

(NTS/FOETS) Mixed Monolayer The (crystalline organosilane/amorphous organosilane) mixed monolayers show a phase-separated structure on the water subphase.*(3-6)* Therefore, it is also expected that the (crystalline NTS/amorphous FOETS) mixed monolayer will form a phase-separated structure. Figure 7 (a) shows the AFM image of the (NTS/FOETS) mixed monolayer prepared on the water subphase at 293 K and transferred onto a silicon wafer substrate at a surface pressure of 20 mN m^{-1}. The (NTS/FOETS) mixed monolayer showed that circular flat-topped domains of ca. 1-3 μm diameter were surrounded by a sea-like matrix in a similar fashion to the (OTS/FOETS) one.*(3-6)* The number and the area fraction of domains increased with an increase of the NTS weight fraction. Figure 7 (c) shows the height profile along the line shown in Figure 7 (a). The height difference between the domain and matrix phase was ca. 1.2-1.4 nm, which corresponds to the difference in their molecular lengths. It follows reasonably that the domain and the matrix phases in the (NTS/FOETS) mixed monolayer are composed of the NTS and the FOETS molecules, respectively. Furthermore, LFM measurement was carried out to investigate the surface mechanical properties of the mixed monolayer surface. Figure 7 (b) shows the LFM image of the (NTS/FOETS) mixed monolayer. The bright and dark parts correspond to the regions with higher and lower lateral force. Figure 7 (b) shows that the magnitude of the lateral force of the FOETS phase was higher than that of the NTS domain. In the case of the (OTS/FOETS) mixed monolayer, the FOETS matrix phase showed a higher lateral force than the OTS one due to higher shear strength between the sample surface and cantilever tip, that might originate from a rigid rod-like conformation of the fluoroalkyl group.*(5)* Correspondingly, it seems reasonable to

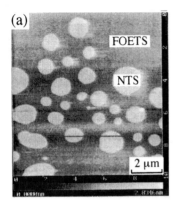
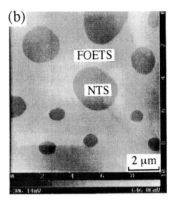
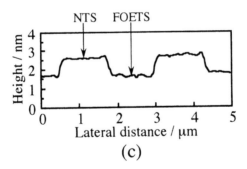

Figure 7 (a) AFM and (b) LFM image of the (NTS/FOETS) mixed monolayer prepared by the Langmuir method and (c) its height profile along the line as shown in (a).

consider that the lateral force of the FOETS matrix phase was higher than that of the NTS domain.

(NTS_{COOH}/FOETS) Mixed Monolayer In order to prepare the mixed monolayer which exhibits a distinct phase-separated structure of the hydrophilic domain and the hydrophobic matrix, the surface of the NTS phase in the (NTS/FOETS) mixed monolayer was oxidized. Figures 8 (a) and (b) show the AFM and the LFM images of the (NTS_{COOH}/FOETS) mixed monolayer prepared by oxidizing the NTS phase in the (NTS/FOETS) mixed monolayer surface. Figure 8 (c) shows the height profile along the line shown in Figure 8 (a). Clearly the surface morphology of the (NTS/FOETS) mixed monolayer was not changed even after oxidation. The height difference between the NTS_{COOH} domain and FOETS matrix phase in the (NTS_{COOH}/FOETS) mixed monolayer was the same as that for the (NTS/FOETS) mixed monolayer. Also, XPS measurement was performed for the (NTS/FOETS) and the (NTS_{COOH}/FOETS) mixed monolayers to confirm the oxidation of the NTS phase. Since the ratio of oxygen/carbon for the (NTS_{COOH}/FOETS) mixed monolayer was larger than that for the (NTS/FOETS) one, it appears that the vinyl end groups of the NTS molecules were oxidized to carboxyl groups. The lateral force of the NTS_{COOH} phase was higher than that of the FOETS phase in the case of the (NTS_{COOH}/FOETS) mixed monolayer as shown in Figure 8 (b), in contrast to the case of the (NTS/FOETS) mixed monolayer. As the NTS_{COOH} phase had hydrophilic carboxyl end groups at the surface, presumably these end groups can form intermolecular hydrogen bonds with neighboring NTS_{COOH} molecules. Therefore, the surface of the outermost NTS_{COOH} phase is expected to show higher shear strength than the case of the NTS one due to difficulty in surface deformation. Also, the magnitude of the lateral force corresponds to the sum of the frictional force and the adhesion force between the sample and the sliding cantilever tip. Guckenberger et al. reported that the enhanced adhesion force due to water capillary force acted as a predominant component of the normal force interacting between the material surface and the cantilever tip.*(22)* Fujihira et al. reported that the adhesion force strongly affected the lateral force in the case of the Si substrate covered partially with OTS.*(23)* Since the adsorbed water layer on the hydrophilic NTS_{COOH} surface might be thicker than that of the hydrophobic surface, the water capillary force interacting between NTS_{COOH} monolayer surface and hydrophilic Si_3N_4 tip could strongly contribute to the adhesion force of the NTS_{COOH} phase. Therefore, it is reasonable to consider that the NTS_{COOH} phase exhibited higher lateral force than the FOETS one due to the formation of intermolecular hydrogen bonding and thicker absorbed water layer as discussed above. As a result, the phase-separated monolayer with a large surface energy gap could be constructed.

In order to investigate influences of the formation of intermolecular hydrogen bonding and the absorbed water layer of the NTS_{COOH} monolayer surface on the magnitude of the lateral force, LFM measurement was carried out for the (NTS_{COOH}/FOETS) mixed monolayer in vacuo, since the adhesion force originating from a water capillary effect might be negligible in LFM measurement in vacuo. Figure 9 shows the LFM image of the (NTS_{COOH}/FOETS) mixed monolayer measured in vacuo. The lateral force of the NTS_{COOH} phase became lower than that of the FOETS phase compared to the case of Figure 8. Therefore, it seems reasonable to conclude from Figure 8 (b) and 9 that an increase in hydrophilicity on the NTS_{COOH} monolayer surface explains why the NTS_{COOH} phase shows a higher lateral force in the (NTS_{COOH}/FOETS) mixed monolayer in air.

(NTS-NTS_{COOH}/FOETS) Ultrathin Film The adsorption behavior of functional molecules depends on the irregularity and the surface roughness as well as wettability, frictional property, and electrical response. Therefore, local height control of the monolayer surface is an important technique to produce the ultrathin film with high functionality. The domain height in the (NTS_{COOH}/FOETS) mixed monolayer can be changed by the chemisorption of the NTS molecules onto the NTS_{COOH}

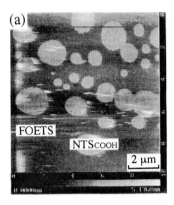
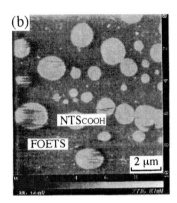
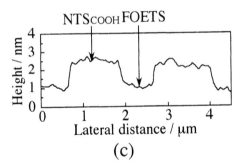

Figure 8 (a) AFM and (b) LFM images of the (NTS$_{COOH}$/FOETS) mixed monolayers measured in air, and (c) its height profile along the line as shown in (a).

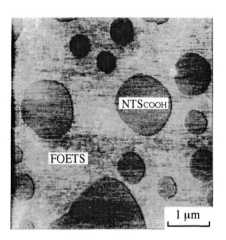

Figure 9　LFM image of the (NTS_{COOH}/FOETS) mixed monolayer measured in vacuum state.

domain. The layer composed of the NTS and NTS_{COOH} molecules is designated as $NTS-NTS_{COOH}$. Figure 10 (a) shows the AFM image of the (NTS-NTS_{COOH}/FOETS) ultrathin film, which was prepared by the chemisorption of the NTS molecules onto the (NTS_{COOH}/FOETS) mixed monolayer from a NTS bicyclohexyl solution. Figure 10 (c) is the height profile along the line shown in Figure 10 (a). The height difference between the domain and matrix phases was ca. 3.8-4.2 nm, which corresponds to the difference between twice the NTS molecular length (2.6 x 2 = 5.2 nm) and FOETS one (ca. 1.2 nm). Therefore, it seems reasonable to conclude that one layer of the NTS molecule was selectively chemisorbed on the NTS_{COOH} monolayer surface with hydrogen bonding formation between carboxyl and silanol groups. Figure 10 (b) shows the LFM image of the (NTS-NTS_{COOH}/FOETS) ultrathin film measured in air. The lateral force of the chemisorbed NTS surface on the (NTS-NTS_{COOH}/FOETS) ultrathin film was lower than that of the FOETS phase such as the (NTS/FOETS) mixed monolayer (Figure 8 (a)). This clearly indicates that the NTS layer is selectively chemisorbed on the NTS_{COOH} surface of the (NTS_{COOH}/FOETS) mixed monolayer and the surface characteristics for the (NTS/FOETS) mixed monolayer and the (NTS-NTS_{COOH}/FOETS) mixed film are comparable with respect to the magnitude of the lateral force.

Figure 11 shows the schematic representation of the formation process of the (NTS-NTS_{COOH}/FOETS) ultrathin film prepared by utilizing the Langmuir and chemisorption methods. A NTS phase in the (NTS/FOETS) mixed monolayer was selectively oxidized to give carboxyl groups. That is, the active reactive site can be introduced in the NTS phase of the (NTS/FOETS) mixed monolayer surface. Furthermore, in the chemisorption process of the NTS molecule onto the (NTS_{COOH}/FOETS) mixed monolayer, the NTS molecule adsorbed preferentially onto the oxidized NTS_{COOH} phase domain. Thus, the domain height in the (NTS/FOETS) mixed monolayer can be controlled by utilizing the NTS molecule with the vinyl end group through the Langmuir and chemisorption methods.

Conclusion

The NTS monolayer which oriented the reactive vinyl groups at the monolayer surface was prepared by the Langmuir method. The NTS monolayer is in a hexagonal crystalline state at 293 K, and the size of the NTS mono-crystalline domain extended to a distance of 10 µm. The stable hydrophilic NTS_{COOH} monolayer was obtained by oxidation of the NTS monolayer. The NTS multilayer film was constructed by repetitions of the processes of oxidation of the vinyl end group and depositions of the NTS monolayers. It was shown from X-ray reflectivity and B-FT-IR measurements that the NTS multilayer film has a well-ordered layer structure.

The (NTS/FOETS) mixed monolayer with phase-separated structure was prepared by the Langmuir method. The hydrophilic domain with the carboxyl groups (NTS_{COOH}) was introduced by oxidization of the vinyl end groups of the NTS phase in the (NTS/FOETS) mixed monolayer. It was also shown that the domain height in the (NTS/FOETS) mixed monolayer can be controlled by utilizing the NTS molecule with the vinyl end group through the Langmuir and chemisorption methods.

Acknowledgment

This study was partially supported by a Research Fellowship of the Japan Society for the Promotion of Science for Young Scientists, and by a Grant-in-Aid for COE Research and Scientific Research on Priority Areas, "Electrochemistry of Ordered Interfaces " (No. 282/09237252), from the Ministry of Education, Science, Sports and Culture of Japan. The FOETS was kindly supplied by Shin-Etsu Chemical Ltd., Co..

 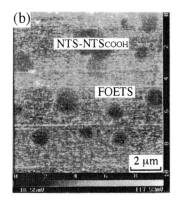

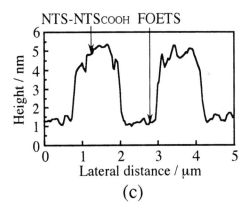

Figure 10 (a) AFM and (b) LFM image of the (NTS-NTS$_{COOH}$/FOETS) ultrathin film and (c) its height profile along the line shown in (a).

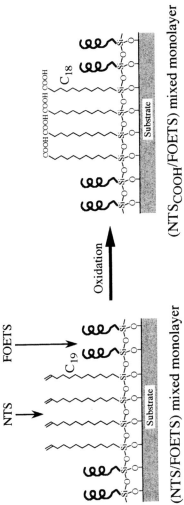

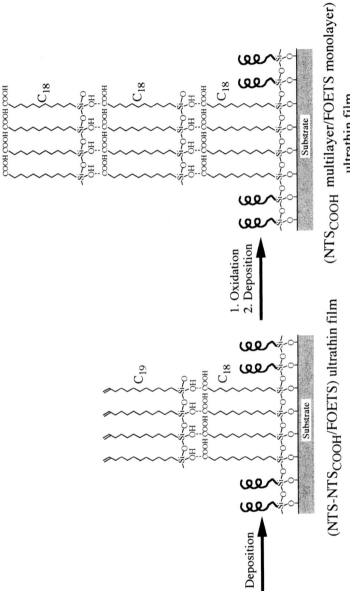

Figure 11 Schematic representation of the domain height control in the (NTS/FOETS) mixed monolayer. NTS layer was built-up on the oxidized NTS phase.

Literatures cited

1. Ulman, A. *An Introduction to Ultrathin Organic Films : From Langmuir-Blodgett to Self-Assembly* (Academic Press, New York, 1991).
2. Sagiv, J. *J. Am. Chem. Soc.* **1980**, *102*, 92.
3. Ge, S. R.; Takahara, A.; Kajiyama, T. *J. Vac. Sci. Technol.* **1994**, *A 12(5)*, 2530.
4. Ge, S. R.; Takahara, A.; Kajiyama, T. *Langmuir* **1995**, *11*, 1341.
5. Kajiyama, T.; Ge, S. R.; Kojio, K.; Takahara, A. *Supramol. Sci.* **1996**, *3*, 123.
6. Takahara, A.; Kojio, K.; Ge, S. R.; Kajiyama, T. *J. Vac. Sci. Technol.* **1996**, *A14(3)*, 1747.
7. Kojio, K.; Ge, S. R.; Takahara, A.; Kajiyama, T. *Langmuir* **1998**, *14*, 971.
8. Takahara, A.; Ge, S. R.; Kojio, K.; Kajiyama, T. In *Scanning Probe Microscopy of Polymers"* Ratner, B. D.; Tsukruk, V. V. Ed.; *ACS Symp. Ser.* **1998**, *694*, 204.
9. Kojio, K.; Takahara, A.; Kajiyama, T. *Rept. Prog. Polym. Phys. Jpn.* **1998**, *41*, in press.
10. Ge, S. R.; Kojio, K.; Takahara, A.; Kajiyama, T. *J. Biomater. Sci. Polym. Edn.* **1998**, *9*, 131.
11. Kojio, K.; Takahara, A.; Kajiyama, T. *Rept. Prog. Polym. Phys. Jpn.* **1995**, *38*, 371.
12. Maoz, R.; Sagiv, J. *Langmuir* **1987**, *3*, 1045.
13. Bain, C. D.; Whitesides, G. M.; *J. Am. Chem. Soc.* **1989**, *111*, 7164.
14. a) Harrick, N. J. *Appl. Spectrosco.* **1977**, *31*, 548. b) Maoz, R.; Sagiv, J.; Degenhardt, D.; Möhwald, H.; Quint, P. *Supramol. Sci.* **1995**, *2*, 9.
15. Wasserman, S. R.; Tao, Y.-T.; Whitesides, G. M. *Langmuir* **1989**, *5*, 1074.
16. Kajiyama, T.; Oishi, Y.; Uchida, M; Morotomi, N; Ishikawa, J.; Tanimoto, Y. *Bull. Chem. Soc. Jpn.* **1992**, *65*, 864.
17. Kajiyama, T.; Oishi, Y.; Uchida, M.; Morotomi, N.; Kozuru, H. *Langmuir* **1992**, *8*, 1563.
18. Kajiyama, T.; Oishi, Y.; Uchida, M.; Takashima, Y. *Langmuir* **1993**, *9*, 1978.
19. Sakurai, K.; Iida, A.; *Adv. X-ray Anal.* **1992**, *35*, 813.
20. MacPhail, R. A.; Strauss, H. L.; Snyder, R. G; Elliger, C. A. *J. Phys. Chem.* **1984**, *88*, 334.
21. Strauss, H. L.; Snyder, R. G.; Elliger, C. A. *J. Phys. Chem.* **1982**, *86*, 5145.
22. Guckenberger, R.; Manfred, H.; Cevc, G.; Knapp, F. H.; Wiegräbe, W.; Hillebrand, A. *Science* **1994**, *266*, 1538.
23. Fujihira, M.; Aoki, D.; Okabe, Y.; Takano, H.; Hokari, H.; Frommer, J.; Nagatani, Y.; Sakai, F. *Chem. Lett.* **1996**, 499.

Chapter 23

An Investigation of the Surface Properties and Phase Behavior of Poly(dimethylsiloxane-*b*-ethyleneoxide) Multiblock Copolymers

Harri Jukarainen[1], Stephen J. Clarson[2], and Jukka V. Seppälä[3]

[1]Leiras Oy, Tykistokatu 4 D, PL 415, 20101 Turku, Finland
[2]University of Cincinnati, College of Engineering,
P.O. Box 210018, Cincinnati, OH 45221
[3]Helsinki University of Technology, Polymer Technology,
PL6100, Helsinki 02015 TKK, Finland

Several different multi-block poly(dimethylsiloxane-b-ethyleneoxide) (PDMS-b-PEO) copolymers were synthesized in order to study their phase behavior and surface properties. The surface properties of the PDMS-b-PEO polymers were studied using water contact angle measurements. Their phase behavior was studied by differential scanning calorimetry (DSC). The water contact angle values were measured for cured PDMS-b-PEO copolymer films. The contact angle values were found to change quite dramatically with extended measurement time for the films made of single phase PDMS-b-PEO copolymers. Initially, single phase materials showed similar values to PDMS, but dropped far below the values of PDMS with extended time. This observation indicates the reorientation or migration of PEO segments on the surface, giving an environment responsive character for these materials. In the case where the film was made of phase separated PDMS-b-PEO copolymer, behavior similar to PDMS was found for the whole range of the time measurements. This indicates that the surface of the film was covered with PDMS blocks and the PEO blocks were not able to migrate or reorient to the surface.

The surface behavior of PDMS copolymers and PDMS containing blends is almost always dominated by their PDMS component because of its low surface energy. Thus for materials where PDMS is present the surface is usually covered with PDMS segments. This quality gives PDMS its wide range of commercial applications on mold release, polishing, defoaming agents, etc. (*1,2*). In the studies (*3,4*) where PDMS film surfaces have been functionalized with covalently bonded hydrophilic groups by plasma polymerization, eventually it has resulted that the hydrophilic groups are buried and a pure PDMS surfaces is restored (hydrophobic recovery), which is not usually desirable. This type of behavior is thought to be mainly due to diffusion of low molecular weight species from the bulk onto the surface or by segmental reorientation (5). In the present study we have prepared films from the single and two phase PDMS-b-PEO copolymers and studied their surface behavior with water contact angle measurements.

Experimental

Two different kinds of α,ω-terminated poly(ethylene glycols) (PEO) were used as a starting material for the PDMS-b-PEO copolymers; commercial samples of PEOs which were subsequently terminated with allyl groups by us (6) and α,ω-divinyl ether terminated PEOs (commercial sample) were used as a prepolymers for the copolymer synthesis. Three different kinds of α,ω-dimethylsilylhydride-terminated PDMS (commercial samples) were used as the other prepolymer for preparing different kinds of multi-block PDMS-b-PEO copolymers. These copolymers were prepared by Pt(0) catalyzed (Karsted's catalyst) hydrosilylation (7,8) reactions using oxygen as a cocatalyst. In order to get vinyl group terminated PDMS-b-PEO copolymers, the polymerizations were carried out in a 10 % molar excess of allyl or vinyl terminated PEOs.

Crosslinked films were prepared from these PDMS-b-PEO copolymers by hydrosilylation reaction between the terminal vinyl groups and multi hydride functional poly(dimethyl-co-methylhydride siloxane) (9), which was a commercial sample from Bayer (CA 730). A Si-H/vinyl ratio of 3 was used. Again Pt(0) was used as a catalyst and ethynyl cyclohexanol as an inhibitor, which prohibited curing before raising the temperature to 115 °C.

DSC (PL) runs were done for liquid samples of PDMS-b-PEO under nitrogen using liquid nitrogen cooling. Cooling and heating rates were 10 °C/min. Water in air contact angle measurements were performed by applying a drop (4±0.2 mg) of distilled water on the planar crosslinked sample film and then taking measurements after 15, 300, 600, 900, 1200, and 1500 seconds. Five parallel measurements were done for each sample and the average result was reported. Naturally, a little evaporation of the water droplets occurred during the course of contact angle measurements, but this did not interfere with the aim of the study, which was a comparison of the different materials. The measured total evaporation times for the droplets were about 35 min for all of the samples for the laboratory condition used. The contact angle device used was a Rame-Hart's goniometer model 100-00

Results and Discussion

The PDMS-b-PEO copolymers prepared for the phase study are presented in Table I. The number of phases of the copolymers were determined according to clarity and DSC thermograms of the copolymers. DSC thermograms are shown in Figure I. The transition values obtained from DSC runs are collected in Table II.

Table I. The PDMS-b-PEO copolymers prepared, their visual clarity, and the determined number of phases

Polymer	M_n PDMS block (g/mol)	M_n PEO block (g/mol)	Clear or opaque	Single or two phase
BP1	6000	270	clear	single
BP2	6000	520	clear	single
BP3	2000	270	clear	single
BP4	760	270	clear	single
BP5	6000	1350	opaque	two

Table II. Glass transition temperatures (T_g), cold crystallization temperatures (T_c), and melting points (T_m) of the PDMS-b-PEO copolymers (from Figure 1).

Polymer	T_g (°C)	T_c (°C)	$T_{m,PDMS}$ (°C)	$T_{m,PEO}$ (°C)
BP1	-127	-88	-51	-
BP2	-125	-93	-49	-
BP3	-112	-	-	-
BP4	-103	-	-	-
BP5	-125	-	-50	37

As all the copolymers except BP5 were clear and did not have the T_g or melting peak of PEO, it was determined that the only gross phase separated polymer was BP5, the rest being single phase.

The reason why BP1 and BP2 had the T_c and T_m typical of PDMS was probably due to the fact that these polymers had the longest PDMS blocks (6000 g/mol) and thus they were able to form the helical conformation which gave these peaks (10). The reason why BP3 and BP4 lacked that behavior was most likely the reduced size of the PDMS blocks and the growing weight proportion of the PEO. This would give more restrictions to the PDMS blocks and thus prohibit the formation of an ordered structure, therefore totally amorphous materials were achieved. This same behavior has been reported for random copolymers of PDMS (11).

Increasing the amount of PEO in the single phase materials (BP1...BP4) had the effect of increasing the T_g. This was according to expectations considering that the T_g of pure PEO lies around -50 °C (12).

In order to study the surface behavior of these copolymers a set of PDMS-b-PEO copolymers was prepared. The size of the PDMS blocks were 6000 g/mol and the size of the PEO blocks were varied. All the copolymers except the one with the PEO block size of 1350 g/mol were single phase. These copolymers were cured and contact angle measurements for these crosslinked films were carried out. Results of the experiments are shown in Figure 2.

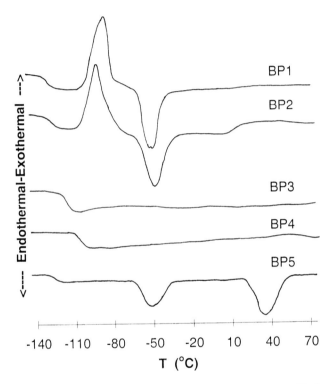

Figure 1. DSC thermograms of the PDMS-b-PEO copolymers. Each curve is labeled with the abbreviation of the copolymer (ref. Table I).

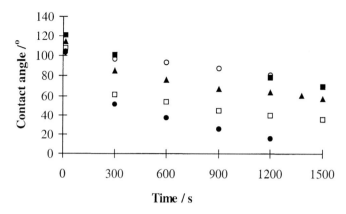

Figure 2. Contact angle values versus time for cured PDMS and different kinds of cured PDMS-b-PEO copolymers where the size of the PDMS block was 6000 for every sample and the size of the PEO block varied; ▲ = 270 g/mol, □ = 330 g/mol, ● = 520 g/mol, ■ = 1350 g/mol. o = PDMS film. All the films except the one marked with filled square (■) were single phase.

From the Figure 2. it can be seen that contact angle values dropped quickly with different rates from 0 to 300 s after which similar declining rates were achieved. The time from 0 to 300 s was related to the time when changes on the surface of the film happened.

The copolymer which had the highest amount of PEO was phase separated and had similar contact angle values with the pure PDMS film. This suggests that no PEO is present at any time on the surface of this copolymer, as could be expected for the phase separated material. Initial water contact angle values (15 s) similar to PDMS were found for the single phase copolymers but with extended time (300 s) their values dropped quite dramatically. This drop increased with increasing size and wt.% of the PEO blocks. This also suggests that for the single phase materials the surface is predominantly PDMS initially, but with extended time in contact with water the hydrophilic PEO blocks or segments migrate or reorient to the surface. This rearrangement of species at the surface causes the declining contact angle values and changing interfacial tension between water and the copolymer.

Conclusions

The number of phases of the multi-block PDMS-b-PEO copolymers can be detected with DSC and the surface behavior of these copolymers is ruled by their phase behavior. Phase separated copolymers were found to behave like PDMS in contact angle measurements implying that only PDMS is present on and near the surface. Single phase copolymers have similar values to PDMS initially, but with extended time contact angle values drop quite dramatically, which suggests the migration or reorientation of the PEO blocks on the surface. This gives a noted environmentally responsive character to these single phase copolymers.

Acknowledgments

It is a pleasure to thank Leiras Oy, Finland for providing the financial support for this study.

References

1. Owen, M.J., *Chemtech.*, **1981**, *11*, 288.
2. Owen, M.J., *Comm. Inorg. Chem.*, **1988**, *7*, 195.
3. Owen, M.J., In *The Analytical Chemistry of Silicones*, A.L. Smith, Ed., John Wiley & Sons Inc., New York, 1991.
4. Owen, M.J., In *Siloxane Polymers*, Clarson, S.J. and Semlyen, J.A. Eds., Prentice Hall, Englewood Cliffs, NJ, 1993.
5. Stasser, J.L. and Owen, M.J., *Polym. Prepr. (Am. Chem. Soc., Div. Polym. Chem.*, **1997**, *38*, 1087.
6. Yang, M-H., Li, L-J., Ho, T-F, *J. Ch. Colloid & Interface Soc.*, **1994**, *3(17)*, 19.
7. Lewis, L.N., Stein, J., Gao, Y., Colborn, R.E., and Hutchins, G., *Platinum Metals Rev.*, **1997**, *41(2)*, 66.
8. Jukarainen, H., Ruohonen, J., Lehtinen, M., Ala-Sorvari, J. and Seppälä, J., *Fin. Pat. Apl.* 973427 (**1997**).
9. Lestel, L., Cheradame, H., and Boileau, S., *Polymer,***1990**, *31*, 1554-58.
10. Clarson, S.J. and Rabolt, J., *Macromolecules*, **1993**, *26*, 2621.
11. Kennan, J.J., Siloxane Copolymers in '*Siloxane Polymers*' (Eds. Clarson, S.J., and Semlyen, J.A.), *Prentice Hall*, Englewood Cliffs, NJ, **1993**, 73.
12. Bailey, F.E.Jr., Koleske, J.V., *Alkylene Oxides and Their Polymers, Marcel Dekker*, New York, **1991**, 191.

Chapter 24

Poly(siloxyethylene glycol) for New Functionality Materials

Yukio Nagasaki[1] and Hidetoshi Aoki[2]

[1]Department of Materials Science and Technology,
Science University of Tokyo, Noda 278–8510, Japan
[2]Research and Development Center, Hokushin Corporation,
Turumi, Yokohama 230–0003, Japan

Poly(siloxyethylene glycol)s, (PSEGs) which consist of alternating oligooxyethylene and oligosiloxane derivatives, were prepared via polycondensation reactions between oligoethylene glycol and bis(diethylamino)dimethylsilane. When the dimethylamino group was used as the leaving group of the electrophilic silyl compound, the polycondensations proceeded smoothly without any degradation. It should be noted that cleavage reactions takes place when dichlorodimethylsilane was used as an electrophile due to the liberated HCl. PSEGs thus prepared possesses alternating hydrophilic and hydrophobic unit in the main chain, which showed unique phase transition phenomena in aqueous media. Especially, the lower critical solution temperature (LCST) can be controlled by their Si-content. The glass transition temperature can also be varied by the Si-content. Thus, the unique structure and characteristic of PSEG derivatives can be utilized as matrix for functional materials such as hydrogels with stimuli-sensitivity, resists with high etching durability and water development and as a matrix for solid electrolytes. This paper reviews the synthesis and applications of PSEG derivatives.

Polymers having an organosilicon moiety show unique characteristics due to the nature of silicon atom.(*1,2*) We have been studying on synthesis of several types of silicon-containing polymers and investigating their characteristics. For example, polystyrene possessing doubly trimethylsilyl groups in each repeating unit, poly[4-{bis(trimethylsilyl)methyl}styrene] (PBSMS), shows fairly high oxygen permselectivity among vinyl polymers.(*3*) Based on this characteristic of PBSMS, we are preparing high performance artificial lung via copolymerization of BSMS with 2-hydroxyethyl methacrylate.(*4*) The same polymer can be applicable as a resist material due to the high dry etching resistance of the organosilicon moiety. PBSMS shows negatively-working electron-beam resist with high resolution parameter.(*5*)

Our alternating polymer, poly(silamine), which consists of alternating 3,3-dimethyl-3-silapentamethylene and N,N'-diethylethylenediamine units in the main chain,(6) shows phase transition phenomena in aqueous media.(7) Because of suitable hydrophilic and hydrophobic balances in the main chain, the polymer is soluble in neutral water at lower pH. With increasing temperature or pH, the solution becomes turbid due to the deprotonation from amino group in the main chain. It should be noted that the rubber elasticity of the polymer drastically changes with the phase transition. Actually, the glass transition temperatures of poly(silamine) with and without protonation were +80 and -80 °C.(8) Such a remarkable change in the glass transition temperature was due to the freezing of molecular motion by the anion binding to the repeating silicon atoms along with protonation of amino groups in the main chain. Thus, the character of silicon segment in the polymer influences the characteristics of the base polymer molecule.

In this paper, we describe another new silicon-containing polymer, poly(siloxyethylene glycol), which consists of alternating oligoether and oligosiloxane linkage in the main chain.(9) The synthesis, characterizations, characteristics in aqueous media are summarized. The applications for an electron-beam resist (10) and a ion conducting polymer are also described.

Experimental Section

Materials: Commercial THF (Wako) was purified in the following way. After THF was pre-dried with KOH for several days, reflux was carried out over lithium aluminum hydride for 5h, followed by distillation. The fraction at 68 °C was collected and stored under Ar atmosphere. Oligo(ethylene glycol)s were dried at 110 °C for 2 d in vacuo. Diethylamine (Wako), dichlorodimethylsilane (Shin-etsu Chemical Co. Ltd.), 1,3-dichlorotetramethyldisiloxane (Shin-etsu Chemical Co. Ltd.) and dichlorodivinylsilane (Shin-etsu Chemical Co. Ltd.) were used as received. Bis(diethylamino)dimethylsilane (DAS), 1,3-bis(diethylamino)tetramethyldisiloxane (DADS) and bis(diethylamino)divinylsilane (DADVS) were prepared by the reaction of diethylamine with the corresponding dichlorosilanes.(11) The boiling points of DAS, DADS and DADVS were 30 - 31 °C / 1 mmHg and 79 °C / 3 mmHg and 63.5 - 64.5 °C / 2mmHg, respectively. Other materials were used as received.

Polymer Synthesis: A typical polymerization was performed in a 100-mL round-bottomed flask with a 3-way stopcock. After 6.19 g of OEG (M_n = 300) (20.6 mmol) was weighed into the flask, the inside of the reactor was degassed sufficiently and filled with Ar gas. 20 mL of THF and 4.17 g of DAS (20.6 mmol) were then added to the flask via syringe and allowed to react for 24 h at 60 °C. After THF and liberated diethylamine were removed by evaporation in vacuo, the obtained polymer was analyzed by GPC. The ^{1}H NMR of the obtained polymer was recorded after the sample was purified by GPC fractionation.

Hydrolytic Stability Test: The stability of PSEG against hydrolysis in aqueous media was estimated as follows: A polymer sample was dissolved in phosphate buffer (1.5 wt.%; pH = 7.0; I = 0.05) at 4 °C. Every few hours, an aliquot of the solution was subjected to measurement of its turbidity after the sample was heated above the

LCST. The turbidity was monitored as a normalized attenuance using the following equation.

$$\text{Normalized Attenuance} = \log(I_R / I_{s,t}) / \log(I_R / I_{s,min})$$

where, I_R and $I_{s,t}$ denote the intensities of reference (phosphate buffer) and sample after t hours reaction, respectively. $I_{s,min}$ represents the minimum intensity of the sample during the experiment.

Resist Processing: For the photo irradiation examination, a THF solution of the polymer sample was prepared with 20 wt.% concentration. Tetramethylmethanetetra(3-mercaptopropyonate) (4TP-5) (1.34 mmol) and benzoinmethylether (BME) (0.21 mmol) were added to the THF solution of the polymer as follows: $[-CH=CH_2]$: $[-SH]$: $[BME] = 1 : 0.2 : 0.04$. The prepared solution was spin-coated at 3000 r.p.m. for 30 s on a silicon wafer surface. The thickness of the resist film thus obtained was about 1μm, and it was irradiated with a 500W high pressure mercury lamp (USHIO INC.) equipped with a heat cut filter (HA-30 type, Kenko). A dose that ranged from 15 to 3240 mJ/cm^2 was employed. In an EB exposure examination, a THF solution of PVSE300 (10 wt.%) was spin-coated at the same conditions on a Si wafer surface. The thickness of the PVSE300 film thus obtained was 0.7 - 0.9 μm, and EB was exposed using a JEOL JSM-5200 scanning electron microscope (SEM) equipped with a Tokyo Technology L&S Pattern Generator LSPG1-1S at several probe currents (3, 10 and 100 pA) and accelerating voltages (2, 5, 10 and 20 kV). A dose that ranged from 0.01 - 44 μC/cm^2 was employed. The PVSE300 film on the Si wafer was developed by soaking in cold water (4 °C) for 2 or 10 minutes after the photo irradiation and EB exposure. The remaining film thickness was measured with a Tencor ALPHA STEP 300. The sensitivity and resolution parameter were calculated from the sensitivity characteristic curve as Dg^{50} and γ value. The Dg^{50} value denotes the exposure dose for a remaining thickness of 50 % of the initial thickness.[12,13] The γ value was calculated from the following equation.

$$\gamma = [2\log(Dg^{50}/Dg^i)]^{-1}$$

where Dg^i represents the minimum dose in which the cross-linking reaction is proceeded by EB.[14]

O$_2$ RIE Resistance: A polymer sample for the O$_2$ RIE experiment was prepared in a similar way as stated above, *viz.*, the film was exposed to an EB dose of 10 μC/cm^2 at 100 pA, 20 kV followed by development in 4 °C water for 2 minutes. The plasma etching experiment was carried out in an oxygen atmosphere in a Model: BP-1, from the Samco International Research Corporation. The radio frequency (Rf) power and gas pressure were 100 W and 0.5 Torr, respectively. The etching time was varied from 5 to 60 minutes. The temperature of the substrate was maintained by heating the sample-plate at 40 °C during the etching. The thickness of the film was measured before and after etching using a ALPHA STEP 300. The etch rate curve was plotted as the reducing film thickness versus etching time.

Ionic conductivity of PSEG gel: PSEG having pendent vinyl groups was mixed with tetrafunctional thiol, pentaerythritol tetrakis(2-mercaptopropionate) (4TP-5), and photoinitiator, 2,2-dimethoxy-2-phenylacetophenone (DMPA), in acetonitrile. Lithium perchlorate was dissolved to the mixture. The ratio for [Li] / [-O-] was in the range of 0.025 - 0.317. After the polymer was crosslinked by UV light (1.6 J/cm^2), the obtained sample was dried for 48 h at ambient temperature.

The obtained sample was cut to disk shape, the diameter of which was 13 mm. The disk was placed between two Pt electrodes and the conductivity was measured in thermoregulated cell from -20 to 100 °C.

Measurement: GPC measurements were performed on a Toso CCPE with a RI detector and TSK-Gel GM$_{H6}$ x 2 + GM$_{HXL}$ x 2 columns. The NMR spectra (^{1}H: 399.65 MHz; ^{13}C: 100.53 MHz) were determined on a JEOL EX400 spectrometer using CDCl$_3$ as a solvent at room temperature. Chemical shifts relative to CHCl$_3$ (^{1}H: δ = 7.26) were employed. Glass transition temperatures of the polymers were determined using a differential scanning calorimeter (DSC) (Mettler TA4000 system) at a heating rate of 20 °C/min from -170 °C to 200 °C. The turbidity of the polymer solution was recorded using Photorode Mettler DP 550. The absorption at 550 nm was monitored.

Results and Discussion

Synthesis and characteristics of poly(siloxyethylene glycol)

Synthesis and Characterization: In this study, PSEGs were synthesized by polycondensation reactions between OEG and a bi-functional silicon monomer (Scheme 2). There are several choices of leaving groups from silicon monomers. Especially, dichlorosilane is the most common organosilicon monomer for the reaction with hydroxyl groups. However, if dichlorosilane were employed to synthesize the PSEG, the main chain of the polymerized product should be decomposed by the liberated acid.(*15*) Therefore, we employed a diethylamino group as a leaving group in the bi-functional silicon monomers. PSEG possessing vinyl groups as pendent, PVSE was also synthesized by a polycondensation reaction between OEG and bis(diethylamino)divinylsilane (DAVS) in the same way as that of PSEG (Scheme 2).

The polycondensation reaction between OEG and diethylaminosilane derivatives proceeded smoothly, and no gel formation took place even in the case of DAVS which possesses vinyl groups as a side chain.

Figure 1 shows the gel permeation chromatography (GPC) profiles of the representative polycondensation products; PSEG(1/7), PSEG(2/7) and PVSE(1/7), where the numbers in parenthesis (m/n) denote the number of Si atoms in siloxane and ethylene glycol units in OEG, respectively. The polymers thus obtained were viscous liquids, and the number average molecular weights of the products were in

Scheme 1

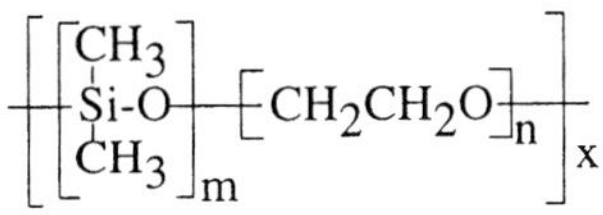

$$\left[\left[\begin{array}{c}CH_3\\|\\-Si-O-\\|\\CH_3\end{array}\right]_m [CH_2CH_2O]_n\right]_x$$

Poly(siloxyethylene glycol), PSEG(m/n)

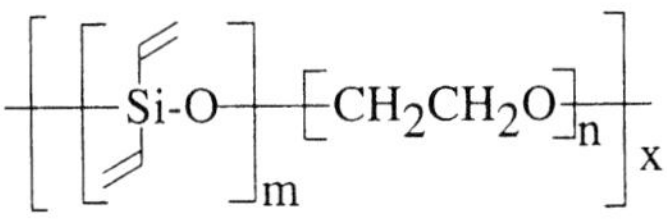

$$\left[\left[\begin{array}{c}\diagup\\-Si-O-\\\diagdown\end{array}\right]_m [CH_2CH_2O]_n\right]_x$$

Poly(divinylsiloxyethylene glycol), PVSE(m/n)

Scheme 2

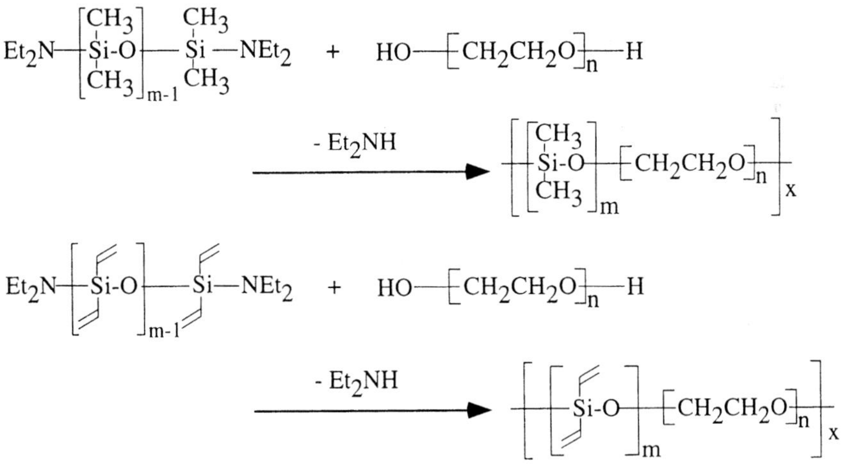

$$Et_2N-\left[\begin{array}{c}CH_3\\|\\Si-O\\|\\CH_3\end{array}\right]_{m-1}\begin{array}{c}CH_3\\|\\Si-NEt_2\\|\\CH_3\end{array} + HO-[CH_2CH_2O]_n-H$$

$$\xrightarrow[\text{- Et}_2\text{NH}]{} \left[\left[\begin{array}{c}CH_3\\|\\-Si-O-\\|\\CH_3\end{array}\right]_m [CH_2CH_2O]_n\right]_x$$

$$Et_2N-\left[\begin{array}{c}\diagup\\Si-O\\\diagdown\end{array}\right]_{m-1}\begin{array}{c}\diagup\\Si-NEt_2\\\diagdown\end{array} + HO-[CH_2CH_2O]_n-H$$

$$\xrightarrow[\text{- Et}_2\text{NH}]{} \left[\left[\begin{array}{c}\diagup\\-Si-O-\\\diagdown\end{array}\right]_m [CH_2CH_2O]_n\right]_x$$

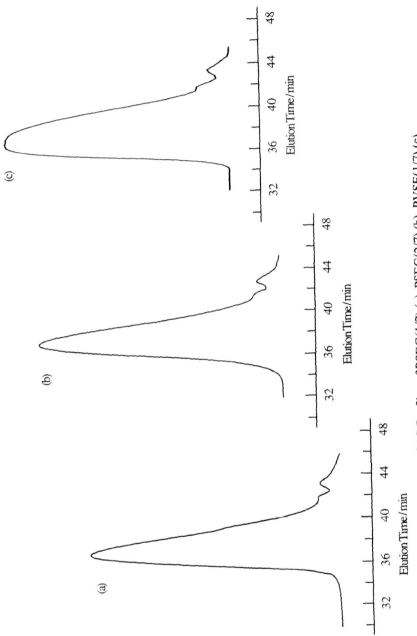

Figure 1. GPC Profiles of PSEG(1/7) (a), PSEG(2/7) (b), PVSE(1/7) (c).

the range of 3,500 - 17,700 with a distribution factor (M_w/M_n) of 1.17 - 2.27. It should be noted that if the polycondensation were carried out using dichlorosilane compounds instead of diaminosilane compounds, a large amount of cyclic oligomers was produced along with linear oligomers to result in a very wide molecular weight distribution with rather low average molecular weight.

To analyze the chemical structure of the obtained polymers, ^{1}H-NMR measurements were carried out. Figure 2 shows the ^{1}H-NMR spectra of the representative products (The same samples as in Figure 1). When diaminodimethylsiloxanes were used as one of the counterparts in the polycondensation reactions, two proton signals appeared at 0.10 ppm and 3.65 ppm, which are assignable to $OSi(CH_3)_2O$ and OCH_2CH_2O protons, respectively (Figure 2a and b). The integral ratios of $OSi(CH_3)_2O$ vs. OCH_2CH_2O in Figure 2a and b were 6/27 and 12/27, respectively, which agreed well with PSEG(1/7) and PSEG(2/7), respectively. Thus, new silicon-containing polymers having alternating oligo(ethylene glycol) and oligosiloxane in the main chain were obtained through polycondensation reactions. As shown in Figure 2c, vinyl protons adjacent to Si atom appeared around 6.0 ppm(*16*) along with oxyethylene protons at 3.5-4.0 ppm. The integral ratio of olefinic protons vs. oxyethylene protons was 6.0/27.4, which agreed well with the alternating structure of OEG with divinylsiloxane in the main chain as shown in Scheme 1. On the basis of these results, polymers having alternating OEG and oligosiloxane with and without a reactive double bond could be prepared easily. The polycondensation conditions and other data for the obtained polymers are summarized in Table 1.

Physicochemical Properties Because PSEG derivatives possess a new repeating unit in the main chain, their physicochemical properties are of interest. From the differential scanning calorimetry (DSC) profiles shown in Figure 3, it was found that the mobility of PSEG derivatives is fairly high as anticipated. For example, the glass transition temperature (T_g) of PSEG(1/9) was -68.6 °C (Figure 3a). With increasing Si content in the polymer, the T_g decreased as shown in Figure 4.

In this way, the mobility of the PSEG series was controllable between poly(ethylene glycol) (PEG) (T_g = -53 °C)(*17*) and silicone (-127 °C).(*18*) The mobility of the PSEG was not affected very much by the substituents on the Si atom in the repeating units. For example, PVSE(1/7), which possesses two vinyl groups on the Si atom in each repeating unit, showed a T_g of -72.5 °C as shown in Figure 3b, which is similar to that of PSEG having the same Si content as PVSE(1/7). Thus, the T_g of the PSEG derivatives was governed only by Si content. Crystallization phenomena, however, were influenced significantly by the substituents on the Si atom in PSEG series. For example, PSEG(1/9) shows a peak assignable to a melting point (Tm) at - 6.4 °C as shown in Figure 3a. However, in the case of PVSE(1/7), there is no melting peak under the same conditions (Figure 3b). In summarizing the T_m data of the PSEG derivatives in Table 2, the polymers having more than 6 OEG repeating units tend to show a crystalline phase to some extent when the polymer possesses dimethyl substituents on the Si atom. The bulky divinyl substituents on the Si atom may prevent the crystallization under the same conditions in the case of

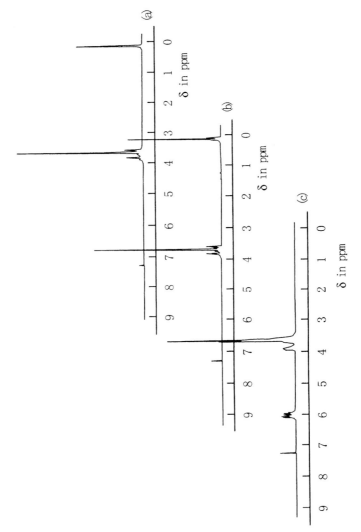

Figure 2. ^{1}H-NMR Spectra of PSEG(1/7) (a), PSEG(2/7) (b), PVSE(1/7) (c).

Table 1. Results of Synthesis of PSEG and PVSE Series through Polycondensation Reactions between Bis(diethylamino)siloxane (BAS) and Oligo(ethylene glycol) (OEG)

Polymer code	m in BAS	[BAS]$_0$ in mmol	n in OEG	[OEG]$_0$ in mmol	THF[a] in ml	Temp in °C	Time in h	10^{-3} M_n[b]	M_w/M_n[b]
PSEG(1/2)	1	20.0	2	20.0	20	reflux	36	14.0	1.50
PSEG(1/3)	1	22.4	3	22.5	20	reflux	36	4.7	1.42
PSEG(1/4)	1	20.8	4	20.8	20	60	24	17.7	1.52
PSEG(1/7)	1	11.1	7	11.1	—	60	48	11.8	1.41
PSEG(1/9)	1	6.5	9	6.4	20	60	24	5.9	1.25
PSEG(1/13)	1	10.0	13	10.0	20	60	24	3.7	1.17
PSEG(2/2)	2	9.9	2	10.0	20	60	24	10.8	1.28
PSEG(2/3)	2	9.9	3	9.7	20	60	24	8.4	1.47
PSEG(2/4)	2	6.2	4	6.1	20	60	24	7.7	1.52
PSEG(2/7)	2	11.2	7	11.1	—	60	48	13.2	1.47
PSEG(2/9)	2	6.5	9	6.4	20	60	24	11.5	1.21
PSEG(2/13)	2	3.7	13	3.8	20	60	24	3.5	1.20
PVSE(1/3)	1	10.0	3	10.0	—	80	48	9.3	2.27
PVSE(1/4)	1	10.0	4	10.0	—	80	48	7.0	2.13
PVSE(1/7)	1	70.0	7	70.0	—	80	96	6.5	2.01

a) THF was used as solvent.   b) Determined from GPC results

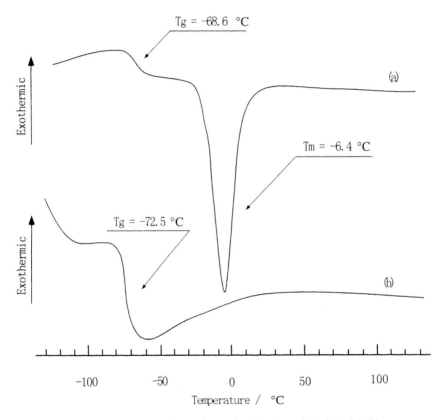

Figure 3. DSC Profiles of PSEG(1/9) (a), and PVSE(1/7) (b)

PVSE(1/7).   Detailed data for the T_g and the T_m values of PSEG derivatives are summarized in Table 2.

PEG is known to be a hydrophilic polymer, which is soluble in aqueous media, due to the etheral oxygen which causes hydrogen bonding with water.   However, PEG is insolubilized from water when the solutions heated to 90 °C due to the negative dissolution entropy.(19)   By the introduction of a hydrophobic group into PEG in a suitable balance, it is anticipated that the phase separation temperature should be lowered.   If the LCST becomes ca. 20 to 50 °C, the possibility of the polymer as a functional material will be expanded.   The alternating units of OEG and the oligosiloxane linkage in the main chain can be anticipated to control the LCST.   Actually, the polymers having relatively higher oxyethylene content were soluble in cold water at neutral pH.   Figure 5 shows the change in transmittance at 550 nm of the PSEG aqueous solution (Phosphate buffer, pH = 7.0; I = 0.05) as a function of temperature.   As seen in Figure 5a, even a trioxyethylene / dimethylsiloxane alternating copolymer, PSEG(1/3), was soluble in cold water. With increasing solution temperature, the transmittance abruptly decreased at 10.5 °C. Thus, the phase separation took place at 10.5 °C in the case of PSEG(1/3).   With increasing number of EG unit in the PSEG(1/n) series, the phase transition points increased.   With PSEG having tetraoxyethylene units in each repeating unit, the LCST became ca. 30 °C.   PSEG(1/2) was not soluble in cold water due to the strong hydrophobicity.

In the case of the PSEG(2/n) series, the hydrophobicity of the polymer increased compared with PSEG(1/n) when compared with the same number of OEG units.   Actually, PSEG(2/3) was not soluble even in cold water.   As shown in Figure 5b, the PSEG(2/n) series was soluble in cold water when more than 6 EG units were present.   The LCSTs of PSEG(2/7) and (2/13) were 13 and 37 °C, respectively. The same phenomenon was observed in the case of PVSE(1/7).   The hydrophobicity of the polymer was higher than that of PSEG(1/7) due to the hydrophobicity of two the vinyl groups on each Si atom instead of two the methyl groups in PSEG(1/7).   The LCST of PVSE(1/7) was 10.5 °C as shown in Figure 5c.

On the basis of these results, it is concluded that the LCST of PSEG derivatives is controllable by the extent of siloxane units in the main chain.   Actually, the LCST was only governed by the Si content in PSEG with dimethylsiloxane repeating units as shown in Figure 6.   In the case of PVSE, however, the LCST was lower as compared with the same Si content as in the PSEG series.   The bulky divinylsiloxane linkage makes the polymer much more hydrophobic than the polymer with dimethylsiloxane repeating units.

Hydrolytic Stability in Aqueous Media It is well-known that an Si-O linkage is susceptible to hydrolytic degradation.(20,21)   Especially, a siloxane bond in the C-O-Si linkage is easy to hydrolyze even at neutral pH.   Because PSEG derivatives possess a C-O-Si linkage in each repeating unit, the hydrolytic stability of the polymers should be analyzed.   The stability of the PSEG and PVSE series against hydrolysis in aqueous media was estimated in the same manner as described in a previously reported paper.(22)

Table 2. Glass Transition Temperatures and Melting Points of PSEG Series and PVSE(1/7)

polymer code	Tg (°C) a)	Tm (°C) a)
PSEG(1/2)	−81.3	
PSEG(1/3)	−78.4	
PSEG(1/4)	−74.7	
PSEG(1/7)	−72.2	−30.4
PSEG(1/9)	−68.6	−6.4
PSEG(1/13)	−64.0	16.6
PSEG(2/2)	−95.4	
PSEG(2/3)	−88.9	
PSEG(2/4)	−85.3	
PSEG(2/7)	−78.6	−33.2
PSEG(2/9)	−73.1	−12.1
PSEG(2/13)	−67.2	10.7
PVSE(1/7)	−72.5	

a) Tg and Tm were determined DSC at a heating rate of 20 °C/min from −170 to 200 °C

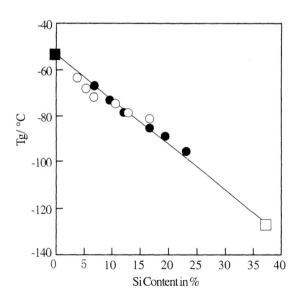

Figure 4. Plots of Glass Transition Temperature versus Silicon Content in the Poly(siloxyethylene glycol): (●) PSEG(1/n); (○) PSEG(2/n); (▲) PVSE(1/7); (■) PEG; (□) PDMS

371

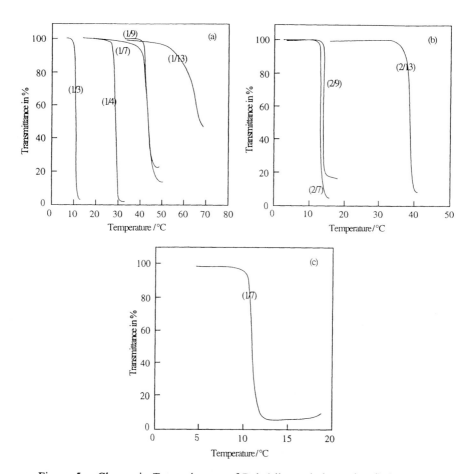

Figure 5. Change in Transmittance of Poly(siloxyethylene glycol) Aqueous Solution (1.5 wt. %) in Phosphate Buffer (pH = 7.0, I = 0.05): (a) PSEG(1/n); (b) PSEG(2/n); (c) PVSE(1/7) (Number in Parentheses Denotes m / n in the Polymers)

Figure 7 shows the time course of the normalized attenuance change as a function of time. In the case of the PSEG(1/n) series shown in Figure 7a, the normalized attenuance decreased rapidly. Especially, in the case of PSEG(1/7) solution, the normalized attenuance decreased within a few hours, indicating that rapid degradation took place even in neutral phosphate buffer. The rate of hydrolysis was found to depend on the hydrophobicity. Actually, phase separation could be observed up to 20 h in the case of PSEG(1/3) solution. The case in which the hydrophobic effect of the polymer showed the most significant influence, is the PVSE homologues. PVSE(1/7) solution showed the phase transition up to a several ten hours, indicating a low hydrolytic rate compared to that of the PSEG series. It should be noted that the PSEG with the same number of OEG units, PSEG(1/7), degraded rapidly up to a few hours.

The hydrophobicity of the polymer significantly influenced the hydrolysis of the PSEG(2/n) series. When dimethyldisiloxane was used as one of the comonomers, no change in the normalized attenuance was observed. Especially, PSEG(2/9) shows the phase transition at more than 60h (no data over 60h). Thus, it is concluded that polymers having fairly high stability were prepared, when the dimethyldisiloxane linkage was utilized for PSEG synthesis.

The lack of a sharp phase transition phenomenon in the case of PSEG(1/3) shown in Figure 5a may be attributable to hydrolysis during the measurement.

Poly(siloxyethylene glycol) as a Negative Working Resist The miniaturization of microelectronic devices has been progressing rapidly for the last twenty years. For example, the accumulation of integrated circuits in an ultra-large-scale-integration (ULSI) progresses four times per every 3 - 4 years.(*23, 24*) For the fabrication of a further integrated pattern, a high performance resist polymer must be developed. The primary demand for the high performance resists is, of course, to improve resolution. One of the important factors to improve resolution is that kind of irradiation source that can be utilized on the resist. In other words, a resist having a high sensitivity toward a rather narrower beam (UV → deep UV → X-ray, EB) is suitable for the integrated pattern.

One of the other important points is to improve the resistance of the polymer against etching. To obtain high resolution patterns, a very thin resist film must be used.(*25*) For this purpose, however, organic polymers do not have enough resistance to oxygen reactive ion etching (O_2 RIE). Organosilicon polymers are being considered as one of the candidates for materials with a high resistance to O_2 RIE.(*26, 27, 28*) Several problems, however, remain for the Si-containing polymers as resists. For example, the organosilicon moiety does not have enough sensitivity, especially toward UV and EB exposure.(*29, 30, 31, 32*) Also, a big disadvantage is that organic solvents must be used as the developing solvent because of the strong hydrophobicity of the organosilicon groups.

As stated above, PVSE 1) is soluble in cold water; 2) has a high Si-content; and 3) it is anticipated to be a negative working resist through the crosslinking reaction of side chains by certain stimuli such as UV and EB, because PVSE possesses two reactive double bonds in each repeating unit. Thus, the resist

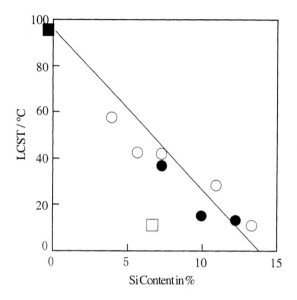

Figure 6. Plots of the LCSTs of Poly(siloxyethylene glycol) versus Silicon Content in the Polymers: (●) PSEG(1/n); (○) PSEG(2/n); (□) PVSE(1/7); (■) PEG

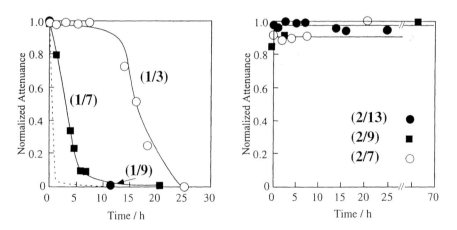

Figure 7. Time Course of the Normalized Attenuance of Polymers in Phosphate Buffer: (a) (○) PSEG(1/3), (■) PSEG(1/7), (●) PVSE(1/7); (b) (○) PSEG(2/7), (■) PSEG(2/9), (●) PSEG(2/13)

characteristics of PVSE are of interest. In this section, the sensitivities of PVSE(1/7), which has an LCST of 10.5 °C, to UV and EB, and the resistance to O_2 plasma etching are described.

Evaluation as photoresist It is known that the polymerizability of vinylsilane compounds in a radical polymerization is not very great.(*33,34*) Actually, no polymerization takes place under the common radical polymerization conditions in the case of trimethylvinylsilane. Therefore, a crosslinking reaction of vinylsilane pendent groups with polythiol compounds was employed in the UV irradiation. When the PVSE(1/7) film coupled with a dithiol compound, di(2-mercaptoethylether), and benzophenone, the sensitivity of the crosslinking reaction was extremely low. Actually, a gel was not observed at UV irradiation of 17,100 mJ/cm². When a tetrafunctional polythiol, tetramethylolmethanetetra(3-mercaptopropyonate) (4TP-5), coupled with benzoinmethyl ether (BME) was used for the crosslinking agent, the sensitivity increased substantially. Figure 8 shows a sensitivity characteristic curve of PVSE(1/7) coupled with 4TP-5/BME. A Dg^{50}, which denotes the dose at 50 % remaining film thickness, was very high (35 mJ/cm²). In the case of the crosslinking reaction, the molecular weight (MW) of the polymer significantly influences the sensitivity, *viz.*, a higher MW makes the crosslinking reaction easier.(*35*) The MW of the PVSE(1/7) used in this study was only 8,000, indicating the extremely high sensitivity of the polymer.

The lower crosslinking efficiency of the dithiol derivatives may be attributed to a cyclization reaction between dithiol and the divinylsilane unit. The γ-value, which is one of the index parameters for resolution, was 1.36, which is moderate compared with other common resists (γ = 1 - 2).

Evaluation as Electron-beam resist Recently, irradiation sources other than UV have been investigated to improve the resolution. Namely, smaller patterns than the UV wavelength have been required. For these requirements, several irradiation sources such as EB, X-ray and ion beam have been employed for the patterning. Thus, the evaluation of PVSE(1/7) with EB exposure was investigated. In the case of the EB exposure, PVSE crosslinked easily without any crosslinking agents. The rather higher energy of the EB makes the reaction of the vinylsilane pendent groups easier.

Figure 9 shows the sensitivity characteristic curve with EB exposure. It was surprising for us that the Dg^{50} of PVSE(1/7) was extremely high (Dg^{50} = 1 μC/cm²). It should be noted that the Dg^{50} of poly(4-chloromethylstyrene), which is a commercially available negative EB-resist, is 6.7 μC/cm².(*36*) On the basis of these results, PVSE(1/7) is one of the candidates for a highly sensitive negative working resist.

Dry etching resistance of PSEG The etching resistance to oxygen plasma was investigated. After crosslinking by EB exposure, the PVSE film was exposed to oxygen plasma (O_2: 0.5 Torr; 100 W). Under the same conditions, the thickness of a diazonaphthoquinone / novolac resin (OFPR-800) decreased rapidly. Actually, the

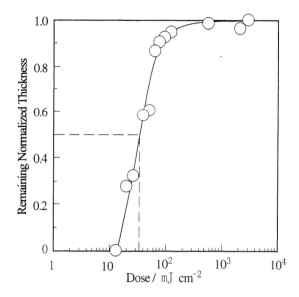

Figure 8. Sensitivity Characteristic Curve of PVSE(1/7) Film against Photo Irradiation (Developed in Water of 4 °C for 2 Minutes) (Reproduced from ref. 10 by courtesy of publishers, Jon Wiley & Sons, Inc.)

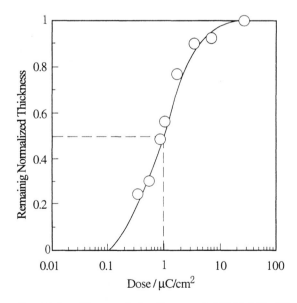

Figure 9. Sensitivity Characteristic Curve of PVSE(1/7) Film against Electron-beam Exposure (Exposed at Accelerating Voltage of 20 kV, Developed in Water of 4 °C for 10 Minutes) (Reproduced from ref. 10 by courtesy of publishers, Jon Wiley & Sons, Inc.)

film of 8000 Å thickness disappeared within 30 min. The etching rate of the OFPR-800 was 220 Å /min as shown in Figure 10. On the other hand, the PVSE film showed fairly high etching resistance. In the early stage of the etching, the film thickness decreased ca. 2000 Å (180 Å /min; Figure 10). After 10 min, however, the etching stopped completely, indicating an effective thin SiO_2 layer formation on the resist surface. From these results, a very thin PVSE film can be applied for EB-resists, which is a big advantage for very fine pattern formation.

Figure 11 shows negative tone images of the PVSE resist with UV- (a) and EB- (b) exposure. The image obtained by cold-water development should be noted again.

Electroconductivity of Poly(siloxyethylene glycol) In the middle of 1970s, Wright pointed out an opportunity of using a polymer matrix as a solid electrolyte.(*37*) Solvent-free polymer electrolytes(*38*) have gained much attention because of their potential applications to variety of area such as high energy batteries, solid state electrochromic displays and photoelectrochemical cells. Poly(ethylene glycol) (PEG) was one of the promising materials as solid electrolyte because of high solubility of cations in PEG due to the chelation of ether oxygen to cations. Its high mobility is one of the other important factor because ion transfer mainly takes place not alone but accompanying with polymer segment surrounding the ions.(*39*) As stated above, PSEG possesses a very flexible chain because it is homologue of PEG and silicone. In addition, OEG segments in PSEG can be anticipated a high chelating ability toward cation species. Thus, the evaluation of PSEG as solid electrolyte is of interest. In this section, the ion conductivity characteristics of PSEG along with lithium perchlorate are described.

Preparation of PSEG Gels and Their Ion Conductivity For preparation of solid matrix, pendent double bonds were introduced to PSEG via copolycondensation reactions. For the polymerization reaction, a mixture of 1,3-bis(diethylamino)tetramethyldisiloxane (DADS) and bis(diethylamino)divinylsilane (DADVS) (5 - 30 mol%) was employed as siloxane source. PSEG(x), where x denotes mol% of DADVS segment, thus obtained were mixed with lithium perchlorate along with crosslinking agent in acetonitrile, then crosslinked by photoirradiation.

Figure 12 shows ion conductivity characteristics as a function of temperature. Other typical polymers for solid electrolyte investigated so far are also drawn in the same figure. As can be seen in the figure, the conductivity of PSEG coupled with $LiClO_4$ was ca. 10^{-4} S cm^{-1} above room temperature. It should be noted that PSEG matrix showed much better conductivity characteristics at temperature lower than 20 °C showed as compared to other polymers. The siloxane linkage between OEGs lowered the glass transition temperature to result in keeping high mobility of the segment-Li complex even at low temperature. Lowered crystallinity of PSEG as compared with PEG also works positively at low temperature.

Thus, PSEG-Li^+ complex is promising as solid electrolyte, especially at low temperature.

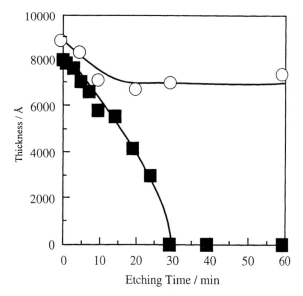

Figure 10. Etching Characteristics of PVSE(1/7) (○) and Diazonaphthoquinone / Novolac Resin (■) Film in an Oxygen Plasma (Radio Frequency Power; 100 W, Gas Pressure; 0.5 Torr, O_2 Flow Rate; 100 mL/min) (Reproduced from ref. 10 by courtesy of publishers, Jon Wiley & Sons, Inc.)

Figure 11(a). SEM Micrograph of Negative Tone Image for Photo Irradiation (Exposed at 92 mJ/cm^2 through a Mask of 70 μm Line) (Reproduced from ref. 10 by courtesy of publishers, Jon Wiley & Sons, Inc.)

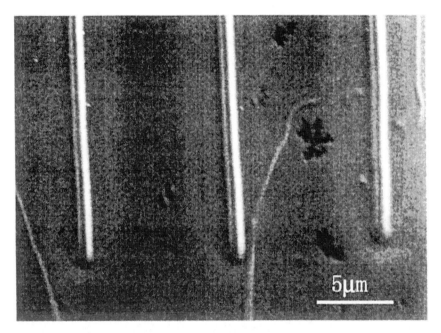

Figure 11(b). SEM Micrograph of L&S Negative Pattern Exposed Using an Electron-beam (Exposure Dose; 2.4 °C/cm², Accelerating Voltage; 20 kV, Developed in Water of 4 °C for 10 Minutes) (Reproduced from ref. 10 by courtesy of publishers, Jon Wiley & Sons, Inc.)

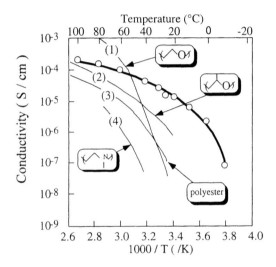

Figure 12. Arrenius Plots of Lithium Cation Conductivity in PSEG(2/13)(vinyl=20mol%)(○) along with other common matrixes. [Li]/[-O-] = 0.075. (1) PEG-LiClO$_4$ (0.083); (2) aliphatic polyester-LiClO$_4$ (0.111); (3) PPG-LiCF$_3$SO$_3$ (0.25); (4) poly(N-methylethylene imine)-LiClO$_4$ (0.125)(40)

Conclusion Poly(siloxyethylene glycol), PSEG, can be synthesized via polycondensation reactions between oligo(ethylene glycol) and diaminosiloxane derivatives to form polymers with narrow molecular weight distributions. PSEG in aqueous media showed the lower critical solution temperature (LCST). The LCST can be controlled by the ratio of EG/Si units. When disiloxane was used as one of comonomer, the resulting PSEG was stable against hydrolysis in neutral buffer.

PSEG with vinyl side groups can be applied as a resist, which can be developed by cold water. Especially, the sensitivity of PSEG irradiated by electron-beam showed extremely high. The PSEG film also showed very high resistance against O_2 dry etching. PSEG coupled with $LiClO_4$ showed high ionic conductivity, especially at low temperature.

On the basis of these results, the PSEG series can be regarded as a new functional polymer which is useful for stimuli-sensitive gels, actuators, negative resists, solid electrolytes, surface modification agents, etc.

References
1 Jones, R. G., ed. "Silicon-containing Polymers", The Royal Society of Chemistry, Cambridge, **1995**
2 Clarson, S. J.; Semlyen, J. A., ed. "Siloxane Polymers", Ellis Horwood-PTR Prentice Hall, Englewood, Cliffs, New Jersey, **1993**
3 Nagasaki, Y.; Kurosawa, K.; Suda, M.; Takahashi, S.; Turuta, T.; Ishihara, K.; Nagase, Y. *Makromol. Chem.*, **1990**, 191, 2103
4 Ito, H.; Nagasaki, Y.; Kato, M.; Kataoka, K.; Tsuruta, T.; Suzuki, K.; Okano, T.; Sakurai, Y. *Artificial Hart*, **1998**, 6, 148
5 Kato, N.; Takeda, K.; Nagasaki; Y., Kato, M. *Ind. Eng. Chem. Res.*, **1994**, 33, 417
6 Nagasaki, Y.; Honzawa, E.; Kato, M.; Kihara, Y.; Tsuruta, T. *J. Macromol. Sci., - Pure & Appl. Chem.*, **1992**, A29, 457
7 Nagasaki, Y.; Honzawa, E.; Kato, M.; Kataoka, K.; Tsuruta, T. *Macromolecules*, **1994**, 27, 5858
8 Nagasaki, TY.; Kazama, K.; Honzawa, E.; Kato, M.; Kataoka, K., Tsuruta, T., *Macromolecules*, **1995**, 28, 8870
9 Nagasaki, Y.; Matsukura, F.; Kato, M., Aoki, H.; Tokuda, T. *Macromolecules*, **1996**, 29, 5859
10 Aoki, H.; Tokuda, T.; Nagasaki, Y.; Kato, M. *J. Polym. Sci., Part A, Polym. Chem. Ed.*, **1997**, 35, 2827
11 "Organosilicon Compounds", Bazant, V.; Chvalovsky, V.; Rathousky, J. **1965**, Academic Press, New York
12 Sugita, K.; Ueno, N. *Prog. Polym. Sci.*, **1992**, 17, 319
13 Jones, R. G. *Trends in Polymer Science*, **1993**, 1, 372
14 Choong, H. S.; Kahn, F. J. *J. Vac. Sci. Technol.*, **1981**, 19, 1121
15 Nagaoka, K.; Naruse H.; Shinohara, I.; Watanabe, M. *J. Polym. Sci. Polym. Lett. Ed.*, **1984**, 22, 659
16 Nagasaki, Y.; Honzawa E., Kato, M.; Tsuruta, T. *J. Macromol. Sci. -Pure Appl. Chem.*, **1992**, A29, 457
17 Miller, W. G.; Saunders, J. H. *J. Appl. Polym. Sci.*, **1969**, 13, 1277
18 Watanabe, M. unpublished data

19 Harris, J. M. *ÒPoly(Ethylene Glycol) Chemistry: Biotechnical and Biomedical Applications"*, **1992**, Plenum Press, New York
20 S. P. Gitto, S. P.; Wooley, K. L. *Macromolecules*, **1995**, 28, 8887
21 Cunico, R. F.; Bedell, L. *J. Org. Chem.*, **1980**, 45, 4798
22 Nagasaki, Y.; Matsukura, F.; Kato, M.; Aoki, H.; Tokuda, T. *Macromolecules*, **1996**, 29, 5859
23 Ogawa, . *J. Photopolym. Sci. Technol.*, **1996**, 9, 379
24 Thompson, L.F. *"Introduction to Microlithography Second Edition"*, Thompson, L. F.; Willson, C. G.; Bowden, M. J. (ed.), **1994**, ACS, Washington, DC, 1994, chap. 1, p. 1.
25 Hasegawa, S.; Iida, Y. *IEEE J. Solid-State Circuits*, **1985**, 20, 15
26 Taylor G. N.; Wolf, T. M.; *Polym. Eng. Sci.*, **1980**, 20, 1087
27 MacDonald, S. A.; H. Ito H.; Willson, G. C. *Microelectronic Engineering*, **183**, 1, 269
28 Sugita, K.; Ueno, N., *Prog. Polym. Sci.*, **1992**, 17, 319
29 Tai, K. L.; Sinclair, W. R.; Vadimsky, R. G.; Moran, J. M. Rand, M. J. *J. Vac. Sci. Technol.*, **1979**, 16(6), 1977
30 Reichmanis, E.; Smolinski G.; Wilkins, Jr., C. W. *Solid State Technol.*, **1985**, 28(8), 130
31 Hofer, D. C.; Miller, R. D.; Willson, C. G. *Proc. SPIE*, **1984**, 469, 16
32 McDonnell, L. P.; Gregor, L. V.; Lyons, C. F. *Solid State Technol.*, **1986**, 29(6), 133
33 Petrov, A. D.; Pollyakova, A. M.; Sakharava, A. A.; Korshak, V. V.; Mironov, V. F. Nikishin, G. J. *Doklady Akad. Nauk., SSSR*, **1954**, 99, 7858
34 Korshak, V. V.; Pollyakova, A. M.; Mironov, V. F.; Petrov, A. D. *Izu. Akad. Nauk., SSSR, Otd. Khim. Nauk*, **1959**, 178
35 Choong, H. S.; Kahn, F. J. *J. Vac. Sci. Technol.*, **1981**, 19, 1121
36 Choong, H. S.; Kahn, F. J. *J. Vac. Sci. Technol.*, **1981**, 19, 1121
37 Wright, P. V. *Br. Polym. J.*,**1975**, 7, 319
38 MacCallum, J. R.; Vincent C. A. Ed. *"Polymer Electrolyte Reviews, 2"* **1990**, Elsevier, Amsterdam
39 Watanabe, M.; Ogata, N. *Br. Polym. J.*, **1988**, 20, 181
40 Armand, M. B. *Ann. Rev. Mater. Sci.*, **1986**, 16, 245

Chapter 25

Polycarbonate–Polysiloxane-Based Interpenetrating Networks

Sylvie Boileau[1], Laurent Bouteiller[1], Riadh Ben Khalifa[1], Yi Liang[1], and Dominique Teyssié[2]

[1]Laboratoire de Recherche sur les Polymères, UMR 7581-CNRS, B.P. 28, 2 rue Henri Dunant, 94320 Thiais, France
[2]Université de Cergy-Pontoise, Neuville sur Oise, Département de Chimie, 5 mail Gay-Lussac, 95031 Cergy-Pontoise Cedex, France

Interpenetrating polymer networks based on polysiloxanes (PS) and polycarbonates (PC) were prepared by the in-situ method : a polysiloxane bearing various proportions of RT crosslinkable -Si(OEt)$_3$ side groups and hydrolyzable phenylcarbonate moieties was mixed with diethyleneglycol bis-allylcarbonate and benzoyl peroxide. After the formation of the PS network at RT, the crosslinking of the PC network was achieved at 100°C. Various chemical modifications of the PS component in the IPN were performed in order to improve the degree of interpenetration as checked by turbidity, DSC and DMA measurements. Kinetics of phenol release was studied on linear PS, single PS networks, semi-IPNs and IPNs of various compositions in buffered medium (pH = 7.5) at 37.5°C. First-order phenol release rate constants decrease on increasing the crosslinking densities of the systems.

Interpenetrating polymer networks (IPNs) are combinations of crosslinked polymers held together by permanent entanglements (*1,2*). Due to their interlocking configuration, the state of phase separation obtained at the end of their synthesis is frozen so that their properties are not influenced by ageing and they are well suited for combination of highly incompatible polymer pairs. A number of IPNs based drug delivery systems have been described where the drug is physically embedded in a gradient type IPN (*2*). We wanted to design a system where the drug delivery would be under chemical control, i.e. a pH-mediated hydrolysis rather than only diffusion controlled in order to prepare a therapeutic lens model. The choice of the two polymeric partners was thus made according to the specific requirements for this application : the optical and mechanical properties where brought by a polycarbonate component, the flexibility and oxygen permeability came from a drug modified polysiloxane component. The drug was covalently bound onto the polysiloxane partner through a spacer containing a carbonate linkage. This function conveniently hydrolyzes at pH 7.5 which corresponds to the pH value in the tear medium.

© 2000 American Chemical Society

Moreover the presence of a carbonate group close to the siloxane backbone was expected to help the compatibilization with the polycarbonate component.

Simultaneous and sequential IPNs based on various polymeric systems have been prepared using polydimethylsiloxane (PDMS) as the host network (*3-8*). These systems include poly(ether-urethane), polystyrene, poly(2,6-dimethyl-1,4-phenyleneoxide), polyacrylic acid, PDMS, polymethylmethacrylate, polyethylene oxide (PEO) ... as the guest network. Some semi-interpenetrating networks (s-IPNs) based either on a linear polymer embedded in a polysiloxane network (*5,9,10*) or on a linear polysiloxane combined with a PEO network (*8*) have also been described. In some cases, PDMS has been replaced by polyaromatic siloxanes such as polydiphenyl or polymethylphenylsiloxanes (*10-12*). The focus of this paper concerns the preparation and properties of IPNs and s-IPNs based on polysiloxanes and poly(diethyleneglycol bis-allylcarbonate) (*13,14*).

Experimental

Polymethylhydrogenosiloxane (PMHS), $\overline{DP_n}$ = 35 (Merck), allyltriethoxysilane (ATES), (ABCR), benzoylperoxide (BPO), (Fluka), dibutyltin dilaurate (DBTDL), (Merck), platinum methylvinylcyclosiloxane complex (PCO 85), (ABCR), were used as received. Diethyleneglycol bis-allylcarbonate (XR 80), (Essilor) was purified on a silica gel column. Butenylphenylcarbonate (BPC) was prepared by reaction of phenyl chloroformate with 3-buten-1-ol under phase transfer catalysis conditions (*13,15*). Precursors PSIa, PSIb and PSIc (Scheme 1) were prepared by cohydrosilylation with PMHS of both ATES and BPC in various relative proportions. The hydrosilylation reactions were carried out in dry toluene at 60°C with a Pt° complex catalyst ([-CH=CH$_2$]/[SiH] = 1.1 ; [Pt]/[SiH] = 5.10^{-4}).

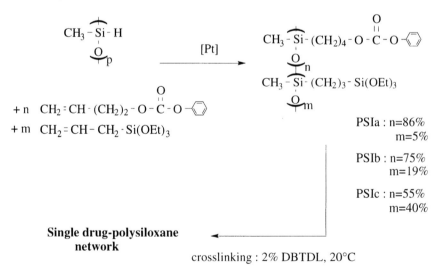

Scheme 1.

The reaction was followed by monitoring the decrease of the SiH IR band at 2160cm^{-1}. The polymers were recovered by precipitation in methanol, purified with carbon black treatment in diethylether, precipitated again in methanol and dried under high vacuum (yields ≈ 80%). As checked by ^{1}H NMR, the final proportions of ATES and BPC grafted onto the polysiloxane main-chain were 0% and 95%, 5% and 85%, 19% and 75%, and 40% and 55% for PS0, PSIa, PSIb and PSIc respectively. The remaining 5% side-functions on each polymer are either unmodified SiH groups (residual signal at 4.7 ppm on the ^{1}H NMR spectrum, see Figure 1), or phenoxy groups (6.95 ppm) resulting from a side-reaction (15).

Single polysiloxane networks PSIaN, PSIbN and PSIcN were obtained by adding DBTDL (2% by weight) which catalyzes the crosslinking at room temperature which occurs through the $Si(OEt)_3$ groups. The polycarbonate network was prepared by free-radical polymerization of XR 80 with BPO as the initiator (3% by weight) at 95°C for 1 h, under nitrogen. The amount of extracted products measured after Soxhlet extraction with CH_2Cl_2 was lower than 5% and the amount of unreacted double bonds as measured by iodometry was equal to 11%. A series of s-IPNs was prepared from various weight proportions (20%, 35%, 50%, 65%) of PSO embedded in a PC network which was cured with BPO (3% by weight) at 95°C for 8h. Finally the IPNs were synthesized by a two step in-situ method. The carbonate monomer, BPO, the modified polysiloxane (PSIa, PSIb or PSIc) and DBTDL were mixed together under nitrogen. The mixture was allowed to react at room temperature until the polysiloxane was crosslinked (15 to 20h) and the temperature was then raised up to 100°C in order to promote the PC network formation (Scheme 2).

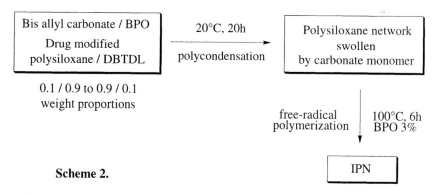

Scheme 2.

^{1}H and ^{13}C NMR spectra were recorded in $CDCl_3$ at 20°C on a Bruker AM 200SY apparatus. The Tg values were taken as the onset point from the second heating curves recorded at a 20°C/min heating rate. DMA measurements were performed on a Perkin-Elmer DMA 7 apparatus with a 10°C/min heating rate and at a frequency of 1 Hz.

The kinetics of the phenol release were studied on ca. 1 g samples immersed in 100 ml pH=7.5 phosphate buffer at 37.5 °C. 1 ml aliquots of the buffer solution were then extracted with diethylether and the phenol concentration was determined by UV spectroscopy on a Perkin-Elmer 544 spectrometer at 273 nm (ε=2200 l mol^{-1} cm^{-1}).

Results and Discussion

A series of IPNs was prepared from PSIa, PSIb and PSIc on one hand and polycarbonate PC on the other hand. The relative weight proportions of the two networks were varied between 10 and 90% and the resulting IPNs showed less than 8% extracted material in all cases. The IPNs prepared from the PSIb and PSIc series were all transparent which is indicative of a satisfactory level of interpenetration considering the fact that the refractive indices of the two partners are different.

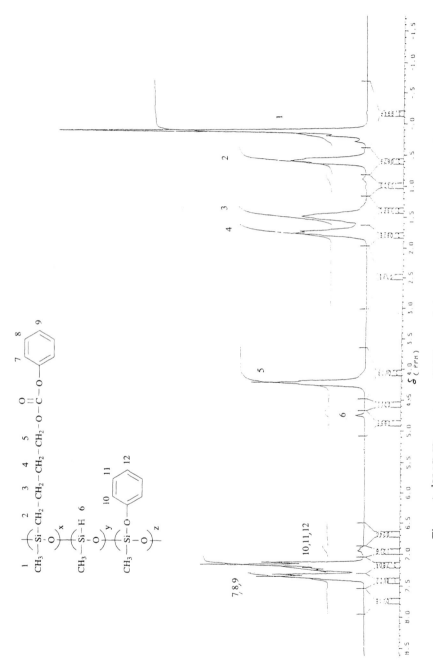

Figure 1. ^{1}H NMR spectrum of PS0 recorded at 250 MHz in CDCl$_3$ at 25°C.

Physicochemical characterization of the IPNs. The IPNs were characterized by thermal analysis, density and turbidity measurements and low angle neutron scattering. Characterization by DSC shows that all investigated IPNs exhibit only one glass transition temperature (T_g) whatever the crosslinking density in the polysiloxane partner and the weight proportion of the polycarbonate partner. However the T_g values plotted as a function of the polycarbonate proportion in the IPN (Figure 2) are unexpectedly lower than those which could be calculated from a theoretical relationship. Thus thermal analyses were performed independently on the two partners of the IPNs. The polycarbonate network alone exhibits a Tg value equal to 70°C. Thermal analysis performed on linear polymethylhydrosiloxanes substituted with increasing amounts of BPC grafts (without ATES) show that the Tg values increase strictly linearly from -123°C (PDMS) to -30°C (100% BPC modified PMHS). This classical behavior probably arises from a decrease in the mobility due to the increasing number of side-chains and their mutual interactions. On the other hand Tg values of single networks prepared from PSIa, PSIb and PSIc increase slightly from -30°C to -20°C in an expected trend corresponding to the increase in the crosslinking density (from 5 to 40% crosslinking groups). The Tg values of the full IPNs which are all lower than the Tg of single polysiloxane networks thus follow a trend which differs from what has been observed in other systems such as polymethylphenylsiloxane-polyurethane IPNs (*11*) or polyacrylate-polyurethane IPNs (*16*).

These observations could arise from a specific interaction due to the particular chemistry of this system : the carbonate groups belonging to the PC network partner in the IPN could compete with the interactions of the neighbouring carbonate groups belonging to the polysiloxane side-chains. They could thus act such as plasticizing elements and induce a decrease in the apparent Tg value when compared to the polysiloxane. It is indeed believed that the single thermal transition observed for the IPNs by DSC is only an apparent Tg which does not fully account for the morphology of the IPN. Tan δ measurement performed by DMA on the IPN synthesized from 30% PSIc and 70% PC reveals two glass transitions at -30°C and +40°C. The lower one, corresponding to the polysiloxane phase is also detected by DSC, while the upper one is not. This could arise from the very small size of the polycarbonate rich domains. This last hypothesis is in agreement with an estimation of the domain size from turbidity measurements at 460 nm. This estimation was made according to an iterative procedure taking into account the volume fraction and the refractive index of each polymer in the IPN (*17*). The sizes of the heterogeneities thus determined for IPNs made from PSIc increase from 30 nm to 55 nm when the weight proportion of PC increases from 10 to 70%. Furthermore, neutron scattering experiments (performed by C. Taupin, CEN Saclay, France) show that the carbonate group domains from either partner in these IPNs are impossible to discriminate, which suggests that there is a high degree of interpenetration. The very strong interactions between the two polymers in the networks were further confirmed by the very low amount of extractible material (as measured after Soxhlet extraction in CH_2Cl_2) in a series of semi-IPNs made from PS0 and PC, which vary from 4 to 16% for 1:4, 1:2, 1:1 and 2:1 weight proportions, where at least 50 to 60% extractible materials would be expected. Finally, the densities of IPNs prepared from PSIa, PSIb and PSIc and various proportions of PC are consistently higher (4% for the 1:1 PS/PC weight composition) than the theoretical values calculated from the volume additivity, showing strong interactions between the two polymers in the IPNs.

Kinetic study of the drug model release. The release of phenol which results from the hydrolysis of the carbonate bond in the side-chain of the polysiloxane component in the IPN was followed as a function of time, by UV spectroscopy. The experimental

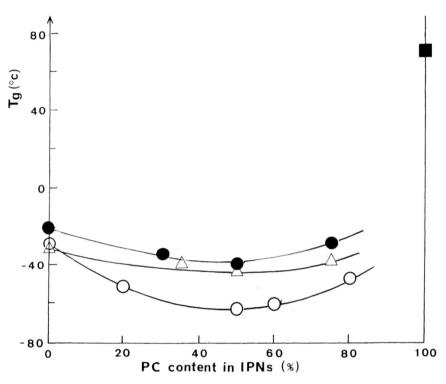

Figure 2. Tg values as a function of polycarbonate (PC) weight proportion in full IPNs with △ PSIa; ○ PSIb and ● PSIc.

conditions, 37.5°C and pH 7.5 buffer, were chosen as model conditions for the eye tear medium.

Blank experiments first performed on the release of phenol physically trapped in an IPN showed that the rate of appearance of the phenol molecule in the external medium is very fast (10h for a 90% release, 20h for a 100% release) compared to the rate of delivery of phenol covalently bound in any of the studied IPNs (20 to 40 days). Thus both the diffusion rate of the buffer into the IPN and the diffusion of the drug model out of the IPN were neglected in a preliminary analysis of the kinetics of the drug release.

The results of typical kinetics are plotted as $Ln([C]/[C_0])$ (concentration of the remaining phenoxy groups on the polysiloxane over the initial concentration) as a function of time (Figure 3). The plots were obtained for a linear 100% BPC modified polysiloxane (PS0), two s-IPNs made from PS0 embedded in a PC network in 1:1 and 1:4 proportions and a full 1:1 IPN made from PSIb and PC. For all tested samples, the plots are linear up to at least 50% or even 70% conversion. Thus (i) the kinetics of the release can be considered as first order with respect to the phenol concentration within 50 or 70% reaction and (ii) the time scale of the release (ca. 25 days) is compatible with a therapeutical program.

The kinetics of the phenol release was also studied on a completely soluble system, i.e. a PMHS modified with 80% PEO grafts ($\overline{DP_n}$ =8) (*18*) and 20% BPC grafts and on insoluble single networks prepared from BPC and ATES modified PMHS. The first order rate constants, k, calculated from the $Ln([C]/[C_0])$ versus time plots are reported in Table I. As expected, k decreases from the water-soluble linear system to the insoluble linear PS0 (columns 1 and 2). This rate constant further decreases in the subsequent crosslinking systems (columns 3 to 5). The crosslinking densities in the s-IPN or in the PSIb single network, PSIbN, seem to have comparable slowing down effects on the phenol release whereas the crosslinking density in the 1:1 PSIb/PC IPN which is of course higher leads to a dramatic decrease in the k value (Table I).

Table I. Kinetics of phenol release from different polymeric systems at pH=7.5

Polymer	water soluble PS[a]	PS0	1:1 PS0/PC s-IPN	PSIbN [b]	1:1 PSIb/PC IPN [b]
$C_0\ 10^4$ (mol/l)[c]	5.0	8.0	5.6	7.7	4.2
$k\ 10^3$ (day^{-1})	80	42	24	25	≈6

[a] PMHS modified with 80% PEO 350 grafts and 20% BPC units (*18*)
[b] $[Si(OEt)_3]$=19 mol%; [c] initial phenol concentration in the medium.

The very large differences in the rates of the phenol release in each of the studied systems suggested that a chosen kinetics of release could probably be obtained by adjusting the parameters which control the rate of hydrolysis of the carbonate bond in the IPN and the release of the phenol out of the system. The accessibility of the carbonate function to the hydrolyzing agent and the diffusion of phenol out of the system will both be controlled by (i) the crosslinking density in each network and (ii) the weight proportions of the two networks in the IPN. Thus a very simple kinetic model was used starting from the following equation (1) expressing the rate constant k

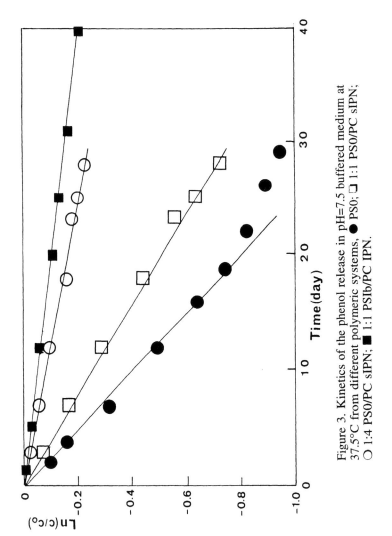

Figure 3. Kinetics of the phenol release in pH=7.5 buffered medium at 37.5°C from different polymeric systems, ● PS0; □ 1:1 PS0/PC sIPN; ○ 1:4 PS0/PC sIPN; ■ 1:1 PSIb/PC IPN.

of the delivery (each term in this equation being determined experimentally on independently synthesized systems as will be described below) :

$$k = k_0 - k_1 \delta_1^{a_1} - k_2 \delta_2^{a_2} \qquad (1)$$

- δ_1 and δ_2 are the respective crosslinking densities in the polysiloxane and the polycarbonate networks. δ_1 values were calculated from the average molecular weight of the monomer unit and the percentage of crosslinking side-chains in the PSI partners. δ_2 values were calculated similarly for the PC networks from the measured density, the molecular weight of the monomer and the percentage of the allyl double bond conversion.
- a_1 and a_2 are coefficients for the polysiloxane and polycarbonate networks respectively.
- The rate constant k_0 of the phenol delivery was determined on a system where the crosslinking density is equal to zero, i.e. the 100% BPC modified polysiloxane PS0. It was found to be equal to $4.2 \ 10^{-2}$ day^{-1}.
- The k_1 and k_2 rate constants were determined by studying the kinetics on systems where either of them could be eliminated in turn. Thus k_2 was determined from the rate constants calculated from the kinetics of the phenol release from a series of s-IPNs made from uncrosslinkable PS0 (δ_1=0) and PC in 1:4, 1:2, 1:1 and 2:1 weight proportions (Figure 4). The plot of those rate constants versus δ_2 being linear, a_2 was found to be equal to 1 and the reduced equation in these systems is :

$$k(\text{day}^{-1}) = 4.2 \ 10^{-2} - 5.4 \ \delta_2 \qquad (2)$$

- Similarly k_1 was determined on systems where δ_2=0, i.e. on single polysiloxane networks formed from PSIa, PSIb and PSIc precursors. The corresponding kinetic plots : Ln ([C]/[C$_0$]) versus time are shown in Figure 5. In this case a_1=0.6 and the reduced equation is :

$$k(\text{day}^{-1}) = 4.2 \ 10^{-2} - \delta_1^{0.6} \qquad (3)$$

In each series PS0 represents the limit system : it is the limit of the s-IPNs series with a zero proportion of PC and it is also the limit of the single polysiloxane network series with a zero crosslinking density.

Combining (2) and (3) relationships and taking x as the weight fraction of carbonate monomer in the IPN and y as the molar fraction of ATES crosslinkable groups in the PSI partner, the final model equation is :

$$k(\text{day}^{-1}) = 4.2 \ 10^{-2} - 8.2 \ 10^{-2} \ x - 4.5 \ 10^{-2} \ y^{0.6} \qquad (4)$$

The validity of this simple model which was derived from measurements on PS0/PC s-IPNs on one hand and on single PSI networks on the other hand was checked on the full IPNs by comparing calculated values to experimentally determined ones. The best fit was obtained between 5 and 40% phenol release showing that truly tailor-made delivery systems can be designed for the specific application.

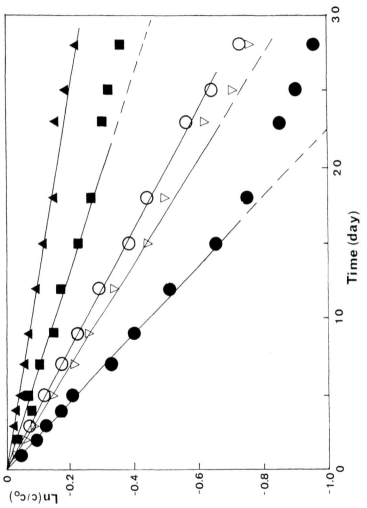

Figure 4. Kinetics of the phenol release in pH=7.5 buffered medium at 37.5°C from sIPNs, ● PS0; ▽ 2:1 PS0/PC; ○ 1:1 PS0/PC; ■ 1:2 PS0/PC; ▲ 1:4 PS0/PC sIPNs

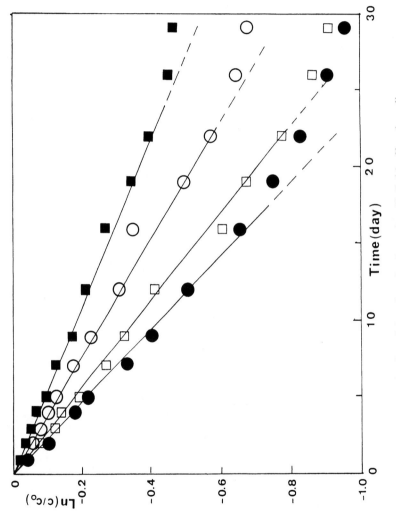

Figure 5. Kinetics of the phenol release in pH=7.5 buffered medium at 37.5°C from single networks obtained from different polysiloxane precursors, ● PS0; □ PSIaN; ○ PSIbN; ■ PSIcN.

References

1. Sperling, L.H. *Interpenetrating Polymer Networks and Related Materials*, Plenum Press, New York, **1981**.
2. Sperling, L. H. in *Interpenetrating Polymer Networks*, Klempner D.; Sperling L. H.; Utracki, L.A. Eds.; Adv. Chem. Ser., ACS Washington DC **1994**, Vol. 239; pp. 3-38
3. Frisch, H.L.; Huang, M.W. in *Siloxane Polymers*; Clarson, S.J.; Semlyen, J.A. Eds., PTR Prentice Hall, Englewood Cliffs, New Jersey, **1993**, pp. 649-667.
4. He, X.W.; Widmaier, J.M.; Herz, J.E.; Meyer, G.C. in *Advances in Interpenetrating Polymer Networks*; Klempner, D.; Frisch, K.C. Eds., Technomic Publishing Co., Lancaster, Pennsylvania, **1994**, pp. 321-356.
5. He, X.W.; Widmaier, J.M.; Herz, J.E.; Meyer, G.C. *Polymer*, **1992**, *33*, 866.
6. Hamurcu, E.E.; Baysal, B.M. *Macromol. Chem. Phys.*, **1995**, *196*, 1261.
7. Miyata, T.; Higuchi, J-I.; Okuno, H.; Uragami, T. *J. Appl. Polym. Sci.*, **1996**, *61*, 1315.
8. Grosz, M.; Boileau, S.; Guégan, P.; Cheradame, H.; Deshayes, A. *Polym. Prep.*, **1997**, *38*(1), 612.
9. Frisch, H.L.; Chen, X. *J. Polym. Sci., Polym. Chem.*, **1993**, *31*, 3307.
10. Gilmer, T.C.; Hall, P.K.; Ehrenfeld, H.; Wilson, K.; Bivens, T.; Clay, D.; Encheszl, C. *J. Polym. Sci., Polym. Chem.*, **1996**, *34*, 1025.
11. Klein, P.G.; Ebdon, J.R.; Hourston, D.J. *Polymer*, **1988**, *29*, 1079.
12. Caille, J.R. Thèse de Doctorat, Univ. P. et M. Curie, Paris 6, **1997**.
13. Ben Khalifa, R. Thèse de Doctorat, Univ. P. et M. Curie, Paris 6, **1994**.
14. Boileau, S. ; Bouteiller, L. ; Ben Khalifa, R. ; Liang, Y. ; Teyssié, D. *Polym. Prep.*, **1998**, *39 (1)*, 457.
15. Yu, J.M.; Teyssié, D.; Boileau, S. *Polym. Bull.*, **1992**, *28*, 435.
16. Frisch, K.C. ; Klempner, D. ; Midgal, S. ; Frisch, H.L. ; Ghiradella, H. *Polym. Eng. Sci.*, **1974**, *14*, 76.
17. Blundell, D.J. ; Longman, G.W. ; Wignall, G.D. ; Bowden, M.J. *Polymer*, **1974**, *15*, 33.
18. Yu, J.M. Thèse de Doctorat, Univ. P. et M. Curie, Paris 6, **1992**.

Chapter 26

High Strength Silicone–Urethane Copolymers: Synthesis and Properties

Emel Yılgör, Ayşen Tulpar[1], Şebnem Kara[1], and Iskender Yılgör

Koç University, Department of Chemistry, İstinye 80860, İstanbul, Turkey

High molecular weight silicone-urethane segmented copolymers were synthesized by the reaction of 4,4'-isocyanatocyclohexylmethane (HMDI), α,ω-hydroxyhexyl terminated polydimethylsiloxane oligomers (PDMS) and 1,4-butanediol (BD). Reactions were conducted in two steps in tetrahydrofuran/dimethylformamide solvent mixture under the catalytic action of dibutyltindilaurate. PDMS oligomers with molecular weights of 900 and 2300 g/mole were used during the synthesis. Hard segment contents of the copolymers were varied between 23 and 50% by weight. Thermal (DSC) and thermomechanical (DMTA) characterization of the copolymers showed the formation of two-phase morphologies. Mechanical strengths of these silicone-urethanes were superior to those of conventional polyether based polyurethanes with similar structures and hard/soft segment compositions.

Block copolymers obtained by the combination of polydimethylsiloxane (PDMS) soft segments and organic hard segments display a unique combination of properties (1). Overall performance of these types of silicone (or siloxane) containing copolymers depends on the type and nature of the organic (hard) segments and the amount and average molecular weight of the PDMS soft segments in the system. As expected, elastomeric properties of these copolymers are determined by the PDMS in the system. On the other hand their mechanical strengths are generally dictated by the type and the nature of the hard segments. Unique properties provided to these copolymers by the presence of PDMS include extremely low glass transition temperatures (-120°C), very low surface energies, improved thermal and oxidative stability, high gas permeabilities, low water absorption, physiological inertness and blood and tissue compatibility (1,2). As a result of these properties, PDMS containing copolymers find applications as specialty elastomers, biomaterials and high performance coatings.

[1]Current address: Chemistry Department, Virginia Polytechnic Institute and State University, Blacksburg, VA 24061.

Preparation and properties of a wide range of well-defined PDMS containing multiphase copolymers have been reported in the literature (1,3,4). Important organic blocks combined with PDMS in the preparation of these block copolymers include polycarbonate, polysulfone, polyurea, polyamide, polyimide, polycaprolactone, poly(methylmethacrylate) and polystyrene (1,3,4). Most of these systems display two phase morphologies due to substantial differences between the solubility parameters of very nonpolar PDMS and moderately to highly polar organic hard segments.

Preparation and characterization of conventional polyether or polyester based polyurethane elastomers with high molecular weights and good mechanical performance have been widely investigated and reported (5,6). Interestingly, there are only a limited number of reports available on the synthesis and characterization of homologous polyurethane systems containing PDMS as the soft segment (7-9). The data published clearly shows that although the formation of copolymers with two phase morphologies (determined by thermal analysis) are observed, the mechanical properties of silicone-urethanes are inferior to those of polyether or polyester based urethane elastomers. Considering the highly flexible nature of PDMS soft segments and achievement of better phase separation in silicone-urethanes leading to stronger hydrogen bonding in the urethane hard segments, one would normally expect better elastomeric properties in silicone-urethanes. We believe poor mechanical properties of silicone-urethanes are mainly due to low molecular weights of the copolymers produced. One of the main reasons for the failure in being able to synthesizing high molecular weight silicone-urethane copolymers is related to the end-group stabilities of commercially available hydroxyalkyl terminated PDMS oligomers, especially the α,ω-hydroxypropyl and α,ω-hydroxybutyl terminated systems. We have recently shown that end-groups on these oligomers are not stable due to the back-biting of silicon (linked to the alkyl bridge) by the hydroxyl oxygen during the vacuum distillation of these oligomers (10). This gives rise to the loss of hydroxyalkyl end-groups and to the formation of stable 5 and 6 membered heterocyclic rings for hydroxypropyl and hydroxybutyl terminated oligomers respectively. Another important reason is the sensitivity of the reaction system to the solvent used. Initially, during the prepolymer formation, the reaction system is fairly non-polar. At this stage tetrahydrofuran (THF) with a solubility parameter of 19.4 $(J/cm^3)^{1/2}$ or 9.48 $(cal/cm^3)^{1/2}$ (11) is a fairly good solvent. However, as the prepolymer is chain extended with butanediol or with other low molecular weight glycols, the polarity and the solubility parameter of the copolymer starts increasing. In order to be able to compensate for this change and increase the polymer molecular weight, polarity of the reaction solvent must also be increased gradually during the chain extension reactions. This can be achieved by the addition of dimethylformamide (DMF) with a solubility parameter of 24.8 $(J/cm^3)^{1/2}$ or 12.12 $(cal/cm^3)^{1/2}$ (11) as a cosolvent into the system, a good solvent for urethane groups.

Recently McGrath and coworkers reported the preparation and characterization of polyurethanes based on 4,4'-diphenylmethanediisocyanate (MDI) and mixed poly(tetramethylene oxide) (PTMO) and PDMS soft segments, where 1,4-butanediol was used as the chain extender (12). They used toluene/DMF (50/50) mixture as the reaction solvent. They reported number average molecular weights of 10,000 to 12,000 g/mole for the copolymers obtained.

In this study siloxane-urethane copolymers were prepared by the reaction of α,ω-hydroxyhexyl terminated PDMS oligomers, 4,4'-isocyanatocyclohexylmethane and 1,4-butanediol, which was used as the chain extender. Influence of the reaction solvent, PDMS molecular weight and hard segment content on the overall molecular weight and thermal and mechanical properties of the products were investigated. Performance of PDMS based polyurethanes were compared with those of PTMO based systems, with similar structures and compositions.

EXPERIMENTAL

Materials

α,ω-Hydroxyhexyl terminated PDMS oligomers with number average molecular weights (Mn) of 900 and 2300 g/mole were kindly supplied by Th. Goldschmidt AG of Essen, Germany. Mn of these oligomers were determined by ^{1}H-NMR spectroscopy by comparing the peaks of hexyl end-group protons to those of the methyl groups attached to silicon, as shown in Figure 1. PTMO oligomer with Mn of 2050 g/mole and 1,4-butanediol (BD) were products of DuPont. 4,4'-isocyanatohexylmethane (HMDI) was supplied by Bayer AG, Leverkusen, Germany. Purity of HMDI, as determined by standard dibutylamine back titration method was better than 99.5%. Reagent grade THF and DMF were purchased from Aldrich. Dibutyltin dilaurate (DBTDL) catalyst was a product of Witco.

Polymer Syntheses

Both PDMS and PTMO based polyurethanes were prepared by following the two-step prepolymer method, which is given in the Scheme I outlined below. Reactions were carried out in 3-neck round bottom flasks equipped with an overhead stirrer, an addition funnel and a nitrogen inlet.

A typical procedure followed in the preparation of siloxane-urethane segmented copolymers is as follows: Calculated amounts of HMDI and PDMS oligomer are introduced into the reaction flask, stirred and heated up to about 60°C. This mixture is not miscible. However, when 0.5 mL of 1.0% DBTDL solution in toluene is added, the mixture turns clear in about one minute indicating a fairly fast reaction between PDMS and HMDI. There is also a dramatic increase in the reaction temperature from 60 to about 90°C, typical for very exothermic urethane formation reaction. Prepolymer formation is followed by FTIR, monitoring the disappearance of the broad hydroxyl peak centered around 3300 cm^{-1}. Prepolymer obtained is then diluted with THF to about 50% solids and heated to reflux temperature of 64.5°C. Chain extender, BD, is dissolved in THF and added dropwise into the reaction mixture. As the system became viscous as a result of the chain extension reaction, it is diluted with THF and DMF to prevent the premature precipitation of the copolymer formed, which is indicated by the formation of a cloudy solution. Reaction was continued until the complete disappearance of sharp isocyanate peak around 2250 cm^{-1} in the FTIR

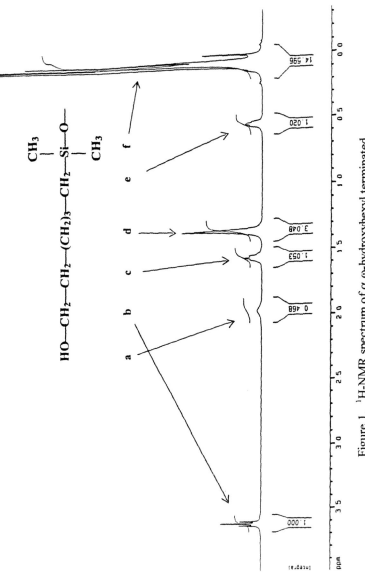

Figure 1. ^{1}H-NMR spectrum of α,ω-hydroxyhexyl terminated polydimethylsiloxane oligomer with Mn=900 g/mol (PDMS-900).

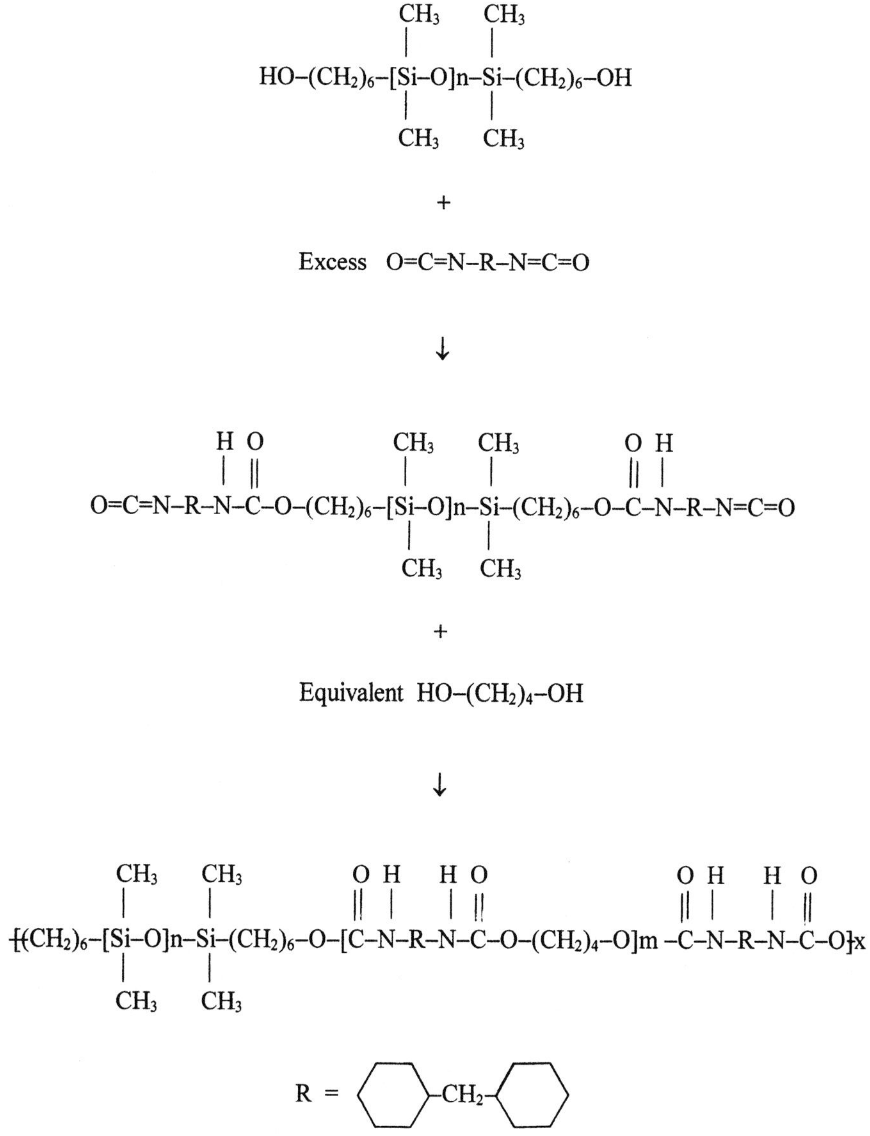

$$\text{HO–(CH}_2)_6\text{–[Si–O]}_n\text{–Si–(CH}_2)_6\text{–OH}$$
with CH$_3$ groups on each Si

$+$

Excess O=C=N–R–N=C=O

$\downarrow$

$$\text{O=C=N–R–N–C–O–(CH}_2)_6\text{–[Si–O]}_n\text{–Si–(CH}_2)_6\text{–O–C–N–R–N=C=O}$$
(with H on N, =O on C of urethane; CH$_3$ on Si)

$+$

Equivalent HO–(CH$_2$)$_4$–OH

$\downarrow$

$$\{(\text{CH}_2)_6\text{–[Si–O]}_n\text{–Si–(CH}_2)_6\text{–O–[C–N–R–N–C–O–(CH}_2)_4\text{–O]}_m\text{–C–N–R–N–C–O}\}_x$$

R = –⟨cyclohexyl⟩–CH$_2$–⟨cyclohexyl⟩–

SCHEME I

Reaction scheme followed in the preparation of PDMS-urethane segmented copolymers spectrum. Polymer produced was coagulated in methanol/water (95/5) mixture, washed several times with methanol, filtered and dried to a constant weight in a vacuum oven at 50°C. Yields were determined gravimetrically.

A similar two-step procedure was used for the preparation of PTMO based polyurethanes, where the solvent used was DMF. Reactions were conducted at 80°C.

A model polyurethane based on HMDI and BD was also prepared in DMF at 80°C by reacting equimolar amounts of each reactant. DBTDL was used as catalyst during the reaction. Product obtained was coagulated in methanol/water (80/20) mixture, washed and dried to a constant weight in a vacuum oven. This sample was used to determine the thermal transitions of the hard segment in DSC.

Table I gives the compositions of silicone-urethane and polyether-urethane segmented copolymers prepared in this study.

Table I. Compositions of the Silicone-Urethane and Polyether-Urethane Reactions

Sample No	Soft Segment Type	Mn (g/mole)	Composition of the reaction mixture (wt.%)		
			Soft Segment	HMDI	BD
PSU-1	PDMS	900	77.4	22.6	–
PSU-2	PDMS	900	61.1	33.6	5.3
PSU-3	PDMS	900	59.0	34.8	6.2
PSU-4	PDMS	2300	78.7	17.8	3.5
PSU-5	PDMS	2300	70.1	24.0	5.9
PSU-6	PDMS	2300	63.3	28.8	7.9
PSU-7	PDMS	2300	56.7	33.3	10.0
PEU-1	PTMO	2050	70.1	24.6	5.3
PEU-2	PTMO	2050	63.2	29.4	7.4
PEU-3	PTMO	2050	56.8	34.1	9.1

Product Characterization

FTIR spectra of starting materials, intermediates and products were recorded on a Nicolet Impact 400D spectrometer. GPC chromatograms were obtained on a Polymer Laboratories, PL 110 instrument equipped with PLgel columns of 500, 10^3 and 10^4 angstroms. THF was used as the solvent and the flow rate was 1 mL/min. Intrinsic viscosities were determined in Ubbelohde viscometers, at 25°C, in THF. Thermal analyses of the products were obtained on a Rheometrics PL-DSC Plus instrument, under nitrogen atmosphere with a heating rate of 10°C per minute. Dynamic mechanical thermal analyses (DMTA) were determined using a Polymer Laboratories DMTA Mk III system, operating in bending mode. Measurements were made simultaneously at frequencies of 1 and 100 Hz, with a heating rate of 5°C per minute. Stress-strain tests were carried out on an Instron Model 4411 Universal Tester, at room temperature, with a crosshead speed of 20 mm/min. For each test a minimum of 5 specimens were used. Films for DMTA and stress-strain tests were prepared by compression molding at 200°C using a Carver, hydraulic press. Transmission micrographs of the solvent (THF) cast polymer films were obtained by using a Leica, Model DM LM, light microscope.

RESULTS AND DISCUSSION

It is well documented that excellent elastomeric properties of thermoplastic, segmented or multiblock polyether or polyester based polyurethanes are a direct result of their backbone composition, the extent of phase separation between their hard and soft segments and the strength of hydrogen bonding in the hard domains (13-15). Extensive phase mixing between polyether or polyester soft segments and urethane hard segments usually lead to poorer mechanical properties, since it effectively reduces the strength of the hydrogen bonding in the system. In that respect, high molecular weight silicone-urethane copolymers can be interesting model polymer systems which may provide valuable information towards a better understanding of the influence of phase separation and hydrogen bonding on the microphase morphology and elastomeric properties of segmented polyurethanes. Extremely nonpolar PDMS soft segments, with very low solubility parameters of 15.6 $(J/cm^3)^{1/2}$ or 7.6 $(cal/cm^3)^{1/2}$ (11) are expected to show almost complete phase separation from the urethane hard segments, which are fairly polar, as indicated by their high solubility parameters around 37.8 $(J/cm^3)^{1/2}$ or 18.5 $(cal/cm^3)^{1/2}$ (16). This will lead to stronger hydrogen bonding in the urethane hard segments in silicone-urethanes, when compared with polyether-urethanes since in the latter system ether oxygen in PTMO soft segments will form weak hydrogen bonding with the urethane groups in hard segments, leading to phase mixing and also effectively reducing the strength of these copolymers. In fact, the dramatic influence of extensive phase separation in silicone-urea systems on the formation of extremely strong thermoplastic elastomers has already been demonstrated (17,18).

Preparation of high molecular weight silicone-urethane copolymers were achieved by using athe conventional two-step, prepolymer technique. During prepolymer formation PDMS oligomers were reacted with excess HMDI in bulk, at 80°C under the catalytic action of DBTDL. Prepolymer mixture was diluted with THF and cooled down to THF reflux temperature of 64-65°C and held constant in this range throughout the chain extension process. For chain extension stoichiometric amount of BD was dissolved in THF (20/100, v/v, BD/THF), introduced into an addition funnel and added dropwise into the reaction mixture. As the urethane hard segment molecular weights grew, polarity of the product increased. In order to prevent premature precipitation of the product, polarity of the solvent was also increased by adding DMF into the reaction flask throughout the chain extension step. This seems to be critical in obtaining high molecular weight segmented siloxane-urethane copolymers. Table II gives various characteristics of the polymers produced. As can be seen from this Table, silicone-urethane copolymers with hard segment contents of up to 50% by weight were prepared in high yields and high molecular weights as indicated by their intrinsic viscosities and GPC chromatograms.

Table II. Characteristics of Siloxane-Urethane and Polyether-Urethane Segmented Copolymers

Sample No	Soft Segment Type	Mn (g/mole)	Hard Segment Content (wt.%)	Yield (wt.%)	$[\eta]^{25°C}$ (dl/g)
PSU-1	PDMS	900	22.6	92	0.90
PSU-2	PDMS	900	38.9	93	0.72
PSU-3	PDMS	900	41.0	93	0.80
PSU-4	PDMS	900	49.9	90	0.68
PSU-5	PDMS	2300	21.3	89	0.67
PSU-6	PDMS	2300	29.9	90	0.73
PSU-7	PDMS	2300	36.7	95	0.79
PSU-8	PDMS	2300	43.3	92	0.81
PEU-1	PTMO	2000	29.9	95	--
PEU-2	PTMO	2000	36.8	96	--
PEU-3	PTMO	2000	43.2	95	--

DSC thermograms of these copolymers all showed two glass transition temperatures, one at around -120°C, due to PDMS soft segments and the other around +70°C due to the urethane hard segments made of BD and HMDI. Silicone-urethane copolymers with high hard segment contents (e. g. PSU-2, PSU-3, PSU-4, PSU-7 and PSU-8) also showed a broad endotherm between 120-150°C due the melting of HMDI-BD hard segments. High temperature DSC thermograms of model polyurethane composed of HMDI-BD and PSU-4 are reproduced in Figure 2. HMDI-BD model polymer shows a Tg at 77°C and a broad melting endotherm between 150 and 170°C whereas Tg for PSU-4 is at 73°C, very close to that of the model system. As can be seen from Figure 2, there is a slight difference in the temperature range of the melting endotherm for the model hard segments and that of PSU-4, which is most probably due to the effect of hard segment molecular weight. However, these results clearly show the formation of two-phase copolymers with good phase separation.

Formation of two-phase morphologies, with good phase separation in these silicone-urethane segmented copolymers can also clearly be seen from the results of DMTA studies. Figure 3 gives the DMTA curves for PSU-4 where Tan δ and Storage Modulus are plotted against temperature. The peaks for PDMS glass transition (-100°C), PDMS melting (-30°C) and hard segment glass transition (60°C) are all clearly visible. After about 80°C, since the sample became too soft it was not possible to observe the transition due to melting of the hard segment in DMTA, which was clearly visible in DSC. Slight shifts in the transition temperatures between DSC and DMTA are typical and are due to differences in the heating rates employed and also due to differences in the nature (static versus dynamic) of these techniques.

Investigation of the tensile behavior of silicone-urethane copolymers and their comparison with those of silicone-urea copolymers (18) or THF based polyether-urethane homologs yielded interesting and somewhat unexpected results. The data on

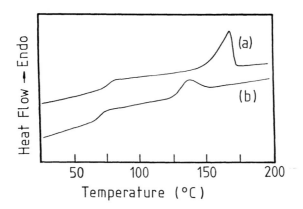

Figure 2. DSC thermograms for (a) model HMDI-BD based polyurethane and (b) PSU-4.

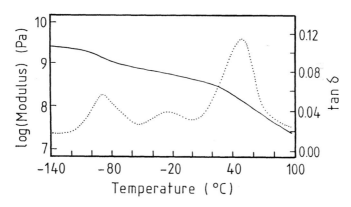

Figure 3. DMTA curves for PSU-4 (———) storage modulus (· · · ·) bending tan δ.

stress-strain tests is given in Table III. Silicone-urethanes with low hard segment contents, such as samples PSU-1 and PSU-5 with hard segment contents of 22.6 and 21.3% by weight, respectively, gave nice films when compression molded or cast from THF solutions, however, they were very weak with almost no tensile strength or mechanical integrity. This is quite different than the tensile test results obtained on silicone-urea copolymers of similar compositions, which show the formation of extremely strong elastomers with very high tensile strengths (17,18). This simple comparison between silicone-urethane and silicone-urea copolymers of similar structure and composition clearly indicate the importance of the strength of hydrogen bonding on the mechanical performance of well phase separated silicone containing thermoplastic elastomers.

Table III. Comparison of the Tensile Properties of Siloxane-Urethane and Polyether-Urethane Segmented Copolymers

Sample No	Soft Segment Type	Mn (g/mol)	Hard Segm. Content (% wt.)	Modulus (MPa)	Ultimate Strength (MPa)	Elong. at Break (%)
PSU-1	PDMS	900	22.6	--	--	--
PSU-3	PDMS	900	41.0	6.75	3.45	400
PSU-4	PDMS	2300	21.3	--	--	--
PSU-5	PDMS	2300	29.9	14.3	11.0	350
PSU-6	PDMS	2300	36.7	35.1	16.4	310
PSU-7	PDMS	2300	43.3	63.9	17.7	250
PEU-1	PTMO	2050	29.9	10.7	3.56	475
PEU-2	PTMO	2050	36.8	23.9	7.68	450
PEU-3	PTMO	2050	43.2	50.9	17.2	425

As expected, the moduli and the ultimate strengths of silicone-urethane copolymers increase with increasing hard segment content in the copolymer. On the other hand there is a small drop in the elongation at break as the amount of the hard segment is increased.

Stress-strain results given in Table III clearly show that for copolymers of similar structures and compositions, silicone-urethanes always show higher tensile moduli and tensile strength when compared with PTMO based polyether-urethanes. This is also very interesting, since at room temperature PDMS is about 145°C above its Tg (-120°C) and approximately 75°C above its melting point (-50°C) (1). On the other hand PTMO is about 90°C above its Tg (-65°C) but 25°C below its melting point (50°C) at room temperature (19). This would intuitively suggest higher tensile strengths for polyether-urethanes due to the contribution of crystalline PTMO soft segments to mechanical performance. However, the experimental results indicate just the opposite. We believe this is due to much better phase separation in silicone-urethanes, which results in more effective intermolecular hydrogen bonding between the urethane hard segments, which effectively determines the modulus and tensile strength of the material. As for the polyether-urethanes, there is usually fairly high

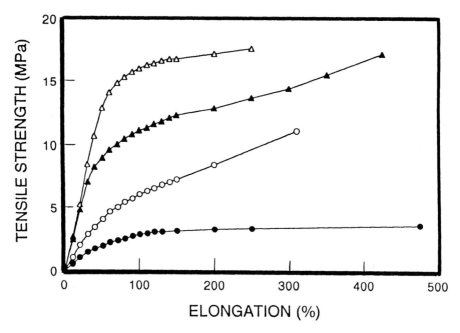

Figure 4. Comparison of the tensile behavior of polydimethylsiloxane and poly(tetramethylene oxide) based segmented polyurethanes with similar compositions and structure. (Δ) PSU-8, (▲) PEU-3, (o) PSU-7 and (•) PEU-2.

amount of phase mixing between urethane and polyether soft segments due to hydrogen bonding between ether linkages in polyethers and -N-H groups of urethanes (6). This effectively reduces the strength of the hydrogen bonding between urethane hard segments and as a result the modulus and overall tensile strength of the polyurethane. Figure 4 gives the stress-strain curves for PSU-7, PSU-8 and PEU-2 and PEU-3 for a better comparison.

Another interesting observation was made in sample PSU-4, which is based on PDMS-900 and has a urethane hard segment content of 49.9% by weight. When THF cast films of PSU-4 were stretched slightly they showed strain induced crystallization, indicated by the formation of regular patterns in the films, which were crystal clear and transparent originally. Figure 5 gives the micrograph of PSU-4 obtained under a light microscope operating in the transmission mode. Figure 5 clearly shows the formation of a very well organized lamellar structure with dimensions of each layer being around 2.5 μm.

CONCLUSIONS

High molecular weight silicone-urethane copolymers with urethane hard segment contents of up to 50% by weight were synthesized in high yields. Thermal, thermomechanical and microscopic characterization of these copolymers clearly indicated the formation of two-phase systems with very good phase separation.

Figure 5. Transmission light micrograph showing the lamellar structure in PSU-4 films formed by strain induced crystallization.

Tensile strengths of silicone-urethanes were higher than those of PTMO based polyether-urethanes of similar structure and compositions. We believe, this is due to better phase separation in silicone-urethanes leading to more efficient and stronger hydrogen bonding between urethane hard segments, the main force, which provides the excellent elastomeric properties to polyurethane elastomers.

REFERENCES

1. Yilgor, I. ; McGrath, J. E. *Adv. Polym. Sci.* **1988**, *86*, 1-87.
2. Noll, W. *Chemistry and Technology of Silicones*, Academic Press: New York, NY, 1968.
3. Noshay, A.; McGrath, J. E. *Block Copolymers: Overview and Critical Survey*, Academic Press: New York, NY, 1978.
4. Burger, C.; Kreuzer, F. -H., in, *Silicone in Polymer Synthesis*, Ed., Kricheldorf, H. R., Springer-Verlag: Berlin, 1996, Chapter 3, pp 113-222.
5. Woods, G.,*The ICI Polyurethanes Book*, John Wiley: New York, 1990.
6. Lelah, M. D.; Cooper, S. L. *Polyurethanes in Medicine*, CRC Press: Boca Raton, Florida, 1986.
7. Kossmehl, G.; Neumann, W.; Schafer, H. *Makromol. Chem.* **1986**, *187*, 1381.
8. Sha'aban, A. K., et al. *Polym. Prepr.* **1983**, *24(2)*, 130.
9. Pascault, J. P.; Chamberlin, Y. *Polym. Commun.* **1986**, 27, 230.
10. Yilgor, E.; Yilgor, I. *Polym. Bull.* **1998**, *40*, 525.
11. Grulke, E. A. in *Polymer Handbook*, Eds. Brandrup, J.; Immergut. H. John Wiley: New York, NY, 1989, Chapter VII, pp VII/519-559.
12. McGrath, J. E.; Wang, L. F.; Mecham, J. B.; Ji. Q. *Polym. Prepr.* **1998**, *39(1)*, 455.
13. Cooper, S. L.; Tobolsky, A. V. *J. Appl. Polym. Sci.* **1966**, *10*, 1837.
14. Marcos-Fernandez, A.; Lozano, A. E.; Gonzales, L.; Rodriguez A. *Macromolecules*, **1997**, *30*, 3584.
15. Abouzar, S.; Wilkes, G. L. *J. Appl. Polym. Sci.* **1984**, *29*, 2695.
16. Van Krevelen, D. W. *Properties of Polymers*; Elsevier: Amsterdam, 1992, Chapter 7, pp 185-225.
17. Yilgor, I.; Sha'aban, A. K.; Steckle, W. P., Jr.; Tyagi. D.; Wilkes, G. L.; McGrath, J. E. *Polymer*, **1984**, *25*, 1800.
18. Tyagi, D.; Yilgor, I.; McGrath, J. E.; Wilkes, G. L. *Polymer*, **1984**, *25*, 1807.
19. Dreyfuss, P. *Poly(tetrahydrofuran)*, Polymer Monographs 8; Gordon and Breach: New York, NY, 1982, Chapter 6, pp-157-190.

Chapter 27

Silicon-Terminated Telechelic Oligomers by Acyclic Diene Metathesis Chemistry

K. R. Brzezinska[1], K. B. Wagener[1], and G. T. Burns[2]

[1]The George and Josephine Butler Polymer Research Laboratory, Department of Chemistry, University of Florida, Gainesville, FL 32611-7200
[2]Advanced Polymeric Systems Laboratory, The Dow Corning Company, Midland, MI 48686

Acyclic diene metathesis (ADMET) polymerization has shown to be a clean route to silicon terminated telechelic oligomers. Methoxydimethylsilane and chlorodimethylsilane terminated telechelic polyoctenamer oligomers have been prepared by ADMET chemistry using Grubbs' ruthenium, or Schrock's molybdenum catalysts. The number average molecular weights (M_n) values of the telechelomers are dictated by the initial ratio of the monomer to the chain limiter. The termini of these oligomers can undergo a condensation reaction with hydroxy terminated poly(dimethylsiloxane) macromonomer, producing an ABA-type block copolymer.

The viability of acyclic diene metathesis (ADMET) polymerization in the synthesis of unsaturated polymers is now well established (1). ADMET polymerization, which is step propagation, condensation type reaction, provides the opportunity to expand metathesis polymerization beyond the well-known ring-opening metathesis polymerization chemistry (ROMP) (2). While continuing to explore the scope of ADMET polymerization in generating various homopolymers and copolymers, we have begun studying ADMET's potential for making telechelomers (Figure 1).

Telechelomers (3) contain functional groups at both ends of the polymer chain and have been the focus of much interest. They have several potential uses, including both theoretical e.g. model network (4) and commercial e.g. liquid rubber (5) applications. Previously, this type of polymer was often prepared by reacting living polymers with suitable terminating reagents in conjunction with difunctional initiators (6,7) or functionally substituted initiators (8). The anionic (9), cationic (10), and most recently, metathesis-based (11,12), living polymerizations are particularly preferred because these routes provide well-defined polymers with a high proportion of functionalities at both ends of the polymer chain. However, these processes are limited in scope since only a few monomers undergo such reactions.

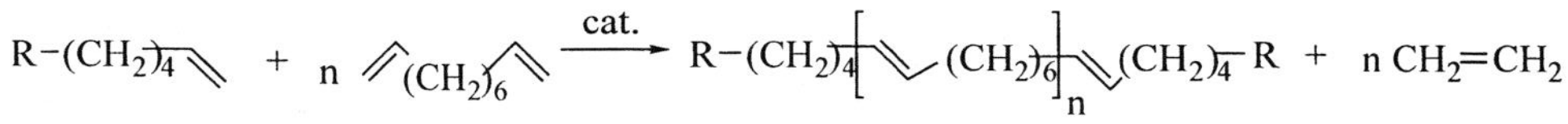

$$R-(CH_2)_4\hspace{-2pt}=\hspace{-2pt} + n =\hspace{-4pt}(CH_2)_6\hspace{-4pt}= \xrightarrow{cat.} R-(CH_2)_4\hspace{-4pt}\left[=\hspace{-4pt}(CH_2)_6\hspace{-4pt}=\right]_n\hspace{-4pt}(CH_2)_4-R + n\ CH_2=CH_2$$

R: -Si(CH$_3$)$_2$OCH$_3$ or   -Si(CH$_3$)$_2$Cl

cat.: RuCl$_2$(=CHPh)(PCy$_3$)$_2$
or Mo(=CH-CMe$_2$Ph)(N-2,6-C$_6$H$_3$-*i*-Pr$_2$)(OCMe(CF$_3$)$_2$)$_2$

Figure 1. Synthesis of telechelic oligomers using ADMET chemistry.

Since ADMET polymerization is an equilibrium step growth reaction (1h), monofunctional reagents can be used to terminate chain propagation and hence, control the molecular weight via end capping (Figure 1). We reported the synthesis of telechelic polybutadiene oligomers (f = 2.0) via ADMET depolymerization using Schrock's molybdenum or tungsten catalysts, where the α,ω-positions of the oligomers are terminated by silane, ester, silyl ether, and imide groups (13). In a similar fashion, Nubel *et al.* reported the synthesis of telechelic polybutadiene oligomers of 5-acetoxy-1-pentene via ADMET polymerization using a catalyst system of $WCl_6/SnMe_4/PrOAc$ (14).

We now report the synthesis of methoxydimethylsilane and chlorodimethylsilane terminated telechelic polyoctenamer oligomers (POCT) using Grubbs' ruthenium, $RuCl_2(=CHPh)(PCy_3)_2$ [Ru] and Schrock's molybdenum, $Mo(=CH\text{-}CMe_2Ph)(N\text{-}2,6\text{-}C_6H_3\text{-}i\text{-}Pr_2)(OC\text{-}Me(CF_3)_2)_2$ [Mo] catalysts (16,15). Additionally, we have shown that these chlorodimethylsilane terminated telechelomers polycondense with a hydroxy terminated poly(dimethylsiloxane) (PDMS) macromonomer, $HOSi(CH_3)_2O\text{-}\{Si(CH_3)_2O\}_xSi(CH_3)_3$. This reaction produces a PDMS-POCT-PDMS block copolymer $(CH_3)_3SiO[Si(CH_3)_2O]_x[CH=CH(CH_2)_6]_y[OSi(CH_3)_2]_xOSi(CH_3)_3$.

The Synthesis of Methoxydimethylsilane Terminated Telechelic Polyoctenamer Oligomers (POCT). Synthesis of the bis(5-hexenyl-methoxydimethylsilane)polyoctenamer was achieved using 5-hexenyl-methoxydimethylsilane as the chain limiter in the ADMET polymerization of 1,9-decadiene. 1,9-Decadiene is well known to polymerize in the presence of Ru (16) or Mo based (15) catalysts, where the resulting hydrocarbon polymer is void of functionality at the termini other than the vinyl group (1). However, if small quantities of functionalized monoolefins are added in the reaction mixture, then telochelomers are produced with the molecular weight determined by the ratio of monomer to terminator (Figure 1). In this study, 5-hexenylmethoxy-dimethylsilane was used as chain limiter. Chain limiter was synthesized from 5-hexenyl-dimethylsilane and methyl orthoformate (17). All polymerizations were done in the bulk state to maximize molar concentration of the olefin, and shift equilibria appropriately for the ADMET reaction (1). Monomer and chain limiter were vacuum transferred to a previously dried and evacuated reaction flask. Catalyst was added to the reaction mixture (monomer / catalyst molar ratio 500:1) and the mixture was stirred for 40 h at 60-85 °C. Ethylene gas evolved was removed under full vacuum.

Integration of the ^1H NMR spectrum (Figure 2) shows that the polyoctenamer sample was at least 98% capped by methoxysilane groups, the remainder being vinyl end groups. Figure 2 shows the ^{13}C NMR spectrum of bis(5-hexenylmethoxydimethylsilane)polyoctenamer. The *cis* and *trans* internal olefinic carbons in the polymer are easily distinguished by ^{13}C NMR spectroscopy since the *cis* olefinic carbons appear at 130.3, and *trans* at 129.8 ppm (1). Quantitative ^{13}C NMR was used to correlate the peak intensities of the resonances to the percentage *trans* stereochemistry for the polymer samples. The telechelic oligomers were determined to be approximately 80% *trans*

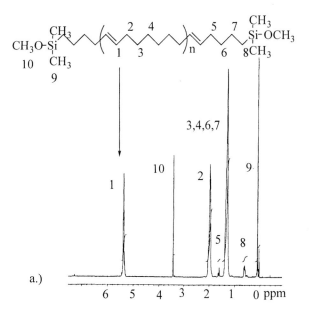
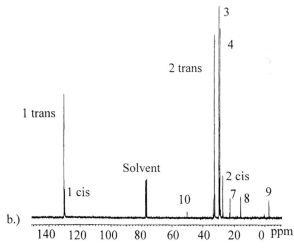

Figure 2. a) ¹H NMR (300 MHz) and b) ¹³C NMR (75 MHz) spectra of bis(5-hexenylmethoxydimethylsilane)polyoctenamer, M_n = 1600 (adapted from ref. 17).

stereochemistry. The ^{29}Si NMR spectrum of bis(5-hexenylmethoxydimethylsilane) polyoctenamer exhibits only one signal at 18.21 ppm and indicates that polymer has been terminated with the proper end group. In the polymer, methoxy end groups were observed by IR spectroscopy at 1900 cm^{-1} and dimethylsilyl end groups are present at 1255 cm^{-1}.

When the chain limiter was added to the reaction mixture, the molecular weight of the bis(5-hexenylmethoxydimethylsilane)polyoctenamers decreased proportionally with the ratio of the monomer to chain limiter. The relationship between the actual measured average molecular weight (by VPO and NMR) and calculated average molecular weight (from ratio of the monomer to chain limiter) is shown in Figure 3. The measured number average molecular weights of bis(5-hexenylmethoxydimethylsilane)polyoctenamers are equal to calculated values within the limit of experimental error. The theoretical M_n values are dictated by the ratio of 1,9-decadiene to chain limiting species. Molecular weight control was quite efficient, and there was no apparent dissociation of the silicon-methoxy bond in the presence of the [Ru] or [Mo] catalyst system.

The Synthesis of Chlorodimethylsilane Terminated Telechelic Polyoctenamer Oligomers. We have also shown that 1,9-decadiene polymerizes in the presence of 5-hexenylchlorodimethylsilane to generate reactive chlorosilyl telechelic oligomers. Using this chain limiter, molecular weight control was also very efficient and there was no apparent dissociation of the silicon-chlorine bond in the presence of the metathesis catalysts. The NMR data found in Figure 4 illustrates the simplicity of this polymerization, where the telechelic oligomers that form are extraordinarily clean in that no other repeat unit is present. Figure 4 shows the ^{1}H NMR spectrum of bis(5-hexenylchlorodimethylsilane)polyoctenamer. The ^{13}C NMR spectrum (Figure 4) of this polymer indicates that it is mostly *trans* (~80%) in its stereochemistry.

Synthesis of the Block Copolymers. The reactive nature of the terminal functional groups was exploited to synthesize ABA-type block copolymers. Silyl chloride groups are able to undergo a facile condensation reaction with hydroxy silicon groups, forming a stable siloxane linkage. Using this chemistry, hydroxyl-terminated poly(dimethylsiloxane) (PDMS) macromonomer [HOSi(CH$_3$)$_2$O{Si(CH$_3$)$_2$O}$_x$Si(CH$_3$)$_3$] was reacted with the chlorosilane-terminated oligomer, producing siloxane-linked ABA-type block copolymers (Figure 5). The HCl gas evolved was removed under reduced pressure. After the addition of PDMS macromonomer was complete, the solution was heated to 60 °C for 4 h under reduced pressure. The copolymer was precipitated into CH$_3$OH and dried *in vacuo* at 50 °C until constant weight was reached.

The ^{1}H NMR spectrum of the copolymer is shown in Figure 6 and shows all of the characteristic resonances exhibited in each individual component of the block copolymer. The ratio between PDMS and polyoctenamer (POCT) segments present in the copolymer is dictated by the comonomer feed ratios, a value which is verified the integration of the methylene peaks at 0.07 and 1.97

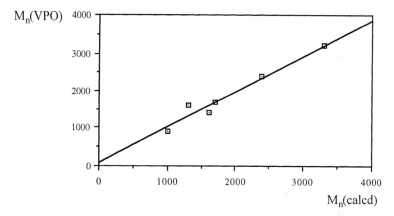

Figure 3. Relationship between the measured M_n (VPO) and calculated M_n (calc.) from the ratio of the monomer to chain limiter (adapted from ref. 17).

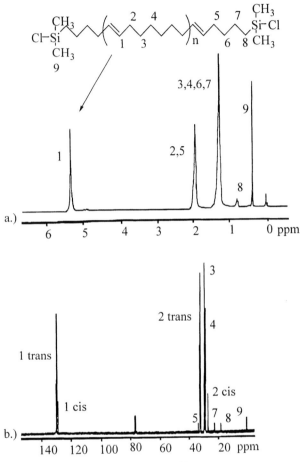

Figure 4. a) ^{1}H NMR (300 MHz) and b) ^{13}C NMR (75 MHz) spectra of bis(5-hexenylchlorodimethylsilane)polyoctenamer, M_n = 1500 (adapted from ref. 17).

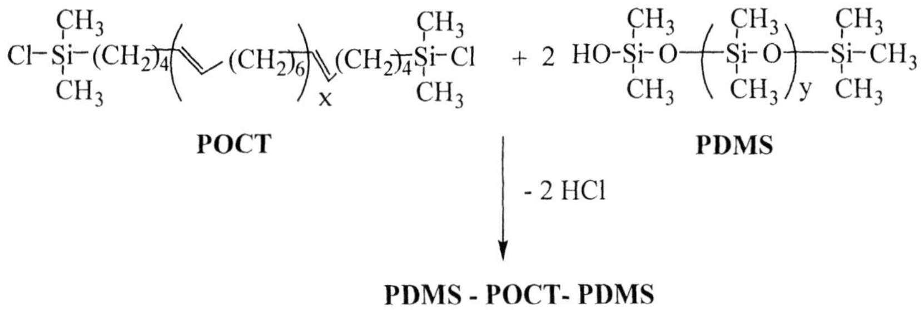

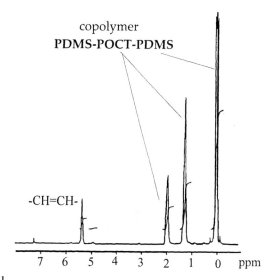

Figure 5. Synthesis of the copolymer: PDMS-POCT-PDMS.

Figure 6. ^{1}H (300 MHz) spectrum of the copolymer (adapted from ref. 17).

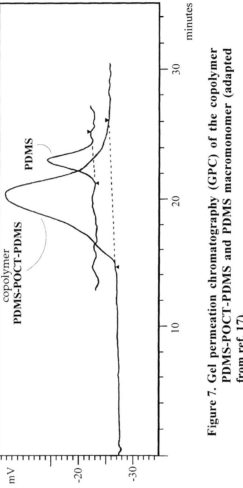

Figure 7. Gel permeation chromatography (GPC) of the copolymer PDMS-POCT-PDMS and PDMS macromonomer (adapted from ref. 17).

ppm. For the sample shown, the PDMS/POCT ratio is approximately 1.8. Good agreement between the theoretical and actual elemental analysis composition (<1%) also support the NMR data for the PDMS/POCT ratio.

Molecular weight analyses also corroborate the formation of a block copolymer. The number average molecular weight for PDMS macromonomer was 3300 and the POCT telechelic oligomer used was 2000. The number average molecular weight (M_n) for the copolymer was determined by VPO to be 8600. Hence, the VPO data support the existance of a PDMS-POCT-PDMS block copolymer. Using GPC, the molecular weight (M_n) for copolymer was determined to be 21000 with M_w/M_n of 1.97. The weight average molecular weight for PDMS macromonomer was 11500 with $M_w/M_n = 1.04$. The GPC curves for these copolymers (Figure 7) are unimodal with no evidence of bimodal behavior. A blend of two homopolymers of PDMS and POCT would show a bimodal distribution. Instead, the unimodel distribution observed suggests the formation of an ABA-type copolymer. Furthermore, the polydispersity for the material synthesized approaches 2.0, typical for copolymer formation.

Acknowledgments. We would like to thank the Dow Corning Company and the Army Research Office for their advice and financial support of this research.

References

1. a.) Wagener, K.B.; Boncella, J.M.; Nel, J.G., *Macromolecules*, **1991**, 24, 2649 b.) Brzezinska, K.; Wagener, K. B. *Macromolecules* **1992**, 25, 2049 c.) Patton, J. T.; Boncella, J. M.; Wagener, K. B. *Macromolecules* **1992**, 25, 3862 d.) Patton, J. T.; Wagener, K. B.; Forbes, M. D. E.; Myers, T. L.; Maynard, M. *Polym. International* **1993**, 32, 411 e.) Smith, D. W.; Wagener, K. B. *Macromolecules* **1991**, 24, 6073 f.) Smith, D. W.; Wagener, K. B. *Macromolecules* **1993**, 26, 1633 g.) O'Gara, D. W.; Portmess, J. D.; Wagener, K. B. *Macromolecules* **1993**, 26, 2837 h.) Wagener, K. B.; Brzezinska, K.; Anderson, J. D.; Younkin, T.R.; Steppe, K.; DeBoer, W, *Macromolecules*, **1997**, 30, 7363 i.) Brzezinska, K.; Wolfe, P.S.; Watson, M. D. and Wagener, K. B. *Macromol. Chem. Phys.* **1996**, 197, 2065.
2. a.) Gilliom, L.R.; Grubbs, R.H., *J. Am. Chem. Soc.*, **1986**, 108, 733 b.) Swager, T.M.; Grubbs, R.H., *J. Am. Chem. Soc.*, **1987**, 109, 894 c.) Schrock, R.R.; Feldman, J.; Cannizo, L.F.; Grubbs, R.H., *Macromolecules*, **1987**, 20, 1488 d.) Wallace, K.C., Schrock R.R., *Macromolecules*, **1987**, 20, 448 e.) Krouse, S.A.; Schrock, R.R., *Macromolecules*, **1988**, 21, 1885.
3. Goethals E.J. *Telechelics Polymers; Synthesis and Applications*, CRS Press Inc., Boca Raton, FL, **1989**.
4. Schnecko, H.; Degler, G.; Dogonowski, H.; Caspary, R.; Anger, G. *Makromol. Chem.* **1978**, 70, 9.
5. Milkovich, R.; Chiang, M. *U.S. Pat.* 3,786,116, **1974**.

6. Tung, L.H.; Lo, G.Y.; Beyer, D.E. *U.S. Pat.* 4,172,100, **1979**.
7. Kennedy, J.P.; Smith, R.A.; Ross, L.R. *U.S. Pat.* 4,276,394, **1981**.
8. Schultz, D.N.; Halasa, A.F.; Oberster, A.E. *J. Polym. Sci., Part A1*, **1974**, 12, 153.
9. Young, R.N.; Quirk, R.P.; Fetters, L. *J. Adv. Polym. Sci.* **1984**, 56, 1.
10. Miyamoto, M.; Sawamoto, M.; Higashimura, T. *Macromolecules*, **1985**, 18, 123.
11. Risse, W.; Grubbs, R.H. *Macromolecules* **1989**, 22, 1558.
12. Schrock, R.R.; Yap, K.B.; Yang, D.C.; Sitzman, H.; Sita, L.R.; Bazan, G.C. *Macromolecules* **1989**, 22, 3191.
13. a.) Marmo, J.C.; Wagener, K.B., *Macromolecules,* **1995**, 28, 2602 b.) Wagener, K. B.; Marmo, J. C. *Macromolecules* **1993**, 26, 2137 c.) Wagener, K.B.; Marmo, J.C., *Macromol. Rapid Commun.*, **1995**, 16, 557.
14. Nubel, P. O.; Lutman, C. A.; Yokelson, H. B. *Macromolecules*, **1994**, 27, 7000.
15. Schrock, R. R.; Murdzek, J. S.; Bazan, G. C.; Robbins, J.; Di Mare, M.; O'Regan, M. *J. Am. Chem. Soc.* **1990**, 112, 3875.
16. Schwab, P. F.; Marcia, B.; Ziller, J. W.; Grubbs, R. H. *Angew. Chem., Int. Ed. Engl.* **1995**, 34, 2039.
17. "*Silicon-Terminated Telechelic Oligomers by ADMET Chemistry: Synthesis and Characterization*" , Brzezinska, K.R.; Wagener, K.B.; Burns, G.T., *J. Polym. Sci., Part A, Polymer Chemistry,* **1999**, vol.37, 849-856, Copyright © 1999, Wiley-Lisa, Inc., a subsidiary of John Wiley & Sons. Inc.

Chapter 28

Hydrogen Bond Interactions and Self-Condensation of Silanol-Containing Polymers in Polymer Blends and Organic–Inorganic Polymeric Hybrids

Eli M. Pearce[1], T. K. Kwei[1], and Shaoxiang Lu[2]

[1]Department of Chemical Engineering, Chemistry and Materials Science, and the Herman F. Mark Polymer Research Institute, Polytechnic University, Six MetroTech Center, Brooklyn, NY 11201
[2]Revlon Research Center, 2121 Route 27, Edison, NJ 08818

A new convenient polymer modification has been developed to synthesize a series of novel silanol-containing polymers by a selective oxidation of Si—H containing precursor polymers with a dimethyldioxirane solution in acetone. The silanol hydrogen bonding interactions in polymer blends as well as the silanol self-condensation to form siloxane semi-interpenetrating polymer networks in miscible polymer blends and organic-inorganic polymeric hybrids are discussed.

Silanol functional groups have been long recognized as reactive intermediates in silicon chemistry.[1] They are formed in the hydrolysis of silanes with various silicon functional groups, such as halosilanes, alkoxysilanes, etc., and then transformed into siloxanes by spontaneous or catalytic condensation. Owing to this tendency, only a limited number of organosilanols have so far been synthesized.

Organosilanols are stronger acids than analogous carbinols.[2-9] They are known to form hydrogen-bonded complexes with phenol, ethers and ketones. On the other hand, the basicity of organosilanol does not prove to be inversely proportional to their acidity as that observed in the case of alcohols. Organosilanols are strongly self-associated through hydrogen bonds even in very dilute solutions. The self-associated silanol hydrogen bonds are characterized by a substantial width and great intensity in the silanol stretching vibration region of infrared spectroscopy.

The combination of organic polymers with inorganic oxides by means of the sol-gel process is one of the most convenient and attractive ways to prepare organic-inorganic hybrids.[10-20] The incorporation of organic polymers into inorganic silicone oxide is of particular interest. Silanol groups produced in hydrolysis tend to further condense to form siloxane networks. On the other hand, silanols or residual silanols on the surface of silicon oxide micro-spheres readily form hydrogen bonds with hydrogen bond acceptor groups in the presence of organic polymers. It is shown that those hydrogen bond interactions between silanol groups and organic polymers are driving forces for formation of these hybrids, and influence the morphology and structure-property behavior of hybrids.

In this chapter, we describe a new convenient polymer modification by the selective oxidation of Si—H containing polymers to synthesize silanol-containing polymers. The studies of silanol-containing polymer blends through silanol hydrogen bonding interactions and the self-condensation of silanols in either miscible hydrogen bonded polymer blends or organic-inorganic polymeric hybrids.

Experimental

Homopolymerization of 4-Vinylphenyldialkyl/arylsilane. 4 g 4-vinylphenyl-dialkyl/arylsilane in 5 ml benzene was introduced into an ampule of 20 ml capacity in the presence of 2,2'-azobisisobutyronitrile (AIBN) (0.2 mole %). The ampule was thoroughly degassed by three freeze-thaw cycles, sealed in an argon atmosphere and then polymerized at 60 °C for 24 h. The resulting polymer was dissolved in methylene chloride and precipitated into methanol twice. A white polymer was obtained with yield greater than 75 % after dried under vacuum at 40 °C for 24 h.

Copolymerization of 4-Vinylphenyldialkyl/arylsilane With Styrene. About 20 g of a mixture of 4-vinylphenyldialkyl/arylsilane and styrene was introduced into an ampule of 25 ml capacity in the presence of AIBN (0.2 mole%). The polymerization procedure was the same as that used for homopolymerization. However, the conversion was limited to about 10% (about 4 h). The copolymer was isolated by precipitation of the methylene chloride solution into methanol and dried under vacuum at 40 °C for 6 h.

Preparation of a Dimethyldioxirane Solution in Acetone. A 500 ml three-necked round-bottomed flask, containing a mixture of twice-distilled water (50 ml), acetone (40 ml), sodium hydrogen carbonate (24 g) and a magnetic stirring bar, was equipped with a gas inlet tube extending into the reaction mixture and an air condenser loosely packed with glass wool. The exit of the air condenser was connected to a condenser filled with dry ice-isopropanol. The bottom of the dry ice condenser was attached in succession to a receiving flask (50 ml) and two cold traps, all being kept in a dry ice-acetone baths. Argon gas was gently passed through the reaction vessel. While applying a slight vacuum (180~220 Torr), 50 g potassium peroxomonosulfate (Oxone) was added quickly in one portion with vigorous stirring at room temperature. The slightly yellow solution (ca. 30 ml) of dimethyldioxirane in acetone (0.06 ~ 0.08 M) was collected in a receiving flask during a 45 minute period.

Polymer Modification of the ≡Si—H Containing Precursor Polymers with a Dimethyldioxirane solution in acetone. To a methyl ethyl ketone solution of ≡Si—H containing precursor polymers or copolymers, a cold solution (ca. -10 °C) of dimethyldioxirane in acetone was quickly added and reacted for 30 min at 0 °C. The mole ratio of dioxirane to polymer was ca. 1.2~1.3. The resulting silanol polymers or copolymers were obtained either in solution and used as is or precipitated into hexane followed by vacuum drying at 40 °C for 24 h.

Preparation of blends, semi-interpenetrating polymer networks (semi-IPNs) and organic-inorganic polymeric hybrids. Blends were prepared by mixing appropriate amounts of each polymer solution in a common solvent while stirring. The resulting blend solutions were stirred at room temperature overnight. Blend films were prepared by solution casting onto glass slides. After the solvent was slowly evaporated at room temperature, all the films were vacuum dried at 80 °C for 3 days unless otherwise specified in the text. In the case of mutual precipitation of two polymers took place while mixing, the precipitates were filtered and washed with chloroform and acetone, respectively, followed by vacuum drying at 80 °C to constant weight.

Thermal analysis. Differential scanning calorimetry (DSC) was performed by means of either the TA 2920 DSC or the Perkin-Elmer DSC-7 calorimeter. Sample weights of 8 ~ 12 mg and a heating rate of 20 °C/min were used. The glass transition temperature of the blend was taken from the second scan unless otherwise specified in the text.

FT-IR spectroscopy. Fourier Transform Infrared Spectroscopy was performed with the use of the Perkin-Elmer 1600 series FT-IR or Digilab FTS-60 spectrometer. A minimum of 64 scans at a resolution of 2 cm^{-1} was signal-averaged. Samples for FT-IR studies were prepared by casting blend solutions onto KBr windows followed by vacuum drying at 80 °C for 3 days. For the precipitated inter-polymer complexes, KBr discs were prepared.

Results and Discussion

Synthesis and Characterization of Silanol-Containing Polymers and Copolymers. Polymer modification is of particular interest when the desired polymer is not readily available from its corresponding monomer by conventional polymerization methods. The conventional method for the synthesis of organosilanols can be accomplished by the hydrolysis of the appropriate substituted silane in the presence of catalysts such as an acid or a base.[1] This synthetic route, however, has some difficulty when applied to the synthesis of silanol polymers which demanded not only high conversion of the functional groups for polymer modification but also resistance to the transformation of silanols to siloxane by self- or catalytic condensation during the preparation.

A new convenient method for the synthesis of silanol polymers was developed by the selective oxidation of corresponding precursor polymers containing Si—H moiety with a dimethyldioxirane solution in acetone.[21,22]. A series of styrene-based silanol polymers and copolymers were synthesized (Scheme 1). The reaction was carried out by an addition of a dimethyldioxirane solution in acetone to the precursor Si—H containing polymer solustion. This reaction resulted in a rapid and selective conversion of the ≡Si—H to ≡Si—OH bonds.

Scheme 1

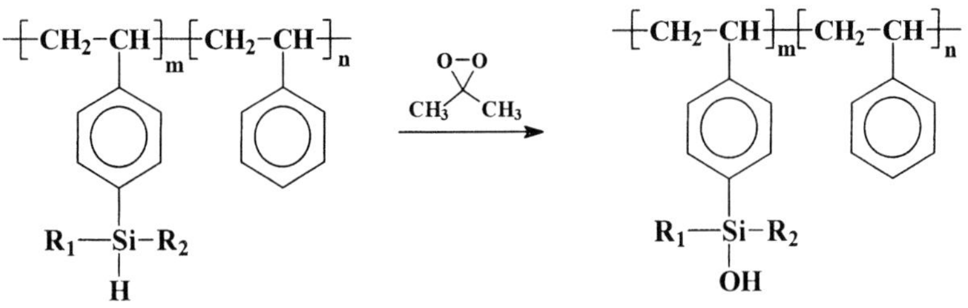

1. $R_1 = CH_3$, $R_2 = CH_3$
2. $R_1 = CH_3$, $R_2 = $ phenyl
3. $R_1 = $ phenyl, $R_2 = $ phenyl

The conversation of the ≡Si—H to ≡Si—OH bonds is evidenced by the infrared spectroscopy as shown in Figure 1. The ≡Si—H stretching vibration band at 2118 cm^{-1} completely disappeared and a new broad band centered at 3367 cm^{-1} with a weak shoulder at 3617 cm^{-1} appeared which are assigned to the self-associated and free silanol stretching vibration bands, respectively.

It was found that the stability of these 4-vinylphenyldialkyl/arylsilanol polymer and their styrene copolymers against self-condensation in the solid state depended largely on the substituents bonded directly to the silicon atom. Although 4-vinylphenyldimethylsilanol polymer and its styrene copolymers obtained in *situ* showed no tendency for self-condensation in solution at room temperature, copolymers underwent self-condensation by forming insoluble siloxane networks after removal of the solvent when the copolymer contained 18 mole % silanol groups or more. When the phenyl group replaced one of the methyl groups, stable 4-vinylphenylmethylphenylsilanol polymer (PVPMPS) and its styrene copolymers (ST-VPMPS) in the solid state were obtained. The increased stability of the polymers against self-condensation in the solid state is attributed to the steric shielding and electron-withdrawing effects of the phenyl substituent bonded directly to the silicon atom. The formation of siloxane crosslinks of PVPMPS at high temperature by the condensation of silanols was characterized by the appearance of a strong Si—O—Si stretching vibration band at 1044 cm^{-1} and a notable reduction in the absorbance of silanol bands in the hydroxyl stretching vibration region (Figure 1).

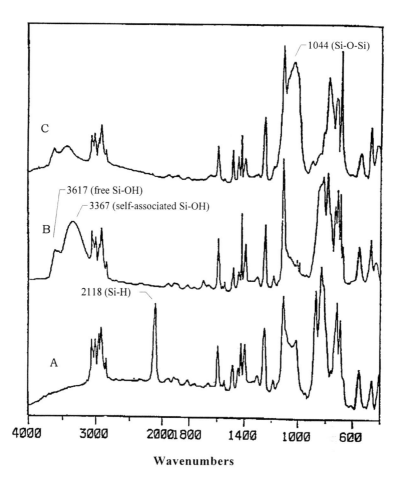

Figure 1. Infrared spectra of (A) poly(4-vinylphenylmethylphenylsilane), (B) poly(4-vinylphenylmethylphenylsilanol) and (C) crosslinked poly(4-vinylphenylmethylphenylsilanol).

The strength of the self-associated silanol hydrogen bonds is measurable through the infrared frequency shifts between the free and self-associated silanol hydrogen bonds in the silanol stretching vibration region. The frequency shifts of the self-associated silanol hydrogen bonds was reduced considerably with increased steric hindrance of the substituents (Figure 2). The infrared frequency shifts were reduced approximately 50 cm^{-1} and 36 cm^{-1} when the methyl groups were replaced by the one and two bulky phenyl groups, respectively.

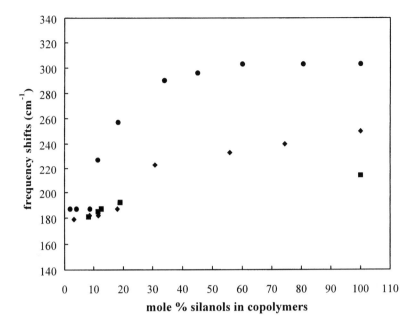

Figure 2. IR frequency shifts in the silanol stretching vibration region. (●) poly(styrene-co-4-vinylphenyldimethylsilanol) (ST-VPDMS), (♦) poly(styrene-co-4-vinylphenylmethylphenylsilanol) (ST-VPMPS) and (■) poly(styrene-co-4-vinylphenyldiphenylsilanol) (ST-VPDPS).

Another novel series of inorganic siloxane-based silanol polymer or copolymers were also prepared by reacting poly(methylhydrosiloxane) (PMHS) or methylhydrosiloxane dimethylsiloxane copolymers with a dimethyldioxirane solution in acetone (Scheme II).[23,24]

Scheme II

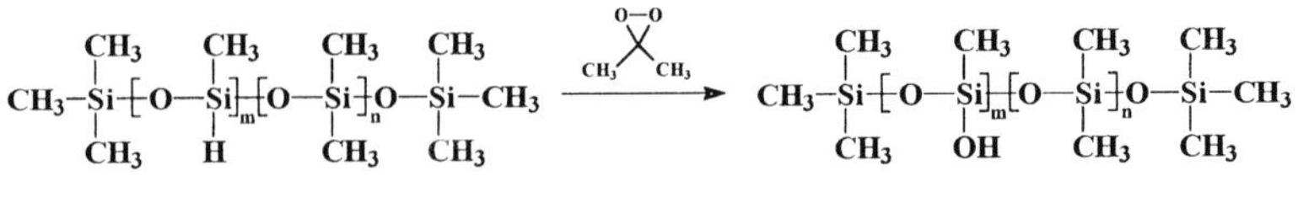

The properties of these siloxane-based silanol polymer and copolymers depended largely on the silanol composition. Polymer materials obtained by the condensation of the silanols led to a wide variety of physical properties ranging from crosslinked elastomer to thermally stable high T_g crosslinked siloxane polymers. Ceramic-like materials were also obtained in a high yield at high temperature under a nitrogen atmosphere.

Silanol Hydrogen Bonds In Miscible Polymer Blends. The introduction of hydrogen bonds between dissimilar polymer chains has been proven to be one of the most effective ways to enhance polymer-polymer miscibility.[25-30] The large negative enthalpy change of hydrogen bonding interactions in the blend provides a driving force for the miscibility. Polymers containing p-(hexafluoro-2-hydroxyl isopropyl) styrene (HFPS) or 4-vinylphenol (VPh) have been extensively studied. It has been demonstrated that as low as few percentage of functional groups are necessary to achieve miscibility between dissimilar polymer chains. On the other hand, the self-associated hydrogen bonds of hydrogen bond donors may compete with hetero-associated hydrogen bonds between the donor and acceptor groups and cause phase separation. The competition and the balance between the strength of the self- and hetero-associated hydrogen bonds determine the phase diagram of the blend.

Studies of ST-VPDMS copolymer blends containing various amount of 4-vinylphenyldimethylsilanol (VPDMS) functional group with poly(n-butyl methacrylate) (PBMA) indicated that as low as 4 mole % VPDMS monomer units in the copolymer are necessary to achieve miscibility with PBMA, which is as effective as p-(hexafluoro-2-hydroxyl isopropyl) styrene (HFPS) or 4-vinylphenol (VPh) in miscibility enhancement. Immiscible blends were also formed when the copolymers contain 34 mole % VPDMS or more. The formation of such a narrow miscibility window is a direct result of strong self-associated silanol hydrogen bonds compared to the hetero-associated hydrogen bond between the silanol and the carbonyl group of PBMA. The infrared spectroscopy investigation revealed that the frequency shifts of the hetero-associated hydrogen bond is about 91 cm^{-1} weaker than that of the sialnol self-associated hydrogen bonds of 180 ~ 300 cm^{-1}.

However, the replacement of one methyl group with a bulky phenyl group shifts the miscibility window to 9 to 56 mole % silanol compositions with PBMA. An electronegative and bulky substituent, such as a phenyl group, would increase the acidity of the silanol and therefore reduce self-associated silanol hydrogen bonds, which makes 4-vinylphenylmethylphenylsilanol a better hydrogen bond donor than 4-vinylphenyldimethylsilanol monomer units. Same consideration can be taken into account when ST-VPDMS copolymers are compared with ST-VPMPS copolymers in terms of the competitive hetero- and self-associated hydrogen bonding interactions in the blends. Infrared spectroscopy studies showed that larger frequency shifts were observed in ST-VPMPS/PBMA than that in ST-VPDMS/PBMA blends in both silanol and carbonyl stretching vibration regions (Table I). The larger frequency shifts suggest stronger hetero-associated hydrogen bond between the silanol and carbonyl groups in ST-VPMPS/PBMA than that in ST-VPDMS/PBMA blends.

Table I. Infrared Frequency Shifts (cm^{-1}) in ST-VPMPS and ST-VPDMS with PBMA Blends

Blend with PBMA	Carbonyl H-bonded silanol group	Silanol H-bonded carbonyl group
ST-VPMPS	107	28
ST-VPDMS	94	25

The shift of the miscibility window would be expected due to the increased strength in hetero-associated hydrogen bonds and the reduced strength in the silanol self-associated hydrogen bonds in ST-VPMPS/PBMA blends.

Studies of blends of ST-VPDMS and ST-VPMPS copolymers with poly(N-vinylpyrrolidone) (PVPr) indicated strong hetero-associated hydrogen bonds between the silanol and the amide carbonyl group of PVPr. Results of infrared frequency shifts in both silanol and amide carbonyl stretching vibration regions are showed in Table II. The hetero-associated hydrogen bonds in ST-VPMPS/PVPr and ST-VPDMS/PVPr blends are almost equal in strength as measured by infrared frequency shifts in both the silanol and amide carbonyl stretching vibration regions. However, the relative strength of the hetero-associated hydrogen bond in ST-VPMPS/PVPr is stronger than that in ST-VPDMS/PVPr due to a notable reduction in the strength of the silanol self-associated hydrogen bond in copolymers.

Table II. Infrared Frequency Shifts (cm^{-1}) in ST-VPMPS and ST-VPDMS with PVPr Blends

Blends with PVPr	Carbonyl H-bonded silanol	Sialnol H-bonded amide carbonyl	Silanol self-association
ST-VPMPS	276	21	187~250
ST-VPDMS	277	20	187~303

The composition dependence of glass transition temperature of blends containing 18 mole % VPMPS and VPDMS monomer units in copolymers are shown in Figure 3 and 4. Positive deviations of glass transition temperature are observed for PVPMPS-18/PVPr blends, whereas, negative derivations of glass transition temperature are found for PVPDMS-18/PVPr blends. The enhancement in glass transition temperature of blends is achieved by the replacement of one methyl group with a phenyl group bonded directly to the silicon atom. This structure-property relationship could be explained by considering the effect of "free volume" changes affected by the competitive hetero- and self-associated hydrogen bonds in the blend. The presence of bulky methylphenylsilanol pendant group tends to increase the "free volume" and therefore lower the T_g of copolymers. However, the silanol self-associated hydrogen bonds enhance the T_g of the copolymer. The influence of these two competitive factors on T_g's of copolymers can be best appreciated from the composition dependence of T_g's of ST-VPMPS copolymers as shown in Figure 5. The silanol self-associated hydrogen bonds are weak at relatively low VPMPS composition. The T_g's of the copolymers decrease with increasing amount of VPMPS monomer units initially, then reached a minimum value at the VPMPS composition range of 12~18 mole %. Further increase in the VPMPS composition resulted in an increase in the strength of silanol self-associated hydrogen bonds that in turn raise T_g's of ST-VPMPS copolymers gradually. In PVPr blends, the formation of the hetero-associated hydrogen bond would result in volume shrinkage at the interaction sites and force the immediate neighboring groups to pack closer. If the net effect of the formation of the hetero-associated hydrogen bond and the cleavage of the self-associated silanol hydrogen results in more efficient close packing of chain segments in the blend, the T_g of the blend would increase. In contrast, if this results in poor packing of chain segments in the blend, the T_g of the blend should decrease. Since PVPMPS-18 copolymer has a relatively large "free volume" to accommodate dissimilar chain segments, blending of ST-VPMPS copolymer with PVPr may result in close-packed hetero chain segments and lead to an incremental increase in T_g of the blends as shown in Figure 3.

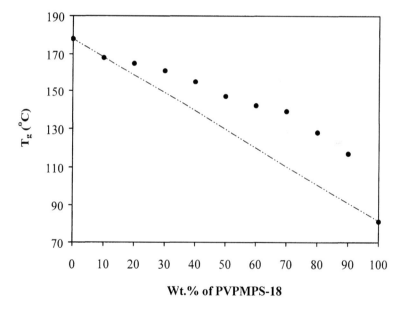

Figure 3. Composition dependence of glass transition temperature of PVPMPS-18/PVPr blends.

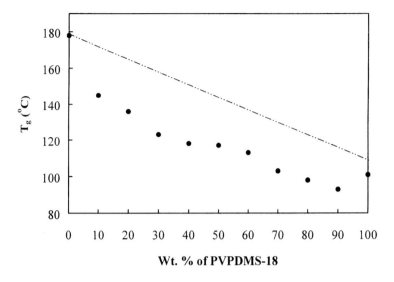

Figure 4. Composition dependence of glass transition temperature of PVPDMS-18/PVPr blends.

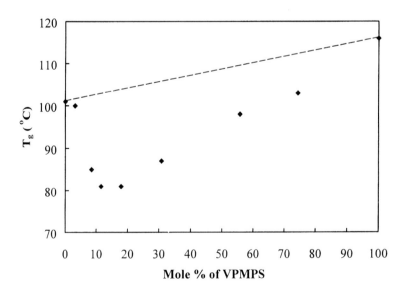

Figure 5. Composition dependence of glass transition temperature of ST-VPMPS copolymers.

Interpolymer Complexes and Semi-IPNs'. The strong inter-polymer hydrogen bonds in ST-VPMPS/PVPr and ST-VPDMS/PVPr blends also resulted in the formation of inter-polymer complexes in the form of precipitates when the copolymer contains 31 mole % VPMPS and 34 mole % VPDMS monomer units (Table III). Further studies of those complexes suggested that semi-interpenetrating polymer networks (IPNs') were formed for those complexes containing 60 mole % or more VPDMS monomer units in ST-VPDMS copolymers. Since the strength of the self-associated silanol hydrogen bonds are little stronger than that of the hetero-associated hydrogen bond in complexes for the copolymer containing 60 mole % VPDMS or more, there is a large fraction of dimethylsilanol pendant group self-associated. Condensation of self-associated silanol groups took place during drying at 80 °C, and transformed those silanols into siloxane crosslink, while the hetero-associated hydrogen bond between silanol and amide carbonyl groups remained almost intact. The formation of siloxane crosslink by the self-condensation of silanol groups greatly enhances the glass transition temperature of PVPDMS-60/PVPr. The non-measurable T_g's of PVPDMS-81/PVPr and PVPDMS-100/PVPr might suggest that the obtained semi-IPNs have a higher crosslink density resulting in a very small heat capacity change (ΔC_p) between the glass and rubbery states.

Table III. Interpolymer complexes of ST-VPMPS/PVPr and ST-VPDMS/PVPr

Blend Code [a]	Solution Appearance	Complex Appearance	T_g (°C)	Calc. T_g (°C) [b]
PVPMPS-31	precipitation	clear	142	132
PVPMPS-56	precipitation	clear	161	138
PVPMPS-74	precipitation	clear	164	141
PVPDMS-34	precipitation	clear	135	147
PVPDMS-60	precipitation	clear	183	150
PVPDMS-81	precipitation	clear	----[c]	157
PVPDMS-100	precipitation	clear	----[c]	163

[a] numeral following PVPMPS or PVPDMS indicates mole % of silanol in the copolymers.
[b] calculated weight-average values.
[c] no T_g s were observed in the temperature range of 50 to 300 °C.

Organic-Inorganic Polymeric Hybrids. The combination of organic polymers with inorganic oxide networks to prepare organic-inorganic hybrids has attracted great research interest. The hybrid can be prepared either by reacting an inorganic alkoxide with a polymer end capped with reactive functional groups to cause cross reaction or by incorporating an organic polymer into an inorganic oxide network in the sol-gel process via catalyzed hydrolysis. In the latter, the incorporation of an organic polymer into silicon oxide network through hetero-associated hydrogen bonding interactions between silanol and the hydrogen bond acceptor group of an organic polymer is of particular interest.

Siloxane polymers are essentially immiscible with almost all other polymers.[31] It has been demonstrated that miscibility with organic polymers can be achieved through interpolymer hydrogen bonding interactions via the modification of siloxane polymer with a strong hydrogen bond donor group, such as 4-hydroxy-4,4-bis(trifluoromethyl) butyl.[32] Since silanol functional groups may either act as a hydrogen bond donor to form inter-polymer hydrogen bonds with organic polymers to enhance miscibility or as reactive sites to condense with themselves to form siloxane crosslink, a variety of organic-inorganic polymeric hybrids were prepared and investigated by using inorganic silanol-containing poly(hydroxymethylsiloxane) (PHMS) and its dimethylsiloxane copolymers (Scheme II). 50/50 w/w Blends of PHMS and its dimethylsiloxane copolymers with poly(N-vinylpyrrolidone), poly(4-vinylpyridine) and poly(ethyloxazoline) indicated that miscibility was accomplished in those polymeric hybrids as evidenced by the clarity of blend films and the presence of a single and composition dependent glass transition temperature. Infrared spectroscopy indicated that inter-polymer hydrogen bonds between silanol and hydrogen bond acceptor groups of organic polymers were present and are responsible for the miscibility in hybrids. Infrared spectroscopy also revealed that condensation of the self-associated silanols took place and converted the self-associated sialnols into siloxane crosslink when the hybrids were vacuum dried at 80 °C.

It is important to mention that properties, such as the T_g of the hybrid, depend largely on the temperature at which the silanol crosslink takes place. For the hybrids vacuum dried at 40 °C for 3 days, a T_g at 54 °C for 50/50 PHMS/PVPr and 40 °C for 50/50 PHMS/PVPy were obtained. When the hybrids were dried in vacuum at 80 °C for 3 days, a higher T_g value was observed. Those results are shown in Table IV. The great dependence of the hybrid properties on the temperature at which the silanol condensation takes place provides evidence for using caution in preparing these hybrids.

Table IV. T_g Values of the 50/50 w/w PHMS/PVPr and PHMS/PVPy Hybrids

Blends	Ratio (w/w)	T_g (°C)	
		drying at 40 °C	drying at 80 °C
PHMS/PVPr	50/50	54	134
PHMS/PVPy	50/50	40	111

Conclusion

A new convenient polymer modification for synthesis of silanol-containing polymer was developed by the selective oxidation of the Si—H bond with a dimethyldioxirane solution in acetone. The oxyfunctionalization of the silane precursor polymers can be utilized to synthesize a wide variety of silanol-containing polymers. Control over the properties of these silanol polymers, such as stability and self-association of silanols, was realized through the placement of different substitute groups bonded directly to the silicon atom. The miscibility in either polymer blends or polymeric hybrids was achieved by the formation of strong inter-polymer hydrogen bonds between the silanols and the hydrogen bond acceptor groups of counter polymers. The solid state crosslink was conducted by the condensation of self-associated silanols in either miscible blends or polymeric hybrids. This new route for solid state crosslink by utilizing the dual nature of the silanol functional groups led to the preparation of intimately mixed semi-interpenetrating polymer networks.

References

1. Noll, W. *Chemistry and Technology of Silicones*, Academic Press, 1968. Chapter 3,
2. West, R., and Baney, R. H., *J. Inorg. Nucl. Chem.*, **1958**, *7*, 297.
3. West, R., Baney, R. H. and Powell, D. L., *J. Am. Chem. Soc.*, **1960**, *82*, 6269.
4. Stone, F. G. A. and Seyferth, D., *J. Inorg. Nucl. Chem.*, **1955**, *1*, 112.
5. Reichstat, M. M., Mioc, U. B., Bogunovic, Lj. J. and Ribnikar, S. V., *J. Mol Struct.*, **1991**, *244*, 283.
6. Mioc, U. B., Bogunovic, Lj. J., Ribnikar, S. V. and Stankovic, N., *J. Serb. Chem. Soc.*, **1989**, *54*, 541.
7. Allred, L., Rochow, E. G. and Stone, F. G. A., *J. Inorg. Nucl. Chem.*, **1956**, *2*, 416
8. Harris, G. I., *J. Chem. Soc.*, **1963**, p. 5978.
9. Kerr, G. T. and Whitemore, F. C., *J. Am. Chem. Soc.*, **1946**, *68*, 2282.
10. Mark, J. E., Jiang, C. Y. and Tang, M. Y. *Macromolecules*, **1984**, *17*, 2613.
11. Tang, M. Y. Mark, J. E. *Macromolecules*, **1984**, *17*, 2616.
12. Clarson, S. J. and Mark, J. E. *Polym. Commun.* **1987**, *28*, 249.
13. Brennan, A. B. and Wilkes, G. L. *Polymer*, **1991**, *32*, 733.
14. Huang, H. H., Orler, B. and Wilkes, G. L. *Macromolecules*, **1987**, *20*, 1322.
15. Wang, B., Wilkes, G. L. Hedrick, J. C. Liptak, S. C. and McGrath, J. E. *Macromolecules*, **1991**, *24*, 3449.
16. Saegusa, T., Chujo, Y. *Proceeding for the 33rd IUPAC Meeting on Macromolecules*, Montreal, **1990**.
17. Chujo, Y. and Saegusa, T. *Adv. Polym. Sci.*, **1992**, *100*, 11.
18. Landry, C. J. T., Coltrain, B. K., Brady, B. K. *Polymer*, **1992**, *33*, 1486.
19. Landry, C. J. T., Coltrain, B. K., Wesson, J. A., Zumbulyadis, N., Lippert, J. L. *Polymer*, **1992**, *33*, 1496
20. Fitzgerald, J. J. Landry, C. J. T. and Pochan, J. M. *Macromolecules*, **1992**, *25*, 3715.
21. Lu, S. Pearce, E. M. and Kwei, T. K. *Macromolecules*, **1993**, *26*, 3514.
22. Lu, S. Pearce, E. M. and Kwei, T. K. *J. Polym. Sci., Polym. Chem. Ed.,* **1994**, *32*, 2597.
23. Lu, S., Melo, M. M., Zhao, J., Pearce, E. M. and Kwei, T. K. *Macromolecules*, **1995**, *28*, 4908.
24. Pearce, E. M., Kwei, T. K., Gao, Y. M. and S. Lu, unpublished results.
25. Pearce, E. M., Kwei, T. K., and Min, B. Y. *J. Macromol. Sci.-Chem.,* **1984**, *A21*, 1181.
26. Fahrenholtz, S., and Kwei, T. K. *Macromolecules*, **1981**, *14*, 1076.

27 Ting, S. P., Pearce, E. M. and Kwei, T. K. *Macromolecules*, **1981**, *14*, 1076.
28 Ting, S. P., Bulkin, B. J., Pearce, E. M. and Kwei, T. K. *J. Polym. Sci., Polym. Chem. Ed.*, **1981**, *19*, 1451.
29 Moskala, E. J., Howe, S. E., Painter, P. C. and Coleman, M. M. *Macromolecules*, **1984**, *17*, 1671
30 Long, L., Pearce, E. M. and Kwei, T. K. *Macromolecules*, **1990**, *23*, 5071.
31 Krause, S. in *Polymer Blends*; Paul D. R., Newman, S., Eds.; Academic Press, New York, **1978**, Vol. 1.
32 Chu, E. Y., Pearce, E. M., Kwei, T. K., Yeh, T. F., Okamoto, Y., *Makromol. Chem., Rapid Comm.*, **1991**, *12*, 1.

Chapter 29

A Review of the Ruthenium-Catalyzed Copolymerization of Aromatic Ketones and 1,3-Divinyltetramethyldisiloxane: Preparation of *alt*-Poly(carbosilane–siloxanes)

William P. Weber, Hongjie Guo, Cindy L. Kepler, Timothy M. Londergan, Ping Lu, Jyri Paulsaari, Jonathan R. Sargent, Mark A. Tapsak, and Guohong Wang

D. P. and K. B. Loker Hydrocarbon Research Institute, Department of Chemistry, University of Southern California, Los Angeles, CA 90089-1661

> Step-growth copolymerization of aromatic ketones and 1,3-divinyl-tetramethyldisiloxane is found to be catalyzed by $(Ph_3P)_3RuH_2CO$. Polymerization occurs by the anti-Markovnikov addition of the C-H bonds, which are *ortho* to the carbonyl group of the aromatic ketone, across the C-C double bonds of 1,3-divinyltetramethyldisiloxane. Prior activation of the catalyst by treatment with a stoichiometric amount of styrene results in formation of ethylbenzene and higher molecular weight copolymers. The scope and current mechanistic understanding of this copolymerization reaction are considered.

Preparation of *alt*-Poly(carbosilane/siloxane) - Previous Methodologies

alt-Poly(carbosilane/siloxanes) have a backbone which contain C-C, Si-C and Si-O-Si bonds. Monomers required for such polymers have been prepared by several approaches. At least two of these involve the formation of Si-C bonds as an essential step. The first involves the formation of Si-C bonds by nucleophilic displacement of chloride from silicon by Grignard or organolithium reagents. For example, 1,4-bis(hydroxydimethylsilyl)benzene, the key monomer for the synthesis of the well studied silphenylene/siloxanes, has been prepared by an in-situ Grignard reaction between 1,4-dibromobenzene and dimethylchlorosilane [1,2], as shown in Figure 1.

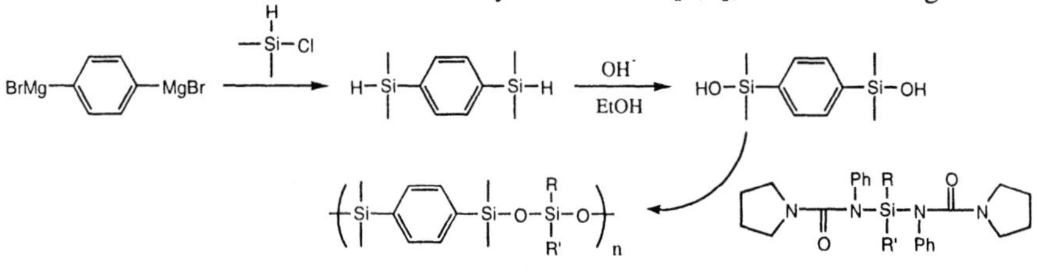

Figure 1

© 2000 American Chemical Society

Alternatively, monomers required for the synthesis of *alt*-poly(carbosilane/siloxanes) have been prepared by formation of Si-C bonds by Pt catalyzed hydrosilation reactions. For example, Pt catalyzed hydrosilation of the C-C double bond of the Diels-Alder adduct of cyclopentadiene and maleic anhydride with sym-tetramethyldisiloxane yields a difunctional monomer 5,5'-bis(1,1,3,3-tetramethyl-1,3-disiloxanediyl)norborane-2,3-dicarboxylic anhydride, see Figure 2. Reaction of this with diamines such as 4,4'-diaminobiphenyl yields a thermally stable poly[imide-*alt*(carbosilane/siloxane)] [3].

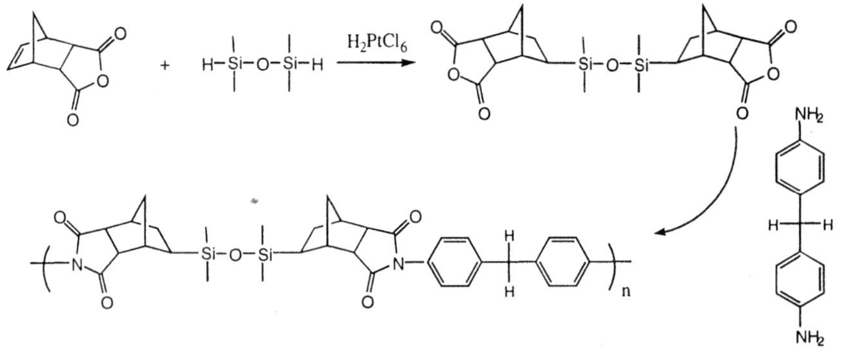

Figure 2

Unsaturated *alt*(carbosilane/siloxane) polymers have been prepared by Wagener's group by use of alicyclic diene metathesis (ADMET) reactions [4]. As shown in Figure 3, the polymer results from the formation of new C-C double bonds. Both the molybdenum and tungsten versions of the Schrock metathesis catalyst are effective [5].

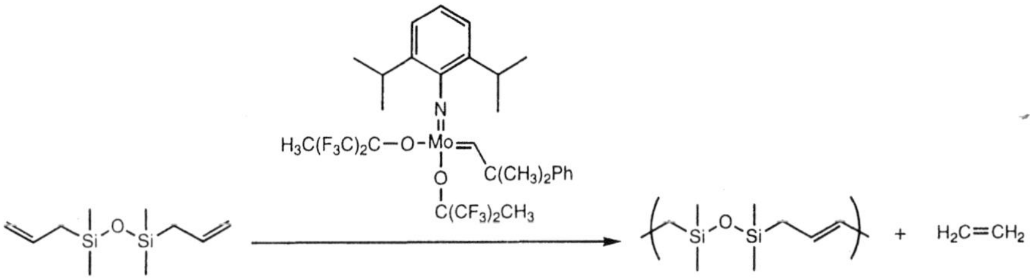

Figure 3

Ru catalyzed synthesis of *alt*-poly(carbosilane/siloxanes)

The reaction, whose scope, limitation and mechanism we are going to review, directly yields *alt*-poly(carbosilane/siloxanes) by the $(Ph_3P)_3RuH_2CO$ (Ru) catalyzed copolymerization of aromatic ketones and 1,3-divinyltetramethyldisiloxane, as shown in Figure 4. The key step in this process involves the ruthenium catalyzed activation of an aromatic C-H bond which is *ortho* to a carbonyl group for anti-Markovnikov addition across the C-C double bond of 1,3-divinyltetramethyldisiloxane. Each time this occurs a C-C bond is formed and the polymer grows. For example, the Ru catalyzed copolymerization of acetophenone and 1,3-divinyltetramethyldisiloxane yields copoly(2-acetyl-1,3-phenylene/3,3,5,5-tetramethyl-4-oxa-3,5-disilaheptanylene), as shown in Figure 4 [6].

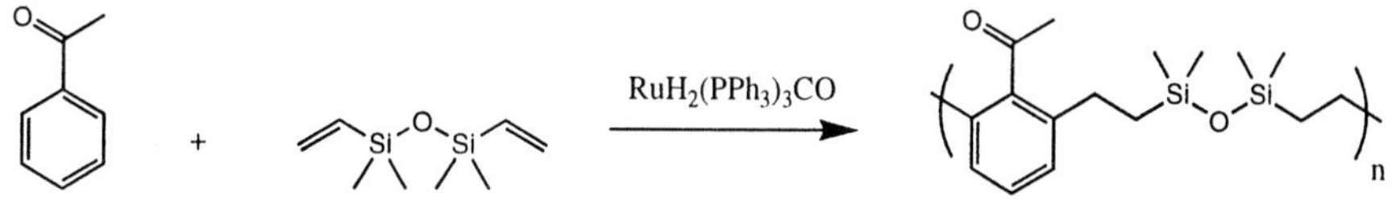

Figure 4

It should be noted that the Ru catalyst is easily prepared by reaction of $RuCl_3$ hydrate with triphenylphosphine, sodium borohydride and formaldehyde [7]. While ruthenium is a rare element, whose natural abundance is reported to be only one tenth that of platinum, its catalytic utility in several different types of polymerization reactions has recently been reported. Among these are ring opening metathesis polymerizations (ROMP) reactions [8], as shown in Figure 5, as well as in isomerization of allyl ethers to reactive propenyl ethers which undergo facile acid catalyzed polymerization [9], as seen in Figure 6. ADMET polymerization can also be achieved using Grubbs Ru-carbene complexes. However, it should be noted that 1,3-divinyltetramethyldisiloxane has been reported not to undergo ADMET type Ru catalyzed polymerization.

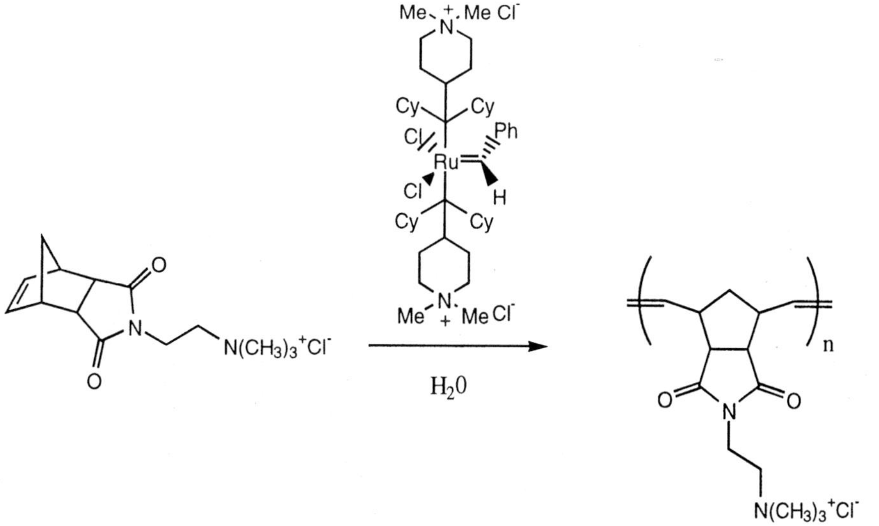

Figure 5

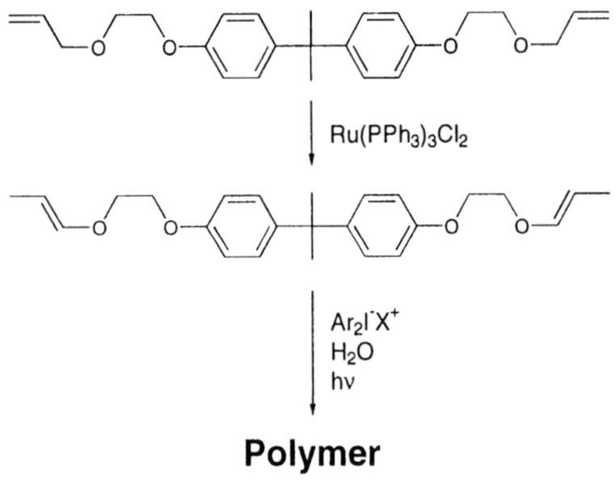

Figure 6

The first examples of this Ru catalyzed reaction in monomer systems were reported by S. Murai, who found that acetophenone and vinyltrimethylsilane undergo Ru catalyzed reaction to yield 2-(2'-trimethylsilylethyl)acetophenone [10,11]. A catalytic cycle, seen in Figure 7, involving coordination of the unsaturated Ru center by the oxygen of the acetyl group followed by insertion of Ru into an adjacent *ortho* C-H bond to form Ru-C and Ru-H bonds has been proposed. Coordination of the C-C double bond of the vinyl silane to the Ru center followed by anti-Markovnikov addition across the C-C double bond regenerates the coordinately unsaturated Ru center and completes the catalytic cycle.

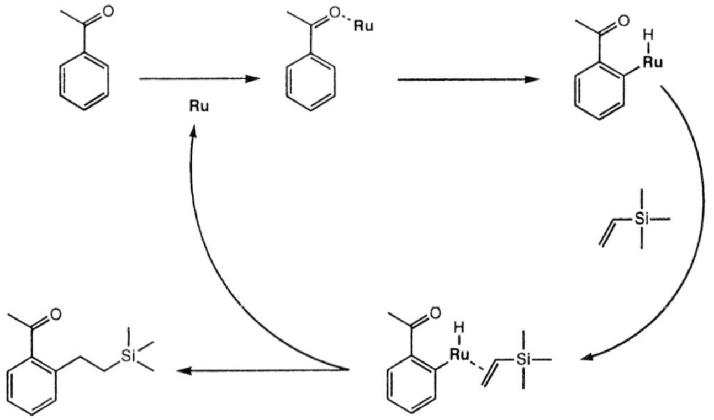

Figure 7

Unfortunately, the reaction fails with most α,ω-dienes. One reason for this is that the Ru catalyst isomerizes terminal α,ω-dienes such as 1,7-heptadiene to internal dienes which are not reactive. Conjugated dienes such as 1,3-butadiene or 2,3-dimethyl-1,3-butadiene are also unreactive. Further C-C double bonds substituted with electron withdrawing groups such those of methyl acrylate, acrylonitrile, or methyl vinyl ketones do not react. So why are the C-C double bonds of vinylsilanes, vinylsiloxanes and styrenes reactive? Perhaps the simplest explanation is that silicon and phenyl groups are not strongly electron withdrawing and do not permit isomerization of adjacent C-C double bonds.

On the other hand, a variety of aromatic ketones have proved to be reactive. Polycyclic aromatic ketones such as anthrone, fluorenone, and xanthone all undergo successful Ru catalyzed copolymerization with 1,3-divinyltetramethyl-disiloxane [12], see Figure 8. It has also been possible to incorporate phenanthrene units, which strongly absorb in ultraviolet and fluoresce, into an *alt*(carbosilane/siloxane) by copolymerization of 2-acetylphenanthrene and 1,3-divinyltetramethyldisiloxane [13]. Initially, the molecular weight of these materials was quite low [12].

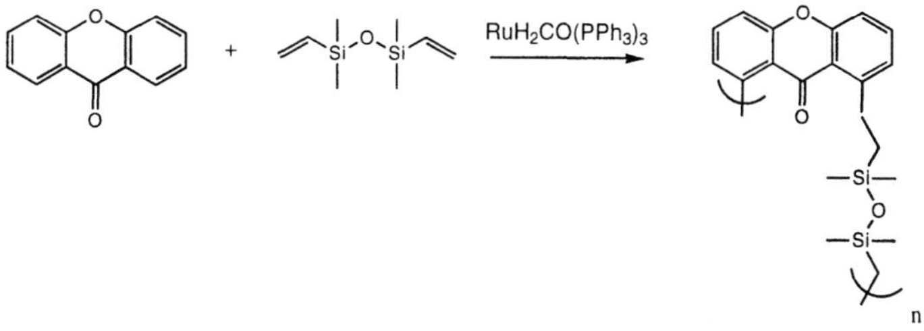

Figure 8

End group analysis by NMR spectroscopy suggested the presence of ethyl groups attached to silicon. This could be explained if the Ru catalyst is capable of catalyzing the hydrogenation of the vinyl groups of 1,3-divinyltetramethyldisiloxane as well as the copolymerization reaction. In fact, we have suggested that loss of hydrogen from the ruthenium center, which creates a site of coordinate unsaturation, may be essential for activating the catalyst for copolymerization [12]. Exact stoichiometric balance is essential in a step-growth copolymerization if one is to achieve high molecular weights. Loss of hydrogen from Ru by hydrogenation of a C-C double bond of 1,3-divinyltetramethyldisiloxane not only activates the catalyst - but also converts a difunctional α,ω-diene into a mono-functional alkene which ultimately becomes an end group. This limits the molecular weight of the copolymer. To overcome this problem, we have treated Ru with a stoichiometric amount of styrene for a few minutes at 135°C. In this way, styrene is converted to ethylbenzene and the activated catalyst is formed. Subsequent addition of an equal molar mixture of an aromatic ketone and 1,3-divinyltetramethyldisiloxane to the activated catalyst yields significantly higher molecular weight copolymers. Solutions of the activated complex are catalytically active for at least twenty-four hours [14].

The activated complex has been characterized by NMR and X-ray crystallography. The signals in the ^{1}H NMR at -8.18 ppm and -6.35 ppm due to the two non-equivalent ruthenium hydrogen bonds of Ru are no longer observed. Further, examination by ^{31}P NMR of the activated catalyst reveals the presence of two non-equivalent triphenylphosphines at 46.21 ppm and 46.78 ppm, as well as one free triphenylphosphine ligand. This suggests that the activated complex is "[Ph$_3$P]$_2$RuCO" or a dimer or trimer species which has this stoichiometry. In this regard, polynuclear ruthenium carbonyl species are well known. Addition of *ortho* acetylstyrene, which has a perfect stoichiometric balance of terminal vinyl groups to *ortho* C-H bonds, does not yield polymer but rather a 1:1 complex of "[Ph$_3$P]$_2$RuCO", see Figure 9. The

"[Ph₃P]₂RuCO", see Figure 9. The structure of this crystalline complex has been determined by X-ray crystallography, as shown in Figure 10. This complex also catalyzes the copolymerization of a 1:1 molar mixture of acetophenone and 1,3-divinyltetramethyldisiloxane. These results suggest that "[Ph₃P]₂RuCO" may be the catalytically active species [15].

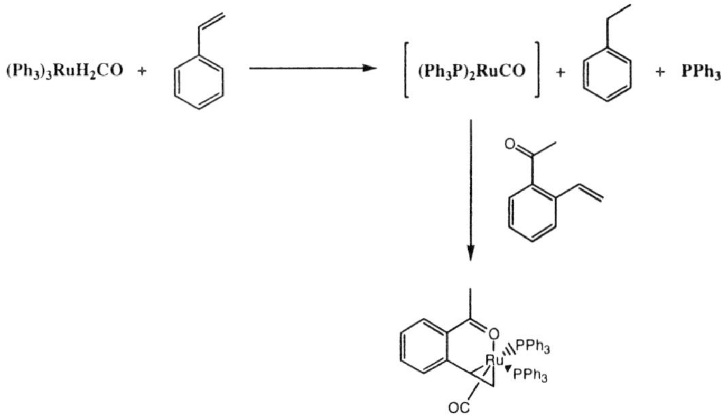

Figure 9

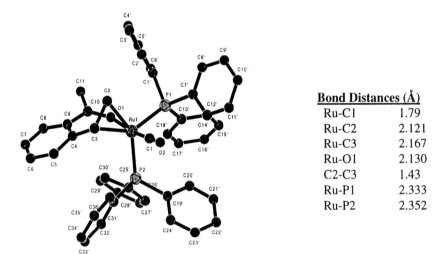

Bond Distances (Å)	
Ru-C1	1.79
Ru-C2	2.121
Ru-C3	2.167
Ru-O1	2.130
C2-C3	1.43
Ru-P1	2.333
Ru-P2	2.352

Figure 10

Hiraki and coworkers have recently reported the X-ray structure of a 1:1 complex of "[Ph₃P]₂RuCO" with N-benzylideneaniline. This complex has both Ru-C and Ru-H bonds formed by insertion of the nitrogen coordinated Ru center into one of the adjacent *ortho* C-H bonds [16]. The relationship of this complex to the catalytic cycle previously proposed should be evident.

We have also reacted activated catalyst "[Ph$_3$P]$_2$RuCO" with dimethyldivinylsilane and obtained a 1:1 complex which demonstrates dynamic NMR spectra depending on temperature. This fluxional complex is also capable of catalyzing the copolymerization of acetophenone and 1,3-divinyltetramethyldisiloxane [17].

Acetophenones substituted with methoxy or phenoxy groups in the *para* position had been shown to undergo successful activated Ru catalyzed copolymerization with 1,3-divinyltetramethyldisiloxane [18]. Activated Ru catalyzed copolymerization of 4-acetylbenzo crown ethers with 1,3-divinyltetramethyldisiloxane provides a synthetic route to *alt*(carbosilane/siloxane) copolymers which incorporate crown ethers, as shown in Figure 11. Polymeric crown ether/lithium perchlorate complexes have been prepared. Of particular note, the sterically congested *ortho* C-H bond, which is between the acetyl group and the crown ether ring, is still reactive [19,20]

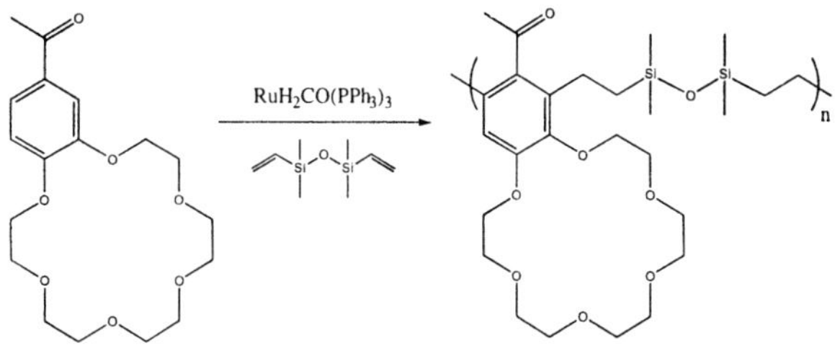

Figure 11

Likewise, activated Ru catalyzed copolymerization of acetophenones which are substituted with dialkylamino groups in the *para* position proceed with facility [21].

We have previously noted that while terminal C-C double bonds of vinylsilanes, vinylsiloxanes and styrenes are reactive, internal C-C double bonds are not. This selectivity has been exploited in the activated Ru catalyzed copolymerization of 4-acetylstilbenes with 1,3-divinyltetramethyldisiloxane. As expected, the internal C-C double bond of the stilbene does not react [22], see Figure 12.

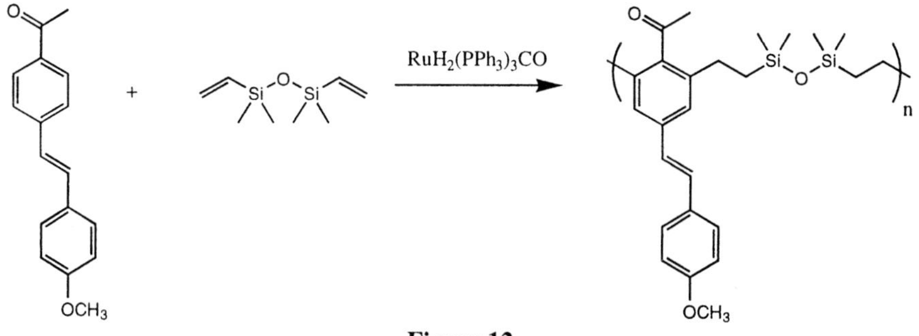

Figure 12

Chemical modification of unsaturated polysiloxanes with pendant or terminal Si-vinyl groups has been achieved by Ru catalyzed reaction with the single reactive *ortho* C-H bond of 2-methylacetophenone. In this way, Si-vinyl groups are converted to 2-(2'-acetyl-3'-methyl-phenethyl) groups [23], see Figure 13. It should be noted that there is considerable current interest in chemical modification of polymers [24,25]. Similar Ru catalyzed reaction of unsaturated polysiloxanes with pendant Si-vinyl groups with the two reactive *ortho* C-H bonds of acetophenone leads to crosslinking.

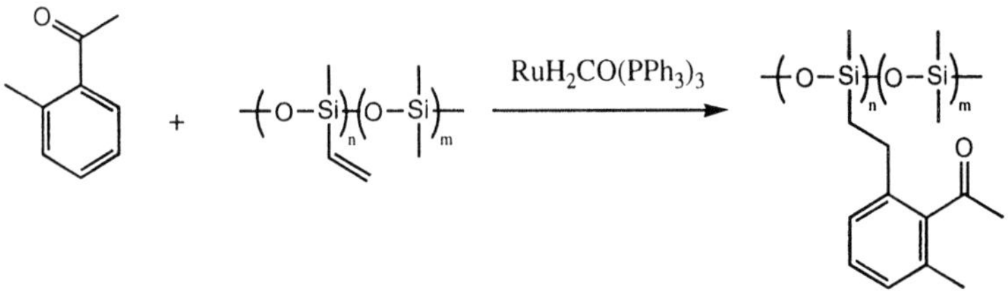

Figure 13

On the basis of the above results, it is surprising that Ru catalyzed reaction between benzophenone which has four potentially reactive *ortho* C-H bonds and 1,3-divinyltetramethyldisiloxane does not yield a crosslinked material - but rather a low molecular weight polymer as well as a cyclic monomer, as shown in Figure 14. Similar results have been obtained in the ruthenium catalyzed reaction with 4-benzoylpyridine and 1,3-divinyltetramethyldisiloxane. On the other hand, 4-acetylpyridine is unreactive. Apparently once one of the *ortho* C-H bond in benzophenone or 4-benzoylpyridine has reacted, the second *ortho* C-H bond in the substituted aromatic ring suffers a significant decrease in reactivity. For this reason, the second *ortho* C-H bond which reacts is usually in the unsubstituted aromatic ring. The formation of cyclic compounds often occurs competitively in polymerization reactions [26].

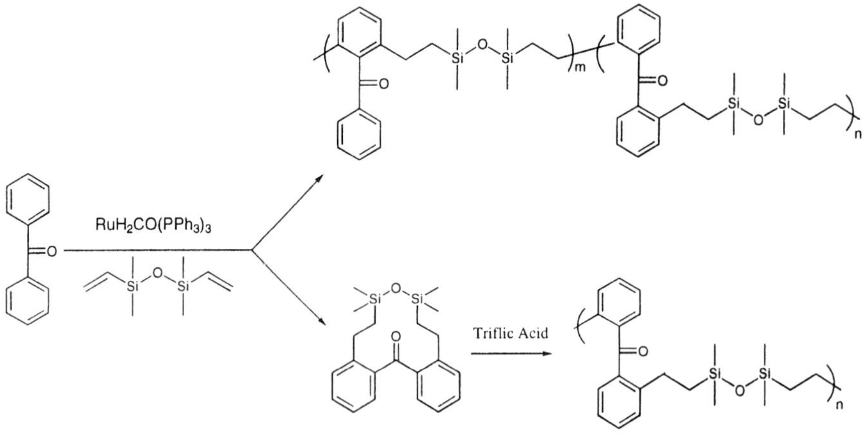

Figure 14

A similar pattern of reactivity is observed with ferrocenyl ketones. Thus acetyl ferrocene is unreactive, while only one of the *ortho* C-H bonds of benzoyl ferrocene undergoes Ru catalyzed anti-Markovnikov addition reaction across the C-C double bonds of 1,3-divinyltetramethyldisiloxane to yield a monomeric product. On the other hand, 1,1'-dibenzoylferrocene undergoes Ru catalyzed copolymerization with 1,3-

divinyl- tetramethyldisiloxane, as shown in Figure 15. The reversible electron chemical oxidation of this material has been studied by cyclic voltametry [27].

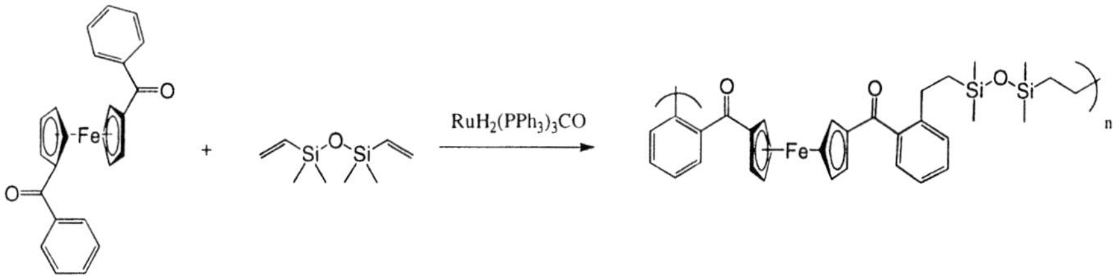

Figure 15

Macromolecules which are highly branched are of substantial interest due to their unusual properties [28-30]. Hyperbranched materials do not have a well-defined architectural structure, by comparison to dendrimers, but rather an irregular pattern of branching. Hyperbranched materials can be prepared in a single step from a monomer which has two mutually reactive functional groups A and B present in a ratio such as AB_2 [31,32]. 1-β-(4'-Acetylphenyl)vinyl-3-vinyl-1,1,3,3-tetramethyldisiloxane is an AB_2 type monomer in that it has two reactive C-H bonds which are *ortho* to the acetyl and a single reactive terminal Si-vinyl group. Reaction of this with activated Ru leads to a hyperbranched material [33], as shown in Figure 16. Similarly, Ru catalyzed reaction of 4-acetylstyrene gives a hyperbranched material [34].

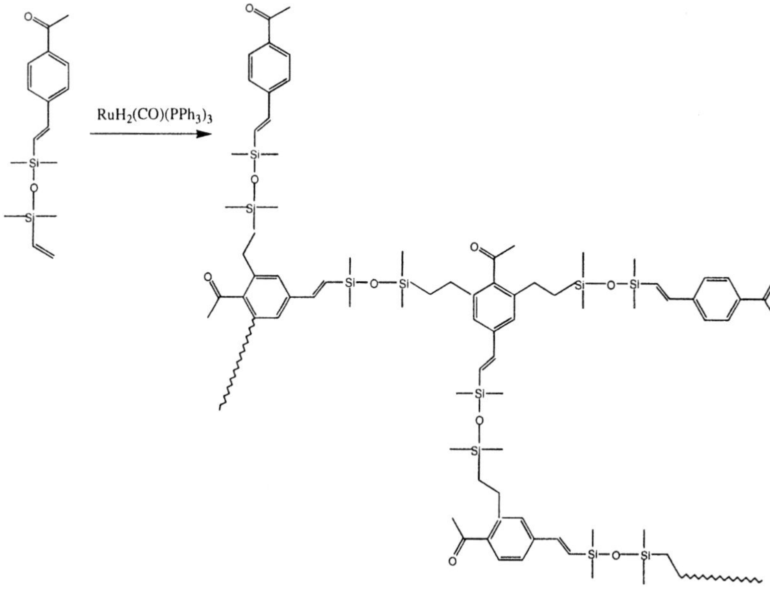

Figure 16

While there is considerable interest in conjugated copoly(arylene/1,2-vinylene)s, there has been much less work on unsaturated cross-conjugated copoly(arylene/1,1-vinylene)s [35]. The Ru catalyzed reaction between α-tetralone and internal acetylenes such as phenyethynyltrimethylsilane has been reported by Murai to yield E/Z-α-(8-α-tetralonyl)-β-trimethylstyrene [36]. We have applied this Ru catalyzed Markovnikov addition of the *ortho* C-H bonds of aromatic ketones across the C-C triple bond of internal acetylene to directly prepare cross-conjugated copoly(arylene/1,1-vinylenes). Thus activated Ru catalyzed reaction of acetophenone and 1,4-bis(trimethylsilylethynyl)benzene directly yields copoly[2-acetyl-1,3-phenylene/α,α'-bis(trimethylsilylmethylene)-1,4-xylenylene], as shown in Figure 17. Similar Ru catalyzed reactions between acetophenone and 1,4-bis(phenylethynyl)benzene are also successful [37].

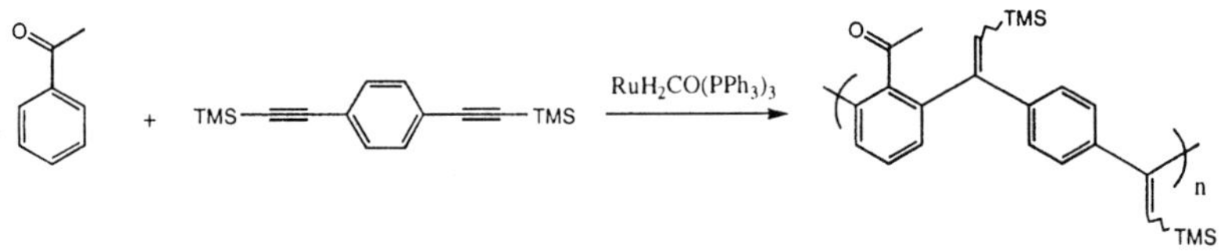

Figure 17

The fluorescence spectra of these cross conjugated materials were performed on a PTI instrument equipped with a model A1010 Xe/Hg lamp and a model 710 photomultiplier defraction detector. Spectra were obtained on methylene chloride, toluene and DMSO solutions which had been degassed by bubbling argon through them for 10 min. Fluorescence quantum yields were determined relative to that of 7-diethylamino-4-methyl coumarin [38]. The fluorescence emission of these materials occurs in the blue, as shown in Figure 18. However, the quantum yield for fluorescence is very low. It is probable that the acetyl group contributes to the low fluorescence.

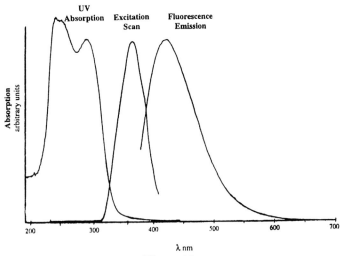

Figure 18

The Ru catalyzed reaction of aromatic ketones with vinyl and acetylenic ketones to yield *alt*-poly(siloxane/carbosilanes) is a reaction which was first reported in 1994. While experimental results to date have helped to elucidate the mechanism, scope and limitations of this reaction, much still remains to be done. It is our hope that further study will demonstrate that the Murai reaction will prove to be a highly versatile and valuable type of step growth polymerization reaction.

References

1. Hani, R.; Lenz, R. W., in "Silicon Based Polymer Science: A Comprehensive Resource", Ed. by Zeigler J. M.; Gordon Fearon, F. W. ACS, Adv. in Chem. Series 224, American Chemical Society, Washington, D. C. **1990**, p 741-752.
2. Merker, R. L.; Scott, M. J., *J. Polymer Science A*, **1964**, *2*, 15.
3. Keohan F. L.; Hallgren, J. E., in Silicon Based Polymer Science: A Comprehensive Resource, Ed. by J. M. Zeigler and F. W. Gordon Fearon, ACS, Adv. in Chem. Series, 224, American Chemical Society, Washington, D. C. **1990**, p 165-179.
4. Cummings, S.; Ginsburg, E.; Miller, R.; Portmess, J.; Smith, D. W. Jr.; Wagener, K., in "Step-Growth Polymers for High Performance Materials, New Synthetic Methods", Ed. by Hedrick, J. L.; Labadie, J. W., ACS Symposium Series 624, American Chemical Society, Washington, D.C. **1996**.
5. Schrock, R. R.; DePue, R. T.; Feldman, J.; Schaverien, C. J.; Dewan, S. C.; Liu, A. H., *J. Am. Chem. Soc.*, **1988**, *110*, 1423.
6. Guo, H.; Weber, W. P., *Polymer Bull.*, **1994**, *32*, 525.
7. Levison, J. J.; Robinson, S. D., *J. Chem. Soc.*, **1970**, A, 2947.
8. Lynn, D.M.; Kanaoka, S.; Grubbs, R.H. *J. Am. Chem. Soc.*, **1996**, *118*, 784.
9. Crivello, J. V.; Kim, W. G., *Pure and Applied Chem.*, **1994**, *A31*, 1005.
10. Murai, S.; Kakiuchi, F.; Sekine, S.; Tanaka, Y.; Sonoda, M.; Chatani, N., *Nature*, **1993**, *366*, 529.
11. Kakiuchi, F.; Sekine, S.; Tanaka, Y.; Kamatani, A.; Sonoda, M.; Chatani, M.; Murai, S., *Bull. Chem. Soc. Jpn.*, **1995**, *68*, 62.
12. Tapsak, M. A.; Guo, H.; Weber, W. P., *Polymer Bull.*, **1995**, *34*, 49.
13. Guo, H.; Wang, G.; Weber, W. P., *Polymer Bull.*, **1996**, *37*, 4231.
14. Guo, H.; Wang, G.; Tapsak, M. A.; Weber, W. P., *Macromolecules*, **1995**, *28*, 5686.
15. Lu, P.; Paulasaari, J.; Jin, K.; Bau, R.; Weber, W. P., *Organometallics*, **1998**, *17*, 584.
16. Hiraki, K.; Koizumi, M.; Kira, S. I.; Kawano, H., *Chem. Letters*, **1998**, 47.
17. Paulasaari, J.; Weber, W. P., unpublished results.
18. Guo, H.; Weber, W. P., *Polymer Bull.*, **1995**, *35*, 259.
19. Wang, G.; Guo, H.; Weber, W. P. *Polymer Bull.*, **1996**, *37*, 169.
20. Wang, G.; Guo, H.; Weber, W. P., *J. Organometal. Chem.*, **1996**, *521*, 351.
21. Guo, H.; Tapsak, M. A.; Weber, W. P., *Macromolecules*, **1995**, *28*, 4714.
22. Londergan, T. M.; Weber, W. P., *Makromol. Rapid Commun.*, **1997**, *18*, 207.
23. Guo, H.; Tapsak, M. A.; Weber, W. P., *Polymer Bull.*, **1994**, *33*, 417.
24. Carraher, C. E. Jr.; Moore, J. A.,"Modification of Polymers", Plenum Press, New York **1983**.
25. Benham, J. L.; Kinstle, J. F. , "Chemical Reactions on Polymers", ACS Symposium Series 363, American Chemical Society, Washington, D. C. **1988**.

26. Kepler, C. L.; Londergan, T. M.; Lu, J.; Paulasaari, J.; Weber, W. P., *Polymer* in press.
27. Sargent J. R.; Weber, W. P., *Polymer Preprints,* **1998**, *39*-I, 274.
28. Tomalia, D.; *Adv. Mater.*, **1994**, *6*, 529.
29. Frechet, J. M., *Science*, **1994**, *263*, 1710.
30. Mekelburger, H. B.; Jaworek, W.; Vogtle, F., *Angew. Chem., Int. Ed. Engl.,* **1992**, *231*, 1571.
31. Kim, Y. H. Webster, O. W., *J. Am. Chem. Soc.*, **1990**, *112*, 4592.
32. Kim, Y. H. Webster, O. W., *Macromolecules*, **1992**, *25*, 5561.
33. Londergan, T. M.; Weber, W. P., *Polymer Bull.*, **1998**, *40*, 15.
34. Lu, P.; Paulasaari, J.; Weber, W. P., *Macromolecules*, **1996**, *29*, 8583.
35. Mao, S. S. H.; Tilley, T. D., *J. Organometal. Chem.*, **1996**, *521*, 425.
36. Kakiuchi, F.; Yamamoto, Y.; Chatani, N.; Murai, S., *Chemistry Letters*, **1995**, 681.
37. Londergan, T. M.; You, Y.; Thompson, M. E.; Weber, W. P., *Macromolecules*, in press **1998**.
38. Murov, S. L.; Carmichael, I.; Hug, G. L., Handbook of Photochemistry 2nd Ed., Marcel Dekker, Inc., New York, New York, **1993**, p 24.

Chapter 30

Block Copolymers Containing Silicone and Vinyl Polymer Segments by Free Radical Polymerization

D. Graiver[1], B. Nguyen[1], F. J. Hamilton[2], Y. Kim[2], and H. J. Harwood[2]

[1]Dow Corning Corporation, Midland, MI 48686-0994
[2]Maurice Morton Institute of Polymer Science, The University of Akron, Akron, OH 44325-3909

Polysiloxanes with terminal hexenyl groups were converted by ozonolysis into polysiloxanes with terminal aldehyde functionality. These were used in combination with copper octanoate, triethylamine, pyridine and triphenylphosphine to initiate the polymerization of styrene, methyl methacrylate and styrene/methyl methacrylate mixtures at 70°C. The polymerizations yielded block copolymers containing polysiloxane and vinyl (co)polymer segments. Depending on how the polymerizations terminated, the block copolymers had triblock or multiblock architectures.

The reactions of the enolates of aldehydes and ketones with copper (II) salts have been known for many years to yield α-acylcarbon-centered radicals which couple to form 1,4-dicarbonyl compounds in nearly quantitative yields (1-3). We have found that the radicals that are intermediates in these reactions are useful for initiating polymerization reactions and have shown that these reactions are very valuable for the synthesis of polymers with useful end-group functionality, block copolymers and graft copolymers (4-8).

In this paper we report the application of this chemistry to novel polysiloxanes with terminal aldehyde functionality. These materials have recently been synthesized in Dow Corning laboratories via the reaction of ozone with polysiloxanes having terminal olefinic unsaturation (8,9). Since unsaturated polysiloxanes can be prepared in considerable variety using conventional polysiloxane technology, and since the ozonolysis process is simple (9,10), these new polysiloxanes are also potentially available in considerable variety. When these materials are used, in conjunction with copper salts, to initiate vinyl monomer polymerizations, a large variety of multi-segment block copolymers containing polysiloxane segments can be synthesized. In this paper we discuss the preparation of multi-block copolymers containing poly(dimethylsiloxane) and polystyrene and/or poly(methyl methacrylate) segments. Block copolymers containing polysiloxane segments are of interest because the polysiloxane blocks can impart impact resistance, toughness, water resistance and valuable surface properties to products containing the block copolymers and because the copolymers can be valuable as thermoplastic elastomers, compatibilizing agents, wetting agents and emulsifiers, etc. Burger and Kreuzer (*11*) have recently reviewed other ways of making such copolymers.

Experimental

Aldehyde-Functional Polysiloxane Synthesis. Poly (dimethylsiloxanes) containing 1-hexenyl-dimethylsiloxane end groups, DC 7691 (n = 200), DC 7692 (n = 100) and DC 7697 (n = 30), were dissolved in hexane or methylene chloride (1 g/ml) and the solutions were cooled to -20°C. Ozone was then bubbled through the solutions and then into a NaI solution to decompose any excess that had not been reacted and prevent its release to the atmosphere. Treatment of the reaction mixtures with ozone was stopped after the NaI solution developed a blue color indicating that all the unsaturation had been consumed. The reaction mixtures were then allowed to warm to room temperature and heated for one hour at 40°C with Zn powder/acetic acid (1:1 molar ratio) to convert the ozonized groups on the polymers to aldehyde groups.

The reaction mixtures were then filtered to remove zinc oxide, washed with water to remove excess acetic acid and then dried. Removal of the solvent by distillation then yielded the aldehyde-functional polysiloxanes. Listed below are the polymers prepared for use in this study. The polymers exhibited characteristic aldehyde absorption at ~ 1733 cm^{-1}.

$$\underset{\text{HC}(CH_2)_4}{\overset{O}{\|}} \underset{|}{\overset{CH_3}{\text{Si}}} (O\underset{|}{\overset{CH_3}{\text{Si}}})_n O \underset{|}{\overset{CH_3}{\text{Si}}} (CH_2)_4 \overset{O}{\overset{\|}{\text{CH}}}$$

Table I. Aldehyde-Functional Poly(dimethylsiloxanes) Employed in This Study

Polysiloxane	n	Parent Polysiloxane	Mn Calc*	Mn GPC**	Mw/Mn
I	30	DC 7697	2522	4800	1.88
II	100	DC 7692	7702	10000	1.81
III	200	DC 7691	15102	20400	1.69

*Calculated from n. **Polystyrene calibration.

Polymerization Procedure. All polymerizations were conducted in one ounce bottles under an argon atmosphere either in bulk or using benzene as a solvent to minimize chain transfer reactions. A typical reaction mixture contained 10 ml benzene that had been distilled from CaH$_2$, copper octanoate (0.1 g, 2.86 x 10^{-4} mole), pyridine (0.5g), triphenylphosphine (0.3g, 1.14 x 10^{-3} mole), triethylamine (0.1g, 1 x 10^{-3} mole), monomer (~ 5 ml) and aldehyde-functional polydimethylsiloxane (0.4 – 2.34 g, 3.1 x 10^{-4} mole of aldehyde groups). After being heated at 70°C for 21 hr., the reaction mixtures were cooled to room temperature, diluted with benzene, and added to methanol to precipitate the polymers. The products were washed with fresh methanol and dried to a constant weight under vacuum at room temperature. They were then analyzed by ^{1}H-NMR and GPC. In some cases, the products were extracted with hexane and reanalyzed by ^{1}H-NMR. Control experiments in which aldehyde-functional polysiloxane was omitted from the formulation or replaced by silicone oil either failed to yield polymer or yielded only small amounts of very high molecular weight homopolymer (Mn ~ 300K). When styrene was polymerized using butyraldehyde as the coinitiator instead of aldehyde-functional polysiloxane but in which silicone oil was present in an amount equivalent to the aldehyde-functional polysiloxane only polystyrene formed. This result is consist with the report that chain transfer constants for polysiloxanes in styrene and MMA polymerizations are very low (*12*).

Decomposition of Polysiloxane Segments in Polydimethylsiloxane-Polystyrene Block Copolymers. To characterize the polystyrene segments in polydimethylsiloxane-polystyrene multi-block copolymers, the copolymers were degraded by a strong acid. This was achieved by dissolving 0.5 g samples of the copolymers in toluene (40 mL) and adding p-toluenesulfonic acid monohydrate (0.3 g). The solutions were refluxed for four hours at 120°C, concentrated to about one-half their volume and poured into methanol to precipitate the hydrolyzed polymers. These were then washed with water and dried. The NMR spectra of the products were equivalent to the spectra of conventional polystyrene samples and contained only very weak signals in the 0 ppm region, where residual polydimethylsiloxane segments would be detected. The polymers were then characterized by GPC. In other experiments, samples of the copolymers (0.5 g) in benzene (40 mL) containing a few drops of concentrated H_2SO_4 were refluxed overnight before the degraded products were isolated. Control experiments indicated that polystyrene segments were also degraded somewhat when H_2SO_4 was used but not when p-toluenesulfonic acid monohydrate was used.

NMR Spectra. ^{1}H-NMR spectra of the copolymers (~ 20 mg) in $CDCl_3$ (~ 0.7 mL) solution containing a small amount of Me_4Si internal standard were recorded at 200 MHz using a Varian Gemini 200 NMR spectrometer, a pulse angle of 90°, an acquisition time of 3.74 sec and a 5 sec delay between pulses. ^{13}C-NMR spectra were recorded at 100 MHz using $CDCl_3$ as the solvent and Me_4Si as the internal reference. A 90° pulse, a pulse delay of 4 sec and an acquisition time of 1.6 sec were employed.

Results and Discussion

General Chemistry. Although polysiloxanes that bear aldehyde functionality have not been described previously, they can be prepared quite easily by ozonolysis of precursors that bear terminal olefinic groups. Hydrosilation can be used to prepare such materials from readily available precursors. The following series of reactions was thus used to prepare polysiloxanes with terminal aldehyde functionality.

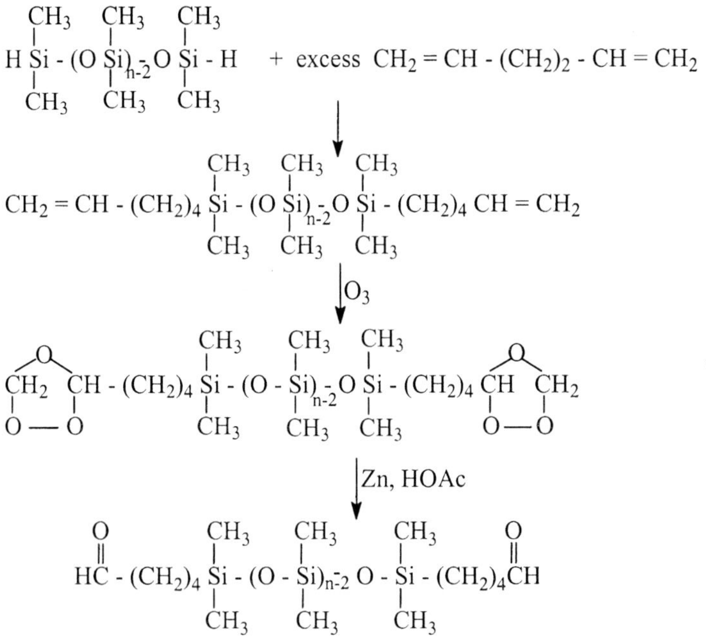

$$\text{H}\underset{\underset{\text{CH}_3}{|}}{\overset{\overset{\text{CH}_3}{|}}{\text{Si}}} - (\text{O}\underset{\underset{\text{CH}_3}{|}}{\overset{\overset{\text{CH}_3}{|}}{\text{Si}}})_{n-2}\text{O}\underset{\underset{\text{CH}_3}{|}}{\overset{\overset{\text{CH}_3}{|}}{\text{Si}}} - \text{H} + \text{excess } \text{CH}_2 = \text{CH} - (\text{CH}_2)_2 - \text{CH} = \text{CH}_2$$

$$\downarrow$$

$$\text{CH}_2 = \text{CH} - (\text{CH}_2)_4 \underset{\underset{\text{CH}_3}{|}}{\overset{\overset{\text{CH}_3}{|}}{\text{Si}}} - (\text{O}\underset{\underset{\text{CH}_3}{|}}{\overset{\overset{\text{CH}_3}{|}}{\text{Si}}})_{n-2}\text{O}\underset{\underset{\text{CH}_3}{|}}{\overset{\overset{\text{CH}_3}{|}}{\text{Si}}} - (\text{CH}_2)_4 \text{CH} = \text{CH}_2$$

$$\downarrow \text{O}_3$$

$$\underset{\underset{\text{O}-\text{O}}{|\quad |}}{\text{CH}_2\;\;\text{CH}}\overset{\overset{\text{O}}{\frown}}{} - (\text{CH}_2)_4 \underset{\underset{\text{CH}_3}{|}}{\overset{\overset{\text{CH}_3}{|}}{\text{Si}}} - (\text{O} - \underset{\underset{\text{CH}_3}{|}}{\overset{\overset{\text{CH}_3}{|}}{\text{Si}}})_{n-2}\text{O}\underset{\underset{\text{CH}_3}{|}}{\overset{\overset{\text{CH}_3}{|}}{\text{Si}}} - (\text{CH}_2)_4 \underset{\underset{\text{O}-\text{O}}{|\quad |}}{\text{CH}\;\;\text{CH}_2}\overset{\overset{\text{O}}{\frown}}{}$$

$$\downarrow \text{Zn, HOAc}$$

$$\overset{\overset{\text{O}}{\|}}{\text{HC}} - (\text{CH}_2)_4 \underset{\underset{\text{CH}_3}{|}}{\overset{\overset{\text{CH}_3}{|}}{\text{Si}}} - (\text{O} - \underset{\underset{\text{CH}_3}{|}}{\overset{\overset{\text{CH}_3}{|}}{\text{Si}}})_{n-2} \text{O} - \underset{\underset{\text{CH}_3}{|}}{\overset{\overset{\text{CH}_3}{|}}{\text{Si}}} - (\text{CH}_2)_4 \overset{\overset{\text{O}}{\|}}{\text{CH}}$$

Except for ozonides obtained from vinylsiloxanes, the ozonides that result from these reactions are surprisingly stable. Figure 1 shows the ^{13}C-NMR spectrum of an ozonide that was prepared from DC 7697 (n = 30). It contains resonances at 1, 18, 23, 29, 41, 94 and 104 ppm. Chemical shifts calculated for the carbon atoms in this compound (1.5(3,4),15.7(1),24(2),26.3(6),30.4(7),97.3(5) and 99.3(6) ppm) correspond reasonably well with those observed. Polysiloxanes containing the ozonide functionality can, in fact, be used as macroinitiators for block copolymer synthesis (*13*).

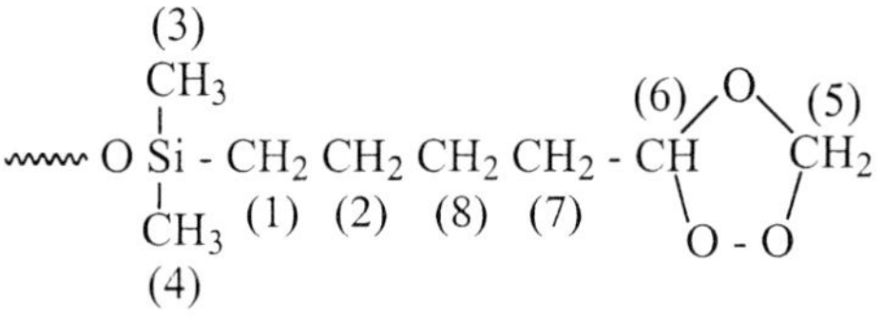

$$\sim\sim\sim\text{O}\underset{\underset{\text{CH}_3\;(4)}{|}}{\overset{\overset{\text{CH}_3\;(3)}{|}}{\text{Si}}} - \underset{(1)}{\text{CH}_2}\;\underset{(2)}{\text{CH}_2}\;\underset{(8)}{\text{CH}_2}\;\underset{(7)}{\text{CH}_2} - \underset{\underset{\text{O}-\text{O}}{\diagdown\quad\diagup}}{\overset{\overset{(6)\diagup\text{O}\diagdown\;(5)}{}}{\text{CH}\qquad\text{CH}_2}}$$

Reaction of ozonized polysiloxanes with zinc and acetic acid at 40° converts the ozonide groups to aldehyde groups. Table I provides information about the aldehyde-functional polysiloxanes that were employed in this investigation.

When solutions of aldehyde-functional polysiloxanes, copper octanoate, triethylamine, pyridine, triphenylphosphine and monomer in inert solvents such as benzene are heated, radicals can be generated at the polysiloxane ends and these can initiate polymerization. This is shown below, where M is a monomer unit.

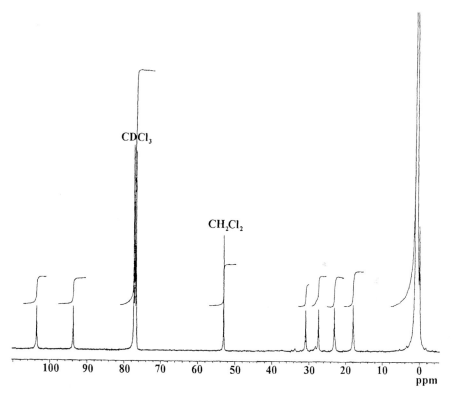

Figure 1. ^{13}C-NMR Spectrum of Ozonide Obtained from DC 7697.

$$\underset{\text{CH}_3}{\overset{\text{O}}{\text{HC}}} - (\text{CH}_2)_4 \underset{\text{CH}_3}{\overset{\text{CH}_3}{\text{Si}}} - (\text{O} - \underset{\text{CH}_3}{\overset{\text{CH}_3}{\text{Si}}})_{n-2} \text{O} - \underset{\text{CH}_3}{\overset{\text{CH}_3}{\text{Si}}} - (\text{CH}_2)_4 \overset{\text{O}}{\text{CH}}$$

$$\downarrow \begin{array}{l} \text{Cu(OCOR)}_2 \\ \text{Et}_3\text{N} \\ \text{Pyridine, } \phi_3\text{P} \end{array}$$

$$\underset{\text{CH}_3}{\overset{\text{O}}{\text{HC}}} - (\text{CH}_2)_4 \underset{\text{CH}_3}{\overset{\text{CH}_3}{\text{Si}}} - (\text{O} - \underset{\text{CH}_3}{\overset{\text{CH}_3}{\text{Si}}})_{n-2} \text{O} - \underset{\text{CH}_3}{\overset{\text{CH}_3}{\text{Si}}} - (\text{CH}_2)_3 \text{CH}\overset{\text{O}}{\text{CH}} \bullet$$

$$\downarrow n\text{M}$$

$$\underset{\text{CH}_3}{\overset{\text{O}}{\text{HC}}} - (\text{CH}_2)_4 \underset{\text{CH}_3}{\overset{\text{CH}_3}{\text{Si}}} - (\text{O} - \underset{\text{CH}_3}{\overset{\text{CH}_3}{\text{Si}}})_{n-2} \text{O} - \underset{\text{CH}_3}{\overset{\text{CH}_3}{\text{Si}}} - (\text{CH}_2)_4 \underset{\text{CHO}}{\text{CH}} - \text{M}n \bullet$$

Depending on the mode of termination of the propagating polymer radicals, and the efficiency of the initiation process, the block copolymers that result can have diblock, triblock or multiblock structures. When methacrylate esters are polymerized, diblock (AB) or triblock (ABA, where A = polyM and B = polysiloxane) copolymers are expected because termination occurs by disproportionation. When acrylates or styrenes are polymerized, multiblock (AB) should be obtained.

Information about the size of the individual blocks in some types of block copolymers can be obtained by degrading the polysiloxane segments with acid and then examining the A segments that remain. For example, when block copolymers containing polystyrene and polysiloxane segments are degraded with toluenesulfonic acid monohydrate, the polysiloxane segments are destroyed and the polystyrene segments can be isolated.

$$[-(\text{polysiloxane}) - (\text{polystyrene})-]_x$$
$$\downarrow \text{RSO}_3\text{H}$$
$$x \text{ polystyrene}$$

Polysiloxane-Polystyrene Block Copolymers. Figure 2 shows GPC curves measured from polymer I (A), a polysiloxane-polystyrene block prepared from it before (B) and after (C) hexane extraction and the hexane extract (D). Very little, if any, of material eluting above 22.0 mL is evident in the B and C chromatograms and it appears that most of the product that is soluble in hexane has a higher molecular weight than polymer I. This result, the low amount of the hexane extract (13 wt%) and the fact that its NMR spectrum indicates it to contain about 30 weight percent polystyrene, suggest that block copolymer formation occurred efficiently. Figure 3 shows the ^{1}H-NMR spectrum of a polysiloxane-polystyrene copolymer in CDCl$_3$ solution that was prepared from polymer I and then exhaustively extracted with hexane. It contains resonances characteristic of polystyrene [δ = 1.0–1.6 (2H, CH$_2$), 1.6–2.0 (1H, CH), 6.2–6.8

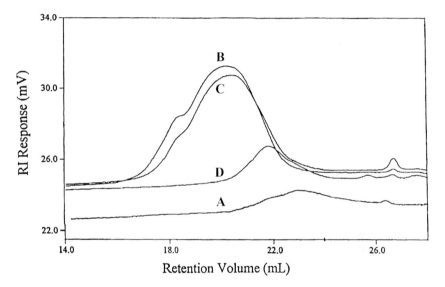

Figure 2. GPC Curves Observed for Parent Polysiloxane (A), a Polysiloxane-Polystyrene Block Copolymer Derived From it Before (B) and After (C) Hexane Extraction and the Hexane Extract (D).

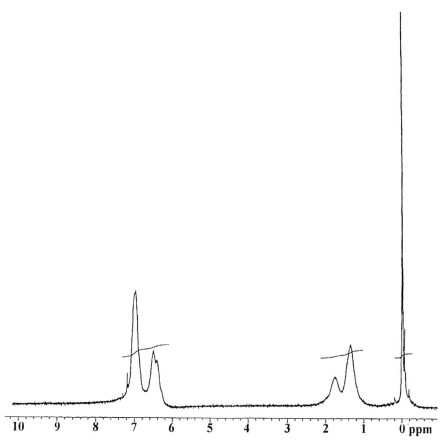

Figure 3. 200 MHz ^{1}H-NMR Spectrum of a Polysiloxane-Polystyrene Block Copolymer.

(2H, o-phenyl-H) and 6.8–7.3 (3H, m + p-phenyl-H) ppm] and polydimethylsiloxane ($\delta = 0$). The relative areas of the polystyrene (A_{st}) and polysiloxane (A_{Si}) resonances and the degree of polymerization (n) of the parent polysiloxane can be used to calculate the average number of styrene units in the polystyrene segments (N) as follows:

$$N = \frac{A_{st}/8}{A_{Si}/(6n)}$$

This equation is based on the assumption that there are as many polystyrene segments in the copolymer as polysiloxane segments which, in turn, is based on the assumptions that each aldehyde group in the parent polysiloxane yields a radical and that termination occurs exclusively by radical combination.

Figure 4 shows the ^{1}H-NMR spectrum of the polystyrene that was obtained after the block copolymer was degraded by heating at 120°C with a solution of p-toluenesulfonic acid in toluene for four hr. The spectrum is almost devoid of polysiloxane resonances at 0 ppm. Control experiments with polystyrene demonstrated that polystyrene was not degraded by this treatment.

Figure 5 compares the GPC curves observed for the parent polysiloxane, the polysiloxane-polystyrene block copolymer, and the polystyrene that was obtained after the polysiloxane segments had been destroyed by treatment with acid. The molecular weight and molecular weight distribution results obtained from such GPC data provide a considerable amount of information about the structures of the copolymer.

Table II provides information about styrene polymerizations that were initiated by polymers I, II, and III in combination with copper octanoate, $\emptyset_3 P$, $Et_3 N$ and pyri-

Table II. Synthesis and Properties of Block Copolymers Containing Polysiloxane and Polystyrene Segments

Starting Polysiloxane	I	I	II	III
Wt. Polysiloxane (g)	0.4	0.4	1.2	2.3
Wt. Polystyrene (g)	5.0	5.0	5.0	5.0
Yield of Copolymer (g)	3.0	2.45	2.26	5.75
% Styrene Polymerized	52	41	21	69
Mn of Copolymer (GPC)	98K	83K	32K	156K
Mw/Mn of Copolymer	5.7	6.8	2.2	3.8
Mn of Copolymer (^{1}H-NMR)	23K	31K	20K	36K
Mn of Copolymer (Calc)	20K	23K	14K	37K
Mn of Recovered PS	14K	----	----	36K
Mw/Mn of Recovered PS	3.7	----	----	3.3

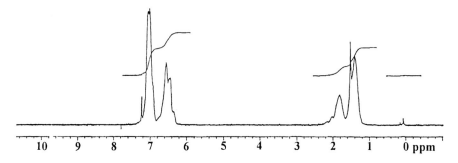

Figure 4. 200 MHz ^{1}H-NMR Spectrum of a Polystyrene Recovered After Degrading a Polysiloxane-Polystyrene Block Copolymer with p-Toluenesulfonic Acid Hydrate.

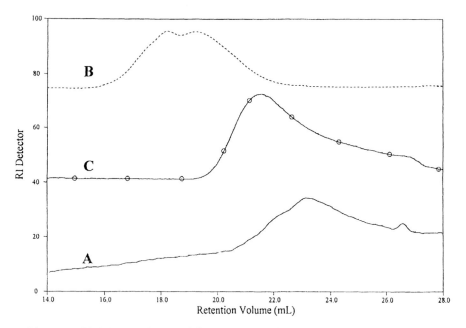

Figure 5. GPC Curves Observed for Parent Polysiloxane (A), a Polysiloxane-Polystyrene Block Copolymer Derived from it (B), and the Polystyrene Obtained After Degrading the Copolymer with Sulfuric Acid (C).

dine. Included in the table are Mn and Mw/Mn values measured for the copolymers by GPC along with values that were obtained for the polystyrenes that were recovered from the copolymers after the polysiloxane segments had been degraded by treatment with H_2SO_4.

In most cases, the molecular weights of the polymers are approximately four times the calculated sums of the molecular weights of the parent polysiloxanes and the polystyrenes recovered from the degraded copolymers. They are also approximately four times the molecular weights that can be calculated from the styrene unit/siloxane unit ratios (determined by NMR or from the styrene conversions and molecular weights of the parent polysiloxanes) of the copolymers, assuming one polystyrene segment per polysiloxane segment. All this information is consistent with the copolymers having structures that can be represented as shown above.

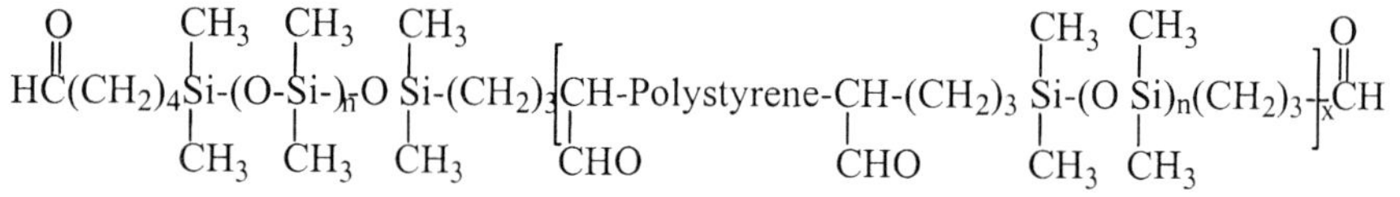

$$\underset{HC}{\overset{O}{\|}}(CH_2)_4\underset{\underset{CH_3}{|}}{\overset{\overset{CH_3}{|}}{Si}}\text{-}(O\text{-}\underset{\underset{CH_3}{|}}{\overset{\overset{CH_3}{|}}{Si}}\text{-})_{n}O\underset{\underset{CH_3}{|}}{\overset{\overset{CH_3}{|}}{Si}}\text{-}(CH_2)_3\left[\underset{\underset{CHO}{|}}{\overset{}{CH}}\text{-Polystyrene-}\underset{\underset{CHO}{|}}{\overset{}{CH}}\text{-}(CH_2)_3\underset{\underset{CH_3}{|}}{\overset{\overset{CH_3}{|}}{Si}}\text{-}(O\underset{\underset{CH_3}{|}}{\overset{\overset{CH_3}{|}}{Si}})_n(CH_2)_3\right]_x\overset{O}{\overset{\|}{CH}}$$

Such structures are believed to result from termination of growing polystyrene segment radicals by combination. Each such termination reaction yields a polysiloxane-polystyrene block copolymer (or a multiblock copolymer) with terminal polysiloxane segments that can also bear aldehyde ends. These, in turn, can initiate the polymerization of other polystyrene segments and ultimately larger multiple segment block copolymers can be formed.

This view of the polymerization process is supported by the fact that the copolymers have multimodal molecular weight distribution curves and Mw/Mn values that increase with styrene conversion, becoming as large as 6.8. In addition, the molecular weights of the polystyrene segments that were obtained by degrading the copolymers with H_2SO_4 followed by GPC analyses were consistent with those expected based on NMR analyses of the block copolymers and the reactant quantities and conversions used to prepare them. As should be expected, the Mw/Mn values of the polystyrene segments were less than those of the parent block copolymers.

Block Copolymers Containing Polysiloxane and PolyMMA Segments. Figure 6 shows the ^{1}H-NMR spectrum of a polysiloxane-polyMMA block copolymer prepared from polymer I. It contains resonances typical of polyMMA [$\delta = 0.7$–$1.3(3H,CH_3)$, 1.3–$2.2(2H,CH_2)$, 3.5–$3.7(3H,OCH_3)$ ppm] and polydimethylsiloxane ($\delta \sim 0$ ppm) segments. Assuming one polysiloxane segment per molecule, the number average molecular weight of the copolymer can be calculated from the relative areas of the resonances due to polyMMA (A_{MMA}) and polysiloxane (A_{Si}) segments as shown below where n is the DP of the parent polysiloxane.

$$Mn = \frac{100\left(\dfrac{A_{MMA}}{8} - \dfrac{14 A_{Si}}{6n}\right) + \dfrac{A_{Si}}{6}\left(\dfrac{75(n-2)+300}{n}\right)}{\dfrac{A_{Si}}{6n}}$$

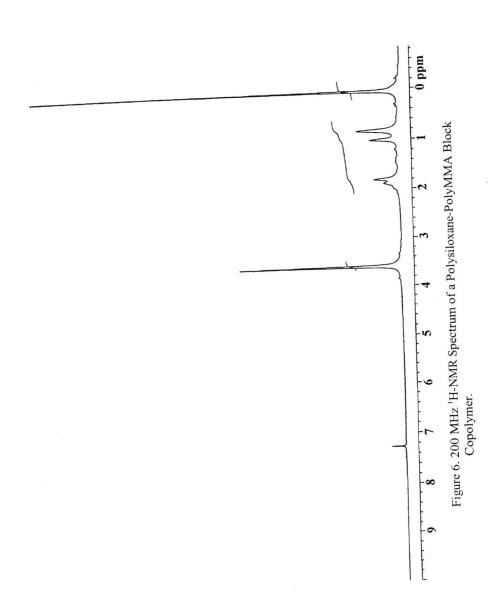

Figure 6. 200 MHz ^{1}H-NMR Spectrum of a Polysiloxane-PolyMMA Block Copolymer.

The weight percentage of MMA units in the copolymers can also be calculated from the A_{MMA} and A_{Si} values, as follows:

$$\text{wt \% MMA} = \frac{100 \left(\dfrac{A_{MMA} - \dfrac{14 A_{Si}}{6n}}{8} \right) \times 100}{\dfrac{100 \left(A_{MMA} - \dfrac{14 A_{Si}}{6n} \right)}{8} + \dfrac{A_{Si}}{6}\left[\dfrac{75(n-2) + 300}{n} \right]}$$

Since poly(methyl methacrylate) radicals terminate by disproportionation, the polysiloxane-polyMMA block copolymers may not have the multiblock structures the polysiloxane-polystyrene block copolymers have. Instead they might be expected to have BAB triblock or AB diblock structures. In accord with this expectation, the polysiloxane-polyMMA blocks prepared (Table III) have narrow molecular weight distributions, Mw/Mn ~ 2, and monomodal GPC curves.

Table III lists Mn values determined for the copolymers by GPC along with values calculated as described above, assuming one polysiloxane segment per copolymer molecule. For copolymers prepared from I (n = 30), the Mn values measured by GPC are nearly twice those calculated from A_{MMA} and A_{Si} values. This indicates that the copolymers contain, on an average, 1.7 – 1.9 polysiloxane segments per molecule. However, for copolymers prepared from II or III, the Mn values determined by GPC and from A_{MMA} and A_{Si} values are rather similar. They very likely have the ABA or AB structures expected. It appears that some process which is dependent on conversion and end group concentration can cause the copolymers to contain more than one polysiloxane segment.

This could indicate involvement of radicals derived from the aldehydes in termination reactions. We have observed that the parent aldehyde-functional polysilox-

Table III. Synthesis and Properties of Block Copolymers Containing Polysiloxane and PolyMMA Segments

Starting Polysiloxane	I	I	I	I	II	III
Wt. Polysiloxane (g)	0.4	0.4	0.4	0.4	1.2	2.3
Wt. MMA (g)	5.0	5.0	5.0	5.0	5.0	5.0
Yield of Copolymer (g)	4.9	5.0	4.2	4.6	4.3	6.8
% MMA Polymerized	89	92	76	84	62	90
Mn of Copolymer (GPC)	81K	78K	82K	74K	128K	99K
Mn of Copolymer (NMR)	43K	46K	48K	40K	124K	79K
Mn of Copolymer (Calc)	30K	31K	30K	29K	27K	44K
Mw/Mn	1.9	2.1	1.9	2.1	2.3	1.9

anes couple to form higher molecular weight polymers upon being exposed to the polymerization conditions in the absence of monomer. This will be the subject of a subsequent publication

Block Copolymers Containing Polysiloxane and Poly(styrene-co-MMA) Segments. Table IV provides results obtained when polymer I was used to initiate polymerizations of styrene, MMA and styrene-MMA mixtures. The ^{1}H-NMR spectra of the products corresponded well with those of conventional polymers and copolymers having the same composition except for the presence of a signal at ~ 0 ppm due to the polysiloxane segments. Exhaustive extraction of the products with hexane failed to remove this component, providing evidence that block copolymers had indeed been formed.

The GPC curves of the products were, with the exception of the polysiloxane-polyMMA block copolymer, all multimodal, and their Mn values, measured by GPC, were approximately 4x those calculated by NMR analysis, assuming one polysiloxane segment per molecule. This indicates that the copolymers containing statistical styrene-MMA copolymer segments had multiblock structures similar to those of copolymers with polystyrene segments.

The Mw/Mn values observed for the copolymers tend to increase with styrene content. This may reflect the fact that termination by disproportionation of propagating methacrylate radicals may be significant when the MMA content of the copolymerization mixture is high but that termination by combination of propagating styrene radicals with other propagating styrene radicals or with propagating MMA radicals will become of increasing importance as the styrene content of the polymerization mixtures increases.

Table IV. Results of Copolymerizations Conducted Using Polymer I

Experiment	1	2	3	4	5
I (g)	0.4	0.4	0.4	0.4	0.4
Styrene (ml)	0	2.0	2.5	3.0	5.0
MMA (ml)	5.0	5.0	2.5	2.0	0
Polymer Yield (g)	4.76	4.62	2.84	3.12	2.66
Mn (GPC)	56.7K	158K	93.3K	102K	83.4K
Mn (NMR)	28.4K	41.8K	20.9K	19.8K	20.8K
Mw/Mn	1.8	2.6	4.0	7.5	6.8
Mole% Styrene in Monomer Feed	0	28	49	59	100
Mole% Styrene in Copolymer Segment (Calc)*	0	33	50	58	100
Mole% Styrene in Copolymer Segment (NMR)	0	31	47	62	100

*Calculated from monomer feed composition, conversion and reactivity ratios of 0.52 and 0.46 for styrene and MMA, respectively.

Conclusions

Hexenyl groups in polysiloxanes react readily with ozone to form rather stable ozonides that can be converted to aldehydes by treatment with zinc and HOAc. This flexible ozonolysis process yields aldehyde-containing siloxanes, cyclic polysiloxanes and polysiloxanes. Polydimethylsiloxanes containing terminal aldehyde functionality that are prepared by this process can be used in conjunction with copper octanoate and organic bases to initiate polymerizations of styrene, MMA, and other monomers and thereby obtain copolymers with polysiloxane and organic polymer segments. The reactions described herein provide the basis for an extremely versatile and diverse route to multiblock copolymers containing polysiloxane segments. Depending on the structure of the starting polysiloxane, block copolymers, graft copolymers, and polymers with star structures can be prepared. In one sense, polysiloxanes with terminal aldehyde functionality can be regarded as multifunctional components of redox initiation systems, but in a perhaps more significant sense, they can be regarded as starting materials for the synthesis of a multitude of multiblock copolymers containing organic polymer and polysiloxane segments. In future publications we will provide more information about the scope of this chemistry and about the properties of the materials that are obtainable with it.

Literature Cited

1. Brackman, W.; Volger, H. C. *Rec. Trav. Chim.* **1966**, *85*, 446.
2. Sayer, L. M.; Jin, S.-J. *J. Org. Chem.* **1984**, *49*, 3498.
3. Jin, S.-J.; Arora, P. K.; Sayer, L. M. *J. Org. Chem.* **1990**, *55*, 3011.
4. Goodrich, S. D. *Investigation of Free Radical Polymerizations Initiated by Anaerobic Copper (II) Oxidation of Enolates Derived From Aldehydes and Ketones;* Ph.D. Dissertation; The University of Akron: Akron, OH, Dec. 1991.
5. Holland, T. V. *Cu(II)-Enolate Redox Initiated Polymerizations: Aspects of End Functionalization and Blocking;* Ph.D. Dissertation; The University of Akron: Akron, OH, May 1996.
6. Goodrich, S. D.; Harwood, H. J. *U. S. Patent* 5,405,913; **April 11, 1995**; *U. S. Patent* 5,470,928; **November 28, 1995**.
7. Harwood, H. J.; Christov, L.; Guo, M.; Holland, T. V.; Huckstep, A. Y.; Jones, D. H.; Medsker, R. E.; Rinaldi, P. L.; Saito, T.; Tung, D. S. *Macromol. Symp.* **1996**, *111*, 25.
8. Holland, T. V.; Goodrich, S. D.; Guo, M.; Harwood, H. J.; Rinaldi, P. L.; Saito, T. *Amer. Chem. Soc., Polym. Chem. Preprints* **1995**, *36(2)*, 91.
9. Graiver, D.; Khieu, A. Q.; Nguyen, B.T. *U.S. Patent* 5,739,246; **April 14, 1998**.
10. Graiver, D.; Khieu, A. Q.; Nguyen, B.T. *U.S. Patent* 5,789,516; **Aug. 4, 1998**.
11. Burger, C.; Kreuzer, F. H. In *Silicon in Polymer Synthesis;* Kricheldorf, H. R., Ed.; Springer Verlag: New York, NY, 1996; pp 113–222.
12. Saam, J. C.; Gordon, D. J. *J. Polym. Sci., Part A-1: Polym. Chem.* **1970**, *8(9)*, 2509.
13. Graiver, D.; Khieu, A. Q.; Nguyen, B.T. *U.S. Patent* 5,789,503; **Aug. 4, 1998**.

Chapter 31

Synthesis of Polymers Containing Silicon Atoms of Regulated Structure

Yusuke Kawakami

Graduate School of Materials Science, Japan Advanced Institute of Science and Technology [JAIST], 1-1 Asahidai, Tatsunokuchi, Ishikawa 923–1292, Japan

> Molecular design and precision synthesis of silicon-containing polymers are described. Polymerizations of substituted silacyclobutanes by phenyllithium and platinum complexes gave poly(carbosilane)s of head to tail regular structure. However, extensive chain transfer seems to have occurred in the polymerization by platinum complexes.
> Syntheses of stereoregular and/or optically active poly(carbosilane)s and poly(carbosiloxane)s from optically active silicon compounds are also shown.

The importance of controlling the chemical structure has become well recognized as necessary for developing the most desirable properties of polymers. It is needless to say that this is also true for silicon-containing polymers. Silicon compounds are often used as protecting or activating groups in organic and polymer syntheses (1,2). However, in many cases, the silicon moiety is not contained in the resulting products. If the silicon components remain in the products, new function will appear as silicon-containing polymers.

Silicon-containing polymers, such as poly(siloxane) (3), poly(carbosilane) (4,5), and poly(silylene) (6) have found practical importance in many areas of applications, but they have been insufficiently studied in relation to their primary structure, stereoregularity, molecular weight, and higher order structure.

In this article, we report the control of the molecular weight and terminal structure, and the stereoregularity of the polymer by ring-opening polymerization and stereospecific hydrosilylation.

Control of Molecular Weight of Polycarosilanes.

There are considerable numbers of reports on the polymerization of 1,1-dimethylsilacyclobutane using platinum complex (7-10) or alkyllithium (11-16). However, there are only few reports on the polymerization of other substituted silacyclobutanes (17, Komuro, K.; Kawakami, Y. Polym. J., in press.), and almost no report on the control of the stereoregularity of polymers. Stereospecific living ring-opening polymerization is considered suitable to control both molecular weight and stereoregularity of the polymer. As a preliminary study, control of the molecular weight and elucidation of the terminal structure of the polymer from 1,1,2-trimethylsilacyclobutane (TMSB) are described.

$$R = H, CH_3$$

By Transition Metal Complex. The results of polymerization of dimethyl-silacyclobutane derivatives using transition metal complexes are summarized in Table I.

Table I. Polymerization of Silacyclobutanes by Transition Metal Catalysts[a]

Run	Monomer	Catalyst	Convesion [%]	M_n^b /10^3	M_w/M_n^b	$M_{n, NMR}$ /10^3	Yield [%]
1	TMSB	PtDVTMDS[c]	quant.	3.7	1.5	2.8	91
2		PtDVTMDS[c,d]	0	---	---	---	---
3		H_2PtCl_6	quant.	4.7	1.6	3.3	95
4		Rh(PPh$_3$)$_3$Cl	0	---	---	---	---
5	DMSB[e]	PtDVTMDS[c,f]	quant.	110	1.3	---	94

[a]In bulk, for 17h, at 80 °C, under Ar, [M]/[Cat] = 3.0×10^4. [b]Estimated by GPC using standard polystyrene. [c]DVTMDS = 1,3-divinyl-1,1,3,3-tetramethyldisiloxane. [d]At room temperature. [e]DMSB = 1,1-dimethylsilacyclobutane, [f]In toluene, [M] = 5.0 mol dm^{-3}, for 30 min.

In contrast to the polymerization of DMSB, lower molecular weight materials were formed in the polymerization of TMSB. This suggests the existence of chain transfer reaction, although molecular weight distribution is reasonably narrow.

The major peaks of ^{29}Si, ^{13}C, ^{1}H NMR spectra were the same as those of the polymers obtained by phenyllithium described in later part. The ^{29}Si spectrum of the polymer by PtDVTMDS is shown in Figure 1. Only one sharp peak is seen at around 5.0 ppm. This fact proves the existence of only one kind of silicon atom in repeating unit. ^{1}H and ^{13}C NMR spectra also support basically the regular head to tail structure.

However, when the ^{1}H NMR spectrum was carefully examined, multiplets at 5.20-5.85 ppm assignable to the olefin protons were seen (Figure 2b). The olefinic protons consisted of basically three protons reflecting the structure at 5.21-5.46, 5.52-5.53 (two singlets), and 5.56-5.86 ppm in the ratio of 63.4 : 2.6 : 18.0. The majority

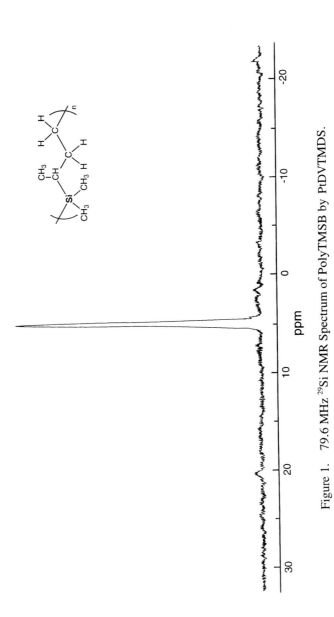

Figure 1. 79.6 MHz ^{29}Si NMR Spectrum of PolyTMSB by PtDVTMDS.

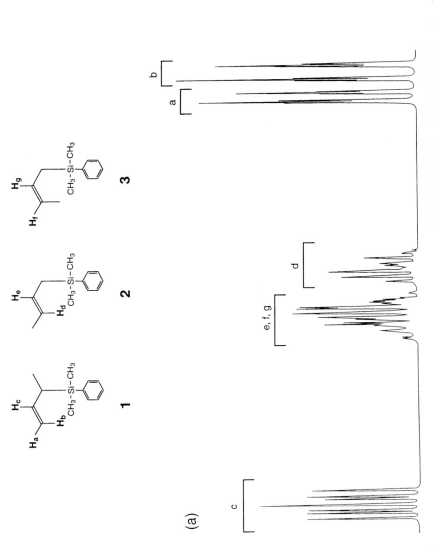

Figure 2. 500 MHz ^{1}H NMR Spectra of Olefinic Region of (a) Model Compounds and (b) PolyTMSB by PtDVTMDS. (*Continued on next page*)

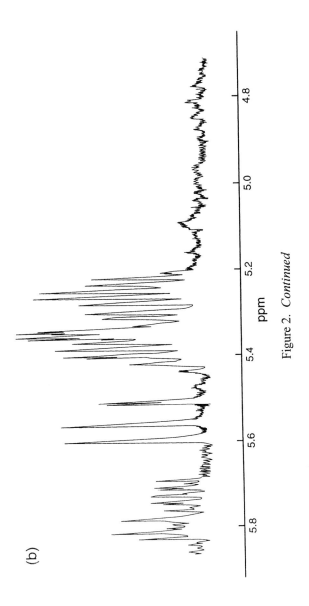

Figure 2. *Continued*

of the olefinic protons of the polymer terminal appeared at quite similar positions with inner olefinic protons of the model compounds (**2, 3**) (Figure 4a) at 5.21-5.46 ppm. These olefinic protons are considered to be the inner type **6** in Scheme 1 (two protons). The singlets at 5.52 and 5.55 are tentatively assigned to olefinic proton of *cis* and *trans* form of the terminal type **5** (one proton each). The signals at 5.57 and 5.93 ($J = 18.0$ Hz), and multiplets at 5.71-5.86 are assignable to terminal vinyl protons of the type **4**.

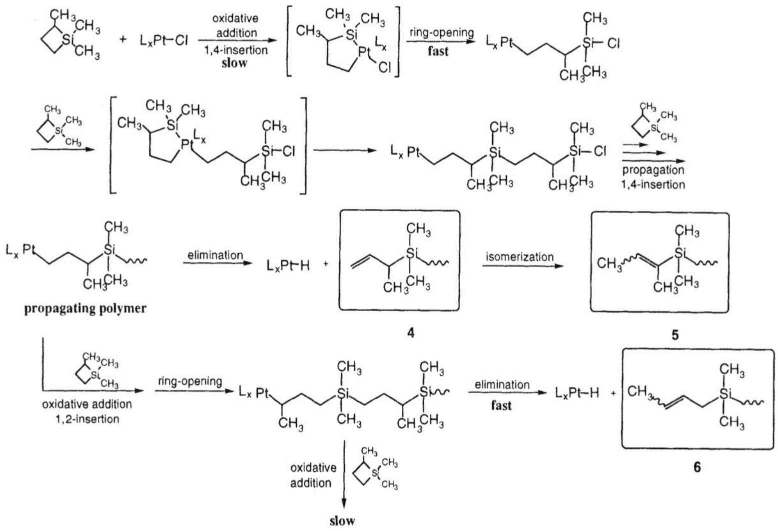

Scheme 1. Proposed Polymerization Mechanism of TMSB by Platinum Catalyst.

The polymer by anionic procedure did not show such signals. The presence of the inner olefinic signals suggests that these olefinic bonds were formed by chain transfer reaction in the polymerization through the β-elimination of the propagating end, and that isomerization of the terminal olefin to inner olefin might have occurred.

Although it is not clear which bond of 1,2- or 1,4- is cleaved in the polymerization of TMSB by platinum catalyst, it may be reasonable to assume 1-platina-2,2,3-trimethyl-2-silacyclopentane is formed as the intermediate which gives growing end through 1,4-cleavage of the ring to give head-tail controlled polymer structure (*18*). The polymerization is considered to proceed according to the following mechanism (Scheme 1): 1) oxidative addition of TMSB to the Pt complex to form platinasilacyclopentane ring (slow), followed by ring-opening (fast) (initiation reaction), 2) repeated formation of platinasilacyclopentane ring and ring-opening (propagation reaction), 3) elimination of platinum hydride (chain transfer reaction) to give the structures **4** and **6**. Isomerization from **4** to **5** may occur thermally and catalytically. When the insertion of TMSB to the growing end occurs in the 1,2-mode, the formed propagating end might be less reactive than that formed *via* 1,4-addition because of the steric factor, and elimination of platinum hydride which gives inner olefinic terminal **6** might be preferred over propagation reaction. The major chain transfer reaction may occur by this reaction path, and the butenyl terminal would be produced mainly in the polymer.

From the ratio of terminal olefinic protons (2H for **6**, 1H for **5**, and 3H for **4**) and $Si(CH_3)_2$ (6H), the degree of polymerization was estimated. Number average molecular weights estimated by ^1H NMR were considerably smaller (25-30 %) than by GPC (Table I).

By Phenyllithium. The reaction of TMSB and phenyllithium in 1 : 1 molar ratio as a model reaction was examined. The ^{29}Si, ^{13}C, ^{1}H, DEPT NMR and GC-Mass spectrometries prove dimethylphenyl(s-buthyl)silane as the initial product (Scheme 2). This fact indicated that phenyllithium attacked the silicon atom of TMSB, followed by 1,4-opening of TMSB-ring.

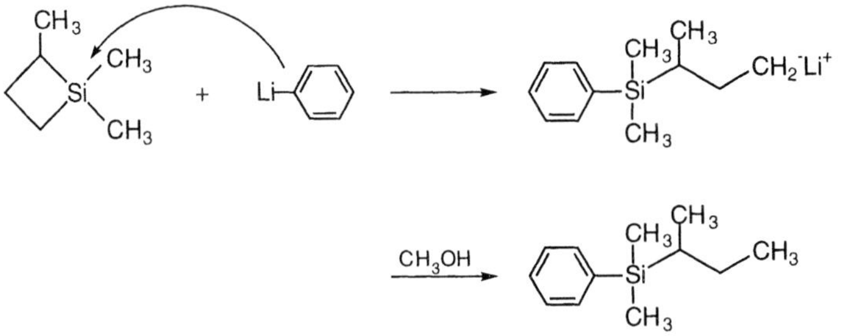

Scheme 2. Initiation Reaction of TMSB by Phenyllithium

The reaction in 1 : 1.2 molar ratio gave 2,3,6,6,7-pentamethyl-2-phenyl-2,6-disilanonane (dimer) and 2,3,6,6,7,10,10,11-octamethyl-2-phenyl-2,6,10-trisilatridecane (trimer) in the reaction mixture. This fact suggests that the propagation reaction is faster than the initiation reaction in THF.

The conditions and results of polymerization of TMSB using alkyllithium reagents as initiators are summarized in Table II.

Table II. Polymerization of TMSB by Alkyllithiums[a]

Run	Initiator	Convesion [%]	M_n[b] /10^3	M_w/M_n[b]	$M_{n, NMR}$ /10^3	M_n calcd /10^3	Yield [%]
1	MeLi	quant.	2.6	1.4	---	2.3	89
2	n-BuLi	quant.	3.6	2.0	---	2.3	95
3	PhLi	quant.	3.3	2.7	2.4	2.4	94

[a]In THF, for 1h, at -40 °C, [M]/[I] = 20, [M] = 2.0 mol dm^{-3}. [b]Estimated by GPC using standard polystyrene.

The number average molecular weight, M_n, of the polymer obtained by phenyllithium was 3.3 × 10^3 (run 3) by GPC analysis. In contrast to hexane-THF mixed solvent suggested by Matsumoto (19, 20), pure THF gave polymers of rather wide molecular weight distribution.

The major peaks of ^{29}Si, ^{13}C, ^{1}H NMR spectra were same to those of the polymer obtained by platinum complex, except for the lack of terminal olefinic signals (Figure 3). All major signals were assigned to the protons or carbons in the repeating unit, and supported by DEPT, ^{1}H-^{1}H COSY, ^{13}C-^{1}H COSY spectra. In ^{1}H and ^{13}C NMR spectra of the polymer, the initiator fragment phenyl group was proved attached to the silicon atom of the polymer terminal (Figure 3c,d). The degree of polymerization calculated from the ratio of phenyl proton and SiCH$_3$ proton in the repeating unit was 20.2. The initiator efficiency was 99 % from the ratio of [M]/[I] and M_n estimated by

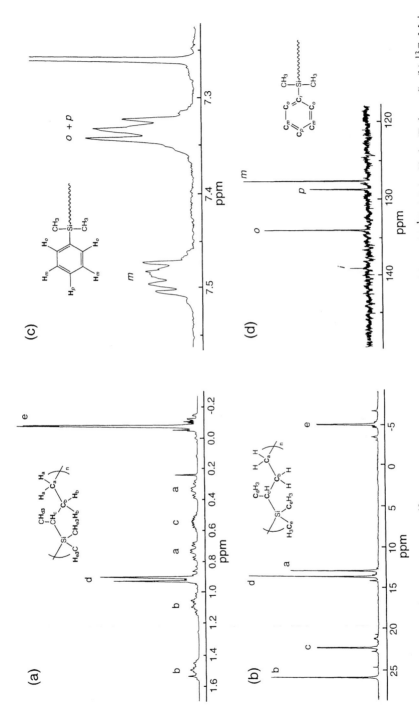

Figure 3. 300 MHz ^{1}H and 75.4 MHz ^{13}C NMR Spectra of PolyTMSB by Phenyllithium. (a) ^{1}H, Main Chain, (b) ^{13}C, Main Chain (Enlarged), (c) ^{1}H, Terminal Aromatic, and (d) ^{13}C, Terminal Aromatic.

^{1}H NMR. On the basis of these data, it is concluded that polymerization of TMSB using alkyllithiums also gave head to tail structure of the polymers.

According to the enlarged spectra (Figure 3a), the SiCH_3 protons (H_e) are split into two singlets. The two CH_2 protons of the repeating unit (H_a and H_b) also split into two multiplets. There is a possibility that the splittings are caused by diad tacticity. However, if such splitting is caused by diad tacticity, the only one pair of signals for each repeating methylene protons strongly suggests the polymer being isotactic, which should not be the case for the polymerization of racemic monomer. These splittings are considered to be enantiotopic splitting by the presence of neighboring asymmetric center (C_c). The SiCHCH_3 carbon (C_c) was also split into two singlets of equal intensity. This splitting might be caused by triad tacticity, but details are not clear at present, since this carbon should be splitted into three singlets instead of two, if this carbon reflects the triad tacticity.

Stereoregular and/or Optically Active Poly(carbosilane)s and Poly(carbosiloxane)s

Poly(carbosilane)s have been of interest as precursors to silicon carbide. Recently, investigations focusing on liquid crystalline behavior of these polymers have been published *(21-24)*. It is very important to control the stereochemistry of Si atom in the repeating units to correlate precisely such behavior with the primary structure of polymers. We designed a synthesis starting from optically active silane *(25,26)*. Methyl(1-naphthyl)phenyl-(-)-menthyloxysilane is a useful and versatile starting material in the synthesis of variety of optically active silicon compounds *(25-29)*. Its optical purity can be determined by diastereomeric splitting in ^{1}H NMR (Figure 4) and HPLC on optically active stationary phase (Figure 5) *(26)*. After the optical resolution by the recrystallization from pentane, Si-(*S*)-form, which has the diastereomer excess (d.e.) of >99 %, is predominantly obtained as a crystalline material. The optical purity of methyl(1-naphthyl)phenylsilane can be also determined on the HPLC.

Optically active allylmethylphenylsilane **10** was synthesized according to the synthetic route shown in Scheme 3 *(26)*.

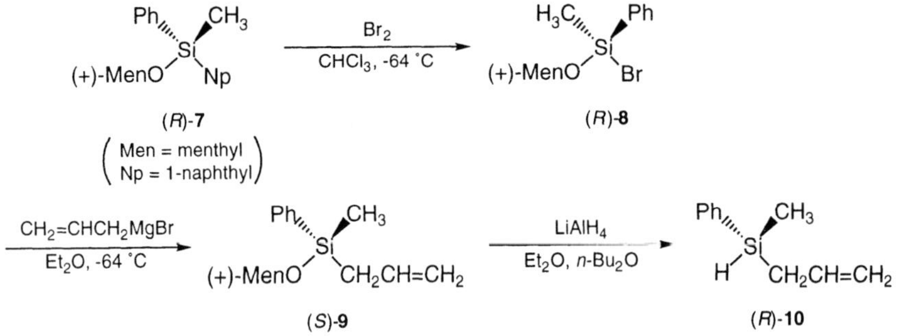

Scheme 3. Synthesis of Optically Active **10**

The d.e. of **8** is estimated by the splitting of the signal at 0.62 and 0.79 ppm. The d.e. of allylated product **9** is estimated by the splitting of the signal at 0.60 and 0.63 ppm. The d.e. of 76.5 % is attained from **8** of d.e. of 78.9 %. Reduction by lithium aluminum hydride gives **10** having $[\alpha]_D^{26}$ = +24.0 (*c* 1.00, pentane). The enantiomer excess (e.e.) of the product could not be estimated at this point. Recently, we found that the introduction of methoxy group at 4-position of naphthyl group improved the d.e. of **8** as high as 91 % (Kawakami, Y.; Oishi, M., JAIST, unpublished data).

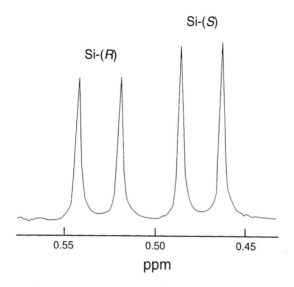

Figure 4. Diastereomeric Splitting of Methyl Groups in Isopropyl of Methyl(1-naphthyl)phenyl-(-)-menthyloxysilane.

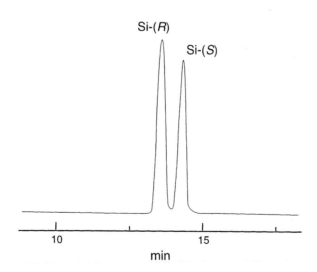

Figure 5. HPLC Separation of Methyl(1-naphthyl)phenyl-(-)-menthyloxysilane. (cellulose carbamates (30), Daicel Chem. Ind., CHIRALCEL OD, 0.46 φcm × 25 cm)

The 500 MHz ^{1}H NMR spectra of the polymer are shown in Figure 6. In the aliphatic region of the spectrum, three types of signals assignable to SiCH_3 (0.13 ppm), α-CH$_2$ (0.70 ppm) and β-CH$_2$ (1.29 ppm) are observed. No signal derived from α-addition is present. In the spectra, the SiCH_3 signal is split into three singlets at 0.120, 0.125 and 0.131 ppm reflecting the triad tacticity. In the figure, the signal at 0.120 ppm becomes relatively stronger and that at 0.131 ppm relatively weaker compared with those of the polymer from racemic monomer. Based on these facts, the signals at 0.120 and 0.131 ppm were assigned to the isotactic and syndiotactic triad, respectively, and that at 0.125 ppm to the heterotactic triad.

The calculated concentration of each triad starting from the optically active monomer with 76.5 % e.e. assuming complete retention of Si stereochemistry in the reduction step by lithium aluminum hydride and in the polymerization is S : H : I = 1.0 : 2.0 : 6.6. The actual concentration of each triad evaluated from Figure 6 is 1.0 : 2.0 : 7.0. This fact proved there was no racemization in allylation and reduction steps to synthesize allylmethylphenylsilane. This also proved the optical purity of allylmethylphenylsilane having $[α]_D^{26}$ = 24.0 is 76.5 % e.e. This value is a little higher than that recently estimated value (-28.89) (*31*).

Although the optical activity of the starting monomer reflects in the isotacticity of the formed polymer *via* selective β-addition, the optical activity itself is lost in the polymer since the formed polymer is *pseudo*-asymmetric.

In order to keep the optical activity in the polymer from the optical active monomer, it is necessary to distribute such optically active repeating unit connected with different constitutional unit in the polymer chain. Optically active vinylhydrodisiloxane, synthesized according to the Scheme 4 (*32*), was chosen as a monomer, and the polyaddition reaction was studied. Attempt to determine the optical purity of (*S*)-**13** by HPLC failed. However, successful separation of two enantiomers of (1-naphthyl)phenylvinylsilane ((*R*)-**14**) by HPLC made it possible to determine the optical purity of monomer (*S*)-**13** by analyzing the optical purity of (*R*)-**14** to be close to 100%.

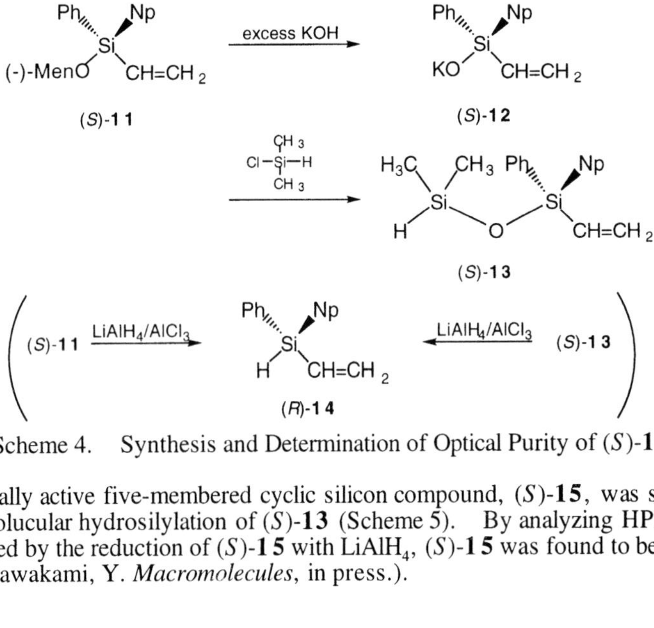

Scheme 4. Synthesis and Determination of Optical Purity of (*S*)-**13**

Optically active five-membered cyclic silicon compound, (*S*)-**15**, was synthesized by intramolucular hydrosilylation of (*S*)-**13** (Scheme 5). By analyzing HPLC of (*R*)-**16** obtained by the reduction of (*S*)-**15** with LiAlH$_4$, (*S*)-**15** was found to be > 98 % e.e. (Li, Y.; Kawakami, Y. *Macromolecules*, in press.).

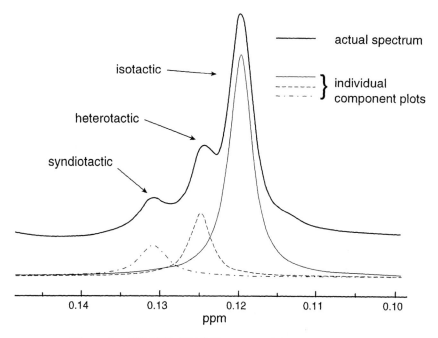

Figure 6. 500 MHz ^{1}H NMR spectra of CH_3 Region of Poly(methylphenylsilylenetrimethylene).

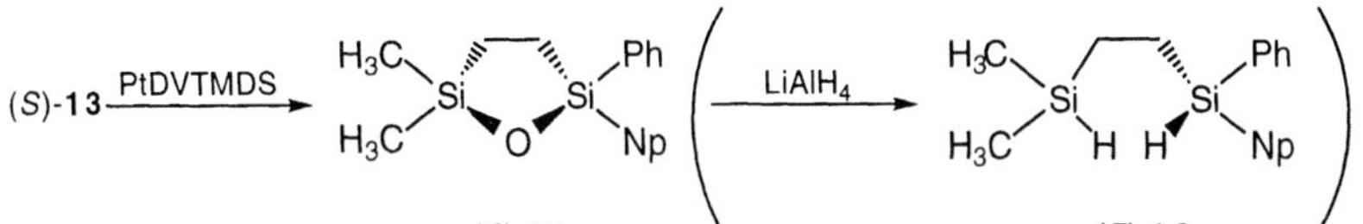

Scheme 5. Synthesis and Determination of Optical Purity of (S)-15

^{13}C NMR spectra of Si(CH$_3$)$_2$ in the polymer obtained from racemic and optically active **13** are shown in Figure 7. Atactic polymer, which was obtained from racemic monomer, showed three peaks (-0.427 ppm, -0.503 ppm and -0.564 ppm), while the polymer from optically active monomer showed mainly two peaks (-0.427 ppm and -0.564 ppm).

There are four possible types of diads for this polymer depending on the asymmetric silicon centers: *S-S, S-R, R-S* and *R-R* (Figure 8). Two methyl carbons in *S-S* (or *R-R*) diad are located in quite different environments and therefore have two chemical shifts. In the case of *S-R* (or *R-S*) diad, two methyl groups are in very similar environments, and their carbon resonances will appear at the similar position, *i.e.*, methyl groups of *S-R* and *R-S* diads will appear as one inseparable peak. Si(CH$_3$)$_2$ in atactic polymer, therefore, would be split into three peaks (the central peak represents the *S-R* and *R-S* diads, two side peaks represent *S-S* and *R-R* diads) in the ^{13}C NMR spectrum, with an intensity of 1 : 2 : 1. While pure isotactic polymer, containing only *S-S* (or *R-R*) diad, would show two peaks of methyl carbons, and pure syndiotactic polymer only one (*S-R* and *R-S* diads). The present polymer showed two distinct peaks of methyl carbons, indicating that the polymer is highly isotactic.

Since we can assume that chirality of the asymmetric silicon atom of the monomer would not change during the hydrosilylation, *i.e.*, present polymer must have the same optical purity with that of the monomer (> 99 % e.e.). Therefore the polymer should be highly optically pure (> 99 % e.e.) and highly stereoregular (isotacticity > 99 %).

However, this method suffered some drawbacks, such as, contamination with cyclic oligomers, the rather low yield (37.8%) and the low molecular weight (M_n = 2.9 × 10^3, polystyrene standard) of the polymer obtained. A considerable amount of cyclic compounds (26.3 %) were also formed. The ring-opening polymerization of (*S*)-**15** using phenyllithium (PhLi) as an initiator is highly regioselective, and provided a new route to obtain a higher molecular weight polymer with a narrow molecular weight distribution in a high yield (*32*).

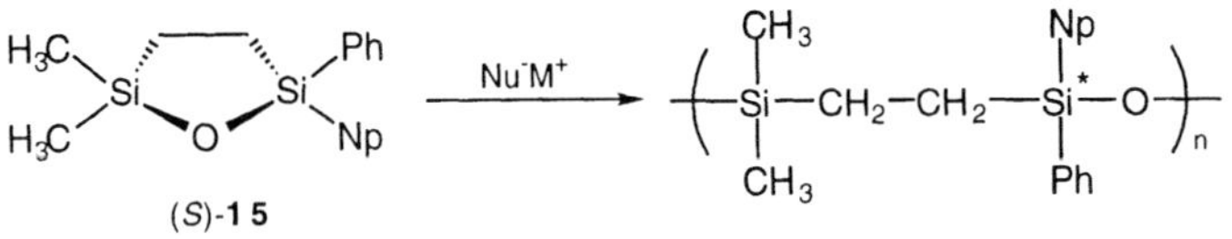

Scheme 6. Anionic Ring-opening Polymerization of (*S*)-**15**

When potassium *tert*-butoxide (*tert*-BuOK) was used as an initiator, a serious chain scission occurred, that resulted in the production of the racemized cyclic polymers with low molecular weight and cyclic dimers.

Conclusion. Approaches to molecular design of primary structure of silicon-containing polymers were illustrated for poly(carbosilane)s with controlled molecular weight and for stereoregular poly(carbosilane) and poly(carbosiloxane).

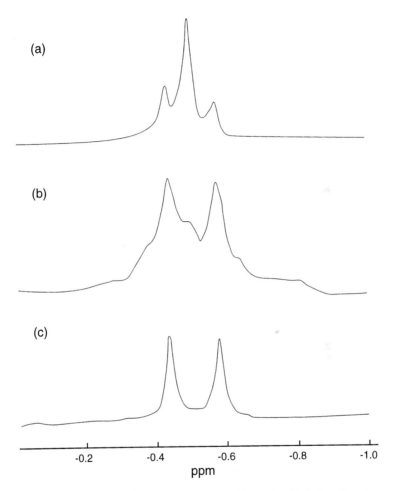

Figure 7. 75.4 MHz ^{13}C NMR Spectra of Si(CH$_3$)$_2$ of Poly[{(1S)-1-(1-naphthyl)-1-phenyl-3,3-dimethyldisiloxane-1,3-diyl}ethylene] Obtained from (a) racemic **13**, (b) (S)-**13**, and (c) (S)-**15**.

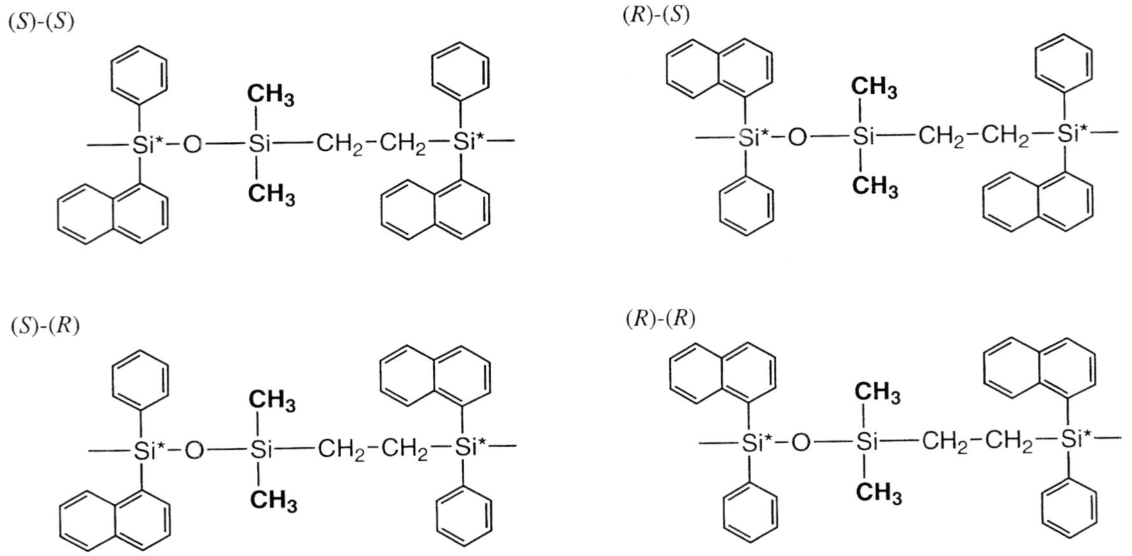

Figure 8. Diad Sequences of Poly[{(1S)-1-(1-naphthyl)-1-phenyl-3,3-dimethyldisiloxane-1,3-diyl}ethylene]

Acknowledgments. The author is grateful to Shin-Etsu Chemical Co., Ltd. for generous donation of organosilicon compounds. Acknowledgements are given to Dr. I. Imae, Mrs. K. Komuro and Y. Li. This work was partially supported by a Grant-in-Aid for Scientific Research (08544538), a Grant-in-Aid for Scientific Research in Priority Areas, *New Polymers and Their Nano-Organized Systems* (10126222) and a Grant-in-Aid for Scientific Research in Priority Areas, *the Chemistry of Inter-element Linkage* (10133220) from the Ministry of Education, Science, Sports, and Culture, Japan.

Literature Cited
1. Greene, T. W.; Wuts, P. G. M. *Protective Group in Organic Synthesis*; John Wiley & Sons: New York, 1991.
2. Kricheldorf, H. R. *Silicon in Polymer Synthesis*; Springer: Berlin, 1996.
3. Smith, A. L. *The Analytical Chemistry of Silicones*; Wiley Interscience: New York 1991.
4. Yajima, S.; Hayashi, J.; Omori, M. *Chem. Lett.* **1975**, 931.
5. Yajima, S.; Okamura, K.; Hayashi, J. *Chem. Lett.* **1975**, 1209.
6. Sakamoto, K.; Obata, K.; Hirata, H.; Nakajima, M.; Sakurai, H. *J. Am. Chem. Soc.* **1989**, *111*, 7641.
7. Nametkin, N. S.; Vdovin, V. M.; Grinberg, P. L. *Izv. Akad. Nauk SSSR Ser. Khim.* **1964**, 1133.
8. Wyenberg, D. R.; Nelson, L. E. *J. Org. Chem.*, **1965**, *30*, 2618.
9. Cundy, C. S.; Eaborn, C.; Lappert, M. F. *J. Organometal. Chem.* **1972**, *44*, 291.
10. Poletaev, V. A.; Vdovin, V. M.; Nametkin, N. S. *Dokl. Akad. Nauk SSSR*, **1973**, *208*, 1112.
11. Nametkin, N. S.; Vdovin, V. M.; Zav'yalov, V. I. *Izv. Akad. Nauk SSSR Ser. Khim.* **1964**, 203.
12. Nametkin, N. S.; Bespalova, N. B.; Ushakov, N. V.; Vdovin, V. M. *Dokl. Akad. Nauk SSSR* **1964**, *209*, 621.
13. Nametkin, N. S.; Vdovin, V. M.; Poletaev, V. A.; Zav'yalov, V. I. *Dokl. Akad. Nauk SSSR* **1967**, *175*, 1068.
14. Liao, C. X.; Weber, W. P. *Polym. Bull.* **1992**, *28*, 281.
15. Xiugao, C.; Weber, W. P. *Macromolecules* **1992**, *25*, 1639.
16. Theuring, M.; Weber, W. P. *Polym. Bull.* **1992**, *28*, 17.
17. Komuro, K.; Toyokawa, S.; Kawakami, Y. *Polym. Bull.* **1998**, *40*, 715.
18. Yamashita, M.; Tanaka, M.; Honda, K. *J. Am. Chem. Soc.* **1995**, *117*, 8873.
19. Matsumoto, K.; Yamaoka, H. *Macromolecules* **1995**, *28*, 17.
20. Matsumoto, K.; Shimazu, H.; Deguchi, M.; Yamaoka, H. *J. Polym. Sci., A, Polym. Chem. Ed.* **1997**, *35*, 3207.
21. Koopmann, F.; Frey, H. *Makromol. Rapid Commun.* **1995**, *16*, 363.
22. Koopmann, F.; Frey, H. *Macromolecules* **1996**, *29*, 3701.
23. Sargeant, S. J.; Weber, W. P. *Macromolecules* **1993**, *26*, 2400.
24. Guo, H.; Volke, R.; Weber, W. P.; Ganicz, T.; Pluta, M.; B-Florjanezyk, E.; Stnczyk, W. *J. Organomet. Chem.* **1993**, *444*, C9.
25. Kawakami, Y.; Takeyama, K.; Komuro, K.; Ooi, O. *Macromolecules* **1998**, *31*, 551.
26. Kawakami, Y.; Takahashi, T.; Yada, Y.; Imae, I. *Polym. J.* **1998**, *30*, 1001.
27. Sommer, L. H.; Frey, C. L. *J. Am. Chem. Soc.* **1959**, *81*, 1013.
28. Sommer, L. H.; Frey. C. L.; Parker, G. A. *J. Am. Chem. Soc.* **1964**, *86*, 3276.
29. Corriu, R. J. P.; Lanneau, G. F.; Leard, M. *J. Organometal. Chem.* **1974**, *64*, 79.
30. Oguni, K.; Oda, H.; Ichida, A. *J. Chromatogr. A* **1995**, *694*, 91.
31. Kobayashi, K. Kato, T.; Unno, M.; Masuda, S. *Bull. Chem. Soc. Jpn.* **1997**, *70*, 1393.
32. Li, Y.; Kawakami, Y. *Macromolecules* **1998**, *31*, 5592.

Chapter 32

Investigation of the Synthesis of Bis-(3-Trimethoxysilylpropyl)fumarate

Peter M. Miranda

General Electric Company Silicones,
260 Hudson River Road, Waterford, NY 12188

This laboratory has investigated and improved the 2-step synthesis of bis-(3-trimethoxysilylpropyl)fumarate. Previously, yields of 40 - 80% of a yellow/brown reaction mixture were common. Now, one can obtain > 95% of a colorless product.
The first step involves the platinum catalyzed hydrosilylation of diallyl maleate (DAM) with trichlorosilane. By purifying the DAM and using Karstedt's Pt catalyst, one obtains the bis-trichlorosilyl intermediate in > 98% yield. The utilization of diallyl fumarate in this step does not provide significant advantage over the maleate isomer.
The second step involves methanolysis of the chlorosilane and the cis/trans isomerization that occurs in the presence of the HCl generated. Previously, addition of methanol to the intermediate resulted in significant amounts of oligomeric siloxane formed by condensation reactions. Modification of the reaction conditions has eliminated this oligomer formation. Also investigated was the replacement of methanol with trimethylorthoformate (TMOF). By utilizing TMOF one eliminates the HCl generation, resulting in exclusive formation of the bis-(3-trimethoxysilylpropyl)maleate isomer.

The focus of this report is to investigate and improve the synthesis[1] of bis-(3-trimethoxysilylpropyl)fumarate and develop a method to synthesize the bis-(3-trimethoxysilylpropyl)maleate isomer. This paper also outlines some of the analytical methods used to characterize and quantify the intermediates and isomers obtained during the synthesis.

The platinum catalyzed hydrosilylation reaction, equation 1, is commonly employed as a 'curing' method in polymeric systems where the oligomers are functionalized with silyl hydride and vinyl groups[2], as exemplified in equation 2.

$$R_3Si-H + H_2C=CHR \xrightarrow{Pt} R_3Si-CH_2-CHHR \tag{1}$$

$$Me_3SiO(\underset{|}{\overset{|}{Si}}O)(\underset{|}{\overset{|}{Si}}O)SiMe_3 + Me_3SiO(\underset{|}{\overset{|}{Si}}O)(\underset{H}{\overset{|}{Si}}O)SiMe_3 \xrightarrow{Pt} \text{Polymer Network} \tag{2}$$

Typically, these curable systems will employ reversible inhibitors, such as diallyl maleate (DAM), which coordinate with the platinum catalyst to prevent premature polymerization. At elevated temperatures, these platinum-inhibitor complexes dissociate and allow for the hydrosilylation reaction to occur.

Of interest is the synthesis of the bis-(3-trimethoxysilylpropyl)fumarate, which is utilized as a low molecular weight polymer additive in a variety of silicone applications[3,4]. Prior to the present investigation, this compound was synthesized through a two-step process starting with commercially available diallyl maleate[5]. This reaction consistently generated a yellow/brown product with erratic yields of < 80% of the theoretical value.

The first step, as shown in equation 3, is a solventless process involving the hydrosilylation of DAM with trichlorosilane (TCS) in the presence of Lamoreaux' catalyst[6] (chloroplatinic acid in excess octanol). As noted above, since DAM inhibits platinum catalyzed hydrosilylation, the reaction requires elevated temperatures (> 40 °C). However, the low boiling point of TCS (31.9 °C) and the thermal instability of DAM necessitate reaction temperatures not to exceed 110 °C. Thus higher levels of platinum had been used to overcome these limitations, but inevitably results in increased coloration and cost of the final product.

$$2\ HSiCl_3 + \begin{array}{c} CO_2CH_2CH=CH_2 \\ \| \\ CO_2CH_2CH=CH_2 \end{array} \xrightarrow[\Delta]{Pt} \begin{array}{c} CO_2CH_2CH_2CH_2SiCl_3 \\ \| \\ CO_2CH_2CH_2CH_2SiCl_3 \end{array} \tag{3}$$

$$6\ CH_3OH + \begin{array}{c} CO_2CH_2CH_2CH_2SiCl_3 \\ \| \\ CO_2CH_2CH_2CH_2SiCl_3 \end{array} \xrightarrow{\Delta} \begin{array}{c} CO_2CH_2CH_2CH_2Si(OMe)_3 \\ \| \\ CO_2CH_2CH_2CH_2Si(OMe)_3 \end{array} + 6\ HCl \tag{4}$$

The second step, as shown in equation 4, involves the methanolysis of the silyl chlorides in the presence of excess methanol. This reaction releases six equivalents of HCl. At later stages of this reaction, high concentration of HCl in the presence of methanol leads to the formation of water as shown in equation 5:

$$CH_3OH + HCl \longrightarrow CH_3Cl + H_2O \qquad (5)$$

In the presence of water and catalytic amounts of acid, the methoxysilanes are hydrolyzed to silanols that then participate in the condensation reactions forming oligomeric and high molecular weight siloxanes (equations 6a-c).

$$\equiv SiOMe + H_2O \xrightarrow{H^+} CH_3OH + \equiv SiOH \qquad (6a)$$

$$\equiv SiOH + \equiv SiOMe \xrightarrow{H^+} CH_3OH + \equiv SiOSi\equiv \qquad (6b)$$

$$\equiv SiOH + \equiv SiOH \xrightarrow{H^+} H_2O + \equiv SiOSi\equiv \qquad (6c)$$

The present investigation has improved both steps of the above mentioned synthesis. Additionally, it has confirmed that the final product based on the aforementioned process is predominantly fumarate isomer based on high field ^{1}H-NMR analysis of the olefinic protons.

Results and Discussion

Hydrosilylation. Historically, the synthesis of the silyl chloride intermediate from DAM and TCS requires between 100-200 ppm of platinum. Often, the hydrosilylation would terminate prior to completion, with yields as low as 40%. To determine whether DAM was responsible for poor reaction results, the diallyl fumarate (DAF) isomer was evaluated as an alternate substrate; DAF is a 'weaker' inhibitor of platinum[7]. The comparison revealed that when the DAF and DAM were distilled prior to use, there were no significant differences in reaction kinetics. It was determined that a sulfur contaminant found in commercial sources of DAM – assumed to be a residual of acid catalyst from the maleate esterification – results in poisoning of the platinum catalyst. Thus with distillation, DAM can now be hydrosilylated with TCS in the presence of only 10 ppm platinum catalyst. This observation was consistently reproduced when utilizing an active platinum catalyst complex. Consequently, the hydrosilylation offers > 99% yield with respect to depletion of either DAF or DAM, and severe coloration is completely eliminated from the final product by reduction of the amount of platinum catalyst required.

Thus, the hydrosilylation reaction was carried out using either purified DAM or DAF with TCS in the presence of Karstedt's catalyst[8,9], and was monitored by gas chromatography (GC). The GC analysis was accomplished by a capillary column (DBS 30 m x 0.32 mm) with injection port at 250 °C and FID at 300 °C; program started at 80 °C for 2', then ramp at 25 °C/min. to 280 °C. The DAF was found to be a slightly faster reaction with all detectable DAF being consumed in ~100 minutes, whereas DAM consumption required ~150 minutes. Both reactions were run at a peak temperature of 90 °C. The fumarate isomer was detected in the synthesis initiated with DAM as a result of trace amounts of water in the reaction. This allows for the formation of HCl from TCS, thus catalyzing the cis/trans isomerization via an enol species.

Figures 1 and 2 show the reaction rates of hydrosilylation. As one observes in Figure 1, initially the trichlorosilylpropylallyl maleate is formed and then further reaction produces the bis-(trichlorosilylpropyl)maleate intermediate. The reaction corresponding to DAF follows a similar course (Figure 2).

The minimal difference (50 minutes) in hydrosilylation reaction rates between DAF and DAM is due to the use of TCS as the hydride reactant. When less reactive hydrides are used – $HSiMeCl_2$ and $HSiMe_2Cl$ – DAF was found to hydrosilylate more readily than DAM, though both DAM and DAF could be driven to completion. It was also observed by ^{1}H-NMR that in a 1:1 solution of DAF and DAM with Karstedt's catalyst, the fumarate was the preferred ligand[10].

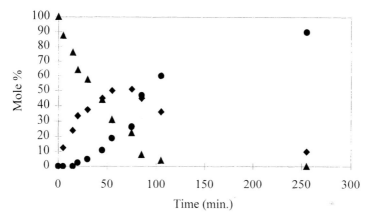

Figure 1. Amounts of DAM (▲), trichlorosilylpropylallyl maleate (♦), and bis-(trichlorosilylpropyl) maleate (●) versus reaction time during the hydrosilylation of DAM with TCS by GC analysis.

Methanolysis and Isomerization. The majority of the isomerization occurs during the methoxylation step of the synthesis that provides the required heat and acid catalysis to cause this reaction. The methanolysis is typically run at 105-120 °C to remove HCl and water as they are produced. Both maleate and fumarate silyl chlorides readily react in the presence of methanol to form bis-(3-trimethoxysilylpropyl)fumarate.

In an alternate synthesis, trimethyl orthoformate (TMOF) is used as the methoxy source[11], as shown in equation 7 below:

$$\equiv SiCl + HC(OCH_3)_3 \longrightarrow \equiv SiOCH_3 + CH_3Cl + HCO_2CH_3 \quad (7)$$

This procedure does not liberate HCl, however methyl chloride and methyl formate are formed as by-products. Thus, no catalyst is produced to affect the isomerization of the product, and no water is formed preventing siloxane formation and condensation.

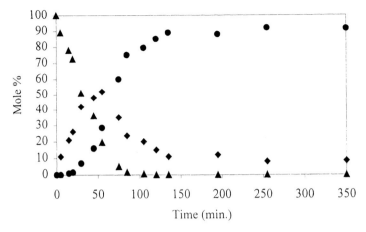

Figure 2. Amounts of DAF (▲), trichlorosilylpropylallyl fumarate (♦), and bis-(trichlorosilylpropyl) fumarate (●) versus reaction time during the hydrosilylation of DAF with TCS by GC analysis.

Hence, TMOF is used to synthesize exclusively bis-(3-trimethoxysilylpropyl)maleate, however, this procedure fails to convert all of the silyl chloride groups. Addition of methanol is successful in removing the residual silyl chloride groups, but does result in some cis/trans isomerization.

Optimization of the methanolysis for the fumarate derivative is achieved by reducing the reaction temperature to 70 °C during the methanol addition while applying a partial vacuum, ~50 mm Hg, to remove HCl and water. Incorporation of this modification resulted in negligible amounts of siloxane formation.

Experimental

Reagents. Diallyl fumarate was obtained from Pfaltz & Bauer and was used without further purification. The impure diallyl maleate (failed in hydrosilylation), trichlorosilane, and trimethylorthoformate were obtained from Aldrich. Distilled DAM was obtained from Bimax Corp. and was utilized without additional treatment. Platinum catalysts and solvents were obtained internally from GE Silicones. All reactions were run under a nitrogen atmosphere.

Hydrosilylation of DAM(DAF). Into a reaction vessel, equipped with a condenser and dropping funnel, one adds 32.61 parts of DAM and 0.006 parts (~10 ppm as platinum) Karstedt's catalyst. The solution is then warmed to 40 °C while stirring. To the dropping funnel, equipped with a pressure equalization arm, is added 67.38 parts of TCS (50% molar excess). Attached to the funnel outlet is Teflon tubing; this allows for TCS to be added below the DAM solution surface.

The TCS is then added at a rate of 0.5 parts/minute, and an exotherm is immediately observed. The reaction temperature is not allowed to exceed 95 °C during the TCS feed; control of temperature can be accomplished by slowing or stopping the TCS addition. After ~1/3 of TCS has been added, the exotherm was moderate and balances with the evaporative cooling of the TCS. At this point, the reaction temperature is maintained at 60-65 °C by external heating. After all the TCS has been added, the reaction is held at reflux (60-65 °C) for two hours.

After two hours, the temperature is increased to 80 °C and the excess TCS is distilled off. Once the overhead stops, 86.25 parts of toluene is added to the reaction vessel, and with a moderate sweep of N_2, the reaction is again stripped at 80 °C until ~26 parts of toluene is removed. The hydrosilylation product was found to be stable in toluene under an N_2 atmosphere.

The product purity is determined by utilizing GC analysis as described earlier. Proton NMR has also been employed for product characterization exemplified by analysis of the DAF hydrosilylation in Figure 3.

Methanolysis. To the bis-trichlorosilane intermediate is added 26 parts of toluene to bring the total amount of toluene in the reaction mixture back to 86 parts. The reaction kettle is then equipped with a reflux condenser and a Dean-Stark trap. The toluene solution is heated to 105-115 °C. To an additional funnel with Teflon tubing on the outlet, is added 51.3 parts of methanol. The methanol is added under the

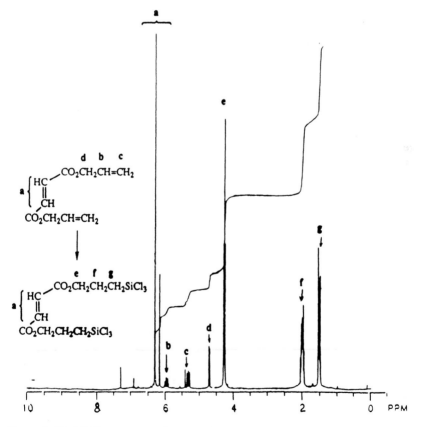

Figure 3. ^{1}H-NMR spectrum of product from the hydrosilylation reaction of DAF. Residual allyl can be detected in this sample.

surface of the hot solution at a rate of 0.2 parts/minute while maintaining a temperature range of 105-115 °C. If the temperature drops below 105 °C, the methanol additional is halted.

As methanol addition proceeds, two phases are collected in the Dean-Stark trap – a lower phase composed of methanol/toluene/HCl/H_2O and an upper phase of toluene/methanol. The lower phase is removed from the trap as the reaction progresses; the upper phase is allowed to return. Upon completion of the methanol addition, the reaction is held at 105 °C for an additional 30 minutes. The reaction solution is then stripped at a vacuum of 100 mm Hg while heating to 110 °C. Once this temperature is reached, the vacuum is removed and the reaction is cooled to room temperature.

Both ^{1}H-NMR and ^{29}Si-NMR have been used to follow the methanolysis reaction, see Figures 4 and 5, respectively.

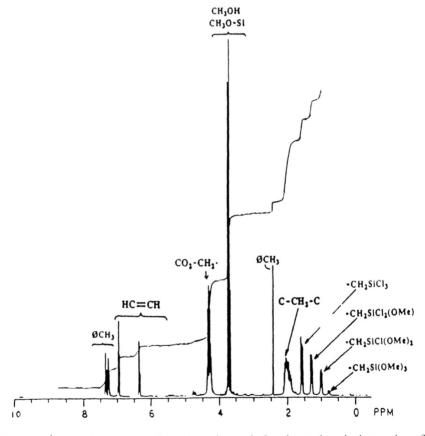

Figure 4. ^{1}H -NMR spectrum of reaction mixture during the methanolysis reaction of bis-(trichlorosilylpropyl) maleate.

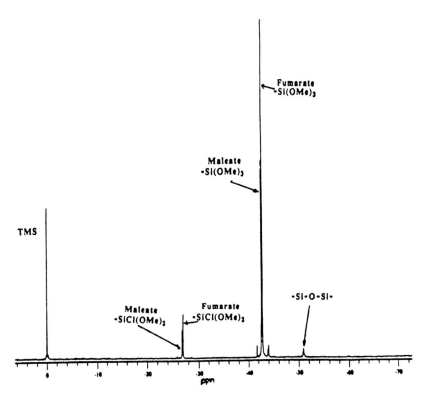

Figure 5. ^{29}Si-NMR spectrum of the product mixture from methanolysis of bis-(trichlorosilylpropyl) maleate. Both residual chlorosilane and siloxane condensation impurities can be detected.

A sample of the product is tested for residual HCl concentration. The residual HCl is neutralized with an equivalent molar amount of triethylamine (TEA). The TEA is added to the product as a 50:50 weight % mixture in methanol. After addition, the product solution is stirred at room temperature for 30 minutes, and then allowed to stand for 1 hour before filtering to remove the salt. The filtrate is then re-stripped at 100 mm Hg to a temperature of 60 °C to remove all traces of methanol.

TMOF Reaction. The procedure followed the methods reported in the literature[11]. A 100% molar excess of trimethylorthoformate is slowly added to the bis-trichlorosilane intermediate. An exothermic reaction occurs with gas evolution. After the addition is completed, the reaction is held at 100-105 °C for 24 hours. If by ^{1}H-NMR the reaction is not completed, a catalytic amount of $AlCl_3$ is added followed by additional heating. The reaction by-products – methyl formate, methyl chloride – and the excess TMOF were then removed *in vacuo*.

Analytical Equipment. The ^{1}H-NMR spectra were acquired by in $CDCl_3$ on a GE model QE-300 NMR spectrometer at 300.15 MHz. The ^{29}Si-NMR were acquired at 59.6 MHz on a Varian XL-300 NMR spectrometer. Chromium III acetylacetonate was added to shorten the T1 values of ^{29}Si, and gated decoupling was used to suppress Nuclear Overhauser Enhancement (NOE). All chemical shifts were referenced to tetramethylsilane.

Acknowledgment.
The author wishes to acknowledge the significant contributions provided by Susan L. Bontems, formerly of GE Corporate Research & Development Center.

References
1. Mitchell, T. D., *U. S. Patent 4,281,145*, 1981.
2. Hardman, B. B.; Torkelson, A., *Kirk-Othmer: Encyclopedia of Chemical Technology*, 3rd Ed.; John Wiley & Sons: New York, 1982, *20*, 922-962.
3. Traver, F. J.; Pawar, P.; Miranda, P. M., *U. S. Patent 5,232,783*, 1993.
4. Smith, A. H.; DeZuba, G. P.; Mitchell, T. D., *U. S. Patent 4,273,698*, 1981.
5. Berger, A.; Selin, T. G., *U. S. Patent 3,759,968*, 1973.
6. Lamoreaux, H. F., *U. S. Patent 3,220,972*, 1965.
7. Lewis, L.N.; Stein, J.; Colborn, R.E.; Gao, Y.; Dong, J., *J. Organometallic Chem.* 1996, *521*, 221-227.
8. Karstedt, B. D., *U. S. Patent 3,775,452*, 1973.
9. Karstedt, B. D., *U. S. Patent 3,814,730*, 1974.
10. Unpublished results.
11. Noll, W., *Chemistry and Technology of Silicones*; Academic Press: New York, 1968, 68-115.

Chapter 33

In Situ Formed Siloxane–Silica Filler for Rubbery Organic Networks

Libor Matějka

Institute of Macromolecular Chemistry, Academy of Sciences of the Czech Republic, Heyrovský Sq. 2, 162 06 Prague 6, Czech Republic

Rubbery epoxide network was reinforced with a silica filler formed in situ by the sol-gel process from tetraethoxysilane (TEOS). Simultaneous or sequential polymerization of organic monomers and TEOS were used to form a microphase separated epoxy-silica hybrid system. Structure evolution during polymerization was followed using small-angle x-ray scattering and the cured networks were characterized by electron microscopy and dynamic mechanical analysis showing the morphology, interphase interaction and mechanical properties. Reaction conditions of the sol-gel process and polymerization procedure determine the reaction mechanism, structure of siloxane-silica regions and interaction between silica and epoxy phases. Modulus increased by two orders of magnitudes at low contents of silica (<10 vol. %) and the reinforcement efficiency increases with increasing interphase interaction. The relationship reaction mechanism – structure – mechanical properties is discussed. Based on the analysis of DMA data, a morphological model of the organic-inorganic hybrid with co-continuous phases was suggested.

Reinforcement of thermoplastics and elastomers with an inorganic filler formed in situ by the sol-gel process has been recently studied by many authors (1-5). These organic-inorganic (O-I) hybrid systems are a new class of composites, called often nanocomposites because of the nanometer-scale size of the inorganic structures formed. An important advantage of the O-I hybrid composite consists in fine dispersion of the inorganic, mostly silica, filler within an organic matrix and there is a possibility of tailoring the material properties by reaction conditions. The reaction mechanism of the sol-gel process involving hydrolysis and condensation of alkoxysilanes to form silica is variable and depends on such factors as catalysis, water content and solvent (6). Mainly the type of catalyst determines the course of the polymerization and the final structure of silica in the organic matrix. However, the ways of controling the structure, morphology and O-I interphase as well as their effects on final properties of the hybrids are still open for investigation.

In our previous works (7-9) siloxane-silica clusters formed by the sol-gel process from tetraethoxysilane (TEOS) were used to reinforce the rubbery epoxide network based on diglycidyl ether of Bisphenol A (DGEBA) and poly(oxypropylene)diamine (Jeffamine® D2000).

$$Si(OEt)_4 \xrightarrow[-EtOH]{n\ H_2O} Si(OH)_n(OEt)_{4-n} \xrightarrow{-H_2O} -O-\underset{\underset{O-}{|}}{\overset{\overset{O-}{|}}{Si}}-O-$$

TEOS

Three polymerization procedures were employed to synthesize DGEBA-D2000-TEOS hybrid involving simultaneous or sequential formation of epoxy and silica networks (7). The sol-gel process was catalyzed with p-toluenesulfonic acid and with a basic catalyst, amine D2000, used as epoxy curing agent. In addition, the O-I hybrid network was prepared by crosslinking a prepolymer D2000 endcapped with alkoxysilane groups (SED2000) using the sol-gel process. The epoxy-silica hybrids synthesized by the following procedures were studied:

(a) one-stage simultaneous polymerization (epoxy-TEOS, ET1) of both systems under basic catalysis of the sol-gel process with amine D2000,
(b) two-stage polymerization (ET2) with acid prehydrolysis of TEOS in the first stage and simultaneous formation of both networks in the second stage where the sol-gel process is base-catalyzed,
(c) two-stage sequential polymerization with preformed epoxy network (E1T2) in the first stage and swelling of the network with TEOS followed by acid-catalyzed TEOS polymerization within the organic network in the second stage,
(d) crosslinking of alkoxy-endcapped prepolymer (SED2000).

Recently we have also followed silica structure evolution during polymerization (8), final structure and morphology as well as mechanical properties of cured networks (9). The effect of reaction conditions, polymerization procedure and interaction between organic and inorganic phases were studied. Small-angle X-ray scattering (SAXS), electron microscopy (SEM) and dynamic mechanical analysis (DMA) were used to characterize the hybrids. The present paper tries to answer an important question relating to morphology and to the interphase. Which type of the morphology is produced? The organic matrix with a dispersed silica filler or interpenetrating silica-epoxy networks?

Experimental

The organic phase of the O-I hybrid studied in the present work consists of a stoichiometric epoxide-amine network prepared by curing DGEBA with oligomeric amine D2000 (M=1970).

Inorganic phase – silica network, was formed by sol-gel process from alkoxysilane precursors – TEOS or alkoxysilane-endcapped prepolymer SED2000. This endcapped prepolymer was prepared by alkoxy functionalization of oligomer D2000 with (3-glycidyloxypropyl)trimethoxysilane (GTMS).

$$H_2N\text{\textasciitilde}\text{\textasciitilde}NH_2 + CH_2CHCH_2O(CH_2)_3Si(OCH_3)_3 \rightarrow \begin{array}{c}(CH_3O)_3SiRRSi(OCH_3)_3\\ N\text{\textasciitilde}\text{\textasciitilde}N\\ (CH_3O)_3SiRRSi(OCH_3)_3\end{array}$$

D2000   GTMS   SED2000

where $R = CH_2CH(OH)\,CH_2O\,(CH_2)_3$

Evolution of silica within the organic matrix was followed by SAXS using a Kratky camera and a linear position-sensitive detector. The structure was described by the size of silica clusters and by fractal dimensions, D_m, characterizing compactness of the object. The fractal dimension was determined from the slope of the linear part of a double logarithmic SAXS plot of the scattered intensity I vs. scattering vector q ($=(4\pi/\lambda)\sin\theta$), where 2θ is the scattering angle.

DMA measurements were performed using a Rheometrics System Four apparatus. Complex shear modulus of rectangular specimens was determined by oscillatory shear deformation at a frequency of 1 Hz.

Results
Silica structure evolution

During simultaneous polymerization of organic epoxy-amine and silica-siloxane systems, the microphase separation takes place. Relative rates of structure growth resulting in gelation of the hybrid and of microphase separation determine the morphology of the microheterogeneous hybrid.

In the base-catalyzed <u>one-stage polymerization,</u> the scattered intensity of the SAXS profile increases gradually and smoothly through the gel point as silica clusters grow and a network is formed (see Fig. 1). The point of gelation was determined visually as a stop of flow of the reaction mixture. It was found that large overlapped clusters are formed from the very beginning of the reaction (8). Figure 2a shows that Guinier radius R_G is concentration-dependent and higher values of R_G were found in dilute solutions indicating an overlap of polysiloxane clusters in the bulk reaction mixture. Consequently, the determined value R_G corresponds to a spatial correlation length within the overlapped clusters where intermolecular interferences contribute to the scattering profile. During the reaction, the Guinier radius approaches the final value of the spatial correlation length in the gel, $R_G \sim 11$ nm. SAXS profiles of the samples in the pregel stage diluted with poly(oxypropylene)glycol PPG1200 (5 % solution) differ from those measured in the bulk (see Fig. 3). Contrary to the smooth intensity curves observed in the bulk samples, the "dilute" profiles are more complex. The shape of the curves may be explained by formation of small subunits as the inner structure of the clusters is screened in the bulk concentrated mixture. The size of these domains increases during polymerization as the deflection point on the curve shifts to lower q. Finally, the inhomogeneities interconnect and fill in the whole cluster and the break disappears. These "subunits" were interpreted as more branched and condensed domains formed in the polysiloxane chain due to a nonrandom branching under basic catalysis. In this case, more branched siloxane groups are the most reactive (6).

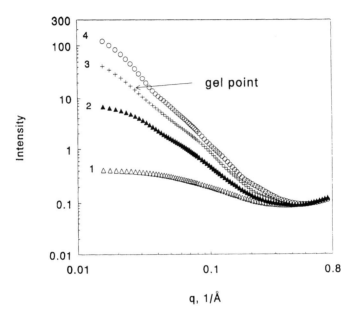

Figure 1. Evolution of the SAXS profiles during a one-stage polymerization of the hybrid DGEBA-D2000-TEOS. t = 1 min (1), 29 min (2), 81 min (3) - gel point, 263 min (4).

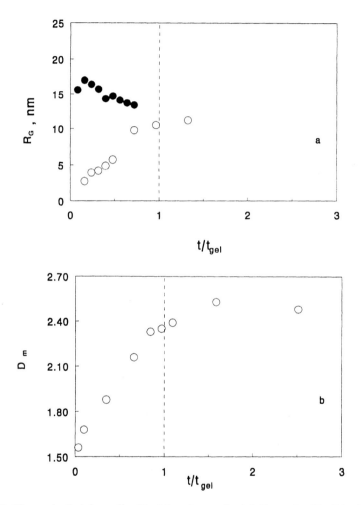

Figure 2. Change in Guinier radius R_G (a) and mass fractal dimension D_m (b) of fractal structures during a one-stage polymerization. ○ reaction mixture, ● 5 % solution in poly(propylene)glycol.

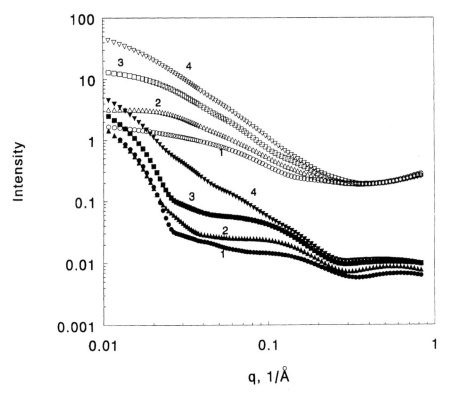

Figure 3. Evolution of the SAXS intensity during a one-stage polymerization of the DGEBA-D2000-TEOS hybrid. ○ reaction mixture, ● dilute solution; reaction time with respect to time of gelation t_{gel}; t/t_{gel} = 0.24 (1), 0.32 (2), 0.48 (3), 0.72 (4).

Therefore, the branched parts of the siloxane cluster grow preferentially. The increasing number and size of the branched domains during polymerization are reflected as a change in the inner structure and growth of the fractal dimension. The fractal dimension gradually the reaction mechanism generally accepted for reaction under base catalysis; the reaction limited monomer-cluster aggregation.

The polymerization is significantly accelerated in a two-stage procedure by acid prehydrolysis of TEOS. Fast hydrolysis takes place and small primary particles, 1.5-2 nm in size, are formed during the first reaction step in the acid medium (see Fig. 4). In the second stage, under base catalysis with D2000, the particles quickly aggregate to form loose clusters of a low fractal dimension, $D_m = 1.7$. The fractal dimension, i.e., the inner structure does not change during the reaction as shown in Fig. 4 (constant slope) in contrast to the one-stage procedure. The open structure of the formed clusters is explained by the diffusion-limited cluster-cluster aggregation reaction mechanism (6) of this very fast reaction. Gelation of silica occurs in 90 s at room temperature compared with 81 min in the case of the one-stage polymerization.

Quite different structure evolution was determined during formation of a network from endcapped prepolymer SED2000. Gelation is much slower and proceeds by formation of SiO_2 cluster crosslinks. In this case, the appearance of an interference maximum was detected in SAXS profiles (see Fig. 5) as a result of phase separation. The maximum indicates a regular arrangement of silica structures in the hybrid in contrast to the systems containing TEOS as a precursor of the silica phase. The position of the interference maximum (q_{max}) depends on the chain length of the endcapped prepolymer and corresponds to correlation length d ($=2\pi/q_{max}$) between crosslinks. No shift of the maximum position with conversion was observed.

Structure and morphology of the epoxy-silica hybrid

The final structure and morphology of hybrids cured at a high temperature were characterized using SAXS and electron microscopy (SEM).

The largest clusters arise in the systems prepared by one-stage polymerization, ET1. The structures are very compact with high fractal dimension, $Dm = 2.7$, as is apparent from the steep slope of the SAXS profile in Fig. 6. SEM shows large silica domains of the size ~100-300 nm (7). In simultaneous two-stage hybrids, ET-2, the two linear parts of the SAXS curves correspond to two length scales of large loose clusters with low fractal dimension composed of smaller and denser particles. According to SEM, the silica clusters are smaller compared with the one-stage case reaching the size 50-100 nm and seem to be co-continuously interconnected. Sequential polymerization with a preformed epoxy network, E1T2, leads to the finest silica structure. The silica domains are small, 10-20 nm in size, and loose, $D_m=1.9$-2.2, as a result of the acid catalysis of the sol-gel process and steric restrictions of the cluster growth within the epoxy network. Structure of networks from endcapped prepolymer SED2000 are characterized by regular arrangement with typical interference maximum.

Polymerization procedures lead to different structures, which is manifested also in mechanical properties of the hybrids.

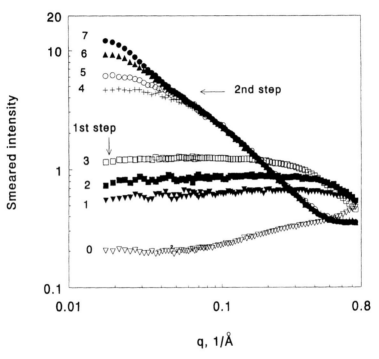

Figure 4. Evolution of the SAXS profiles during a two-stage polymerization of the hybrid DGEBA-D2000-TEOS. t = 0 (0), 1st step - 1 min (1), 5 min (2), 1h (3); 2nd step - 50s (4), 90s (5)- gel point, 4 min (6), 10 min (7).

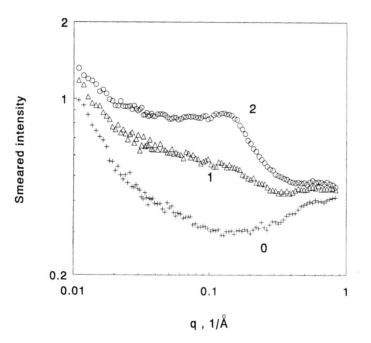

Figure 5. Evolution of the SAXS profiles during crosslinking of endcapped prepolymer SED 2000 by sol-gel process. t = 0 (0), 3h (1) - gel point, 20 h (2).

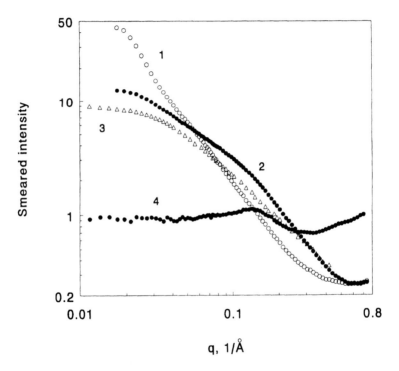

Figure 6. SAXS intensity profiles of the hybrid networks. 1 - ET1, 2 - ET2, 3 - E1T2, 4 - crosslinked SED2000.

Mechanical properties

A significant reinforcement of the rubbery epoxide network with the in-situ formed silica was found by DMA (see Fig. 7). The siloxane-silica domains are in the glassy state and serve as a hard filler with much higher modulus compared with the rubbery network. A relatively low silica fraction (<10 vol. %) leads to an increase in modulus by more than two orders of magnitudes. The modulus rises with increasing silica content depending strongly on the way of polymerization. A weak reinforcement was observed in the case of crosslinked SED2000 where the organic epoxy-amine junctions in the network were replaced by inorganic siloxane-silica clusters. The efficiency of the reinforcement increases in the series: crosslinked SED2000 < one-stage polymerization, ET1 < two-stage process with acid prehydrolysis of TEOS, ET2 < sequential polymerization with the preformed epoxy network, E1T2.

In addition to the increase in modulus, the hybrids show a decrease in the magnitude of the loss factor, tan δ peak at T = –30 °C corresponding to the glass transition temperature of the epoxy network. Moreover, a new peak of tan δ appears at higher temperature (see Fig. 8) which is attributed to the relaxation of the epoxy chains immobilized by interaction with hard silica domains. According to the Wilkes morphological model (10) of the O-I hybrids, our microphase-separated system consists of the organic rubbery epoxy phase, inorganic glassy silica phase and the interphase involving the bound immobilized epoxy chains. The part of interphase which is in glassy state contributes to the effective volume of the silica filler. Approximate fractions of free and bound epoxide chains were estimated from the lowering of the low-temperature tan δ peak reflecting the decrease in amount of free flexible chains (9). The fraction of the bound epoxides is a measure of the interphase epoxy-silica interaction. The modulus increases with enhancing interaction and the highest reinforcement was achieved in sequential interpenetrating networks E1-T2. For a given content of silica, the highest modulus, the most significant decrease in the low-temperature maximum of tan δ of free epoxide chains and the largest shift of the peak to higher temperature were observed in this case.

The analysis of DMA results shows that theoretical models of a composite with a hard filler dispersed in a soft matrix do not account for the observed increase in the modulus. The experimental moduli in Fig. 9 are much higher compared with the theory of the Kerner-Nielsen (11) model (curve 1) (eq.1).

$$G_C/G_M = (1+ABv_f)/(1-B\Psi v_f) \tag{1}$$
$$A = (7-5v_M)/(8-10\ v_M)$$
$$B = ((G_f/G_M)-1)/((G_f/G_M)-A)$$
$$\Psi = 1 + v_f(1-v_{max})/v^2_{max}$$

where G_C, G_M, G_f are moduli of the composite, matrix and filler, respectively, v_M is the Poisson constant of the matrix, v_f is volume fraction of the filler and v_{max} is maximum packing fraction of the filler.

The following input parameters were used for the Kerner-Nielsen model: v_M = 0.5 for rubbery matrix; v_{max} = 0.6 for random packing of spheres; modulus of the rubbery epoxy network matrix $G_M(=G_E) = 2.2 \times 10^6$ Pa; modulus of incompletely condensed siloxane-silica domains was taken from literature data (12) on xerogels prepared from TEOS, $G_f(=G_{Si}) = 4 \times 10^9$ Pa.

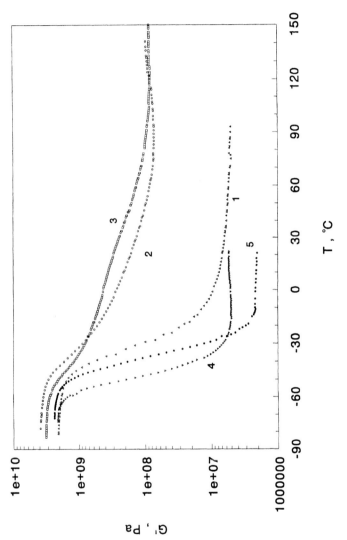

Figure 7. Dynamic shear modulus of the hybrid networks as a function of temperature; 5-13 vol.% SiO$_2$. 1 (△) ET1, 2 (○) ET2, 3 (□) E1T2, 4 (▽) crosslinked SED2000, 5 (●) DGEBA-D2000.

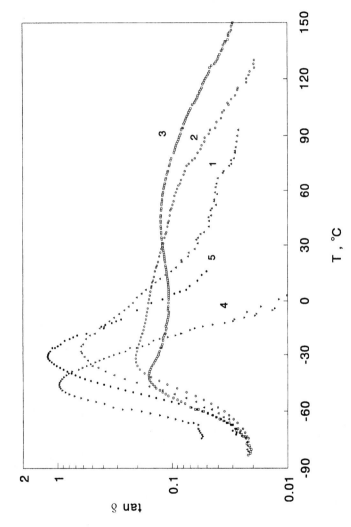

Figure 8. Loss factor tan δ as a function of temperature at frequency ω=1 Hz. 1 (△) ET1, 2 (○) ET2, 3 (□) E1T2, 4 (▽) crosslinked SED2000, 5 (●) DGEBA-D2000

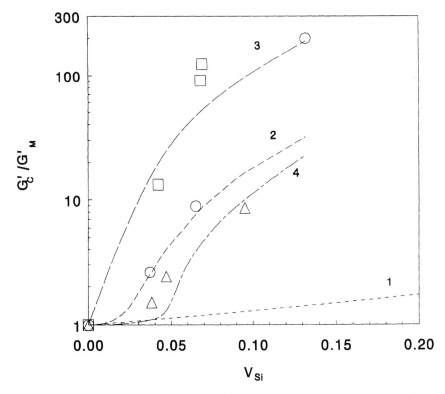

Figure 9. Relative modulus of the O-I hybrid as a function of the silica phase volume fraction. Curves - theoretical models, $G_{Si} = 4 \times 10^9$ Pa, $G_E = 2.2 \times 10^6$ Pa, $G_{Eg} = 2 \times 10^9$ Pa; 1 Kerner-Nielsen model (eq. 1): $v_M = 0.5$, $v_{max} = 0.6$, 2 Equivalent box model (eq. 2b), $K = 0$, $t = 2.0$, $v_{cr} = 0$, 3 EBM, $K = 10$, $t = 2.0$, $v_{cr} = 0$, 4 EBM, $K = 0$, $t = 2.0$, $v_{cr} = 0.02$. Experimental results: △ ET1, ○ ET2, □ E1T2, ● DGEBA-D2000

An agreement with experimental results was obtained by taking into account the increased effective fraction of the filler, v_{eff}, due to the glassy interphase of the bound epoxide layer and assuming a co-continuous morphology of the epoxy-silica hybrid network. Mechanical properties in dependence on the phase continuity are treated by parallel and series models for bicontinuous morphology and discontinuous phases, respectively. The equivalent box model (EBM) developed by Takayanagi (13) (eqs 2-5) and Davies model (14) (eq. 6) were used to compare the experimental data with the theory (9).

According to the EBM model, the modulus is determined mainly by parallel (continuous) phases of the system. In our case, the model involves three phases: silica phase (Si), glassy epoxide phase (Eg) and rubbery epoxide phase (Er)

$$G_C \cong (v_{Si})_p G_{Si} + (v_{Eg})_p G_{Eg} + (v_{Er})_p G_E \cong (v_{Si})_p G_{Si} + (v_{Eg})_p G_{Eg} \quad (2)$$
as $G_E \ll G_{Si}, G_{Eg}$

where $(v_i)_p$ is volume fraction of a component i in parallel phase; $G_{Eg}(= 2 \times 10^9$ Pa) is glassy modulus of the epoxide - used for the immobilized epoxide layer in glassy state.

Parameter K characterizing immobilization of epoxy chains was introduced in the calculation in order to differentiate the interphase interaction and thickness of the immobilized glassy layer in various hybrid systems: $K = v_{Eg}/v_{Si}$

$$G_C \cong (v_{Si})_p (G_{Si} + K G_{Eg}) \quad (2a)$$

The fraction of a phase combined in parallel can be calculated using the percolation theory (15)

$$(v_{Si})_p = [(v_{Si}-v_{Sicr})/(1- v_{Sicr})]^t \quad (3)$$

where v_{cr} is critical volume fraction for a phase continuity of a component, t is a universal critical exponent. $v_{Sicr} = 0$ was taken for permanent chemical epoxy-silica IPNs or $v_{Sicr} = 0.02$ for the most heterogeneous one-stage ET1 network.

$$G_C = (v_{Si})^t (G_{Si} + K G_{Eg}) \quad (2b)$$

The theoretical curves of the EBM model in Fig.9 were calculated using the critical exponent t = 2.0 (16) and various values of parameter K corresponding to systems with very strong epoxy-silica interaction (curve 3) and without interaction (curve 2). Lower continuity of the one-stage ET1 network was assumed by taking $v_{Sicr} = 0.02$ (curve 4).

Modulus of systems containing two continuous phases may also be predicted by the model of Davies (14). The theoretical curve in Fig.10 was calculated considering the effective volume fraction of the hard phase, v_{eff}, consisting of silica and glassy immobilized epoxide phases.

$$G_C^{1/5} = v_{Er} G_E^{1/5} + v_{eff} G_{Si}^{1/5} \quad (4)$$

where v_{Er} is volume fraction of unbound rubbery epoxide phase.

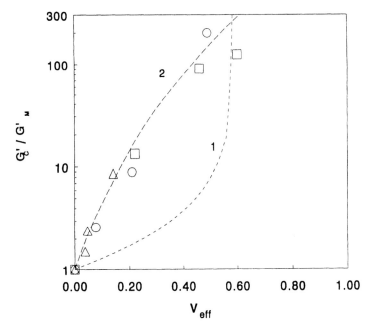

Figure 10. Relative modulus of the O-I hybrid as a function of the effective volume fraction of the hard phase, v_{eff}. Curves - theoretical models, $G_{Si} = 4 \times 10^9$ Pa, $G_E = 2.2 \times 10^6$ Pa, 1 Kerner-Nielsen model (eq. 1): $v_{max} = 0.6$, $v_M = 0.5$, 2 Davies model (eq. 4): $v_{eff} = v_{Si} + v_{\Sigma g}$. Experimental results: △ ET-1, ○ ET-2, □ E1-T2, ● DGEBA-D2000.

Discussion and conclusions

Structure and morphology of the epoxy-silica hybrid systems is determined by (a) reaction mechanism of the sol-gel process in formation of the silica structure, (b) phase separation during the polymerization, and (c) interaction between epoxy and silica phases. These factors are interrelated a close relationship exists as shown below. Also, there is a relationship between the structure and morphology and mechanical properties.

(a) The reaction mechanism of the sol-gel process may be controlled by the way of polymerization, catalytic conditions and reaction medium. The monomer-cluster and cluster-cluster types of aggregation lead to large compact silica structures in the case of base-catalyzed, one-stage procedure, ET1, and to smaller loose clusters in the acid-catalyzed, "sequential" hybrid, E1T2. The open structure of the silica clusters in the two-stage ET2 hybrid proves the crucial role of the early polymerization stage which was catalyzed by an acid while the second stage proceeded under basic catalysis.

(b) Phase separation and its relative rate with respect to the rate of polymerization govern the morphology of the "simultaneous" systems. The very fast polymerization and early gelation in the second step of the two-stage hybrid, ET2, are the reason for a lower extent of phase separation, which is quenched by gelation of silica. Therefore, silica domains are smaller than in the one-stage simultaneous polymerization.

(c) The polymerization of the organic and inorganic systems is independent because of different reaction mechanisms. However, grafting between the epoxy-amine and siloxane polymers occurs by condensation of C-OH from the epoxy system with SiOH of the polysiloxane. The grafting between the phases, affects morphology and properties of the hybrid. A fast and complete hydrolysis of TEOS under acid catalysis leads to a high concentration of silanol groups. This fact promotes grafting to the epoxy network and strengthens the interphase interaction. As a result, a better phase miscibility and formation of a more homogeneous hybrid with smaller silica domains could be expected. This is the case of the acid-catalyzed sequential hybrid E1T2 and of the first stage in the ET2 system. Both these hybrids show a very strong interphase interaction and high modulus, E1T2 being the most reinforced system. On the contrary, in the base-catalyzed one-stage polymerization, the hydrolysis is relatively slow compared with the condensation resulting in a low content of SiOH, small extent of grafting, weak O-I interaction and low modulus.

(d) Figures 9 and 10 show that the theoretical bicontinuous models describe the mechanical behaviour of the reinforced systems both with weak and strong interphase interaction assuming an enhanced effective volume fraction of the hard phase. The analysis of mechanical properties proves that microphase-separated epoxy-silica hybrid forms co-continuous IPN. Grafting of silica-siloxane structures to the epoxide network results in formation of an immobilized epoxide layer with increased T_g. Sufficient shift of T_g then brings about vitrification of a part of this organic layer which contributes to the reinforcement. The results prove the continuity of the hard phase and not of the silica structure. Therefore, one can imagine an O-I network involving silica domains connected by glassy epoxide regions to form a continuous hard phase.

Acknowledgments

The author is greatly indebted to the Grant Agency of the Czech Republic for financial support of this work (project No.203/98/0884).

References
1. Mark, J.E.; Jiang, C.; Tang, M.Y. *Macromolecules* **1984**, 17, 2613.
2. Wilkes, G.L.; Orler, B.; Huang, H.H. *Polym. Prepr* **1985**, 26, 300.
3. Landry, C.J.T.; Coltrain, B.K.; Brady, B.K. *Polymer* **1992**, 33, 1486.
4. Lam, T.M.; Pascault, J.P. *Trends Polym. Sci.* **1993**, 3, 317.
5. Ikeda,Y.; Hashim, A.S.; Kohjiya, S. *Bull. Inst. Chem. Res.*, Kyoto Univ. **1995**, 72, 406.
6. Brinker, C.J.; Scherer, G.W. in *Sol-Gel Science*, Academic Press, San Diego, 1990.
7. Matějka, L.; Dušek, K.; Pleštil, J.; Kríz, J.; Lednický, F. *Polymer* **1998**, 40, 171.
8. Matějka, L.; Pleštil, J.; Dušek, K. *J. Non-Cryst. Solids* **1998**, 226, 114.
9. Matějka, L.; Dukh, O.; Kolarík, *Polymer,* in press.
10. Wilkes, G.L.; Brennan, A.B.; Huang, H.; Rodrigues, D.; Wang, B. *Mater. Res. Soc. Symp. Proc.* **1990**,171, 15.
11. Nielsen, L.E. in *Mechanical Properties of Polymers and Composites*. Marcel Dekker, Inc, New York, 1974.
12. Murtagh, M.J.; Graham, E.K.; Pantano, C.G. *J. Am. Ceram. Soc.* **1986**, 69, 775.
13. Takayanagi, M. Mem. Fac. Eng. Kyushu Univ. **1963**, 23, 41.
14. Davies, W.E.A. *J. Phys.* **1971**, D 4, 1176.
15. DeGennes, P.G. *J. Phys. Lett.* (Paris) **1976**, 37, L1.
16. Stauffer, D. in *Introduction to Percolation Theory*, Taylor and Francis, Philadelphia 1985.

Chapter 34

From a Hyperbranched Polyethoxysiloxane Toward Molecular Forms of Silica: A Polymer-Based Approach to the Monitoring of Silica Properties

V. V. Kazakova[1], E. A. Rebrov[1], V. B. Myakushev[1], T. V. Strelkova[1], A. N. Ozerin[1], L. A. Ozerina[1], T. B. Chenskaya[1], S. S. Sheiko[2], E. Yu. Sharipov[1], and A. M. Muzafarov[1]

[1]Institute of Synthetic Polymeric Materials, Russian Academy of Sciences, 117393 Profsoyuznaya Ul., 70, Moscow, Russia
[2]Organische Chemie-III, Universität Ulm, Albert-Einstein-Allee 11, Postfach 4066, D-89069, Ulm, Germany

Synthesis of a new modification of silica soluble in THF is described. At the first synthetic step, a hyperbranched polyethoxysiloxane (HBPES) is synthesized by heterofunctional condensation using triethoxysilanol previously generated in reaction mixture by neutralization of correspondent sodium salt with acetic acid. At this step, the process was monitored by IR spectroscopy, SEC, and ^{29}Si NMR spectroscopy. At the second step, hydrolysis and intramolecular condensation involving silanol groups is carried out to yield silica sol macromolecules. A SAXS method was used to determine the size and fractal coefficient of trimethylsilated derivatives and silica sols obtained. An atomic-force microscopy imaging of silica sol supported on a mica substrate showed the silica sol particles to be predominantly spherical in shape. Prospects for theoretical, experimental and practical applications of silica sols are discussed.

Complexity of the silica chemistry stems from the high functionality of starting reagents in the reaction mixture involving a large variety of chemical processes: hydrolysis, homo- and heterofunctional condensation, cyclization, chemical aggregation as well as a purely physical aggregation of the particles formed. The process can be further complicated by the formation, at an early step, of a solid-state phase, with the resulting loss of solubility capable of changing the course of chemical reactions. Traditionally, the control over chemical reactions for preparation of silica with tailored properties is exerted empirically by choosing appropriate conditions. On the whole, this approach has led to remarkable results. Researchers have gained expertise in preparing a wide range of silicas – from ultradisperse particles (*1*) to mesoporous silica with controllable pore structure (*2-3*).

The polymer chemistry and silica chemistry, at first sight so basically unlike in many aspects, in recent decades have been sharing spheres of common interests in an ever-increasing number. One reason for this is that various types of silica are used traditionally as fillers for polymer composites. This is to say that commercial properties of a range of polymers depend on the structural properties of silica, its specific surface, concentration of functional groups, and size, shape and degree of ordering of silica particles. Interpenetration of these two chemical disciplines was prompted by the emergence of sol -gel processes and associated therewith interest in hybrid composite materials (4-5). However, the formative processes of silica particles continued, as before, to be monitored on an empirical basis. Main factors involved in the control over the reactive medium were temperature, concentration, proportion of components, and the type of the polymer template used; in fact, the only response to the effect of these factors was a specified property (or a set of them) of the materials formed. Even the latest achievements in the preparation of controllable mesoporous silicas (3) have produced no change in the actual state-of-the-art. The use of surface-active substances (SASs) has brought about merely a change in the shape and size of the template on which hydrolysis and condensation of orthosilisic acid esters were carried out. Despite the significant results obtained in the field, the experimental techniques were in fact indiscriminate in their effect on the structure of silica species formed and provided no means to monitor the silica structure at a molecular level.

An alternative approach to the synthesis of silicas with tailored properties is based on a strategy used in the synthesis of hyperbranched polymers. Extension of this strategy to the chemistry of silica precursors will make it possible to obtain end products with the desired properties, which can be achieved by the structural monitoring of silica macromolecules. The progress that has been accomplished over the last decade in the chemistry of dendrimers, or cascade polymers (6-8), and their irregular analogs, hyperbranched polymers (9-12), in particular, organosilicon polymers (13-16) provides reason to believe in such a possibility.

In the present work, the approaches commonly employed in the molecular structural monitoring of hyperbranched polymeric systems have been extended to polyethoxysiloxanes and products of their hydrolytic polycondensation.

Experimental

1. Materials. Methods. Organic solvents: toluene, hexane, tetrahydrofuran, ethanol (analytical-reagent grade) were dehydrated by boiling and distilling over calcium hydroxide. The gaseous NH_3 was dried by passing through an alkali-packed column. Tetraethoxysilane and hexamethyldisilazan(high-purity grade) were used as-received. Nuclear magnetic resonance (NMR) spectra were measured on a "Bruker WP-200-SY" specrometer, with a working frequency of 200.13 MHz for ^1H nuclei. ^{29}Si-NMR-spectra were taken using a proton pulse suppression technique, with no account for the Overhauser effect, with a delay of 30 s. какой Д-растворитель использовался^1H NMR spectra of the products studied were measured using 20% solutions in CCl_4, and ^{29}Si NMR-spectra, using 50% solutions in THF or toluene. Tetramethylsilane was used as the reference.

Gas-liquid chromatographic (GLC) analysis was done on a 3700 chromatograph (Russia). A thermal conductivity cell was used for detecting signals; the carrier gas

was helium; columns 2 m ¥ 3 mm were used; the stationary phase was SE-30 (5%) supported on a "Chromaton-H-AW" sorbent. Chromatograms were processed using a "CI-100A" computer-controlled integrator (Czechia).
Analysis of polyethoxysiloxanes and silica sols by the gel-permeation chromatography (GPC) method was done on a GPC chromatograph (Czechia). The detector was a refractometer; a 4×250 mm column was used; the sorbent was Silasorb-600 (7.5 μm) treated with hexamethyldisilazan; the eluents were toluene and THF.
Infrared (IR) absorption spectra of the products were measured using a "Bruker IFS-110" IR spectrometer.

SAXS - measurements. SAXS curves were measured using a home-made KRM-1 diffractometer with a slit scheme for the primary beam collimation (CuKa-line, Ni filter, scintillation detector). The scattering coordinate was measured in terms of the scattering vector modulus $s = 4\pi \sin\theta/\lambda$, where θ is the difraction angle, and $\lambda = 0.1542$ nm is the radiation wavelength. The scattering intensity was measured in the range of $s = 0.07$ to 4.26 nm^{-1}. The experimental data were processed as described in *(17-18)*.
Structural studies of silica sols films were carried out using atomic force microscopy (AFM) method. The films, were prepared by casting solutions on a rapidly spinned substrate - mica . The film surface was scanned at room temperature and atmospheric pressure using a Nanoscope III atomic-force microscope (Digital Instruments, California). Silicon tips with a constant of elasticity of 50 N/m and a tip radius smaller that 10 nm, vibrated at a resonance frequency of about 320 kHz, were used.

2. Synthesis of a hyperbranched polyethoxysiloxane. Using a method described in (in print.), a specimen of hyperbranched polyethoxysiloxane was prepared: the solution of 25.4 g (0.125 mole) of sodiumoxytriethoxysilane (preliminary prepared by interaction of 5.0 g of sodium hydroxide and 26.1 g of tetraethoxysilane) was added dropwise to the solution of 7.9g (0.132 mole) of acetic acid in 160 ml of dry toluene with stirring at 0 °C. Reaction mixture was stirred for another two hours at room temperature. The precipitated CH$_3$COONa was filtered and washed with dry toluene on the filter. The solvent was evacuated at room temperature. 21.7 (82%) of triethoxysilanol was prepared and used for the hyperbranched polyethoxysiloxane preparation. Probe of prepared triethoxysilanol was dissolved in dry THF and analized by ^{29}Si NMR (THF,39.76 MHz): $\delta = -78.49$ (s,1Si). 17.2 g(1.01 mole) of dry liquid ammonia was added to solution preliminary cooled at -30°C of 20.0g (0,111 mole) of triethoxysilanol in 15 ml of absolute ethanol. The temperature of reaction mixture was slowly increased to room condition. After three hours of stirring volatile products were evacuated at 133.3 Pa) at 25°C, 14.0 g (94%) of transparent colorless oil like liquid was obtained: n$^{20}_D$ 1.4116, d$^{25}_4$ 1.2030. Found, %: Si 23.50, C 30.41, H 6.42. ^{29}Si NMR (toluene, 39.76 MHz): δ -88.42 . (OSi(OEt)$_3$), δ -96.12. (O$_2$Si(OEt)$_2$), δ -103.02 (O$_3$Si(OEt)), δ -105 - 106 (O$_4$Si).
Trimethylsilylated derivatives were prepared by hyperbranched polyethoxysiloxane probe treatment with excess of trimethyltrifluoroacetoxysilane. Mixture was reflaxed

10 hours, then it was cooled and volatile products was removed in vacuum. Rest was dissolved in toluene with addition of a few drops of hexamethyldisilazan and reflaxed another 5 hours. Solution was cooled, volatile products evacuated, resin like substance was dissolved in toluene or THF and used for analysis of molecular mass distribution.

3. Hydrolysis of polyethoxysiloxane. A mixture of 10 g (0.555 mole) of water and 2 drops of concentrated HCl was added dropwise to a mixture of 5 g (0.037 mole) of polyethoxysiloxane and 557 ml of THF with stirring at 20 °C. The course of the reaction was monitored by sampling portions of the mixture for analysis. The solvent was removed by evaporation in vacuum, and the mixture was inspected over time by IR spectroscopy until the absorption bands $n = 2950$ cm^{-1} (corresponding to the stretching C–H mode of the ethoxy group) and $d = 1480$ cm^{-1} (bending C–H mode of the same group) disappeared completely. The conversion of the ethoxy groups having been completed, the reaction solution was colourless and transparent.

4. Condensation of the polyethoxysiloxane hydrolysate. 1% solution of the polyethoxysiloxane hydrolysate in THF was boiled at 65 °C for 8 hours. To remove ethanol and water (formed as side products of condensation), 200 ml of an aseotropic mixture THF + ethanol + water was removed by distillation in three successive runs (each time, a fresh portion of THF was added). Next, 386 ml of solvent was driven off from the mixture to reach a 3% concentration of the reaction product in THF. The IR spectrum of the product exhibited an absorption band at $n = 3400$ cm^{-1} corresponding to the O–H stretching mode of the hydroxyl group. Elemental analysis of a 9.3% solution of polyethoxysiloxane hydrolysate in THF gives (%): Si 2.57; OH 0.94.

Results and Discussion

Synthesis of the hyperbranched polyethoxysiloxane. Viewed from a standpoint of the chemistry of dendrimers and hyperbranched polymers, triethoxysilanol (regarded in (20) as a primary product of hydrolysis) is no more than a reactant AB$_3$ according to the Flory condition. This signifies that, by generating this product under the conditions of a heterofunctional condensation, one can direct the reaction such that a hyperbranched polyethoxysiloxane is formed, that is, to make the process structurally selective. It is known that, in hyperbranched polymers, cyclization is a minor contributor to the molecular structuring because of the paucity of A-type functionalities. In other words, with allowance made for structural imperfection of the hyperbranched polymer and for the fact that proportions of the dentritic, linear and end chains depend on a number of factors, it is possible in principle to obtain an end product with desired properties by monitoring structure, rather than process parameters, of the polymer formed.

It was shown in (14) that tetraethoxysilane derivatives can be used as building blocks in the buildup of a hyperbranched structure; successive steps in the synthesis of such a polymer have been discussed in (19). The main steps of this process are shown in Scheme 1. It was also discussed that scheme 1 illustrate an idealized development of the processes while real process include a number of side reactions. At the same time the obtained product characterization gave support to consider this scheme as a main.

Scheme 1

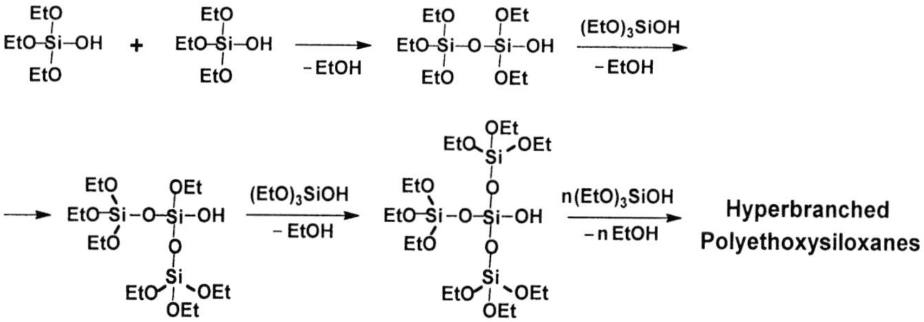

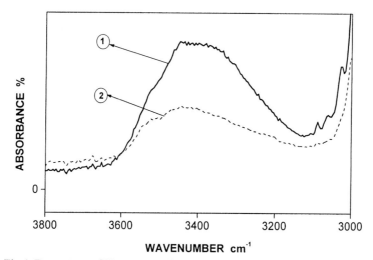

The process in question can be monitored conveniently by following changes in the intensity of the absorption bands of the hydroxyl stretching mode in the IR spectra.

Fig.1 Fragments of IR-spectra of triethoxysilanol (1) and product of its polycondensation (2).

IR absorption spectra of the system under study are shown in Fig. 1: curve 1 refers to the beginning of the process, and curve 2, to its termination. As the condensation proceeds and the ethanol formed is removed from the system, the intensity of IR absorption in this region decreases virtually to zero. The hyperbranched structure of the polymer synthesized has been proven by the ^{29}Si NMR spectra (Fig. 2), the molecular-weight range as determined by the GPC method and the viscosity characteristic of spherically-shaped rigid particles. The GPC curves of hyperbranched polyethoxysiloxane shown at figure 3 (1), together with polystyrene standard (2). Although linear standards did not suit for hyperbranched polymers molecular mass determination this comparison shows that molecular mass of examined polymer at list not smaller of standard value.

The signals in the ^{29}Si-NMR spectra were assigned with reference to the literature data and the spectra of a number of model compounds (21) (m -88 ppm - monosiloxy-substituted units; m - -95 ppm -disiloxy-substituted units; m at -100-105 ppm tri- and tetrasiloxy-substituted units). The ^{29}Si NMR spectral data (Fig. 2) make it possible to determine the proportions of various structural units as constituent parts of the molecule of the end product: $\{(EtO)_3SiO_{0.5}\}_a\{(EtO)_2SiO\}_b\{(EtO)SiO_{1.5}\}_c\{SiO\}_d$, where a = 1.00, b = 1.65, c+d = 2.26}. The last formula shows that the share of dendritic units (for the system in question, tri- and tetrasiloxy substituted silicon atoms) is quite high, which conforms with the high degree of branching of the polymer.

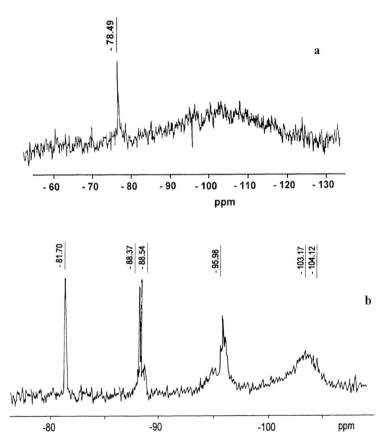

Fig.2 ^{29}Si NMR spectra of triethoxysilanol in THF solution (a), and hyperbranched polyethoxysiloxane (b)

Thus, the hyperbranched polyethoxysiloxane, characterized as a realistic polymeric species, is a basic unit ideally suited for obtaining new molecular forms of silica. Here the molecular forms of silica are understood to mean silica species which are particles of definite size and which may be regarded as individual macromolecules with a specified molecular structure amenable to characterization by instrumental methods of analysis.

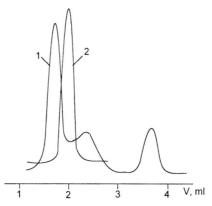

Fig.3 SEC elugrams of trimethylsilylated hyperbranched polyethoxysiloxane (1) and polystyrene MM - 60 000 standart (2)

Synthesis of molecular silica sols. The very fact of using polyethoxysiloxanes for preparation of silica is not by any means a revelation – this technique has been known since long and described in the literature (*22*); however, two points should be emphasized here. For one thing, we have described the starting polyethoxysiloxane as a polymeric entity with defined structure and defined molecular parameters, that is, with characteristics that are seldom reported in the literature for such compounds. For another, it is suggested implicitly that we must change over from polyethoxysiloxane to silica retaining the molecular structure of the former during the course of this polymer-to-analog transition.

In Scheme 2, exemplified by a fragment of the molecular structure, hydrolysis and the subsequent intramolecular cyclization are shown. The only alternative to the last process, meaning abnormal high functionality of polymer under study, is crosslinking. And no crosslinked polymer were detected during series of the experiments.

Scheme 3 reflects the behavior of a molecular particle on its conversion from polyethoxysiloxane to silica. Feasibility of the polymer-to-analog transformation involving a hyperbranched polyethoxysiloxane was demonstrated by us (the first to our knowledge) during the course of a reaction of blocking of functional groups (scheme 4) using trimethyltrifluoroacetoxysilane as blocking agent(*14*). It was established that trimethyltrifluoroacetoxysilane which exerted the least perturbing effect on the molecular structure and simultaneously provided for a sufficient degree of protection was an exceptionally suitable agent for that purpose. The degree of

Scheme 2

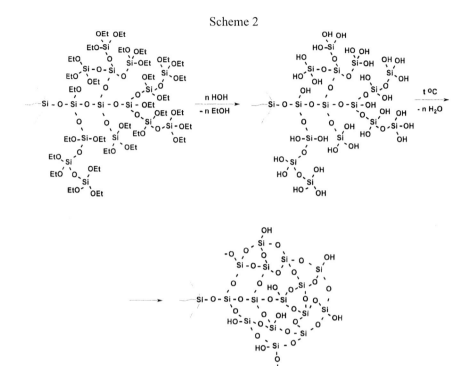

conversion could be monitored with ease and a high accuracy by measuring the ratio of the integrated intensities of the proton signals from trimethylsilyl and ethoxy groups in the ^{1}H NMR spectra; however, the question as to what an extent the molecular skeleton was affected by the process could be answered only on a semiquantitative level, from a comparison attesting to the fact that the macromolecule retained its size unchanged before and after the blocking by the SEC method.

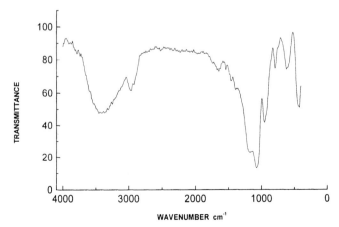

Fig.4 IR-spectrum of silica sol dried in vacuum

The same can be said about the hydrolysis and intramolecular condensation of the hyperbranched polyethoxysiloxane. On the one hand, the completeness of hydrolysis is supported by the IR spectroscopic data (Fig. 4). It is seen in Fig. 4 that virtually no absorption is observed in the region of 3000 cm^{-1} characteristic of the stretching C–H modes whereas a strong absorption is observed in the region of 3700 – 3200 cm^{-1} attesting to the presence of hydroxyl groups.

On the other hand, very little can be said about the condensation processes that may proceed concurrently with the hydrolysis process. Analogously, little can be said about the change which the structure of silica sols could have sustained after prolonged boiling in a dilute solution.

The stabilization of silica sols that was carried out through eliminating numerous hydroxyl groups formed by hydrolysis was expected to lead to a substantial intramolecular condensation with a very little involvement (if any) of intermolecular reactions. Was it really the case for the actual situation?

Scheme 3

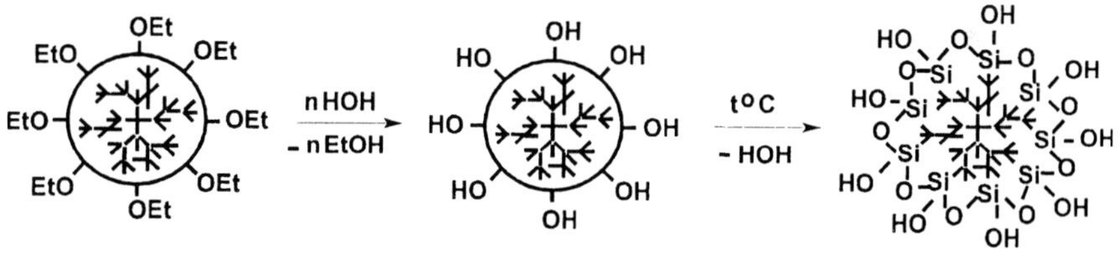

Scheme 4

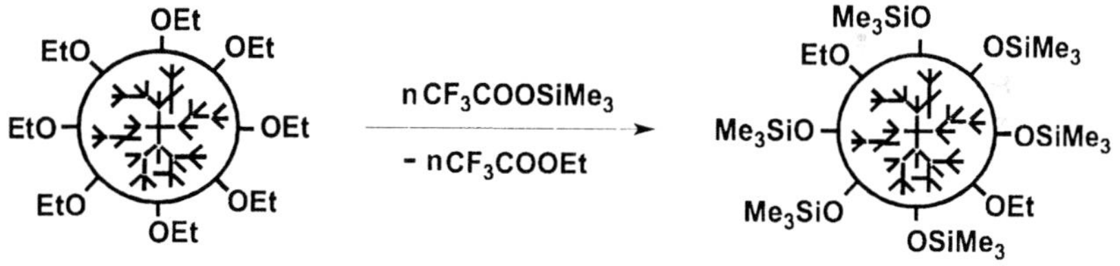

SAXS measurements. A whole range of problems relevant to the systems of interest could be solved by use of the small-angle X-ray scattering method. Dilute solutions of the hyperbranched polyethoxysiloxane, its trimethylsilyl derivative (a product of its blocking) and a silica sol (a product of hydrolytic condensation of the starting hyperbranched polyethoxysiloxane) were studied under conditions as specified above. The relevant results are presented in Fig. 5 and 6.

The scattered data were interpreted in terms of the scattering from a dilute system composed of macromolecular formations – particles with a characteric size of the order of several nanometers. The radius of gyration, found to be $R_g = 2.5 - 3.0$ nm, and the fractal dimensions of scattering particles, $d_f = 1.6 - 1.8$, calculated using data of scattering curves (Fig.5) were identical for all specimens studied, that is, the characteristic dimensions and the density of packing of the molecular structure (fractal dimension) showed no significant change on transition from hyperbranched polyethoxysiloxane to its protected analog or silica sol.

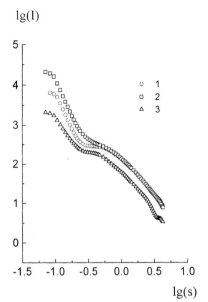

Fig. 5. Log-log plot of the experimental SAXS curves of hyperbranched polyethoxysiloxane (1); Trimethyl-silyl-terminated of (HBPES) (2); silica sol (3). The set of SAXS curves are normalized at $s=1 nm^{-1}$.

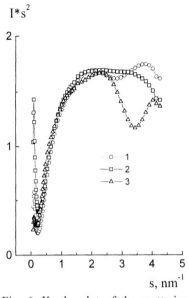

Fig. 6. Kratky-plot of the scattering curves of hyperbranched polyethoxysiloxane (1); Trimethyl-silyl-terminated of (HBPES) (2); silica sol (3). The set of SAXS curves are vertically displaced for clarity.

Thus, the SAXS data for hyperbranched polyethoxysiloxane and its derivatives have provided evidence for a genetic link between the starting hyperbranched polyethoxysiloxane and the products originated from its protection and hydrolytic condensation. This result is important for two reasons: first, it shows that one can use a protected derivative in the analysis of molecular-weight characteristic and, second, that virtually no processes of molecular aggregation are involved in the hydrolytic condensation. Finally, another result of importance is to be mentioned: one can evaluate the volume fraction of particles of larger size which are present in the system along with the small-sized particles. This is done by measuring the scattered power from scattered intensity curves plotted in the Kratky coordinates $I^*s^2 - s$ (Fig. 6). As found, the volume fraction of large-sized particles with $Rg > 20$ nm does not exceed 5%. The very fact that the formation of such particles could be established in the course of the experiment argues for the potential utility of the method employed. One will infer from this evidence, on the one hand, that polyethoxysiloxanes of larger size can be obtained in principle and, on the other hand, that no chemical transformations involving an intermolecular aggregation take place under the actual conditions, considering that the fraction of large-sized particles is retained at a constant low level throughout, starting from the primary hyperbranched polyethoxysiloxane and ending in silica sol (Fig. 6, curves 1 – 3).

AFM measurements. Now we shall dwell in some detail on the silica sols with a view to form an idea what in fact they may be. For one thing, they represent a purely inorganic system soluble in a dehydrated organic solvent. The nature of this solubility is readily enough conceived of: it bears on the property of the numerous hydroxyl groups to be solvated by the THF. The solvating solvent having been removed or replaced by a non-solvating one (hexane, toluene), the silica sol converts immediately (and, to be noted, irreversibly) to a silica gel. At the same time, the silica sol is quite stable in the state of solvation, and its shelf life can last through years with no visible change. From the data of functional and elemental analysis, the silica sol is represented by an empirical formula $(Si_2O_{3.5}OH)_n$; 90% of its constituent hydroxyl groups are accessible to substitution. Attempts were made to visualize silica sol molecules. To that end, silica sol was cast from a very dilute solution onto the fresh surface of mica, and the substrates thus prepared were examined by atomic force microscopy. In the photomicrograph in Fig. 7, spherical features, adsorbed on the surface of mica, are clearly visible.

Separate silica sol macromolecules are aggregated into irregularly shaped particles, which is not quite unexpected considering that the silica sol particles and highly functionalized and their hydroxyl groups are capable of forming hydrogen bonds. The adsorbed silica sol particles are 6 to 10 nm in size, which is in agreement with data obtained from SAXS experiments.

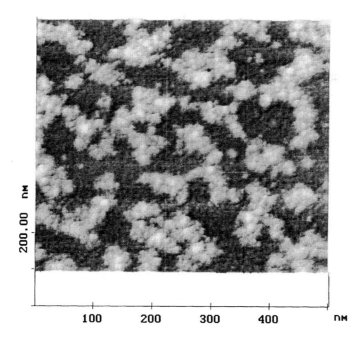

Fig.7 Scanning force photomicrograph of silica sol on mica.

As is seen, the AFM method allows one to get a glimpse into the shape and size of silica sol particles as well as to gain an understanding of the high surface activity inherent in these molecular formations.

Conclusion

The chemical transformations studied by a variety of instrumental techniques and the results obtained provide the grounds to believe that silica sols are silicon species possessing an individual molecular structure, unlike the majority of known silicas which are supramolecular formations. The individualization of these species was rendered possible owing to the fact that among various processes involved in the synthesis, two main processes – formation of a molecular structure and intramolecular cyclization, could be identified and separated in space and time. This could be achieved owing to the use of general synthetic approaches that were developed for hyperbranched systems, and to the choice of conditions specific of the given case; ultimately, the two aforementioned processes could be conducted avoiding aggregation or supramolecular organization of the molecular particles formed. In this context, several directions open up where the use of the new silica form holds great promise.

For one thing, the use of silica sols exhibiting uncommon molecular properties can lead to production of new types of silicas owing to the possibility to monitor the size of silica sol particles and their aggregative behaviour. Polymer-to-analog conversions involving alkoxides of various metals hold promise for preparing new compounds and materials. Molecular silica-based particles, soluble in anhydrous and hydroxyl-free organic solvents, are exceptionally convenient material for chemical transformations and, properly modified, can be used for preparation of nanometer-scale aggregates. They can be effectively used in homogeneous immobilization of various organometallic compounds under anhydrous conditions. This field of research in the chemistry of silica sols augurs well for the development of novel catalytic systems. Finally, the new silica sols whose aggregative properties are amenable to easy control can be used, as components of polymeric templates, for synthesis of molecular particles with needed dimensions and functionalities.

It stands to reason that, in order to implement the potentiality of silica sols in full measure, much work needs to be done at all stages of their preparation, especially in what concerns precise control over the dimensions of starting hyperbranched polyethoxysilanes and the effect of hydrolysis conditions on the parameters of target silica sols. It is felt, however, that our tentative steps in this direction have met with some success.

References

1. Stober, W.; Fink, A.; Bohn, E. *J. Colloid and Interface Sci.* **1968**, *26*, 62.
2. Tanev, P.T.; Pinnavaia, T.J. *Chem. Mater.* **1996**, *8 (8)*, 2068.
3. Kresge, C.T.; Leonowicz, M.E.; Rorh, W.J.; Vartuli, J.C.; Beck, J.S. *Nature.* **1992**, *359*, 710.

4. Wen, J.; Wilkes, G.L. *Chem. Mater.* **1996,** *8 (8),* 1668.
5. Brinker, C.J.; Scherrer, G.W. *Sol-Gel Science, the Phisics and Chemistry of Sol-Gel P rocessing;* Publisher: Academic Press; San Diego, CA, 1990.
6. Tomalia, D.A.; Durst, H.D. *Top. Curr. Chem.* **1993,** 165,193.
7. Newkome, G.R. *Advances in Dendritic Macromolecules.Greenwich;* Publisher: JAI Press, 1996. V.1-3.
8. Hawker,C.J.; Frechet,M.J. *J.Chem.Soc.Chem.Commun.* **1990.**1010.
9. Kim, Y.H.; Webster, O.W. *Polym. Prepr.* **1988,** *29(2),* 310.
10. Voit, B. I. *Acta Polymer.* **1995,** 46, 87.
11. Bochkarev, N. and Katkova, M.A. *Russ. Chem. Rev.* **1995.** 64, 1035.
12. Malmström E. and Hult A. *J. Macromol. Sci. - Rev. Macromol. Chem. Phys..* **1997,** C37, 555.
13. Muzafarov, A.M.; Rebrov, E.A.;.Papkov , V.S. *Russian Chemical Reviews.* **1991,** *60(7),* 1596.
14. Kazakova, V.V.; Myakuchev, V.D.; Strelkova, T.V.; Gvazava ,.N.G.; Muzafarov, A.M. *Doklady RAS.* **1996,** *349(4),* 486.
15. Mathias, L. J. and Carothers, T. W. *J. Am. Chem. Soc.* 113, 4043 (1991).
16. Rebrov, E.A.; Muzafarov, A.M.; Papkov, V.S.; Zhdanov, A.A. *Dokl. Akad. Nauk SSSR..* **1989,** *309(2),* 367.
17. Feigin, L.A.; Svergun, D.I. *Structure Analysis by Small-Angle X-ray and Neutron Scattering;* Publisher: Plenum Press: New York, 1987.
18. Martin, J.E.; Hurd, A.J. *J.Appl.Cryst.* **1987,** 20, 61.
19. Kazakova, V.V.; Myakuchev, V.D.; Strelkova, T.V.; Muzafarov, A.M. Polymer Science. **1999,** *41 (3),* 283.
20. Blaaderen, A.; Vrij, A. *Colloid and Interface Sci.* **1993,** *156,* 1.
21. Liepins, E.; Zismane, I; Lukevics, E. *J. Organomet. Chem.* **1986,** *306,* 167.
22. Iler, R.K. T*he Chemistry of Silica: Solubility, Polymerization, Colloid and Surface Properties and Biochemistry of Silica;* Publisher: Mir: Moscow, Russia, 1982. Vol 1, 244.

Chapter 35

Synthesis, Characterization, and Evaluation of Siloxane-Containing Modifiers for Photocurable Epoxy Coating Formulations

Mark D. Soucek, Shaobing Wu, and Srinivasan Chakrapani

Department of Polymers and Coatings,
North Dakota State University, Fargo, ND 58105

Reactive modifiers for photocurable epoxide coatings were synthesized by reacting caprolactone polyols with tetraethyl orthosilicate (TEOS). The structures of the TEOS functionalized polyols were characterized using IR, ^{1}H-NMR, and ESI-FTMS spectroscopy. The resulting siloxane functionalized polyols were used as reactive modifiers for UV-curable cycloaliphatic diepoxide coatings. The crosslinking reactions were monitored using IR and ^{29}Si-NMR spectroscopy. Based on the curing behavior and spectroscopic data, possible crosslinking reactions were postulated. The coatings were evaluated with respect to curing rate, gel content, and adhesion to aluminum substrate. Tensile properties of the cured films were measured. The modified polyols functioned effectively as adhesion promoters without affecting the coating properties of polyol-epoxide coatings, when used as partial replacements of the polyols at the level of 10 wt % of the parent polyol.

Environmental regulations have significantly impacted coatings research. The current emphasis is on low volatile organic compound (VOC) coatings formulations. Photocuring offers the advantages of rapid cure, low VOC, energy efficiency, low capital investment and space requirement. There are two principal modes of UV coatings, free radical, and cationic. Free radical polymerizations are inhibited by atmospheric oxygen [1]. However, photoinitiated cationic polymerizations are not affected by ambient oxygen [2]. In addition, there are toxicity and sensitivity problems with monomers used typically in photocurable free radical coatings. The monomers used in photocurable cationic coatings, epoxies and vinyl ethers, are typically environmentally benign. These advantages favor photocurable cationic coatings.

Epoxide monomers offer a wider latitude in formulations that enable tailor coating properties for specific requirements. There are two classes of epoxides which are used as monomers in UV-curable coatings, Bisphenol-A (BPA) epoxides, and cycloaliphatic epoxides. The cycloaliphatic epoxides react more readily under acid condition, and therefore are preferred. Once irradiated, the super acid generated by the photoinitiator rapidly homopolymerizes the cycloaliphatic epoxides. Unfortunately, the resultant homopolymer is brittle and has low impact resistance. As a consequence, flexible polyols are added into the formulation to improve the toughness of the coating.

Taking into account reports cited above, we formulated a strategy to incorporate flexibile units which also function as adhesion promoters. The flexible units or "soft segments" were modified with siloxane groups affording a higher functionality availible for additional crosslinking reactions. The siloxane groups were used to endcap caprolactone polyols. These siloxane modified polyols were isolated, spectroscopically characterized, and formulated into coatings. The gel content, adhesion and mechanical properties of the coatings were evaluated.

Experimental

Materials. Caprolactone polyols (Tone 0201: $\overline{M_n}$ = 530, and Tone 0305: $\overline{M_n}$ = 540), cycloaliphatic diepoxide (UVR-6105), triarylsulfonium hexafluoroantimonates (UVI-6974) were provided by Union Carbide Corporation. Surfactant L-7604 was procured from Witco Corporation. Caprolactone polyols were dried and stored with molecular sieves (4Å). Tetraethyl orthosilicate (TEOS, 98 %) and xylene (99 %) were purchased from Aldrich. Xylene was purified according to reported literature [3]. Aluminum panels were purchased from Q-Panel Lab Products. All the materials were used as received unless specified otherwise.

Instrumentation. A Nicolet Magna-IR 850 Series II spectrometer was used to record IR spectra. The IR spectra were obtained by spreading liquid samples onto KBr crystals, and then exposing the crystals to UV radiation. A JEOL GSXFT 270 MHz was used to record ^{1}H-NMR and ^{29}Si -NMR spectra for all the compounds. ^{1}H NMR and ^{29}Si-NMR spectra were obtained in CDCl$_3$ with chemical shifts (δ) referenced to tetramethylsilane. Elemental silicon was analyzed by Galbraith Laboratories. Instron Universal Tester (Model 1000) was used to measure the tensile properties of the cured coatings. Pull-off adhesion tester (Elcometer, Model 106) was used to evaluate the adhesion of the coatings to aluminum substrates.

Siloxane Functionalization of Tone Polyols. Tone polyols were siloxane functionalized by reaction with tetraethyl orthosilicate in refluxing xylene in presence of dibutyltin dilaurate catalyst. We have reported details of this procedure in an earlier communication (Wu, S.; and Soucek, M.D., *Polym. Commun.*, in press).

Coating Formulations. Four types of two component formulations were employed. One component was always UVR-6105 (cycloaliphatic diepoxide). The second component consisted of the diol or triol or their corresponding functionalized derivatives. These served as control formulations to evaluate the suitability of functionalized diol or triol derivatives as partial replacements of the diol or triol in three component formulations.

Two types of three component formulations used consisted of UVR-6105, the unmodified diol or triol and functionalized diol or triol derivatives. All the formulations mentioned above are shown in Table I.

All the coatings were cast on aluminum panels and glass plates using a wire wound rod (No. 16), and cured in a ultraviolet (UV) Processor (RPC, 2 × 300 W/in medium pressure mercury UV lamps, and 50 fpm). The film properties were tested 24 h after UV exposure. The gel content (ASTM D2765), pull-off adhesion (ASTM D 4541), pencil hardness (ASTM D3363), crosshatch adhesion (ASTM D3359), were measured according the standard methods as indicated, respectively. The tensile properties were measured according to ASTM D2370-82. Five measurements were obtained for each sample and the average values are reported.

Table I. Three component Formulation used in the Coatings

Exptl. No.	Cyclo aliphatic Diepoxide (g)	Diol or Triol (unmodified) (g)	Functionalized Diol or Triol (g)
1	10.00	0.00	0.00
2	8.00	2.00	0.00
3	7.00	3.00	0.00
4	6.00	4.00	0.00
5	5.00	5.00	0.00
6	6.00	3.60	0.40
7	6.00	3.00	1.00
8	6.00	2.00	2.00
9	6.00	1.60	2.40
10	6.00	0.80	3.20
11	7.00	2.70	0.30
12	7.00	2.25	0.75
13	7.00	1.50	1.50

* All the formulations contained 4 wt.% of the photoinitiator and 0.5 wt.% of the wetting agent.

The cure response of the coatings under UV light was evaluated using maximum surface-cure and through-cure rates. The surface-cure was determined by lightly touching the coating with a cotton ball within 1 second after the UV exposure. The coating was considered surface-cured if no fibers of the cotton stuck onto the coating. The through-cure was determined by scraping the coating with a thumb within 1 second after UV exposure. The coating was considered through-cured if the coating was not removed by the thumb on the substrate.

Spectroscopic Investigation of the Crosslinking Reactions. The crosslinking reactions of the coatings were monitored using IR and ^{29}Si -NMR. The coating formulation used for the IR spectroscopy consisted of 40 wt % di-TEOS functionalized polyol and 60 wt % cycloaliphatic diepoxide, and the coating formulation for the ^{29}Si -NMR consisted of 40 wt % di- or tri-TEOS functionalized polyol and 60 wt % cycloaliphatic diepoxide. The formulations were spread on KBr plates as thin films. After UV exposure, the IR spectra were recorded at desired time intervals. During the interval, the coated KBr plates were stored in a dark dry box.

Results And Discussion

Siloxane groups are widely used in coatings as versatile modifiers to improve interfacial properties such as wetting, spreading, and adhesion [4]. Under acidic conditions, siloxane groups can react with water to afford silanol groups which can then condense to form siloxane crosslinks [5]. The siloxane addition reaction with epoxy was previously reported in the Epoxy/Silanol/Curing/Acrylic coating systems known as ESCA coatings [6]. The siloxane functionalization of the hydroxyl groups would eliminate hydrogen bonding resulting in a decrease in viscosity. By functionalizing the polyols with siloxane moieties, reactive groups are also introduced, resulting in a reactive diluent.

Synthesis and Characterization of Reactive Diluents. The reaction of the caprolactone diol or triol with tetraethyl orthosilicate (TEOS) afforded di-TEOS and tri-TEOS functionalized polyols as shown in Equations 1 and 2, respectively.

$$R\begin{matrix}-O+\overset{O}{\underset{}{C}}\!\!\sim\!\!\sim\!\!\sim\!\!O+_n H \\ -O+\underset{O}{\overset{}{C}}\!\!\sim\!\!\sim\!\!\sim\!\!O+_n H\end{matrix} + 2\text{ Si}-(OEt)_4 \xrightarrow{\text{Cat.}} R\begin{matrix}-O+\overset{O}{\underset{}{C}}\!\!\sim\!\!\sim\!\!\sim\!\!O+_n Si-(OEt)_3 \\ -O+\underset{O}{\overset{}{C}}\!\!\sim\!\!\sim\!\!\sim\!\!O+_n Si-(OEt)_3\end{matrix} + 2\text{ EtOH} \qquad (1)$$

Caprolactone diol 0201   TEOS   S-di

$$R_1\begin{matrix}-O+\overset{O}{\underset{}{C}}\!\!\sim\!\!\sim\!\!\sim\!\!O+_n H \\ -O+\underset{O}{\overset{}{C}}\!\!\sim\!\!\sim\!\!\sim\!\!O+_n H \\ -O+\underset{O}{\overset{}{C}}\!\!\sim\!\!\sim\!\!\sim\!\!O+_n H\end{matrix} + 3\text{ Si}-(OEt)_4 \xrightarrow{\text{Cat}} R_2\begin{matrix}-O+\overset{O}{\underset{}{C}}\!\!\sim\!\!\sim\!\!\sim\!\!O+_n Si-(OEt)_3 \\ -O+\underset{O}{\overset{}{C}}\!\!\sim\!\!\sim\!\!\sim\!\!O+_n Si-(OEt)_3 \\ -O+\underset{O}{\overset{}{C}}\!\!\sim\!\!\sim\!\!\sim\!\!O+_n Si-(OEt)_3\end{matrix} + 3\text{ EtOH} \qquad (2)$$

Caprolactone triol 0305   TEOS   S-tri

For the di-TEOS functionalized product (S-di), the Na^+ IDS mass spectrum followed the formula; (449 + 114n) Da. The 114 Da was due to the -($COCH_2CH_2CH_2CH_2CH_2O$)- repeating unit of the caprolactone diol. The 449 Da was the summation of the siloxane group (162×2 Da), sodium ion (23 Da), and the caprolactone branching unit (102 Da). For the tri-TEOS functionalized product (S-tri) the Na^+ IDS mass spectrum followed the formula; (640 + 114n) Da for the tri-TEOS functionalized polyol. The 114 Da was due to the -($COCH_2CH_2CH_2CH_2CH_2O$)- repeating unit of the caprolactone diol, and the 640 Da were the contribution from the siloxane group (162×3 Da), sodium ion (23 Da), and the caprolactone branching unit (131 Da). The ^{1}HNMR and IR and elemental analyses were in agreement with the mass spectral data. Detailed information on the characterizations have been reported by us in an earlier communication (Wu, S.; and Soucek, M.D., *Polym. Commun.*, in press).

Curing Reaction Studies. Surprisingly, the TEOS in presence of photogenerated super acid did not cure into films. However, in combination with the epoxy monomer, well-cured films were obtained. Infrared and ^{29}Si-NMR were used to study the curing reactions of these coating formulations. Ether peaks at about 1100cm-1 and epoxy peaks at 789-810cm-1 were monitored for these studies. The epoxy peaks disappeared with time and the ether peaks increased in intensity with time. An ^{29}Si-NMR study revealed the appearance of new siloxane bond formation. As a consequence, a curing mechanism based on the incorporation of the siloxane groups was proposed (Schemes 1&2). For a detailed discussion on the spectral studies, the readers are referred to our another communication (Wu, S.; and Soucek, M.D., *Polym. Commun.*, in press).

Curing Behavior. Curing behavior of coatings could be of two types - surface and bulk or thorough cure. If the reactivity of the monomer towards UV radiation is very rapid, then surface will cure faster than the bulk. This would then adversely affect the penetration of the radiation and result in lower cure rate of the bulk. Figures 1 & 2 show surface and through cure rates. As the epoxy content of the two component systems is increased, the surface cure rate increases linearly. When the second component is siloxane functionalized polyol, surface cure rate suffers a little, due to the fact that the siloxane groups do not contribute to crosslinks as effectively as the hydroxyl groups. In fact, in the three component system, when tone diol is replaced partly by functionalized diol, the surface cure rate drops, though marginally. For the two component systems, the surface curing rate achieves a maximum at 60/40 composition of epoxy to diol (modified or unmodified). As the surface cure rate increase, the through cure rate decreases. Conversely, in the three component system,

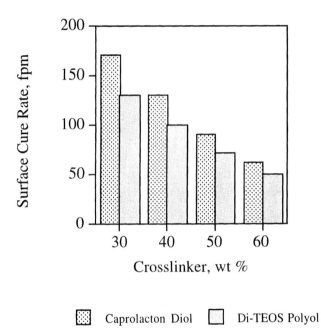

Figure 1. Effects of the Crosslinkers on the Surface Cure Rate of the Polyol or Siloxane Polyol/Epoxide Coatings.

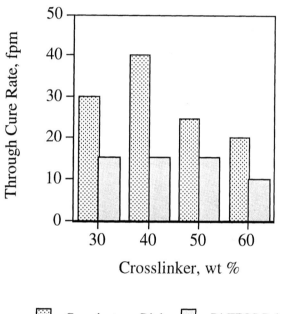

Figure 2. Effects of the Crosslinkers on the Through Cure Rate of the Polyol or Siloxane Polyol/Epoxide Coatings.

as the diol is replaced with an equal amount of the siloxane modified derivative, the surface cure rate decreased and the through cure rate increased (Figure 3).

Mechanical Properties. Figures 4 & 5 show the tensile modulus of the two component systems as a function of the crosslinker concentration. The tensile modulus of all the coatings decreased with increasing concentration of the crosslinker. The polyols and the corresponding TEOS functionalized polyols both functioned as flexible modifiers, decreasing the hardness of the epoxy coatings. This is corroborated by the tensile strength data (Figures 6 & 7). The tensile strength values of all of the coatings dropped with increasing crosslinker content.

The tensile elongation of the two component coatings as a function of the crosslinker is shown in Figures 8 & 9. The elongation of the coatings increased with increasing crosslinker concentration. Values of the siloxane modified polyol containing coatings were consistently lower than those of the unmodified polyol containing coatings. This is probably due to the fact that the siloxane modified polyols are capable of furthering crosslinks due to siloxane condensation reactions.

Scheme 1. Proposed Condensation Reactions of the TEOS Functionalized Polyols under Cationic UV Conditions.

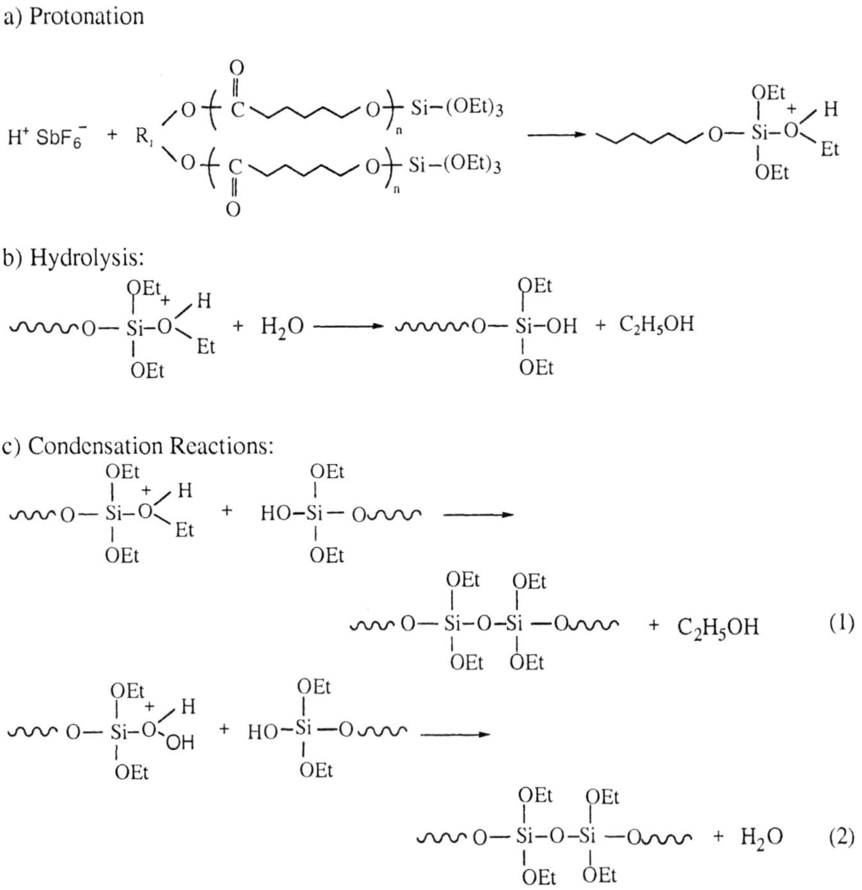

a) Protonation

b) Hydrolysis:

c) Condensation Reactions:

$$\quad (1)$$

$$\quad (2)$$

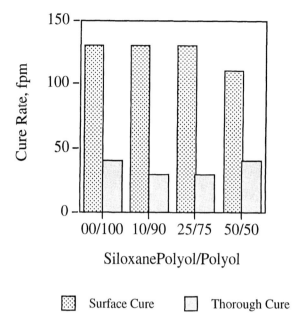

Figure 3. Surface and Thorough Cure Rate of the Di-TEOS Functionalized Polyol Modified UV Coatings as a Funtction of the Ratio of Siloxane Polyol to the Parent Polyol.

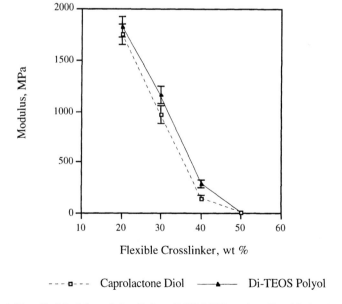

Figure 4. Tensile Modulus of the diol or di-TEOS Functionalized Polyol/Epoxide Coatings as a Function of the Crosslinker.

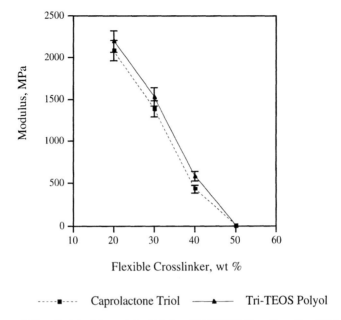

Figure 5. Tensile Modulus of the Triol or Tri-TEOS Functionalized Polyol/ Epoxide Coatings as a Function of the Crosslinker.

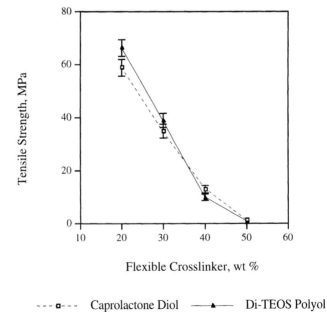

Figure 6. Tensile Strength of the Diol or Di-TEOS Functionalized Polyol/ Epoxide Coatings as a Function of the Crosslinker.

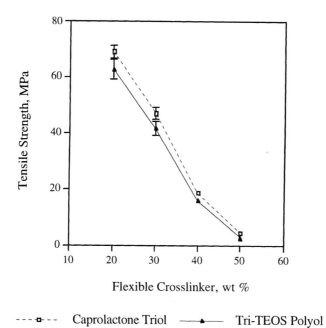

Figure 7. Tensile Strength of the Triol or Tri-TEOS Functionalized Polyol/ Epoxide Coatings as a Function of the Crosslinker.

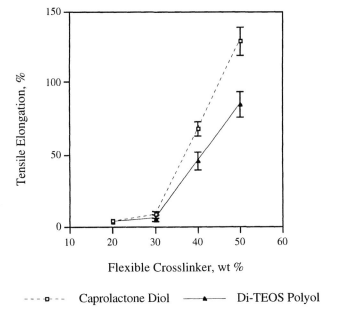

Figure 8. Tensile Elongation of the diol or di-TEOS Functionalized Polyol/Epoxide Coatings as a Function of the Crosslinker.

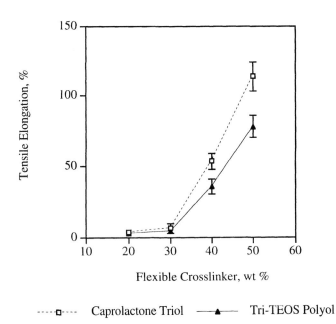

Figure 9. Tensile Elongation of the Triol & Tri-TEOS Functionalized Polyol/Epoxide Coatings as a Function of the Crosslinker.

Scheme 2. Possible Crosslinking Reactions of the TEOS Functionalized Polyols with Cycloaliphatic Epoxide under Cationic UV Conditions.

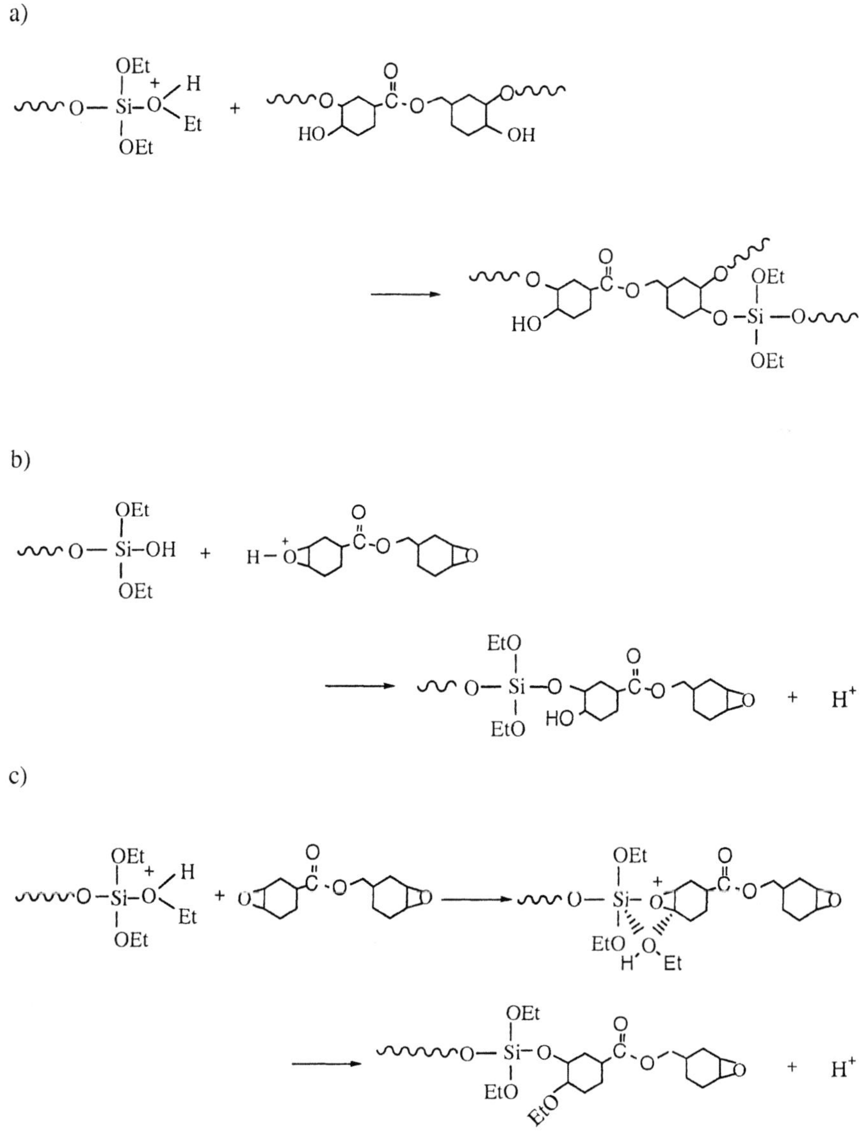

a)

b)

c)

When the three component formulations are considered, as shown in Figures 10 & 11, the trend is a monotonous, but a marginal increase in tensile modulus of the samples with increasing replacement of the unmodified diol with the siloxane functionalized diol. This when, correlated with the gel content suggests that there, probably occurs intrachain crosslinks than interchain crosslinking.

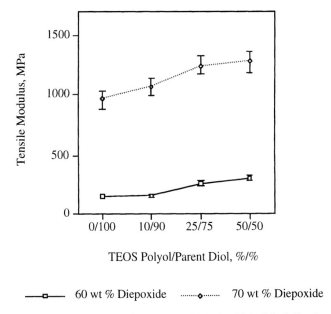

Figure 10. Tensile Modulus of the Di-TEOS Polyol Modified Coatings as a Function of the Ratio of the Siloxane Polyol to the Diol.

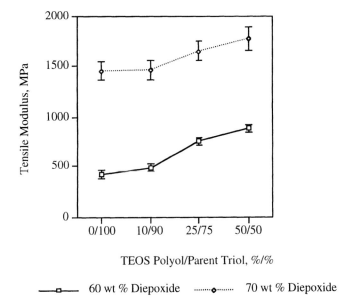

Figure 11. Tensile Modulus of the Tri-TEOS Polyol Modified Coatings as a Function of the Ratio of the Siloxane Polyol to the Triol.

Gel Content. Gel content is a measure of extent of cure, and has implications in terms of crosslink density. The measure gel content is also can be related to mechanical properties. Gelling occurs when the substrate has affinity for the solvent, but the crosslinks present in the substrate prevent dissolution. However, when the gel content is high, swelling is decreased due to entropic limitations. The gel content data is presented in Figures 12a and 12b. The gel content corrletes well with the mechanical properties. An increase in tensile strength and modulus is reflected in the increasing gel content.

In the cationic polymerization of epoxy resins, hydroxyl groups terminate the propagating chain end and initiate new propagating centers [7,8]. However, beyond a certain concentration, the hydroxyl groups have a detrimental effect because the chain transfer is increased and cationically reactive epoxy content is decreased. This will lower the crosslink density or increase coating swellability (Figures 12a and 12b). Thus, an optimum concentration of hydroxyl groups would help in achieving right properties. This is also what has been observed in tensile property measurements.

Siloxane modified polyols undergo reaction with the hydroxyl groups of the unmodified polyols and also with the hydroxyl groups of the epoxy polymer chain. Statistically, due to their concentration, reaction of the siloxane groups with the hydroxyl groups would be more favored. Thus, in the three component system, when more and more of the polyols are replaced with the modified polyols, this probability should decrease. Thus, the coatings with siloxane modified diols should exhibit an optimum gel content and should decrease with decreasing polyol content. Indeed, this is what has been observed (Figures 12 a and b).

Adhesive Properties

Pull-off Adhesion. Adhesive properties of the coatings were tested with pull-off adhesion test to give an idea about the affinity of the coatings to metal substrates. Particularly, it was of interest to evaluate the siloxane-functionalized polyols as replacements for polyols. Figures 13 & 14 represent the results of these tests. When the epoxide content was 70 wt % of the composition (in the three component systems) adhesive properties were poorer than when it was only 60wt.%. This is probably because of lack of sufficient polar functional groups which are contributed by the Tone polyols. When the epoxide content was 60 wt % of the composition, replacement of 10 wt % of the polyol with siloxane functionalized polyol yielded significant improvement in pull-off adhesion property. Further increase of functionalized polyol did not show any further improvement. This could be attributable to the lack of reactivity of alkoxy siloxane groups to yield Si-O-Si bonds coupled with the a simultaneous decrease of hydroxyl functionality. This suggests less network formation resulting in a relatively weaker interfacial adhesion. Thus, a 10 wt % replacement of the unmodified polyol with siloxane functionalized polyol seems to give optimum adhesive properties.

Conclusion

Tone polyols were siloxane functionalized by reacting with tetraethyl orthosilicate in presence of dibutyltin dilaurate catalyst. The synthesized siloxane functionalized polyols did not undergo self-condensation when exposed to UV radiation in presence of photoacid generators. However, the combination of siloxane functionalized polyols and cycloaliphatic diepoxides formed cohesive crosslinked films. Partial replacement of polyols with the siloxane modified polyols enhanced adhesion of the

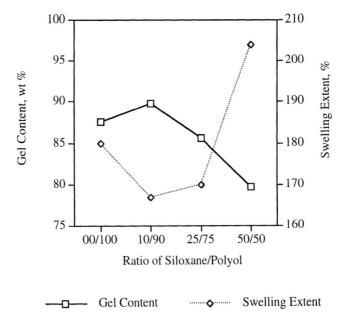

Figure 12a. Gel Content and Swelling Extent of Di-TEOS Functionalized Polyol Modified Coatings Containing 60 wt %.

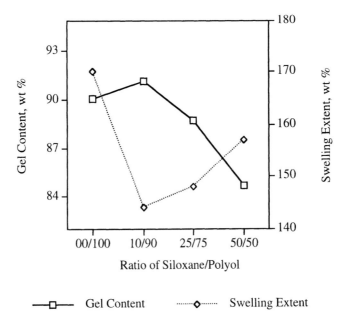

Figure 12b. Gel Content and Swelling Extent of Tri-TEOS Functionalized Polyol Modified Coatings Containing 60 wt %.

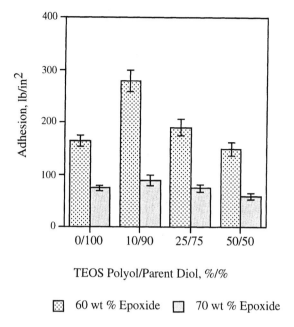

Figure 13. Pulloff-Adhesion of the Di-TEOS Polyol Modified Coatings as a Function of the Ratio of the Siloxane Polyol to the Diol.

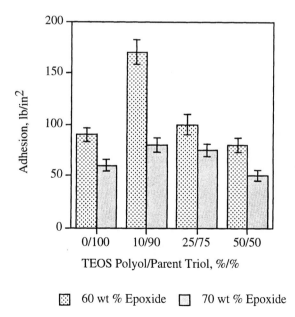

Figure 14. Pulloff-Adhesion of the Tri-TEOS Polyol Modified Coatings as a Function of the Ratio of the Siloxane Polyol to the Triol.

coatings to aluminum metal substrate and flexibilized the coatings. The coatings formulated with the siloxane functionalized polyols exhibited marginally higher tensile modulus than the corresponding unfunctionalized caprolactone polyol coatings.

Acknowledgment

The authors would like to thank Dr. Marek W. Urban for the providing of the IR spectroscopy instrument.

Literature Cited

1. Decker, C.; and Morel, F., *Proc. of the ACS: Polymeric Materials Science and Technology*, **1997**,. 76, pp 70.

2. Crivello, J. V. *Advances in Polym. Sci.* 62, Springer-Verlag, Berlin, Heidelberg **1984**, 62, pp 27.

3. Perrin, D. D.; Armarego, W. L. F.; and Perrin, D. R., *Purification of Laboratory Chemicals*, Pergamon: Great Britain, **1980**, Chapter 1.

4. a) Chen, M. J.; Osterholtz. F. D.; Pohl, E. R., *J. Coat. Tech.*, **1997**, 69, pp43. b) Adamson, A. W. and Gast, A. P., *Physical Chemistry of Surfaces*, John Wiley and Sons: New York, 1997, Chapters 7 and 8.

5. Brinker, C. J. and Scherer, G. W., *Sol-Gel Science*, Academic: New York, **1990**, Chapters 1-3.

6. a) Takahashi, T.; Sano, S. and Ishihara, M., *Waterborne, High Solids, and Powder Coatings Symposium*, Univ. of Southern Mississippi: Hattiesburg, **1994**, pp606. b) USP No. 4191713 and 4518726. c) Murata, K. and Isozaki, O., *Japan Finishing Tosou-Gijutsu*, **1987**, 1, pp133.

7. Koleske, J. V., *Federation of Societies for Coatings Technology: Cationic Radiation Curing*, Philadelphia, **1986**.

8. Manus, P. J. M., *Polymers Paint Colour J.*, **1991**, 181, pp56.

Chapter 36

TOSPEARL: Silicone Resin for Industrial Applications

Robert J. Perry[1] and Mary E. Adams[2]

[1]General Electric Company Silicones,
260 Hudson River Road, Waterford, NY 12188
[2]General Electric Company CRD, P.O. Box 8, Schenectady, NY 12301

Tospearl particles are crosslinked siloxanes made by the controlled hydrolysis and condensation of methyltrimethoxysilane. Their spherical nature, narrow particle size distribution and chemical and thermal stability make them ideal for use in wear resistance, antiblocking and light diffusing applications.

Tospearl is a spherical silicone resin particle made by the controlled hydrolysis and condensation of alkyl trialkoxysilanes (equation 1) and produced by Toshiba Silicones, a joint venture of GE Silicones. The first report of these materials appeared in 1985 when the aqueous amine catalyzed condensation of methyl trimethoxysilane was described (1). Since the initial disclosure, a number of other patents have issued in this area (2-8). The product is a three-dimensional network that is intermediate between inorganic and organic particles. This paper describes the physical and chemical properties of this network as well as its utility in a number of applications.

$$\text{RO-Si(Me)(OR)-OR} + H_2O \xrightarrow[-ROH]{NH_3} \left[\text{HO-Si(Me)(OH)-OH} \right] \xrightarrow{-H_2O} \text{Me-Si(OH)(O-)-O-Si(Me)(OH)-O-Si(O-)(Me)-O-} \longrightarrow \{-\text{O-Si(Me)(O-)-O-}\} \quad (1)$$

Preparation

There are a number of grades of Tospearl produced with sizes ranging form submicron to 12μm. Each of these materials has its own characteristic production process but the common thread is that reaction to form the Tospearl particle occurs at the silane/water interface (Figure 1). As the silane hydrolyzes, it becomes soluble in the aqueous phase where it undergoes condensation. As the silanol condenses, it becomes insoluble and precipitates. The particle size and distribution are affected by base concentration, stirring rate, paddle shape, vessel size, temperature and solids concentration. When the correct balance between all these parameters is achieved, monodisperse, spherical beads form.

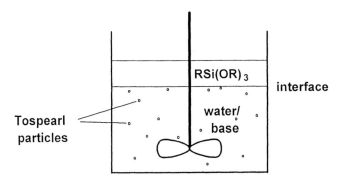

Figure 1. Tospearl kettle.

Figure 2 shows the effect of stirring rate and ammonia concentration on the particle size. The stirring rates increase from 1 to 4 (a range of 2-60 rpm) and at the slower rates smaller particles are formed. As the rate increases, the particles become larger. At a constant stirring rate, the particles grow to a larger size at lower ammonia concentrations.

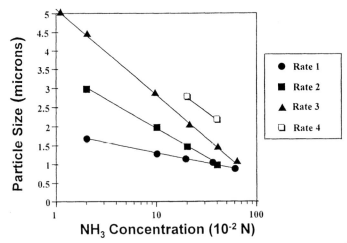

Figure 2. Effect of Stirring Rate and NH_3 Concentration on Particle Size.

In a separate set of experiments, Figure 3 illustrates the effect the methyltrimethoxysilane/ water ratio has on the particle size. The largest particles were formed at high silane/water rations at a given base concentration. However, a maximum was seen in the base range examined.

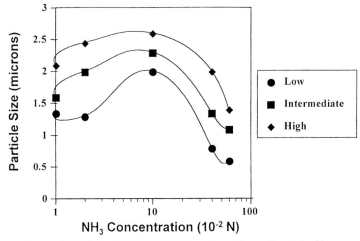

Figure 3. Effect of MeSi(OMe)₃/water ratio on Particle Size.

Figure 4 outlines the process for product isolation. After the particles are formed, the reaction mixture is filtered to remove gel from the suspension. The filtrate is then subjected to another, finer filtration and the resulting filter cake is dried at ~200°C to give the isolated resin. During drying, some fusing of particles occurs resulting in a loose network structure. This is broken up by jet milling the clusters; a process in which aggregates are propelled at a conical ceramic plate at high speeds.

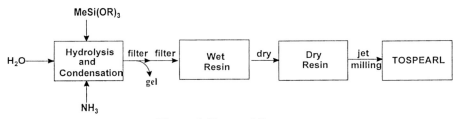

Figure 4. Tospearl Process

Properties

Table 1 shows some of the characteristics of the various grades of Tospearl that can be produced by this method. Diameters from 0.5m to 12 m can be obtained with the smaller particles having high specific surface areas and low bulk specific gravities. With the exception of 2000B and 240, all the particles have narrow particle spheres. Tospearl 2000B is still spherical, but has a broader size distribution centered at 6m. The amorphous 240 grade has a rounded, irregular shape. These materials are pH neutral and have low moisture content.

Table 1. Characteristics of Commercial Tospearl Grades

Tospearl	Diameter (μm)	Bulk sp. gr.	Specific Surface Area (m²/g)
105	0.5	0.25	70
120	2.0	0.35	30
130	3.0	0.36	20
145	4.5	0.43	20
3120	12.0	0.46	18
240	4.0 (amorphous)	0.17	35

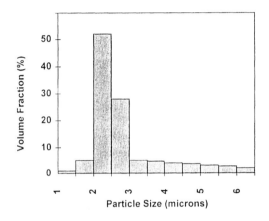

Figure 5. Particle Size Distribution for Tospearl.120.

Electron micrographs of these materials are shown in Figure 6. One can see the spherical nature of the particle as well as the mono-disperse nature of the beads.

As expected for a material that is primarily inorganic, the thermal stability of the silicone resin spheres is extremely high, especially when compared to fine organic particles, as shown in Figure 7. Tospearl 120 was stable to 400+°C in air and at 800°C, there was only 10% weight loss. The weight loss seen at 450 °C is attributed to further condensation within the particle with loss of water and/or methanol.

Tospearl particles are also readily dispersed in a variety of solvents. Table 2 shows the dispersion viscosity of Tospearl 120 in polar and nonpolar solvents. In polar ketone, alcohol and ether solvents, the viscosities are quite low. When nonpolar or chlorinated solvents are used, the dispersion viscosities rise by an order of magnitude. The polar solvents more readily associate with the residual silanol groups on the Tospearl particles effectively isolating them from self-association. In non-polar solvents, self-association is possible and the viscosity rises accordingly.

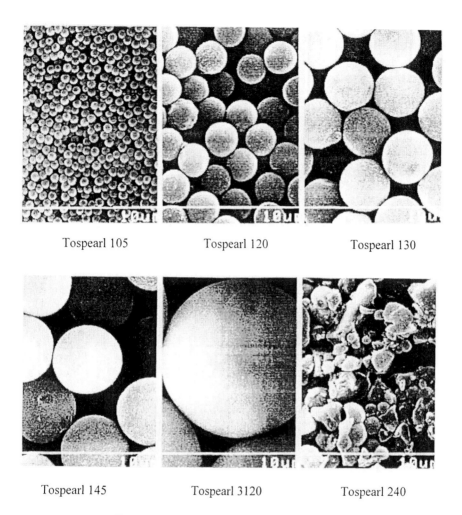

Tospearl 105 Tospearl 120 Tospearl 130

Tospearl 145 Tospearl 3120 Tospearl 240

Figure 6. Micrographs of Tospearl Particles.

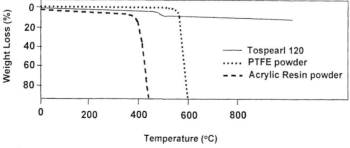

Figure 7. Thermogravimetric Analysis of Tospearl 120 and Organic Resin Fine particles.

Table 2. Dispersion Viscosities of Tospearl 120 at 50 wt %.

Solvent	Dispersion Viscosity (cP)
Ketone-based	5-10
Ester-based	15-20
Alcohol-based (C1-2)	10
Alcohol-based (C3-4)	20-28
Glycol-ether-based	20-25
Diacetone alcohol	25
Tetrahydrofuran	20
Aromatic Hydrocarbon-based	220-230
n-Hexane	370
Kerosene	280
Perchloroethylene	260
Chloroform	640

Applications

Tospearl particles have found widespread use in a number of varied applications. These materials improve wear, water and chemical resistance and modify the slip properties of rubber surfaces. In the paint and ink arena, Tospearl has been used as a dispersing agent and to adjust viscosity as well as to prevent blocking and pigment hardening and as luster control. The cosmetic industry uses these particles to improve the feel of foundations, to prevent the caking in lipsticks and in hand and body creams. Figure 8 gives a brief overview of the particle sizes commonly used in a variety of applications. Table 3 also shows the specific types of Tospearl that have been utilized in lubrication, anti-blocking and light diffusing environments. The "A" designation indicates a Tospearl grade suitable for personal care applications.

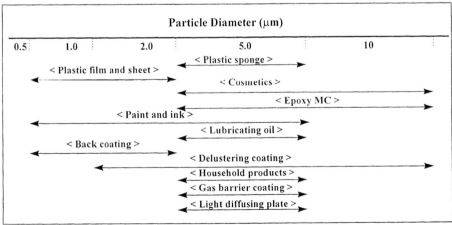

Figure 8. Application vs. Particle Size.

Table 3. Function and Applications of Tospearl particles.

Function	Application	Product
Matting	Paint, Ink, SHC	Tospearl 120, 130, 145, 240
Lubrication	Back coat for color ribbon	Tospearl 105, 120
	Gravure ink	Tospearl 105, 120
	Gas barrier coating	Tospearl 120, 130, 145
	Car wax	Tospearl 120, 130, 145
	Organic photoconductor drum	Tospearl 120
	OPP Film	Tospearl 120, 130, 145
	Cosmetics (foundations)	Tospearl 120A, 130A, 145A
	Weatherstrip coating	Tospearl 120, 130, 145
Anti-blocking	Gas barrier coating	Tospearl 120, 130, 145
	Paint, ink	Tospearl 120, 240
Light Diffusing	Liquid crystal display	Tospearl 120, 130, 145
	Acrylic resin	Tospearl 120, 130, 145
	Polycarbonate resin	Tospearl 120, 130, 145
	PET resin	Tospearl 120, 130, 145
Flowability	Cosmetics (foundations)	Tospearl 120A, 130A, 145A
Anti-agglomeration	Cosmetics (foundations)	Tospearl 120A, 130A, 145A

Tospearl has also been used to increase the abrasion resistance of weather stripping (Table 4). Untreated EPDM base rubber abraded easily. When treated with a curable silicone, the abrasion resistance increased by an order of magnitude and the friction was halved, but a glossy rather than the desired matte finish was obtained. Addition of Tospearl 120 or 145 improved the abrasion resistance another 6-8 fold and the friction coefficient decreased 4-8 times. In addition, the rubber took on a matte appearance. If the particles were too large, as with Tospearl 3120, the abrasion resistance decreased.

Table 4. Results of Treating Weatherstripping with Tospearl.

Weatherstrip Condition	Appearance	Friction Coefficient		Abrasion Resistance*
		STATIC	DYNAMIC	
Untreated	matte	2.1	2.0	500
Treated	glossy	1.5	1.3	5,000
Treated w/120	matte	0.4	0.3	30,000
Treated w/145	matte	0.2	0.2	40,000
Treated w/3120	matte	0.1	0.1	20,000

*number of strokes

Tospearl has been used as a plastic film additive in polypropylene (PP). A specific example is shown in Table 5 where Tospearl 120 was incorporated into a single PP sheet. One third less Tospearl was required when compared to silica to obtain the same anti-blocking properties. In addition, the static slip was half as large.

Table 5. Evaluation of Tospearl in a Single PP Sheet.

Properties	New Formulation	Old Formulation
Base Resin	Homo PP	Homo PP
Antiblocking Agent	Tospearl 120	Silica
Particle Size	2 μm	1 μm
Particle Content	240 ppm	900 ppm
Thickness	30 μm	30 μm
Haze	1.4%	1.8% (ASTM D 1003)
Static Slip (Film/Film)	0.25	0.50 (ASTM D 1894)
Antiblocking Property	800 g/10 cm2	780 g/10 cm2

In a coextruded film (Figure 9) with the same Tospearl loading as silica, the resulting material had a two-fold increase in anti-blocking properties and a 70% reduction in static slip (Table 6).

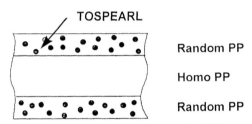

Figure 9. Coextruded Polypropylene Film.

Table 6. Evaluation of Tospearl in a Coextruded PP Film.

Properties	New Formulation	Old Formulation
Core Layer	Homo PP	Homo PP
Skin Layer	Random PP	Random PP
Antiblocking Agent	Tospearl 120	Silica
Particle Size	2 μm	3 μm
Particle Content	2000 ppm	2000 ppm
Thickness	22 μm	22 μm
haze	1.8%	1.6% (ASTM D 1003)
Static Slip (Film/Film)	0.15	0.48 (ASTM D 1894)
Antiblocking Property	150 g/10 cm2	360 g/10 cm2

Table 7 shows a polycarbonate (PC) resin which was treated with Tospearl to provide an improved light diffusing sheet (9). At only 40% of the barium sulfate level, the Tospearl treated sheet allowed more light through and the light was better dispersed than the conventional additive. The mechanical integrity of the PC sheeting was also retained as shown by the mechanical measurements.

Table 7. Evaluation of Tospearl in a Polycarbonate Film.

Properties	New Formulation	Old Formulation
Polycarbonate Resin	100 parts	100 parts
Antiblocking Agent	Tospearl 120	$CaCO_3$
Particle Content	0.2 parts	2 parts
Thickness	1mm	1 mm
Izod (kg.cm/cm$_2$)	89.0	80.0
Tensile Strength (kg/cm$_2$)	630	630
Haze	92.4%	80.2%
Light Transmission	75.9%	70.5%

Table 8 also shows a light diffusing application using acrylic sheeting. Again, more light is transmitted and is more diffuse (10).

Table 8. Evaluation of Tospearl in a PMMA Sheet.

Properties	New Formulation	Old Formulation
Base Resin	PMMA	PMMA
Antiblocking Agent	Tospearl 120	$BaSO_4$
Particle Content	1.0%	2.5%
Thickness	1 mm	1 mm
Total luminous transmittance	95.6%	83.7% (ASTM D 1003)
Diffusing luminance	90.3%	78.4%
Parallel luminance	5.0%	5.1%

In addition to the examples noted above, these spherical silica particles have found use as additives in other PMMA (11-13), PP (14-15), PET (16), PVC (17) and PE (18) formulations.

All applications described thus far have utilized untreated Tospearl particles. Although extensive condensation occurs during the synthesis of these particles, some silanol groups remain on the surface. The following two examples illustrate the ease with which the Tospearl surface can be modified.

The surface of Tospearl 120 can be rendered more hydrophobic by treatment with hexamethyldisilazane (HMDZ) or trimethylchlorosilane (TMSCl) as shown in Table 9. Hydrophobicity is determined by the amount of material that is "wetted" by the solvent mixture and settles out during centrifugation. 45% of the untreated particles are wetted by a 60/40 methanol/water mixture. After hydrophobizing the surface, less than 5% were "wetted" (19).

Table 9. Treatment of Tospearl 120 with Hydrophobizing Agents.

Methanol/Water Ratio	HMDZ 25°C/15 h	HMDZ 160°C/6 h	TMSCI 25°C/16 h	Untreated
60/40	3	0	4	45
80/20	50	30	51	100

In another application, the surface charge could be altered by treatment with aminoethylaminopropyltrimethoxysilane as seen in Table 10 (20). The greater the amine loading on the surface, the greater the positive charge. Numerous other examples of surface treatments have also been reported (21-25).

Table 10. Treatment of Tospearl with Aminoethylaminoproptltrimethoxysilane.

Surface Treating Agent Solution	Example 1	Example 2	Example 3	Untreated
Aminosilane (parts)	1	2	10	-
Methanol (parts)	19	18	10	-
Contact Charge (µC/g)	+150	+270	+330	-950

Summary

The spherical silicone resin particles described above are made by the hydrolysis and condensation of methyltrimethoxysilane. These materials are chemically inert, thermally stable and available in a variety of sizes with a narrow particle distribution. Specific examples have illustrated how a number of diverse industries have used these particles for wear resistance, anti-blocking properties and light diffusing ability and how the surface characteristics can be modified. Other applications can also be favorably impacted by these unique materials. The options are only limited by your imagination.

Acknowledgments

We thank H. Kimura, A. Takagi, M. Yoya, M. Nishida, M. Iwasaki, T. Sugito, Y. Kasahara, M. Matsumoto, A. Horne, J. Russell and K. Murthy for supplying valuable reference and application information.

References

1. Kimura, H., US 4,528,390 to Toshiba Silicone Co., Ltd., 7/9/85.
2. Jpn. Kokai Tokkyo Koho, JP 61221520 to Toshiba Silicone Co., Ltd., 9/19/86.
3. Jpn. Kokai Tokkyo Koho, JP 61247345 to Toshiba Silicone Co., Ltd., 10/29/86.
4. Jpn. Kokai Tokkyo Koho, JP 62129840 to Toshiba Silicone Co., Ltd., 5/28/87.
5. Jpn. Kokai Tokkyo Koho, JP 63069712 to Toshiba Silicone Co., Ltd., 3/25/88.
6. Jpn. Kokai Tokkyo Koho, JP 05239365 to Toshiba Silicone Co., Ltd., 2/26/91.
7. Jpn. Kokai Tokkyo Koho, JP 03047840 to Toshiba Silicone Co., Ltd., 2/28/91.
8. Jpn. Kokai Tokkyo Koho, JP 05140314 to Toshiba Silicone Co., Ltd., 6/8/93.
9. Ohtsuka, U.; Fugiguchi, T.; Oishi, K. US 5,352,747, to GE Plastics, Japan, Ltd., 10/4/94.
10. Jpn. Kokai Tokkyo Koho, JP 9316002 B to Asahi Chemical Co., Ltd., 1993
11. Jpn. Kokai Tokkyo Koho, JP 3207743 to Misubishi Rayon., 11/11/91.

12. Jpn. Kokai Tokkyo Koho, JP 3273046 to Misubishi Rayon., 12/4/91.
13. Jpn. Kokai Tokkyo Koho, JP 3294348 to Misubishi Rayon., 12/25/91.
14. Mizuno, H.; Fujiwara, K. US 4,769,418 to Misubishi Petrochemicals Co., Ltd., 9/6/88.
15. Jpn. Kokai Tokkyo Koho, JP 9020845 to Mitsui Toatsu Chemical Co., Ltd., 1/21/97.
16. Etchu, M.; Murooka, H., US 5,620,774 to Teijin Ltd., 4/14/97.
17. Jpn. Kokai Tokkyo Koho, JP 1103440 to Kohjin Co., Ltd., 4/20/89.
18. Jpn. Kokai Tokkyo Koho, JP 9059457 to Mitsui Toatsu Chemicals Co., Ltd., 3/4/97.
19. Saitoh, K.; Kimura, H., US 4,895,914 to Toshiba Silicones Co., Ltd., 1/23/90.
20. Kimura, H.; Takagi, A., US 4,871,616 to Toshiba Silicones Co., Ltd., 10/3/89.
21. Sumida, H.; Kimura, H., US 4,652,618 to Toshiba Silicones Co., Ltd., 3/24/87.
22. Saito, K.; Kimura, H., US 4,996,257 to Toshiba Silicones Co., Ltd., 2/26/91.
23. Saito, K.; Kimura, H., US 5,034,476 to Toshiba Silicones Co., Ltd., 7/23/91.
24. Saito, K.; Kimura, H., US 5,106,922 to Toshiba Silicones Co., Ltd., 4/21/92.
25. Saito, K.; Kimura, H., US 5,204,432 to Toshiba Silicones Co., Ltd., 4/20/93.

Chapter 37

Oxygen Gas Barrier PET Films Formed by Deposition of Plasma-Polymerized SiOx Films

N. Inagaki

Laboratory of Polymer Chemistry, Faculty of Engineering, Shizuoka University, 3-5-1 Johoku, Hamamatsu 432-8561, Japan

In order to prepare SiOx-deposited PET films with a high oxygen gas barrier capability, SiOx depositions from the plasma polymerization of silane compounds have been investigated from viewpoint of the chemical composition of the deposited SiOx films and oxygen permeation rate through the SiOx-deposited PET films. Tetramethoxysilane (TMOS) was suitable as a starting material for the SiOx deposition from the plasma polymerization. Oxidation and etching processes by selfbias effects were effective in the SiOx deposition. The SiOx film deposited on PET film surfaces from the TMOS/O_2 mixture (60 mol% O_2) was mainly composed of Si-O-Si networks with a carbonized carbon component as a minor product. The SiOx-deposited PET film showed good oxygen gas barrier properties. The oxygen permeation rate was 0.10 cm^3/m^2-day-atm, which corresponded to an oxygen permeability coefficient of 1.4×10^{-17} cm^3-cm/cm^2-sec-cmHg for the SiOx film itself. This permeability coefficient is three orders lower than that of conventional gas barrier polymeric films such as Eval and Saran.

Silicon oxide film (SiOx) is an interesting material in the field of the food and pharmaceutical technologies as well as the microelectronics technology. SiOx film possesses a high gas barrier property besides high thermal stability and high electrical insulation. Therefore, SiOx-coated polyester (PET) film is a possible material used for packing to protect foods or medicines from the deterioration by oxidation.

Usually, SiOx film is synthesized by the sol-gel method using alcohoxysilanes such as tetraethoxysilane (TEOS) or by the plasma chemical vapor deposition (plasma CVD) methods using a mixture of TEOS and oxygen (1). In these processes, TEOS is hydrolyzed and polycondensed into Si-O-Si networks. Building-up the complete Si-O-Si networks and eliminating carbonaceous components from the deposited SiOx film requires a high temperature of more than 500°C (2). However, this process is difficult to apply directly the preparation of the SiOx-deposited PET films because of low thermal-resistance. The SiOx film deposition should proceed at lower temperatures than the glass transition temperature (about 70°C) of the PET film.

A possible process for the SiOx film deposition at low temperatures below 70°C may be a special plasma CVD method which involves some elimination of carbonaceous compounds. In the plasma CVD, we believe that SiOx films are formed in two reactions: (1) the bond scission of Si-O-C bond in TEOS to form Si• or SiO• radicals, and (2) the recombination between two radicals to form Si-O-Si linkage. The repetitious combination of the two reactions leads to the Si-O-Si network, and as a result, a SiOx film is deposited. On the other hand, fragments eliminated from TEOS, ethyl or ethoxy radicals, also are recombined to form carbonaceous compounds and incorporated into the SiOx film. In the conventional plasma CVD process, the carbonaceous compounds in the deposited SiOx film are eliminated by the pyrolysis of the SiOx film at high temperature (500°C) (2). Therefore, in the SiOx-deposition on the PET film surface, some special process for the elimination of the carbonaceous compounds instead of the pyrolysis treatment should be investigated.

In this study, we have investigated three possible processes for the elimination of the carbonaceous compounds from the deposited SiOx films: (1) Choice of silane compounds used as a starting material for the SiOx film deposition by the plasma CVD, (2) Oxidation processes by oxygen plasma, and (3) Etching processes occurring on an electrode surface.

We have discussed the possibility of these processes from the viewpoint of chemical composition of the deposited SiOx films and the oxygen gas barrier capability of the SiOx-deposited PET films.

Experimental.

Materials. Biaxially-stretched poly(ethylene terephthalate), PET, film (Trade name BOPET, T_m 256°C, T_g 67°C) kindly provided in a size of 500 mm wide and 38 μm thick by Toyobo Co. Ltd. was used as a substrate for plasma polymer deposition. The PET film surface was rinsed with acetone in an ultrasonic washer prior to the plasma polymer deposition. Tetraethoxysilane (TEOS), Triethoxysilane (TrEOS), tetramethoxysilane (TMOS), dimethyldimethoxysilane (DMDMOS), and tetramethylsilane (TMS) purchased from Petrarch Systems Co. were used as silanes for plasma polymerization without purification. The purity of these silanes was 98 - 99%.

Plasma Polymerization Reactor for SiOx Deposition. A commercial vacuum deposition apparatus (Ulvac Co., Japan; model EBH6), which had a bell-jar chamber (400 mm diameter, 590 mm high) and a vacuum system with a combination of a rotary pump (320 lit/min) and a diffusion pump (550 lit/sec), was incorporated into a special reactor for SiOx film deposition. A diode planar electrode made of stainless steel for glow discharge, inlets of the silane vapors and oxygen or argon gas, a rolling machine for reeling the PET film (variable reeling rates of 33 to 994 mm/min), a thickness monitor (Ulvac Co., Japan; model CRTM-1000) for monitoring the SiOx deposition rate and a pressure gauge were installed in the bell-jar chamber. An electrical energy input system for initiating a glow discharge (Samuco Co., Japan; model RFG-200), a digital voltmeter with a matching box (Samuco Co., Japan, model MU-2) for measuring the selfbias voltage and a flow rate controlling system for the silane vapors and oxygen and argon gases were annexed to the reactor. The diode electrode contained a circular electrode (electrode A) with a dimension of 380 mm diameter and a rectangular electrode (electrode B) with a dimension of 50 mm wide and 100 mm long. The relative area of the two electrodes is 22.7 : 1. The gap between the electrodes was 60 mm. The electrode A was grounded and the electrode B was connected to an electrical power generator at 13.56 MHz (Samuco Co., Japan; model RFG-200). The PET film was placed between the electrodes and the film was reeled by the rolling machine. The gas flow rate controlling system was a combination of a metering needle valve (Nupro Co., model BM-4BMG) for the TMOS and TEOS vapors and a mass flow controller (Estec Co., Japan; model SEC-400 MARK3) for oxygen and argon gases. A schematic diagram of the plasma reactor used in this study is shown in Figure 1.

SiOx Deposition by Plasma Polymerization. The PET film (110 mm wide) ran between the electrode A and B and was set up on reels of the rolling machine. Two positions where the PET film between the electrode A and B were used for the SiOx deposition. One position was within a 5 mm distance from the electrode B surface (deposition on the electrode surface), and the other position was in the midpoint between the electrode A and B (30 mm far from the electrode B surface). The silane was poured into a reservoir and air dissolved in the silane was removed by a repeated freezing-fusion procedure. The reservoir was kept in temperature-controlled oven at 60°C to increase the vapor pressure of the silane.
Air in the reaction chamber was displaced with argon and the reaction chamber was evacuated to approximately 0.13 Pa. Afterwards, the silane vapor whose flow rate was adjusted to a given flow rate (15 - 6 cm^3/min) by the metering valve and oxygen gas adjusted by the mass flow controller (0 - 9 cm^3/min), were blown off from the inlets to the reaction chamber. The total flow rate of the silane vapor and oxygen gas was kept a constant of 15 cm^3/min. The plasma CVD of the silane/O_2 mixtures was performed at a system pressure of 39.9 Pa at rf powers of 40 - 100 W. The deposition rate of SiOx films was determined from reading of the thickness monitor (model CRTM-1000). After the SiOx deposition, the reaction chamber was evacuated to less than 0.13 Pa for 30 min and argon gas introduced into the reaction chamber at atmospheric pressure. The PET film was taken out from the reaction chamber and served as specimens for x-ray photoelectron spectroscopy (XPS), scanning electron microscopy (SEM) and oxygen permeation rate measurements.

X-ray Photoelectron Spectra of deposited SiOx Films. XPS spectra of the SiOx films deposited on the PET were obtained on a Shimadzu ESCA K1 using a non-monochromatic MgK$_\alpha$ photon source. The anode voltage was 12 kV, the anode current was 20 mA and the background pressure in the analytical chamber was 1.5×10^{-6} Pa. A size of the x-ray spot was 2 mm diameter and a take-off angle of photoelectrons was 90° with respect to the sample surface. The smoothing process of the spectra were not done. The spectra were decomposed by fitting a Gaussian-Lorentzian mixture function (80 : 20 mixture ratio) to an experimental curve using a nonlinear, least-squares curve-fitting program, ESCAPAC, supplied by Shimadzu. The sensitivity factors (S) for the core levels were $S(C_{1s}) = 1.00$, $S(Si_{2p}) = 0.87$, and $S(O_{1s}) = 2.85$. C/Si and O/Si atomic ratios of the deposited SiOx films were estimated from the spectral intensity and the sensitivity factor of corresponding core levels within an experimental error of ± 0.05.

Oxygen Permeation Rate through SiOx-deposited PET film. A gas barrier tester (Mocon Co., model OX-TRAN 2/20) was used for measuring the oxygen permeation rate (in cm^3/m^2-day-atm) through SiOx-deposited PET films (area of 78 mm diameter) at 30°C and a relative humidity of 70 %RH. Three to five specimens were used for the measurement and the oxygen permeation rate was determined from an average of the measurements within an experimental error of 0.1 cm^3/m^2-day-atm. From the oxygen permeation rate, the oxygen permeability coefficient for the deposited SiOx film in $(STP)cm^3$-cm/cm^2-sec-cmHg was estimated under assumption of the two layer model.

The SiOx-deposited PET film is a laminated structure consisting of the PET and SiOx film layers. For the two layers model, the following equations (1) and (2) are given (3),

$$d = d_1 + d_2 \tag{1}$$

$$\frac{d}{P} = \frac{d_1}{P_1} + \frac{d_2}{P_2} \tag{2}$$

where the subscript 1 and 2 designate the original PET film layer and the SiOx film layer deposited on the PET film surface, respectively. P and d designate the oxygen permeability coefficient and the film thickness, respectively. For the original PET film, P_1 is 2.5×10^{-12} $(STP)cm^3 \cdot cm/cm^2/s/cmHg$ and d_1 is 38 μm. For the SiOx film, d_2 is 100 nm. Equation (2) is modified into equation (3).

$$\frac{1}{P} = \frac{1}{P_1} + \frac{d_2}{d_1 + d_2} \frac{1}{P_2} \tag{3}$$

The oxygen permeability coefficient, P_2, for the SiOx film in the SiOx-deposited PET film can be calculated from equation (3).

Results and Discussion.

Atomic Composition of Deposited SiOx Films. We have investigated three processes for the elimination of carbonaceous compounds from the deposited SiOx films. The factors are (1) choice of silane compounds used as a starting material for the SiOx film deposition by the plasma CVD, (2) oxidation reactions by oxygen plasma and (3) etching reactions occurring near an electrode surface by the selfbias.

Choice of Silanes for SiOx Film Deposition. Plasma polymerization of silanes results in plasma polymer films which contain carbonaceous components such as Si-C and CH_2-CH_2 groups within SiOx networks. In order to form SiOx networks without the carbonaceous components, carbonaceous components should be excluded from the deposited SiOx films. Many researchers have followed a heating procedure for the elimination of the carbonaceous components from the deposited SiOx films but this procedure is not applicable to SiOx-deposited PET films because of poor thermal-resistance of the film. In this study, a silane that was suitable for the SiOx deposition with less carbonaceous component was investigated. Five silanes, TEOS, TrEOS, TMOS, DMDMOS, and TMS, which contained different C/Si atomic ratio of 8 to 4 and different bond structure (Si-O-C and Si-C bonds) between Si and C atoms, were used for the plasma polymerization. Table I compares the C/Si atomic ratio in the deposited SiOx films from the five silanes. The C/Si atom ratio in the deposited SiOx films, as shown in Table I, depends on the C/Si atomic ratio and the bond structure in the starting silanes: Silanes with a small C/Si atom ratio deposit SiOx films with a low carbon content, and silanes with Si-O-C bonds also deposit SiOx films with a low carbon content. From this viewpoint, TMOS is preferable to TEOS, TrEOS, DMDMOS, and TMS as a silane for SiOx deposition with a less carbonaceous component, although TMOS is not yet a satisfactory material for the SiOx deposition. The SiOx film deposited from TMOS reveals a C/Si atom ratio of 1.5.

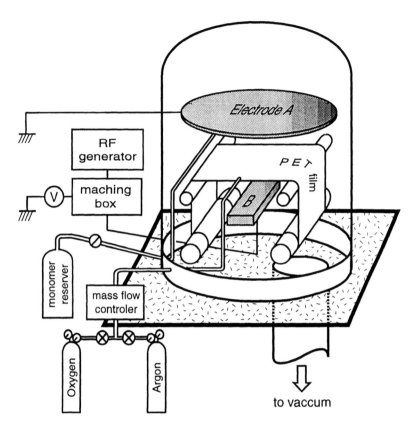

Figure 1 Schematic Presentation of Reaction Chamber for SiOx Deposition.

Table I. Atom Composition of Deposited SiOx Films from Silanes

Silane Compounds	Atom Composition			
	in Stating Compounds		in Deposited SiOx	
	C/Si	O/Si	C/Si	O/Si
$(C_2H_5O)_4Si$	8	4	3.2	1.9
$(C_2H_5O)_3SiH$	6	3	1.9	1.4
$(CH_3O)_4Si$	4	4	1.5	1.5
$(CH_3O)_2Si(CH_3)_2$	4	2	1.7	1.1
$Si(CH_3)_4$	4	0	2.5	0.3

Oxidation Reactions for SiOx Film Deposition. Oxidation processes by the oxygen plasma were investigated for the elimination of carbonaceous components from deposited SiOx films. TMOS was mixed with oxygen and the mixture was plasma-polymerized for SiOx film deposition. Table II shows the C/Si and O/Si atom ratios in the deposited SiOx films as functions of the oxygen concentration in the $TMOS/O_2$ mixture. The mixing of oxygen gas, as shown in Table II, leads to large decreases in the C/Si atom ratio of the deposited SiOx films. When an oxygen gas of 20 mol%O_2 was mixed with TMOS, the C/Si atom ratio in the deposited SiOx films decreases from 1.4 to 0.44 at an rf power of 40 W and from 1.5 to 0.39 at 60W On the other hand, changes in the O/Si atom ratio by mixing with oxygen gas is small. The O/Si atom ratio decreases from 1.7 to 1.5 at an rf power of 40 W, and from 1.8 to 1.4 at 60 W, even when the oxygen concentration increases from 0 to 60 mol%. From the comparison, we conclude that the oxidation process contributes to the SiOx deposition with less carbonaceous components.

Etching Reactions by Selfbias for SiOx Film Deposition. In a glow discharge at 13.56 MHz frequency, the electrode A is grounded, and the electrode B is connected to the electrical power generator. Electrons and ions are oscillated by changing the polarity of the electrical field. As a result, the ion sheath is formed around the electrode B, due to much lower speed of ion species than electrons, and the electrode B is charged negatively (selfbias) against the electrode A which is grounded (4). The selfbias reaches from a few tens to hundreds of volts, depending on operating conditions. As a result, the ion bombardment occurs powerfully in the ionsheath region. From this viewpoint, we expect that light carbonaceous components incorporated into the SiOx films will be sputtered out from the deposited SiOx films, and dense SiOx films with less carbonaceous component will remain.

Figures 2 and 3 show typical results of the selfbias voltage as functions of rf power and the $TMOS/O_2$ mixture composition. The selfbias voltage increases linearly with increasing both oxygen concentration in the $TMOS/O_2$ mixture and rf power. The selfbias voltage reaches from - 36 to -125 volts at oxygen concentrations of 0 - 60 mol% and rf powers of 40 - 100 W. Therefore, we expect that etching actions will occur powerfully on the electrode B. Table III compares the C/Si and O/Si atom ratios between the SiOx films deposited under selfbias and under no selfbias. The SiOx deposition under the selfbias leads to large decreases in the C/Si atom ratio. The C/Si atom ratio for the SiOx films deposited at rf powers of 40 and 60 W decreases from 1.4 to 0.90 and from 1.5 to 1.2, respectively. However, the O/Si atom ratio is negligibly influenced by the selfbias. The O/Si atom ratio is 1.6 - 1.8. Mixing of oxygen gas with TMOS leads to further decreases in the C/Si atom ratio. When an oxygen concentration of 20 mol% is mixed with TMOS, the C/Si atom ratio for the SiOx films deposited at rf powers of 40 and 60 W decreases from 0.90 to 0.39 and from 1.2 to 0.45, respectively. A minimum C/Si atom ratio of 0.24 and 0.31 reaches at an oxygen concentration of 40 mol% at 40 and 60 W, respectively.

From this result, we conclude that the etching process by the selfbias is effective in eliminating carbonaceous compounds from the deposited SiOx films. A combination of the selfbias and oxygen gas-mixing is more effective for the carbon elimination.

Chemical Composition of Deposited SiOx Films. The chemical composition, especially the carbonaceous and silicone components, was investigated by XPS (C1s and Si2p) spectra. Six SiOx films deposited from the $TMOS/O_2$ mixtures of 0, 20, and

Table II. Atom Composition of Deposited SiOx Films from TMOS/O2 Mixtures

TMOS/O$_2$ Mixtures O$_2$ Concentration (mol%)	RF Power (W)	Atom Composition in Deposited SiOx Films	
		C/Si	O/Si
0	40	1.4	1.7
20	40	0.44	1.5
40	40	0.30	1.5
60	40	0.25	1.5
0	60	1.5	1.8
20	60	0.39	1.4
40	60	0.42	1.4
60	60	0.49	1.4

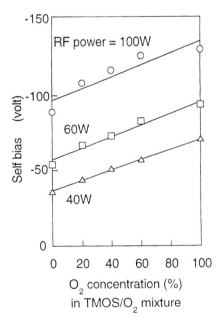

Figure 2   Selfbias Voltage at a Constant RF Power as a Function of Oxygen Concentration in TMOS/O$_2$ Mixture.

40 mol%O_2 at an rf power of 60 W under selfbias and under no selfbias were used as specimens for the XPS measurements. Figures 4 and 5 compare C1s and Si2p spectra among three SiOx films deposited from the TMOS/O_2 mixtures of 0, 20, and 40 mol%O_2 at an rf power of 60 W under no selfbias. The SiOx film deposited from TMOS (no oxygen mixing) shows an intensive and complex C1s spectrum on which a maximum peak appears at 286.9 eV (due to C-O groups whose relative concentration is 83 %) and a shoulder appears at 285 eV (due to CH groups whose relative concentration is 17 %) (5). The Si2p spectrum shows a maximum peak appears at 103.4 eV (due to SiO_2 groups). The full width at half-maximum (FWHM) value of the Si2p spectrum is 1.79 eV. On the other hand, for the two SiOx films deposited from the TMOS/O_2 mixtures of 20 and 40 mol%O_2, the C1s spectra distributes from 283 to 286 eV and a maximum peak appears at 284.3 - 284.7 eV (due to carbonized carbon). The Si2p spectra show a peak at 103.4 eV (due to SiO_2) (5) whose FWHM value is 1.65 eV which is smaller than that for the SiOx film deposited from TMOS (1.79 eV). This comparison indicates that the chemical composition of the carbonaceous compounds incorporated into the deposited SiOx film is distinctly altered by the mixing of oxygen gas. The carbonaceous compound in the SiOx film deposited from TMOS consists mainly of C-O and CH groups, while the film deposited from the TMOS/O_2 mixtures is composed of carbonized carbons. Furthermore, the Si-O-Si networks in the SiOx films deposited from the TMOS/O_2 mixture may grow more than those deposited from TMOS, because of a small FWHM value.

The chemical composition of the SiOx films deposited under the selfbias was analyzed with XPS. Figures 6 and 7 show typical C1s and Si2p spectra for the SiOx films deposited under the selfbias. The C1s spectrum for the SiOx film deposited from TMOS under the selfbias shows narrow distribution which contains a peak at 287.0 eV due to C-O groups and a small shoulder near 285 eV due to CH groups whose relative concentration is estimated to be 87 and 13 %, respectively. The relative concentration of the C-O and CH components in the SiOx film deposited under the selfbias corresponds almost to those in the SiOx film deposited under no selfbias (83 % for the C-O component and 17 % for the CH component), although the SiOx film deposited under the selfbias, as shown in Table III, possesses lower carbon content (C/Si atom ratio = 1.2) than that deposited under no selfbias (C/Si = 1.5). This comparison means that the etching actions by the selfbias removed predominantly carbonaceous components, independently of the kind of their components (C-O and CH components), from the SiOx film surface. The Si2p spectrum is symmetrical and narrow distribution whose FWHM value is 1.68 eV, which is smaller than that for the SiOx film deposited under no selfbias (1.79 eV) (shown in Figure 5). The oxygen-mixing with TMOS produces great changes in the C1s spectrum but a small change in the Si2p spectrum. When an oxygen of 20 and 40 mo% is mixed with TMOS, the C1s spectra becomes very weak, and the peak position of the spectra shifts from 287 to 284.5 eV due to carbonized carbons (Figure 6). The Si2p spectra become intense and peak appears at 103.4 eV due to SiO_2 groups (Figure 7). For discussion of the selfbias effects on the chemical composition, four SiOx films, that is, each of the two SiOx films deposited from the TMOS/O_2 mixtures of 20 and 40 mol%O_2 under the selfbias and under no selfbias, were compared. Their C/Si atom ratio is 0.45 and 0.31 for the SiOx films deposited under the selfbias, and 0.39 and 0.42 for the SiOx films under no selfbias. Their C1s and Si2p spectra are shown in Figures 4 and 5 for the SiOx films under no selfbias, and in Figure 6 and 7 for the SiOx films under the selfbias. All four SiOx films shows very weak C1s spectra, but intense Si2p spectra. There is no difference in C1s and Si2p spectra between the SiOx films deposited under the selfbias and under no selfbias. All C1s

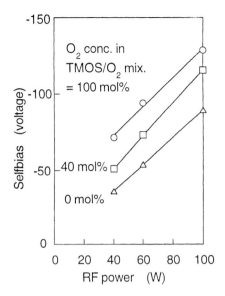

Figure 3 Selfbias Vantage at a Constant Oxygen Concentration in TMOS/O$_2$ Mixture as a Function of RF Power.

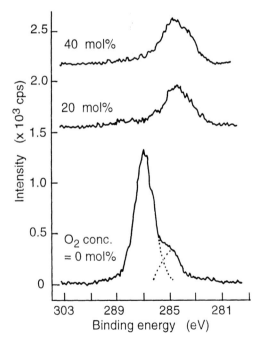

Figure 4 XPS (C1s) Spectra of Deposited SiOx Films as a Function of Oxygen Concentration in TMOS/O$_2$ Mixture.

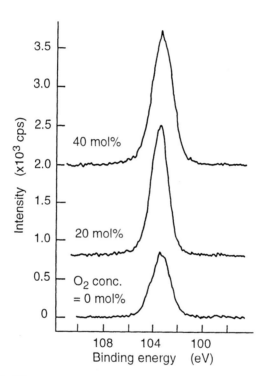

Figure 5   XPS (Si2p) Spectra of Deposited SiOx Films as a Function of Oxygen Concentration in TMOS/O_2 Mixture.

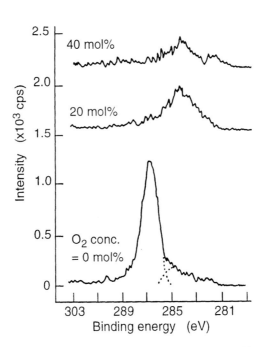

Figure 6 XPS (C1s) Spectra of SiOx Films deposited under Selfbias as a Function of Oxygen Concentration in TMOS/O_2 Mixture.

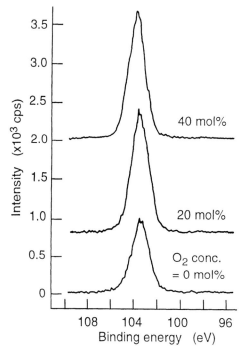

Figure 7   XPS (Si2p) Spectra of SiOx Films deposited under Selfbias as a Function of Oxygen Concentration in TMOS/O_2 Mixture.

Table III.   Atom Composition of SiOx Films Deposited under Selfbias

O_2 Conc. in TMOS/O_2 Mix. (mol%)	RF Power (W)	Atom Composition in SiOx Films			
		Deposited under Selfbias		Deposited under no Selfbias	
		C/Si	O/Si	C/Si	O/Si
0	40	0.90	1.6	1.4	1.7
20	40	0.39	1.7	0.44	1.5
40	40	0.24	1.5	0.30	1.5
60	40	0.35	1.5	0.25	1.5
0	60	1.2	1.7	1.5	1.8
20	60	0.45	1.4	0.39	1.4
40	60	0.31	1.6	0.42	1.4
60	60	0.59	1.3	0.49	1.4

spectra show a peak at 284.3 - 284.7 eV (carbonized carbon groups). All Si2p spectra show a peak at 103.4 eV (due to SiO_2 groups). Their FWHM value of the Si2p spectra is small (1.59 - 1.65 eV), indicating that Si-O-Si networks in the SiOx films are grown up more highly, because of narrow FWHM value of 1.59 - 1.65 eV.

From this result, we conclude that there is no difference in the chemical composition between the SiOx films deposited under the selfbias and under no selfbias, although the C/Si atom ratio is different between these SiOx films.

Oxygen Gas Barrier Properties of SiOx-deposited PET Films. SiOx films (100 nm thickness) were deposited from the $TMOS/O_2$ mixtures on the surface of the PET films under the selfbias and under no selfbias and the SiOx-deposited PET films were provided for the evaluation of the oxygen gas barrier properties. Table IV shows typical results of the oxygen permeation rate (in a dimension of cm^3/m^2-day-atm) at 30°C at a relative humidity of 70% through the SiOx-deposited PET films. The oxygen permeation rate of the original PET film is 44 cm^3/m^2-day-atm, which corresponds to an oxygen permeability coefficient of 2.5×10^{-12} cm^3-cm/cm^2-sec-cmHg. All SiOx-deposited PET films, as shown in Table IV, show lower oxygen permeation rate than the original PET films. This comparison indicates that the SiOx film deposition is effective in improving the oxygen gas barrier properties of the PET films, and that the oxidation process by the oxygen plasma (the oxygen concentration in the $TMOS/O_2$ mixture) and the etching process by the selfbias (the SiOx deposition under the selfbias or under no selfbias) produce strong influences on the magnitude of the oxygen permeation rate for the SiOx-deposited PET films.

When the oxygen concentration in the $TMOS/O_2$ mixture was changed from 0 to 60 mol%, the SiOx-deposited PET films showed a large decrease in the oxygen permeation rate. For example, the oxygen permeation rate, as shown in Table IV, is 0.60 and 0.10 cm^3/m^2-day-atm for the SiOx-deposited PET films from $TMOS/O_2$ mixtures of 40 and 60 mol%O_2, and at an rf power of 60W, respectively, compared with 27 cm^3/m^2-day-atm for the SiOx-deposited PET films from TMOS alone at an rf powers of 60W. Another SiOx-deposited PET film from the $TMOS/O_2$ mixture of 60 mol%O_2 at an rf power of 100W also shows a low oxygen permeability rate of 0.34 cm^3/m^2-day-atm. Furthermore, the etching process by the selfbias also influenced the oxygen permeation rate. For example, the SiOx films deposited under the selfbias from $TMOS/O_2$ mixtures of 40 and 60 mol%O_2 at an rf power of 60 W on the PET film surface, as shown in Table IV, show extremely high oxygen permeation rate of 21 and 14 cm^3/m^2-day-atm compared with the SiOx-deposited PET films under no selfbias (0.60 and 0.10 cm^3/m^2-day-atm). The SiOx films deposited under the selfbias, as discussed in a previous section, are almost same in the chemical composition as those deposited under no selfbias. Both SiOx films consist mainly of Si-O-Si network with small amounts of carbonized carbon groups. Therefore, we believe that a large difference in the oxygen gas barrier capability between the SiOx films deposited under the selfbias and under no selfbias may be due to physical factors such as cracks and voids in the SiOx films rather than chemical factors such as the chemical composition of the deposited SiOx films. Figure 8 shows SEM pictures of the SiOx-deposited PET films under the selfbias and under no selfbias from the $TMOS/O_2$ mixtures of 60 mol%O_2 at an rf power of 60 W. The surface of the SiOx-deposited PET film under no selfbias is smooth, and has no defect such as crack and hole, while the surface of the SiOx-deposited PET film

SiOx film deposited under no selfbias
from TMOS/O_2 mixture of 60 mol%O_2

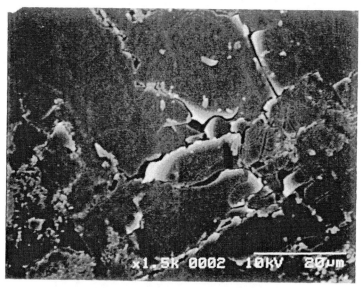

SiOx film deposited under the selfbias
from TMOS/O_2 mixture of 60 mol%O_2

Figure 8   SEM Pictures of SiOx Films deposited under Selfbias and under no Selfbias.

under the selfbias is not smooth but defective. There are many cracks in the SiOx film. We believe that the extremely low oxygen gas barrier capability of the SiOx-deposited PET films under the selfbias may be due to the formation of cracks in the SiOx films. The cracks may be formed during the elimination process of carbonaceous compounds from the SiOx films by the etching actions of the selfbias.

Finally, using equation (3), the oxygen permeability coefficient for the deposited SiOx film itself was estimated and the result of the estimation is listed in Table V. The SiOx film deposited under no selfbias from the TMOS/O_2 mixtures of 60 mol%O_2

Table IV. Oxygen Permeation Rate through SiOx-deposited PET Films

RF Power (W) $(cm^3/m^2\text{-}day\text{-}atm)$	Selfbias	Oxygen Con. (mol%)	Oxygen Permeation Rate in TMOS/O_2 Mix.
60	no	0	27
60	no	20	27
60	no	40	0.60
60	no	60	0.10
60	yes	0	29
60	yes	20	28
60	yes	40	21
60	yes	60	14

Table V. Oxygen Permeation Coefficient for SiOx Films

RF Power (W) $cm^3\text{-}cm/cm^2\text{-}sec\text{-}cmHg)$	Selfbias	Oxygen Con. (mol%)	Oxygen Permeation Coefficient in TMOS/O_2 Mix. (x 10^{-14}
60	no	0	1.5
60	no	20	1.5
60	no	40	8.4×10^{-3}
60	no	60	1.4×10^{-3}
60	yes	0	2.1
60	yes	20	1.7
60	yes	40	3.9×10^{-1}
60	yes	60	3.2×10^{-1}

at an rf power of 60 W shows very small oxygen permeability coefficient of 1.4×10^{-17} cm^3-cm/cm^2-sec-cmHg, which is 860 - 6360 times smaller than conventional polymers with good gas barrier capability, Eval F (hydrolyzed ethylene-vinylacetate copolymer, 1.2×10^{-14} cm^3-cm/cm^2-sec-cmHg) and Saran (polyvinylidene chloride, $3.0 - 8.9 \times 10^{-14}$ cm^3-cm/cm^2-sec-cmHg) (6). Therefore, the SiOx film deposited under no selfbias from TMOS/O_2 mixtures of 60 mol%O_2 at an rf power of 60 W is a good material with high oxygen gas barrier capability. We believe that the SiOx-deposited PET films will practically serve as packing materials in the food and pharmaceutical industries, because of the low oxygen permeation rates of less than 1 cm^3/m^2-day-atm.

From these results, we conclude that the SiOx deposition from the TMOS/O_2 mixture under no selfbias gives a good oxygen gas barrier capability. The oxygen gas barrier capability of the deposited SiOx films is superior to that of the conventional barrier materials such as Eval F and Saran.

Conclusions.

Three factors for the SiOx deposition by plasma polymerization of silanes were investigated. The three processes are (1) choice of silane compounds used as a starting material for the SiOx film deposition by the plasma CVD, (2) oxidation processes by the oxygen plasma, and (3) etching processes occurring near an electrode surface. The chemical composition of the deposited SiOx films and the oxygen gas barrier capability of the SiOx-deposited PET films were investigated. Results are summarized as follows.
1. TMOS is suitable as a starting material for the SiOx deposition from the plasma polymerization.
2. The oxidation process by the oxygen plasma and the etching process by the selfbias are effective in the SiOx deposition with less carbonaceous compounds.
3. The SiOx films deposited under no selfbias from the TMOS/O_2 mixture give good films with a high oxygen gas barrier capability, but the SiOx films deposited under the selfbias give poor films. The oxygen permeation rate for the SiOx-deposited PET films under no selfbias was 0.10 cm^3/m^2-day-atm, which corresponds to an oxygen permeability coefficient of 1.4×10^{-17} cm^3-cm/cm^2-sec-cmHg for the SiOx film itself.

We believe that the SiOx-deposited PET films may practically used as packing materials in the food and pharmaceutical industries.

REFERENCES

1. Inagaki, N., *Chapter 10.4 Materials for Surface Modification*, in *Jikken Kagaku Koza* (in Japanese); The Chemical Society of Japan; 29, Kobunshi Zairyo; Maruzen, Tokyo, 1993; pp. 477.
2. Adams, A. C., *VLSI Technology*; Sze, S. M.; McGraw-Hill, New York, 1988; pp. 233.
3. Rogers, C. E., *Chapter 2 Permeation of Gases and Vapours in Polymers, Polymer Permeability*; Comyn, J; Elsevier, London, 1985; pp. 11.
4. Chapman, B., *Glow Discharge Process*; Wiley, New York, 1980; pp. 143.
5. Beamson, G. and Briggs, D., *High Resolution XPS of Organic Polymers*,; Wiley, New York, 1992.
6. Ashley, R. J., *Chapter 7 Permeability and Plastics Packing*, in *Polymer Permeability*; J. Comyn; Elsevier, London, 1985; pp. 269.

Author Index

Adams, Mary E., 533
Aoki, Hidetoshi, 359
Auner, Norbert, 115
Babb, David A., 308
Backer, Michael, 115
Ben Khalifa, Riadh, 383
Boileau, Sylvie, 383
Bouteiller, Laurent, 383
Brzezinska, K. R., 408
Burnell, Tim, 180
Burns, G. T., 408
Carpenter, John, 180
Chakrapani, Srinivasan, 516
Chaudhury, M. K., 322
Chenskaya, T. B., 503
Chojnowski, J., 20
Chu, H. K., 170
Clarson, Stephen J., 353
Cosgrove, Terence, 204
Crivello, James V., 284
Cypryk, M., 20
Dagger, Anthony C., 38
de Leuze-Jallouli, 241
Diamanti, Steven, 270
Dvornic, Petar R., 241
Fortuniak, W., 20
Fu, Zhidong, 164
Garasanin, Tania, 204
Godovsky, Yu. K., 98
Gordon, Glenn V., 204
Gottlieb, M., 194
Graiver, D., 445
Guo, Hongjie, 433
Hamilton, F. J., 445
Hannington, Jonathan P., 204
Harwood, H. J., 445
Hay, P. J., 81
Hu, J., 226
Inagaki, N., 544
Ji, Junmi, 308
Jukarainen, Harri, 353

Kajiyama, Tisato, 332
Kara, Şebnem, 395
Kawakami, Yusuke, 460
Kazakova, V. V., 503
Kaźmierski, K., 20
Kepler, Cindy L., 433
Kickelbick, Guido, 270
Kim, Y., 445
Kojio, Ken, 332
Kress, J. D., 81
Künzler, J., 296
Kwei, T. K., 419
Leung, P. C., 81
Lewis, Larry N., 11
Liang, Yi, 383
Londergan, Timothy M., 433
Lu, Ping, 433
Lu, Shaoxiang, 419
Makarova, N. N., 98
Malik, Ranjit, 284
Mark, James E., 1
Matějka, Libor, 485
Matisons, J. G., 128
Matukhina, E. V., 98
Matyjaszewski, Krzysztof, 270
Miller, Peter J., 270
Miranda, Peter M., 476
Muzafarov, A. M., 214, 503
Myakushev, V. B., 503
Nagasaki, Yukio, 359
Nakagawa, Yoshiki, 270
Narayan-Sarathy, Sridevi, 308
Neilson, Robert H., 308
Nguyen, B., 445
Owen, Michael J., 241, 322
Ozark, R., 296
Ozerin, A. N., 503
Ozerina, L. A., 503
Pacis, Cristina, 270
Partchuk, T., 194
Paulsaari, Jyri, 433

Pearce, Eli M., 419
Pernisz, Udo, 115
Perry, Robert J., 533
Perz, Susan V., 241
Pollack, Steven K., 164
Provatas, A., 128
Rebrov, E. A., 214, 503
Roberts, Claire, 204
Sargent, Jonathan R., 433
Schmidt, Randall G., 204
Semlyen, J. Anthony, 38
Seppälä, Jukka V., 353
Serth-Guzzo, Judy, 180
Sharipov, E. Yu., 503
She, H., 322
Sheiko, S. S., 503
Smith, Dennis W., Jr., 308
Son, D. Y., 226
Soucek, Mark D., 516
Stein, Judith, 180

Stepto, R. F. T., 194
Strelkova, T. V., 503
Takahara, Atsushi, 332
Tapsak, Mark A., 433
Tawa, G. J., 81
Taylor, D. J. R., 194
Tebeneva, N. A., 214
Teyssié, Dominique, 383
Truby, Kathryn, 180
Tulpar, Ayşen, 395
Turner, Michael J., 204
Wagener, K. B., 408
Wang, Guohong, 433
Weatherhead, Ian, 204
Weber, William P., 433
Wiebe, Deborah, 180
Wu, Shaobing, 516
Yılgör, Emel, 395
Yılgör, İskender, 395

Subject Index

A

Ab initio electronic calculations. *See* KOH-catalyzed ring-opening polymerization of cyclic siloxanes

Ablative oils
 foul release performance, 186–187
 incorporation into silicone network, 185
 mechanisms for oil retention in RTV11 topcoat, 183–184
 synthesis of diphenyldimethylsiloxane oil, 184f
 See also Biofouling release coatings

Acid catalysis, Chojnowski method, 43–44

Acrylate terminated silicones. *See* UV curable silicones

Acyclic diene metathesis (ADMET)
 alt-poly(carbosilane/siloxane), 434
 polymerization using Ru-carbene complexes, 435, 436f
 See also Ruthenium catalyzed synthesis of *alt*-poly(carbosilane/siloxanes); Telechelic oligomers, silicon terminated

Adhesive properties
 pull-off adhesion of di-TEOS and tri-TEOS functionalized polyol modified coatings as function of ratio of siloxane polyol to diol, 531f
 siloxane-modified epoxy coatings, 529
 See also Modifiers for photocurable epoxy coating formulations

Adsorption on surfaces, hydrogenated cyclic and linear PDMS, 57

Aerosil silica particles
 dispersion in polymer melts, 206, 210
 See also Reinforced poly(dimethylsiloxane) melts

Aggregation. *See* 18-Nonadecyltrichlorosilane (NTS) monolayer

Ahmad–Rolfes–Stepto (A–R–S) theory of gelation. *See* Gelation studies

Alanine
 differential scanning calorimetry (DSC) results for amino acid functional siloxanes, 161t
 evaluating efficacy and generality of synthetic procedure, 141
 See also Amino acid functional siloxanes

Amino acid functional siloxanes
 α,ω-amino acid functional siloxanes containing tetramethyldisiloxane (products 6-10), 135–137
 amino acids for evaluation of efficacy and generality of synthetic procedure, 141
 ^{13}C NMR spectrum of siloxane (4), 143, 145f
 ^{13}C NMR spectrum of siloxane (6), 149, 151f
 COSY NMR spectrum of siloxane (4), 143, 146f, 147
 2D COSY NMR spectrum of siloxane (6), 149, 152f
 deprotection of *t*-butyl ester groups (products 21-35), 140
 deprotection of *t*-butyl groups for final amino acid tetramethyldisiloxanes, 156
 deprotection step of *t*-butyl ester or *N*-*t*-BOC groups from parent amino acid, 141, 143
 differential scanning calorimetry (DSC) results, 161t
 DSC trace of siloxane (23), 157, 160f
 experimental, 130, 133–140
 experimental reagents, 130, 133

Fourier transform infrared (FTIR) spectroscopy method, 133
FTIR analysis of unprotected telechelic siloxanes, 156–157
grafting amino groups onto pendant hydridosiloxanes by hydrosilylation, 129–130
^{1}H NMR spectrum of *t*-butyl derivative of siloxane (24), 156, 158*f*
^{1}H NMR spectrum of siloxane (4), 143, 144*f*
^{1}H NMR spectrum of siloxane (6), 147, 149, 150*f*
^{1}H NMR spectrum of siloxane (24), 156, 159*f*
hydrosilylation of allylcarboxy amino acid *t*-butyl esters (1-5), 147
monitoring hydrosilylation reaction by disappearance of Si–H resonances, 147
needs of polymeric medical implants, 129
percent α and β adduct for siloxanes (6-20) [ratio of CH_2CH_2 vs. $CHCH_3$], 155*t*
polymers containing poly[(methylhydrogen)-*co*-(dimethyl)] siloxane (products 16-20), 138–140
polymers containing poly(methylhydrogen)siloxane (PMHS) (products 11-15), 137–138
preparation of allylcarboxy functionalized amino *t*-butyl ester acids (products 1-5), 133–135
preparation of poly[(methylhydrogen)-*co*-(dimethyl)] siloxanes, 133
preparation results, 143–157
quantities and yields for amino acid siloxanes obtained after deprotection of *t*-butyl ester groups by trifluoroacetic acid, 140*t*
reactivity of three siloxanes, 141
separations of amino acid enantiomers using stationary phases, 130, 131*f*
^{29}Si NMR spectrum of siloxane (6), 149, 153*f*
side reactions during hydrosilylation, 155–156
structures of tetramethyldisiloxane and poly(methylhydrogen)siloxane, 142*f*
synthetic scheme, 130, 132
TGA method (thermogravimetric analysis), 133
TGA of siloxane (25), 157, 160*f*
thermal analysis results, 157
typical 2D HETCOR NMR spectrum of siloxane (4), 147, 148*f*
typical hydrosilylation reactions, 149, 154*f*, 155
Antifouling coatings
environmental risk to marine organisms, 181
See also Biofouling release coatings
Applications
polysiloxanes, 5–6
Tospearl particles, 538–542
See also Tospearl particles
Aromatic ether linkage. *See* Fluorosilicones with perfluorocyclobutene (PFCB) aromatic ether linkage
Atom transfer radical polymerization (ATRP)
alternative to ionic techniques, 271–272
ATRP from silsesquioxane initiators, 278, 279
halogen atom transfer by transition metal species, 271
kinetic plot for ATRP of styrene initiated from benzyl chloride-POSS (polyhedral oligomeric silsesquioxane), 280*f*
molecular weight plot for ATRP of styrene initiated from benzyl chloride-POSS, 281*f*
summary of architectures available through ATRP, 273*f*
See also Hybrid polymers from atomic transfer radical polymerization

(ATRP) and poly(dimethylsiloxane) (PDMS)
Atomic force microscopy (AFM). *See* 18-Nonadecyltrichlorosilane (NTS) monolayer; Silicas

B

Barnacle adhesion testing
 data for various coatings, 182*f*
 technique, 181
 See also Biofouling release coatings
Base catalysis
 Brown and Slusarczuk method, 43
 See also KOH-catalyzed ring-opening polymerization of cyclic siloxanes
Bimodal model networks, elastomeric, 5
Biofouling release coatings
 ablative networks and tethered incompatible oils, 183–184
 analysis of radio-labeled samples containing ^{14}C poly(diphenyldimethylsiloxane) oil, 192
 barnacle adhesion data for various coatings, 182*f*
 barnacle adhesion testing, 181
 ^{14}C radio-labeled oil study, 187–188
 condensation moisture cure system of silicone topcoat RTV11, 181
 easy release properties of silicones, 181
 experimental design for ^{14}C radio-labeled study, 192, 193*f*
 foul release performance of ablative and tethered oils, 186–187
 General Electric foul release coatings with RTV11 topcoat, 181, 182*f*
 incorporation of ablative and tethered oils into silicone network, 185
 material balance for fresh water, 189*t*
 material balance for marine water system, 189*t*
 mechanisms for oil retention in RTV11 topcoat, 183–184
 preparation of ^{14}C poly(diphenyldimethylsiloxane) oil, 191–192
 preparation of silicone coatings containing ^{14}C poly(diphenyldimethylsiloxane) oil, 192
 radio-labeled study results, 188
 RTV11 chemistry, 182*f*
 silicone coatings minimizing adhesive strength of biofouling attachment, 181
 summary of ablative and tethered diphenyldimethylsiloxane oils, 183*f*
 synthesis of ^{14}C-labeled poly(diphenyldimethylsiloxane), 188*f*
 synthesis of ablative diphenyldimethylsiloxane oil, 183, 184*f*
 synthesis of bis-allyl-terminated diphenyldimethylsiloxane, 190–191
 synthesis of bis-chlorosilane-terminated diphenylsiloxanedimethylsiloxane, 190
 synthesis of bis-triethoxy-terminated diphenyldimethylsiloxane, 191
 synthesis of tethered diphenyldimethylsiloxane oil, 183, 184*f*, 191
 traditional antifouling coatings, 181
Bis-(3-trimethoxysilylpropyl)fumarate
 alternate synthesis using trimethyl orthoformate (TMOF), 479–480
 analytical equipment, 484
 ^{13}C NMR spectrum of reaction mixture from methanolysis of bis-(trichlorosilylpropyl)maleate, 483*f*
 experimental reagents, 480
 ^{1}H NMR spectrum of product from hydrosilylation reaction of diallyl fumarate (DAF), 481*f*

^{1}H NMR spectrum of reaction mixture during methanolysis reaction of bis-(trichlorosilylpropyl)maleate, 482f
hydrolysis of methoxysilanes to silanols, 478
hydrosilylation, 478–479
hydrosilylation method of DAM and DAF, 480–481
hydrosilylation of diallyl maleate (DAM) with trichlorosilane (TCS), 477
Karstedt's catalyst, 478
methanolysis and isomerization, 479–480
methanolysis method, 481–482
methanolysis of silyl chlorides with excess methanol, 477–478
polymer additive in silicone applications, 477
reaction rates of hydrosilylation of DAM and DAF, 479f, 480f
TMOF reaction method, 484
Bis-(3-trimethoxysilylpropyl)maleate. *See* Bis-(3-trimethoxysilylpropyl)fumarate
Block copolymers
polysiloxanes, 5
reactive homopolymers, 4
synthesizing ABA-type, 412, 417
unique properties with hard/soft segment compositions, 395
See also Silicone–urethane copolymers; Telechelic oligomers, silicon terminated
Block copolymers by free radical polymerization
α-acylcarbon-centered radicals useful for initiating polymerizations, 445, 459
aldehyde-functional poly(dimethylsiloxanes) (PDMS) in study, 446t
aldehyde-functional polysiloxane synthesis, 446
block copolymers containing polysiloxane and poly(methyl methacrylate) (PMMA) segments, 455, 457–458
block copolymers containing polysiloxane and poly(styrene-*co*-MMA) segments, 458
block structures possible, 450
^{13}C NMR spectrum of ozonide from polysiloxane DC 7697, 449f
calculated versus measured molecular weights of the polymers, 455
calculating average number of styrene units in PS segments, 453
calculating number average molecular weight of copolymer containing PMMA segments, 455
calculating weight percentage of MMA units in copolymers, 457
copolymerization results using parent polysiloxane I, 458t
copolymer structure, 455
decomposition of polysiloxane segments in PDMS–polystyrene (PS) block copolymers, 447
gel permeation chromatography (GPC) curves for polysiloxane, polysiloxane–polystyrene block copolymer before and after hexane extraction, and the hexane extract, 451f
general chemistry, 447–450
generating radicals at polysiloxane ends, 448, 450
GPC curves for polysiloxane, polysiloxane–PS block copolymer, and polystyrene after degradation of copolymer, 454f
^{1}H NMR of recovered PS from polysiloxane–PS block copolymer degradation, 454f
^{1}H NMR spectrum of polysiloxane–PMMA block copolymer, 456f
^{1}H NMR spectrum of polysiloxane–PS block copolymer, 452f
NMR spectroscopy method, 447
PDMS containing 1-hexenyl-dimethylsiloxane end groups (DC

7691, 7692, and 7697), 446
polymerization procedure, 446–447
polysiloxane–PS block copolymers, 450, 453, 455
polysiloxane segments imparting important properties, 445–446
reaction of ozonized polysiloxanes with zinc and acetic acid, 448
synthesis and properties of block copolymers containing polysiloxane and PS segments, 453*t*
synthesis and properties of polysiloxane–PMMA copolymers, 457*t*
Brewster angle infrared (IR) spectroscopy. *See* 18-Nonadecyltrichlorosilane (NTS) monolayer
Brown and Slusarczuk base catalysis method for cyclic PDMS, 43
reaction mechanism, 43*f*
Bulk viscosity, hydrogenated cyclic and linear PDMS, 53

C

^{13}C nuclear magnetic resonance (NMR), small per-deuterated dimethylsiloxanes, 64
Caprolactone polyols. *See* Modifiers for photocurable epoxy coating formulations
Carbo-electro reduction process, chemical grade silicon, 11, 13
Cationic curing silicones, advantages and limitations, 170
Cationic polymerization of monomers with silsesquioxane (T_8) core
cationic photopolymerization method, 295
characterization of monomers, 287, 291
gel permeation chromatography (GPC) traces of functionalized T_8monomers, 290*f*
hydrosilation of 4-vinylcyclohexene oxide with T_8H, 285
hydrosilation of T_8H with 1,2-epoxy-5-hexene for octafunctional epoxide monomer III, 285–286
key starting material for multifunctional monomers (T_8H), 285
model compound I by hydrosilation of 1-octene, 285
monomer bearing eight epoxycyclohexyl groups, 286–287
octafunctional monomer bearing 1-propenyl ether groups (IX), 287
photoinitiated cationic polymerization, 291
preparation method for monomer V, 294–295
preparation method of T_8H, 293
reaction yielding octafunctional epoxy monomer V with T_8 core, 286
real-time infrared (RTIR) spectroscopy study of monomer conversion as function of irradiation time, 292*f*
steric hindrance in hydrosilation of T8H, 286
synthesis of 1-vinyloxy-2(2-propenoxy)ethane VIII, 294
synthesis of 4(2-vinyloxyethoxy)cyclohexene IV, 294
synthesis of monomers, 285–287
thermal analysis, 291, 293
thermal analysis method, 295
thermal analysis of polymers from T_8 monomers, 293*t*
typical NMR spectra for model compound I, 288*f*
typical NMR spectra for monomer III, 289*f*
Cationic polymerizations, UV coatings, 516
Cationic ring-opening polymerization of cyclotrisiloxanes
analytical procedure, 35

arrangements of siloxane groups in open chain monomer units, 24
back biting reactions, 22
calculations in terms of Markov chain theory, 35
chain transfer to another chain, 22
chain transfer to terminal Me$_3$SiO siloxane unit, 28, 30
chemoselectivity, 21–30
comparing initiators in polymerization, 27t
comparison of triad contents and regioselectivity factor for polymerization of model monomer with various cationic and anionic initiators, 33t
concentration of SiOH end groups with addition of CF$_3$SO$_3$SiMe$_3$, 28
determination of sequencing in terminal monomer unit, 35
1,1-diphenyl-3,3,5,5-tetramethylcyclotrisiloxane as model monomer, 21
effect of phenyl groups on oxonium ion formation, 31–32
end group information from MALDI TOF analysis, 28
experimental chemicals, 34
fragment of MALDI TOF mass spectrogram of polymer from model monomer, 29f
initiation, 26
monomer synthesis, 34
polymerization conditions, 24
polymer synthesis for MALDI TOF experiment, 35
polymer synthesis for sequencing studies, 34
propagation, 26
quantum mechanical calculations, 31–32
regioselectivity, 30–32
regioselectivity factor, 32
regioselectivity in polymerization of model monomer, 33t

^{29}Si NMR spectrum of polymer from model monomer, 25f
side reactions leading to siloxane chain cleavage, 22, 24
simplified mechanism, 24, 26
size exclusion chromatogram (SEC) of polymer from model monomer, 23f
source of poor regioselectivity, 31
transfer of terminal unit to end of another chain, 22
Ceramics, sol-gel, silica-type materials, 6
Chemistry of silicones
common inhibitors, 18
fillers, 16
formation of elemental silicon, 11, 13
hexamethyldisiloxane (MM), 14, 15
hydrosilylation reaction, 17
Karstedt catalyst, 17–18
methyl chlorosilane reaction (MCS), 13
octamethylcyclotetrasiloxane (D$_4$), 15
platinum-catalyzed addition cure, 17–18
proposed MCS mechanism, 13–14
ring-opening polymerization reaction, 15–16
scheme for conversion of sand to Pt-curable silicone formulation, 12f
siloxane polymers from methylchlorosilanes, 14–16
trichlorosilane (TCS) reaction, 16
Chojnowski acid catalysis
polymerization method, 43–44
reaction mechanism, 44f
Coatings. *See* Modifiers for photocurable epoxy coating formulations
Columnar mesomorphic behavior. *See* Cyclolinear polyorganosiloxanes (POCS)
Contact angles. *See* Thin film poly(dimethylsiloxane) (PDMS)
Contact lens
oxygen permeability, 296–297

poly(dimethylsiloxane) (PDMS), 3
requirements for candidate polymers, 296
See also Fluorosiloxane based hydrogels
Copolymeric dendrimers. *See* Radially layered copolymeric dendrimers
Copolymers
 AB diblock, 277
 ABA triblock, 272, 275–276
 graft, 278
 polysiloxanes, 5
 See also Block copolymers by free radical polymerization; Hybrid polymers from atom transfer radical polymerization (ATRP) and poly(dimethylsiloxane) (PDMS); Multi-block copolymers; Ruthenium catalyzed synthesis of *alt*-poly(carbosilane/siloxanes); Silicone–urethane copolymers; Telechelic oligomers, silicon terminated
Cross-conjugated copolymers
 copoly(arylene/1,1-vinylenes), 441–442
 fluorescence spectra, 442*f*
Crosslinked siloxanes. *See* Tospearl particles
Crosslinking
 curing behavior of siloxane-modified epoxy coatings, 519, 520*f*, 521*f*, 522
 curing reaction studies of siloxane-modified epoxy coatings, 519
 degree in elastomeric networks, 4–5
 possible mechanism in halogenated poly(carbosiloxane)s, 232, 235, 236
 spectroscopic investigation of crosslinking reactions, 518
 See also Gelation studies; Halogenated poly(carbosiloxane)s; Modifiers for photocurable epoxy coating formulations
Crown ethers, incorporation into alt(carbosilane/siloxane) copolymers, 439

Curable silicones
 cationic, 170
 free radical curing of acrylate and acrylamide silicones, 171
 photohydrosilylation using photolabile platinum catalysts, 171
 thiol-ene crosslinking of mercaptosiloxanes and vinylsiloxanes, 170–171
 See also UV curable silicones
Curing of coatings
 behavior of epoxy coatings, 519, 522
 possible crosslinking reactions of tetraethyl orthosilicate (TEOS) functionalized polyols with cycloaliphatic epoxide under cationic UV conditions, 527
 proposed condensation reactions of TEOS functionalized polyols under cationic UV conditions, 522
 reaction studies of siloxane-modified epoxy coatings, 519
 See also Modifiers for photocurable epoxy coating formulations
Cyanate ester resins
 applications, 165
 polymerization process, 164
 See also Telechelic siloxanes
Cyclic poly(dimethylsiloxanes)
 adsorption on surfaces of hydrogenated cyclic and linear PDMS, 57
 analysis of small per-deuterated dimethylsiloxanes, 63–69
 Beltzung reaction scheme for per-deuterated D_4 synthesis, 58*f*
 Brown and Slusarczuk base catalysis, 43
 bulk viscosity of hydrogenated cyclic and linear PDMS, 53
 ^{13}C NMR of small per-deuterated dimethylsiloxanes, 64
 catalytic preparation of both linear and cyclic per-deuterated PDMS, 72
 characterization of hydrogenated cyclic PDMS fractions, 47–50

Chojnowski acid catalysis, 43–44
comparing NMR results with gas liquid chromatography (GLC) traces for hydrogenated cyclic PDMS, 50
comparison between theoretical and experimental data for macrocyclization equilibrium constants for PDMS, 41f
condensations leading to phenyl-terminated linear oligomers, 76f
density and refractive index of hydrogenated cyclic and linear PDMS, 54, 55f
deuterated cyclic PDMS, 57–59
dimensions of rings and chains in dilute solution by small angle neutron scattering (SANS), 52
electron ionization mass spectroscopy (EIMS), low resolution, 50, 51f
entrapment of rings into networks, 57, 58f
fractionation of deuterated cyclic PDMS, 73, 74t
gas chromatography/ammonia chemical ionization mass spectroscopic (GC/CI–MS) analysis of small per-deuterated dimethylsiloxane cyclics, 69, 70f, 71t
gas chromatography/electron ionization mass spectroscopic (GC/EI–MS) analysis of small per-deuterated dimethylsiloxane cyclics, 65, 66f, 67f, 68t, 69
gel permeation chromatography (GPC), preparative, 45–47
glass transition temperature of hydrogenated cyclic and linear PDMS, 54
GLC of hydrogenated cyclic PDMS, 47
GLC traces showing range of individual ring sizes, 47f, 48f
^{1}H NMR of small per-deuterated dimethylsiloxanes, 63

hydrogenated cyclic PDMS, 39–41
infrared (IR) of small per-deuterated dimethylsiloxanes, 64–65, 66f
investigations of hydrogenated cyclic and linear PDMS, 52–57
Jacobson and Stockmayer cyclization theory, 39–41
monitoring triflic acid reaction progress using GLC, 61, 62f
peak assignments for EIMS spectrum of hydrogenated cyclic PDMS, 52t
peak assignments for GC/CI–MS spectra of small per-deuterated cyclic dimethylsiloxanes, 71t
peak assignments for GC/EI–MS of small per-deuterated cyclic dimethylsiloxanes, 68t
preparation of deuterated cyclic PDMS, 60–62
preparation of hydrogenated linear and cyclic PDMS, 42–44
preparation of larger per-deuterated siloxanes, 69, 72
preparation of silyl triflates from silanes and oligosilanes with phenyl groups, 60f
preparation of small cyclic per-deuterated dimethylsiloxanes, 61, 62f
preparative GPC during fractionization of deuterated cyclic PDMS, 73f
preparative GPC instrument, 45, 47
production of silyl triflates and silanol intermediates, 75f
proposed use of triflic acid to produce siloxane oligomers, 59f
reaction mechanism proposal, 73, 75
reaction of phenyl-terminated oligomers leading to cyclic or linear products, 76f
reasons for studying ring macromolecules, 41–42
ring-chain equilibration in PDMS, 39
schematic of preparative GPC

instrument, 46f
separation of linear and cyclic PDMS, 44
^{29}Si NMR chemical shift of hydrogenated cyclic and linear PDMS, 56
^{29}Si NMR of hydrogenated cyclic PDMS, 49–50
^{29}Si NMR of small per-deuterated dimethylsiloxanes, 64
structural comparison of linear and cyclic polymers, 42f
synthesis of per-deuterated dimethylsiloxanes, 60f
use of triflic acid for phenyl group removal from dimethyldiphenylsilane, 59–60
See also Poly(dimethylsiloxane) (PDMS)
Cyclic trapping, reactive homopolymers, 5
Cyclization theory, Jacobson and Stockmayer, 39–41
Cyclolinear polyorganosiloxanes (POCS)
ability to form columnar liquid crystalline (LC) phases, 99
affects of varying homopolymer microtacticity, 111
characteristics of mesophase oligomers with various end groups, 106t
columnar polymesomorphism of POCSs, 100, 103
comparing molecular conformation of POCS with decamethylcyclohexasiloxane fragments (PMCS-6) with PMCS-4, 108–109
controversy in classification of polymer mesophases, 99
1D-monolayered mesophase, 105, 108–112
1D-monolayer polymesomorphism, 109, 111
2D columnar mesophase, 100, 103
determining ability of molecules to form mesophase structures, 109
driving forces for mesophase development in PMCS-6, 108
effect of chemical structure of organic side groups and siloxane backbone on structural parameter d_m of 2D-columnar mesophase, 102f
effect of chemical structure on temperature transitions of POCS exhibiting 2D-columnar mesophase, 101t
effect of chemical structure on temperature transitions of PMCS-6, 106t
experimental, 99–100
general formula of POCS, 98–99
influence of microtacticity of PMCS-6 on ability to form mesophase, 105
mesophase of PMCS-6 with 1-D positional long-range order, 108
mesophase oligomers and block-copolymers, 111–112
model of structural organization, 109
POCS displaying columnar mesomorphic behavior, 112
probable arrangement of PMCS-6 in mesophase, 112
recording thermal transitions with differential scanning calorimetry (DSC), 99–100
reversible polymesomorphic transition within columnar phases, 103
role of length of macromolecules in mesophase behavior, 111
self-assembling of PMCS-6 into 1D-monolayered mesophase, 110f
self-organization into 2D-columar mesophase, 104f
self-organization of PMCS-6 into 1D-mesophase, 112–113
sensitivity of structural organization of mesophase to chemical nature of end groups, 111–112
temperature dependencies of interlayer distance d_m for PMCS-6

homopolymers with various microtacticity, 107*f*
tendency to nanoscale phase separation on inter- and intramolecular levels, 108
tendency toward ordering in columar POCS, 103
transformation of long-range order during polymesomorphic transition, 111
transforming noncrystallizable mesomorphic PMCS-6 into mesomorphic glasses below glass transition temperature, 105, 108
X-ray diffraction patterns of PMCS-6 in mesomorphic state, 107*f*
X-ray diffraction patterns of POCS homopolymers with cyclotetrasiloxane fragments in mesomorphic state with 2D-hexagonal molecular packing, 101*f*
Cyclosiloxanes. *See* Photoluminescence of phenyl- and methyl-substituted cyclosiloxanes
Cyclotrisiloxanes. *See* Cationic ring-opening polymerization of cyclotrisiloxanes

D

Dendrimers
architecturally driven property differences between dendrimers and corresponding linear polymers, 247*t*
copolymeric, 248
emergence of physical properties, 243, 248
generalized structure, 242–243
molecular characteristics of ethylenediamine core poly(amidoamine) (PAMAM) dendrimers, 246*t*
preparation of radially layered poly(amidoamine-organosilicon) (PAMAMOS) copolymeric, 249
schematic of homopolymeric, 244*f*
schematic of radially layered and segmented ordered copolymeric, 245*f*
surface properties of PAMAMOS, 255, 257
See also Radially layered copolymeric dendrimers
Density, hydrogenated cyclic and linear PDMS, 54, 55*f*
Deuterated cyclic poly(dimethylsiloxane)
analysis of small cyclic per-deuterated dimethylsiloxanes, 63–69
Beltzung reaction scheme, 58*f*
fractionation, 73, 74*t*
polymerization of deuterated D_4, 57–59
preparation, 60–62
preparation of small cyclic per-deuterated dimethylsiloxanes, 61, 62*f*
synthesis of per-deuterated dimethylsiloxanes, 60*f*
See also Cyclic poly(dimethylsiloxanes); Per-deuterated siloxanes
Diallyl fumarate (DAF). *See* Bis-(3-trimethoxysilylpropyl)fumarate
Diallyl maleate (DAM). *See* Bis-(3-trimethoxysilylpropyl)fumarate
Diblock copolymers
living anionic PDMS with attachable initiator and species capable of initiating atom transfer radical polymerization, 277
See also Hybrid polymers from atomic transfer radical polymerization (ATRP) and poly(dimethylsiloxane) (PDMS)
Diene isomerization, ruthenium catalysis, 436–437
Dimethyldioxirane
modification of polymers with ≡Si–H groups, 421
preparation in acetone, 420
reaction scheme, 424
See also Silanol-containing polymers

Dimethyldiphenylsilane
  removing phenyl groups with triflic acid, 59–60
  See also Trifluoromethanesulfonic (triflic) acid
Dimethyl fumarate, inhibitor with Karstedt's catalyst, 18
1,1-Diphenyl-3,3,5,5-tetramethylcyclotrisiloxane
  model monomer, 21
  synthesis, 34
  See also Cationic ring-opening polymerization of cyclotrisiloxanes
Disiloxane, subject of electronic structure calculations, 82–83
Drug delivery systems
  design with interpenetrating networks, 383–384
  kinetic study of drug model release, 387, 389, 391
  See also Interpenetrating polymer networks (IPNs)

E

Elastomeric networks, reactive homopolymers, 4–5
Elastomers, reinforcement with inorganic filler, 485
Electroconductivity, poly(siloxyethylene glycol), 376
Electron-beam resist
  evaluation of poly(siloxyethylene glycol), 374
  sensitivity characteristic curve of film against electron-beam exposure, 375f
  See also Resist
Electron ionization mass spectrometry (EIMS)
  hydrogenated cyclic PDMS fractions, 50, 51f
  peak assignments for hydrogenated cyclic PDMS fraction, 52t

Electronic calculations. *See* KOH-catalyzed ring-opening polymerization of cyclic siloxanes
Epoxide monomers
  classes in UV-curable coatings, 516–517
  See also Modifiers for photocurable epoxy coating formulations
Epoxide networks. *See* Rubbery organic networks
Equivalent box model (EBM)
  comparing experimental with theory, 499
  relative modulus of organic-inorganic hybrid as function of silica phase volume fraction, 498f
  See also Rubbery organic networks
Etching reactions, selfbias for silicon oxide (SiOx) film deposition, 549
Ether linkage. *See* Fluorosilicones with perfluorocyclobutene (PFCB) aromatic ether linkage

F

Fillers
  polysiloxane materials, 16
  surface treatment, 16
  See also Rubbery organic networks
Flexibility, siloxane homopolymers, 2–3
Flory–Stockmayer (F–S) theory. *See* Gelation studies
Fluorosilicones with perfluorocyclobutane (PFCB) aromatic ether linkage
  di-olefin intermediates for hydrosilation, 317
  di-olefin TFVE monomers for hydrosilation polymerization, 317f
  first PFCB siloxane polymer, 311f
  hydrosilation polymerization of TFVE di-olefin monomers, 318f
  intermediate strategy for preparing

siloxane functionalized PFCB polymers, 312–313
linear cyclopolymerization of siloxanyl monomers, 313f
linear PFCB fluorosilicones, 313
mono-olefin intermediates for hydrosilation, 314–315
mono-olefin trifluorovinyl ether (TFVE) intermediates for pendant hydrosilation, 314f
network PFCB fluorosilicones, 314–320
PFCB chemistry, 309–310
PFCB fluorosilicone networks 11'-14', 319–320
PFCB fluorosilicone networks 4'-6', 316–317
PFCB fluorosilicone networks from mono-olefin derived copolymers, 316f
PFCB silicones from phenols, 311–313
PFCB siloxane/organosilane networks, 319f
polymerization of tris-vinyl ether (TVE) monomer, 310f
selected properties of di-olefin hydrosilation oligomers, 318t
selected properties of linear PFCB fluorosilicones, 313t
selected properties of mono-olefin hydrosilated copolymers, 315t
selected thermal properties of fluorosilicone networks, 317t
selected thermal properties of PFCB fluorosilicone networks, 320t
TFVE derived silicone copolymers via hydrosilation of mono-olefins, 315f
TFVE functional silicones 11-14, 318
TFVE functional silicones 4-6, 315–316
thermal polymerization of trifluorovinyl aryl ethers, 309f
traditional PFCB materials, 310
Fluorosiloxane based hydrogels
analytical techniques, 300
copolymers of methacrylate end-capped fluoro-substituted siloxanes, 297
equation for determining water content, 300
experimental materials, 298
general procedure for synthesis of methacryloxypropyl tris(3-(2,2,3,3,4,4,5,5-octafluoropentoxy)propyldimethylsiloxy)silane (Tris(F)), 298
hydrogel formulation, 303, 305–306
kinetics of polymerization, 305
limitations to overcome before designing hydrogels based on PDMS, 297
mechanical and physical property results for films cast from Tris(F), Di(F), and Mono(F), 305t
methods for development with high oxygen permeability, 297
modified synthetic procedure for Mono(F) preparation, 304f
oxygen permeability, 296–297
procedure for synthesis of 1-(methacryloxypropyl)-3-(3-(2,2,3,3,4,4,5,5-octafluoropentoxy)propyl)tetramethyldisiloxane (Mono(F)), 298–300
structure of Tris(F), Di(F), Mono(F), and FSi, 301f
synthesis, 300, 303
synthetic procedure for Tris(F) preparation, 302f
varying dimethylacrylamide (DMA) content in hydrogels, 305–306
Fouling
soft and hard, 180
See also Biofouling release coatings
Free radical curing, acrylate and acrylamide silicones, 171
Free radical polymerizations
UV coatings, 516
See also Block copolymers by free radical polymerization
Fumed silica, trichlorosilane (TCS)

reaction, 16
Functionalized polymers, reactive homopolymers, 4–5

G

Gas chromatography/ammonia chemical ionization mass spectroscopy (GC/CI–MS), small per-deuterated dimethylsiloxane cyclics, 69, 70f, 71t
Gas chromatography/electron ionization mass spectroscopy (GC/EI–MS), small per-deuterated dimethylsiloxane cyclics, 65, 66f, 67f, 68t, 69
Gas liquid chromatography (GLC)
comparing NMR results with gas liquid chromatography (GLC) traces for hydrogenated cyclic PDMS, 50
hydrogenated cyclic PDMS fractions, 47, 48f
methods for silicas, 504–505
monitoring triflic acid reaction progress using GLC, 61, 62f
Gas-phase calculations, KOH-catalyzed ring-opening polymerization, 91–94
Gel content
siloxane-modified epoxy coatings, 529, 530f
See also Modifiers for photocurable epoxy coating formulations
Gel permeation chromatography (GPC)
fractionation of deuterated cyclic PDMS, 73f, 74t
preparative, 45–47
preparative instrument, 45, 47
schematic of preparative instrument, 46f
Gelation studies
α_c for non-stoichiometric polymerization mixtures, 195
α_c product of critical extents of reaction at gelation, 195

Ahmad–Rolfes–Stepto (A–R–S) analysis of PDMS gelation data, 199–203
analysis of PDMS data as plots of $v^{3/2}\lambda_{b0}$ versus $1/c_{b0}$, 201f
A–R–S gelation criterion, 199
chemical structures of reactants for PDMS, 196f
determining stiffness parameter values (b) for each PDMS-forming system, 200, 202t
experimental, 195
gel points as functions of reactive-group dilution $1/c_{b0}$, 198f
gel points for polymerization reactions, 195, 197
illustrating "internal" and "external" concentrations of reactive groups around chosen group, 198f
molar mass of linear PDMS chains and numbers of skeletal bonds, 196t
number of skeletal bonds (v) in linear subchain forming smallest look structures for PDMS non-linear polymerizations, 196t
PDMS network formation, 198f
plot of b versus $1/v$ for PDMS-forming systems, 202f
pre-gel intramolecular reaction, 197
quantifying competition between intermolecular and intramolecular reaction, 197, 199
ring-forming parameter, λ_{b0}, 199
sensitivity of b from gel points, 203
theory of non-linear polymerizations and analysis of experimental data, 195, 197–199
Truesdell function, $\varphi(1,3/2)$, 199
values of b by A–R–S theory and chain-conformational analyses, 202t
Glass transition temperature
composition dependence of polymeric blends, 426–427, 428f
hydrogenated cyclic and linear PDMS, 54

poly(dimethylsiloxane) (PDMS), 2–3
See also Silanol-containing polymers
Glasses, polymer modified, silica-type materials, 6
Glycine
  differential scanning calorimetry results for amino acid functional siloxanes, 161*t*
  evaluating efficacy and generality of synthetic procedure, 141
  See also Amino acid functional siloxanes
Graft copolymers, PDMS macroinitiators, 278
Grignard reactions, *alt*-poly(carbosilane/siloxane), 433

H

^{1}H nuclear magnetic resonance (NMR), small per-deuterated dimethylsiloxanes, 63
Halogenated poly(carbosiloxane)s
  ^{1}H NMR of product from 3-Cl_2/tetramethylcyclotetrasiloxane (D_4) reaction, 233*f*
  ^{1}H NMR spectra of chlorinated polymers 3-Cl_2 and 3-Cl(Me), 230*f*
  crosslinking reaction of $SiMe_2(OSiMe_2H)_2$ and polymer 3-Cl_2 with UV radiation, 238–239
  experimental materials, 235
  gel permeation chromatography (GPC) molecular weight data, 229*t*
  hydrosilylation polymerization scheme, 228
  infrared spectrum of crosslinking reaction during course of reaction, 234*f*
  insertion of siloxanes into polymer 3-Cl_2, 233
  insertion reaction of $[SiMe(Vi)O]_4$ (DVi_4) into polymer 3-Cl_2 and subsequent crosslinking reaction, 238
  insertion reactions of $[SiMe2O]_4$ (D_4) into polymer 3-Cl_2, 238
  modification scheme of polymer 3-Cl_2, 231
  polymer insertion reactions, 232, 235
  polymer modification reactions, 229, 232
  polymer synthesis and characterization, 227
  possible crosslinking mechanism, 232, 235, 236
  potential monomers for hydrosilylation polymerization, 228
  preparation of polymer 3-Cl_2, 235, 237
  preparation of polymer 3-Cl(Me) from polymer 3-Cl_2, 237
  preparation of polymer 3-Cl(Me) via hydrosilylation polymerization, 237–238
  preparation of polymer 3-Cl($SiMe_2H$) from polymer 3-Cl_2, 237
  preparation of polymer 4-Cl_2, 237
  reaction between polymer 3-Cl_2 and hexamethyltrisiloxane, 234
  structure of intramolecular cyclization product isolated during polymerization, 231*f*
Hartree–Fock (HF) method. *See* KOH-catalyzed ring-opening polymerization of cyclic siloxanes
Hexamethylcyclotrisiloxane (D_3). *See* KOH-catalyzed ring-opening polymerization of cyclic siloxanes
Hexamethyldisiloxane (MM)
  chain stopper, 15
  hydrolysis/condensation of Me_3SiCl, 14, 15
Homopolymers
  flexibility, 2–3
  permeability, 3
  reactive, 4–5
  siloxane-type, 2–4
  surface and interfacial properties, 4
Hybrid polymers from atom transfer radical polymerization (ATRP) and poly(dimethylsiloxane) (PDMS)

AB diblock copolymers, 277
ABA triblock copolymers, 272, 275–276
ATRP as alternative to ionic techniques, 271–272
ATRP from silsesquioxane initiators, 278
ATRP of styrene from PDMS macroinitiator, 275
ATRP of vinyl monomers from PDMS macroinitiators, 277t
conversion of 1,3,5,7-tetramethylcyclotetrasiloxane into ATRP initiator containing benzyl chloride moieties, 278
copolymerization of PDMS with tougher materials, 270–271
graft copolymers, 278
kinetic plot for ATRP of styrene from difunctional PDMS macroinitiator, 275f
kinetic plot for ATRP of styrene initiated from benzyl chloride-POSS (polyhedral oligomeric silsesquioxane), 280f
living anionic PDMS terminated with attachable initiator containing silyl chloride moiety and species capable of ATRP initiation, 277
molecular weight plot for ATRP of styrene from difunctional PDMS macroinitiator, 276f
molecular weight plot for ATRP of styrene initiated from benzyl chloride-POSS, 281f
motivation for synthesis of di- and triblock copolymers, 272
polymerization of styrene from benzyl chloride-POSS, 278
schematic of halogen atom transfer by transition metal species, 271
schematic of silsesquioxanes, 271
SEC traces of ATRP of styrene initiated from tetrafunctional cyclotetrasiloxane, 279f
series of polymerizations from benzyl chloride and 2-bromoisobutyryloxy terminal PDMS macroinitiators, 276, 277
silsesquioxane initiator, 279
star polymers, 278
summary of architectures available through ATRP, 272, 273f
synthesis of ABA triblock copolymers from terminal functionalized PDMS, 274

Hydrogels. *See* Fluorosiloxane based hydrogels
Hydrogen bond interactions. *See* Silanol-containing polymers
Hydrogenated cyclic poly(dimethylsiloxane) (PDMS)
adsorption on surfaces, 57
Brown and Slusarczuk base catalysis, 43
bulk viscosity, 53
characterization of fractions, 47–50
Chojnowski acid catalysis, 43–44
comparison between theoretical and experimental data for macrocyclization equilibrium constants for PDMS, 41f
density and refractive index, 54, 55f
dimensions in dilute solution by small angle neutron scattering (SANS), 52
electron ionization mass spectroscopy (EIMS), low resolution, 50, 51f
entrapment of rings into networks, 57, 58f
gas liquid chromatography (GLC), 47
glass transition temperature, 54
GLC traces of fractions with range of individual ring sizes, 47f, 48f
Jacobson and Stockmayer cyclization theory, 39–41
peak assignments for EIMS spectrum, 52t
preparation, 42–44
ring-chain equilibration in PDMS, 39
separation from linear PDMS, 44
^{29}Si NMR chemical shift, 56
^{29}Si NMR spectroscopy, 49–50
See also Cyclic poly(dimethylsiloxanes)

Hydrogenated linear
  poly(dimethylsiloxane) (PDMS)
  adsorption on surfaces, 57
  bulk viscosity, 53
  Chojnowski acid catalysis, 43–44
  density and refractive index, 54, 55f
  dimensions in dilute solution by small
    angle neutron scattering (SANS), 52
  glass transition temperature, 54
  preparation, 42
  separation from cyclic PDMS, 44
  ^{29}Si NMR chemical shift, 56
Hydrophobic monolayers
  organotrichlorosilane monolayers by
    Langmuir method, 332–333
  See also 18-Nonadecyltrichlorosilane
    (NTS) monolayer
Hydrosilylation polymerization
  organosilicon polymers, 226–227
  See also Halogenated
    poly(carbosiloxane)s; Organosilicon
    linear polymer with
    octaorganooctasilsesquioxanes (T_8)
    in main chain
Hydrosilylation reactions
  adducts of Markovnikov and anti-
    Markovnikov addition, 149, 155
  diallyl maleate (DAM) with
    trichlorosilane, 477, 478–479, 480–
    481
  di-olefin trifluorovinyl ether (TFVE)
    monomers for, 317
  grafting amino acids bearing
    unsaturated vinyl or allyl groups,
    129–130
  mono-olefin TFVE intermediates for,
    314–315
  optically active five-membered cyclic
    silicon compound by intramolecular,
    470, 472
  platinum-catalyzed, 17–18
  platinum catalyzed curing method in
    polymeric systems, 477
  alt-poly(carbosilane/siloxane), 434
  polymerization of TFVE di-olefin
    monomers, 318f

  side reactions, 155–156
  TFVE functional silicones, 315–316
  typical reaction, 154f
  See also Amino acid functional
    siloxanes; Cationic polymerization
    of monomers with silsesquioxane
    (T_8) core
Hydroxyl-terminated chains, reactive
  homopolymers, 4
Hyperbranched materials, preparation
  with ruthenium catalysts, 441
Hyperbranched polyethoxysiloxane
  condensation method of
    polyethoxysiloxane hydrolysate, 506
  hydrolysis method of
    polyethoxysiloxane, 506
  infrared (IR) absorption spectra of
    system, 507–508
  main steps of synthesis, 507
  ^{29}Si NMR spectra, 508f
  size exclusion chromatography (SEC)
    of trimethylsilylated, 509f
  synthesis of molecular silica sols,
    509–511
  synthetic method, 505–506
  synthetic results, 506–509
  See also Silicas

I

Impact modifiers for cyanate ester
  resins. *See* Telechelic siloxanes
Industrial applications. *See* Tospearl
  particles
Infrared spectroscopy (IR), small per-
  deuterated dimethylsiloxanes, 64–65,
  66f
Inhibitors, platinum-catalyzed
  hydrosilylation reaction, 17–18
Inorganic-organic composites,
  dendrimer-based, 266
Inorganic polymers
  desirable in hybrid materials, 270
  See also Hybrid polymers from atom
    transfer radical polymerization

(ATRP) and poly(dimethylsiloxane) (PDMS)
Interfacial properties, polysiloxanes, 4
Interpenetrating polymer networks (IPNs)
drug delivery systems, 383–384
equation of rate constant for simple kinetic model, 389, 391
experimental, 384–385
glass transition temperatures (Tg) as function of polycarbonate (PC) weight proportion in IPN, 388f
^{1}H and ^{13}C NMR methods, 385
^{1}H NMR spectrum of PSO (benzoylperoxide modified polysiloxane), 386f
kinetics of phenol release from different polymeric systems at pH=7.5, 389t
kinetics of phenol release in pH=7.5 buffered medium from different polymeric systems, 389, 390f
kinetics of phenol release in pH=7.5 buffered medium from single networks from different polysiloxane precursors, 391, 393f
kinetics of phenol release in pH=7.5 buffered medium from s-IPNs (PSO embedded in PC network), 391, 392f
kinetics study of phenol release, 385
kinetic study of drug model release, 387, 389, 391
kinetics of phenol release on completely soluble system, 389
monitoring reaction by decrease in Si-H IR band, 384–385
obtaining single polysiloxane networks, 385
PC network formation, 385
physicochemical characterization, 387
poly(dimethylsiloxane) (PDMS) as host network, 384
schematic of hydrosilylation reactions, 384
series of prepared IPNs, 385
See also Silanol-containing polymers
Ion conductivity

characteristics as function of temperature, 380f
poly(siloxyethylene glycol) gels, 376
test method, 362

J

Jacobson and Stockmayer cyclization theory
comparison between theoretical and experimental data for macrocyclization equilibrium constants for PDMS, 41f
molar cyclization equilibrium constants, 39–40

K

Karstedt's catalyst
hydrosilylation reaction, 17
hydrosilylation using diallyl fumarate or diallyl maleate with trichlorosilane (TCS), 478–479
inhibitors, 18
octafunctional epoxy monomer with silsesquioxane core, 286
octasubstituted silsesquioxane, 285
star polymer by atom transfer radical polymerization, 278
synthesis of ABA triblock copolymers from terminal functionalized PDMS, 272, 274
Kerner–Nielsen model
experimental comparison to theory, 495
relative modulus of organic-inorganic (O-I) hybrid as function of silica phase volume fraction, 498f
See also Rubbery organic networks
Kinetics study. See Interpenetrating polymer networks (IPNs)
KOH-catalyzed ring-opening polymerization of cyclic siloxanes
ab initio electronic calculations along gas-phase reaction path for

octamethylcyclotetrasiloxane (D_4) and hexamethylcyclotrisiloxane (D_3), 95–96
basis sets, 94–95
computational method, 83, 85
determining Hartree–Fock (HF) optimized geometries, 83
disiloxane in electronic structure calculations, 82–83
effect of basis set superposition errors on relative energies between reactants and addition complex, 95
electronic structure, 83, 85
focusing on thermodynamically controlled reaction, 83
gas-phase calculations, 91–94
gas-phase reaction path of D_4 and D_3, 88f
general trend of variation of geometries and atom charges along D_3 reaction path, 93–94
Hartree–Fock gas-phase reaction path of D_3, 90f
KOH-D_4 adduct structures for initial step along reaction path, 91
mechanism, 82
nature of 5-coordinate Si species, 92–93
optimized geometries for D_3 and D_4 rings, 84t
reaction paths for ring-opening of D_4 and D_3 in gas-phase, 85
solvation calculations, 94
solvation model, 85
stability of insertion and ring-opened products, 93
structures of addition complexes with D_3, 89f
structures of adducts of KOH and D_4, 86f
structures of critical points along D_4 reaction path, 87f
synthesis of linear siloxane polymers, 82
thermochemistry of addition reactions to siloxane rings, 90t

L

Langmuir method
 organosilane monolayers for multilayer film preparation, 333
 See also 18-Nonadecyltrichlorosilane (NTS) monolayer
Lateral force microscopy (LFM). See 18-Nonadecyltrichlorosilane (NTS) monolayer
Leucine
 DSC results for amino acid functional siloxanes, 161t
 evaluating efficacy and generality of synthetic procedure, 141
 See also Amino acid functional siloxanes
Liquid crystalline behavior. See Cyclolinear polyorganosiloxanes (POCS)

M

Macroinitiators
 silsesquioxanes, 278
 See also Hybrid polymers from atom transfer radical polymerization (ATRP) and poly(dimethylsiloxane) (PDMS)
Macromolecules
 dendritic polymers, 242
 reasons for studying ring, 41–42
 See also Radially layered copolymeric dendrimers
Marine biofouling
 problem for ships, 180
 See also Biofouling release coatings
Materials, new. See Poly(siloxyethylene glycol) (PSEG)
Mechanical properties. See Modifiers for photocurable epoxy coating formulations
Medical applications, polysiloxanes, 5
Melts, polymer. See Reinforced poly(dimethylsiloxane) melts

Mesophases. *See* Cyclolinear polyorganosiloxanes (POCS)
Methacrylate functionalized fluorosiloxy silanes. *See* Fluorosiloxane based hydrogels
Methacrylate polymerizations
  atom transfer radical polymerization of vinyl monomers from PDMS macroinitiators, 277*t*
  initiation by PDMS containing in-chain silylpinocolate moieties, 272
  See also Hybrid polymers from atom transfer radical polymerization (ATRP) and poly(dimethylsiloxane) (PDMS)
Methylchlorosilane reaction (MCS)
  critical factors, 13
  mechanism lacking at molecular level, 13–14
  proposed mechanism, 14*f*
  typical reaction from elemental silicon, 13
Methylchlorosilanes, siloxane polymers from, 14–16
Methyl-substituted cyclosiloxanes. *See* Photoluminescence of phenyl- and methyl-substituted cyclosiloxanes
Methyl trimethoxysilane
  hydrolysis and condensation for Tospearl production, 533
  See also Tospearl particles
Model networks, bimodal, 5
Models
  relative modulus of organic-inorganic (O-I) hybrid as function of silica phase volume fraction, 498*f*
  See also Rubbery organic networks
Modifiers for photocurable epoxy coating formulations
  adhesive properties, 529
  coating formulations, 517–518
  curing behavior, 519, 522
  curing reaction studies, 519
  effects of crosslinkers on surface cure rate of polyol or siloxane polyol/epoxide coatings, 520*f*
  effects of crosslinkers on through cure rate of polyol or siloxane polyol/epoxide coatings, 521*f*
  experimental materials, 517
  gel content, 529
  gel content and swelling extent of di-tetraethyl orthosilicate (di-TEOS) and tri-TEOS functionalized polyol modified coatings containing 60 wt%, 530*f*
  instrumentation, 517
  mechanical properties, 522, 527
  possible crosslinking reactions of TEOS functionalized polyols with cycloaliphatic epoxide under cationic UV conditions, 527
  proposed condensation reactions of TEOS functionalized polyols under cationic UV conditions, 522
  pull-off adhesion, 529
  pull-off adhesion of di-TEOS and tri-TEOS functionalized polyol modified coatings as function of ratio of siloxane polyol to diol, 531*f*
  reaction of caprolactone diol or triol with TEOS, 519
  siloxane functionalization of tone polyols, 517
  spectroscopic investigation of crosslinking reactions, 518
  surface and through cure rate of di-TEOS functionalized polyol modified UV coatings as function of ratio of siloxane polyol to parent polyol, 523*f*
  synthesis and characterization of reactive diluents, 518–519
  tensile elongation of diol or di-TEOS functionalized polyol/epoxide coatings as function of crosslinker, 526*f*
  tensile elongation of triol or tri-TEOS functionalized polyol/epoxide coatings as function of crosslinker, 526*f*
  tensile modulus of diol or di-TEOS

functionalized polyol/epoxide coatings as function of crosslinker, 523f
tensile modulus of di-TEOS polyol modified coatings as function of ratio of siloxane polyol to diol, 528f
tensile modulus of triol or tri-TEOS functionalized polyol/epoxide coatings as function of crosslinker, 524f
tensile modulus of tri-TEOS polyol modified coatings as function of ratio of siloxane polyol to diol, 528f
tensile strength of diol or di-TEOS functionalized polyol/epoxide coatings as function of crosslinker, 524f
tensile strength of triol or tri-TEOS functionalized polyol/epoxide coatings as function of crosslinker, 525f
Molar cyclization equilibrium constants, Jacobson and Stockmayer, 39–41
Molecular sponges, dendrimer-based, 266
Multi-block copolymers
contact angle values versus time for cured PDMS and various cured PDMS-b-PEO copolymers, 357f
copolymer PDMS-b-PEO starting materials, 354
crosslinked film preparation from PDMS-b-PEO copolymers by hydrosilylation, 354
differential scanning calorimetry (DSC) method, 354
DSC thermograms for PDMS-b-PEO copolymers, 356f
effect of increasing PEO amount, 355
experimental, 354
glass transition temperatures, cold crystallization temperatures, and melting points for PDMS-b-PEO copolymers, 355t
PDMS-b-PEO copolymers, visual clarity and number of phases, 355t
phase separation behavior, 357
phase study, 354–355, 357
surface behavior, 355
Multilayer films. *See* 18-Nonadecyltrichlorosilane (NTS) monolayer
Murai reaction. *See* Ruthenium catalyzed synthesis of *alt*-poly(carbosilane/siloxanes)

N

Nanocomposites. *See* Rubbery organic networks
Nanoscopic reactors, dendrimer-based, 266
Negative working resist, poly(siloxyethylene glycol), 372, 374
Networks
elastomeric, reactive homopolymers, 4–5
entrapment of rings, 57, 58f
perfluorocyclobutane (PFCB) fluorosilicones, 316–317, 319–320
See also Fluorosilicones with perfluorocyclobutene (PFCB) aromatic ether linkage; Gelation studies; Interpenetrating polymer networks (IPNs); Radially layered copolymeric dendrimers; Rubbery organic networks
New materials. *See* Poly(siloxyethylene glycol) (PSEG)
Non-linear polymerizations, theory, 195, 197–199
Non-medical applications, polysiloxanes, 5–6
18-Nonadecyltrichlorosilane (NTS) monolayer
AFM (atomic force microscopy) and lateral force microscopy (LFM) methods for surface morphology and mechanical behavior of monolayer, 334
AFM and LFM images of (carboxylated NTS

(NTS$_{COOH}$)/FOETS) mixed monolayers in air, 346f
AFM and LFM images of NTS/FOETS ([2-(perfluorooctyl)ethyl]trichlorosilane) mixed monolayer by Langmuir method, 344f
AFM and LFM images of NTS-NTS$_{COOH}$/FOETS ultrathin film, 349f
Brewster angle infrared (IR) measurement, 334–335
Brewster angle IR spectra for 1- to 5-layer NTS films by Langmuir deposition, 341f
characterization of monolayer and multilayer film, 334–335
characterizing layered structure of multilayer film by X-ray reflectivity (XR), 339, 343
construction of NTS multilayer film, 339, 343
electron diffraction (ED) studying molecular aggregation of monolayer, 334
LFM image of NTS$_{COOH}$/FOETS mixed monolayer in vacuum, 347f
molecular aggregation state in NTS monolayer, 335, 339
NTS$_{COOH}$/FOETS mixed monolayer, 345
NTS/FOETS mixed monolayer, 343, 345
NTS monolayer preparation, 333–334
NTS-NTS$_{COOH}$/FOETS ultrathin film, 345, 348
oxidation of NTS monolayer, 334
π-A isotherm for NTS monolayer on water subphase at 293 K, 338f
schematic of domain height control in NTS/FOETS mixed monolayer, 350f, 351f
schematic of NTS multilayer film formation by Langmuir method, 336f, 337f
surface free energy measurement, 334

XPS C1s spectra for NTS and NTS$_{COOH}$ monolayers, 340f
XR and fitting curves for NTS monolayer on silicon wafer substrate, 340f
XR and fitting curves for NTS multilayer film on silicon wafer substrate, 342f
XR method, 335
X-ray photoelectron spectroscopy (XPS) method, 334
Nucleophilic substitution reactions, reactions of silane and related compounds, 175

O

Octamethylcyclotetrasiloxane (D$_4$) isolation by distillation, 15
See also KOH-catalyzed ring-opening polymerization of cyclic siloxanes
Octaorganooctasilsesquioxanes. See Organosilicon linear polymer with octaorganooctasilsesquioxanes (T$_8$) in main chain
Optically active poly(carbosilane)s and poly(carbosiloxane)s. See Silicon-containing polymers
Organic-inorganic (O-I) hybrid systems advantage, 485
See also Rubbery organic networks
Organic-inorganic polymeric hybrids combining organic polymers with inorganic oxides by sol-gel process, 419–420
See also Silanol-containing polymers
Organic polymers functionality in terms of block copolymers, 271
See also Hybrid polymers from atom transfer radical polymerization (ATRP) and poly(dimethylsiloxane) (PDMS)
Organization. See Cyclolinear polyorganosiloxanes (POCS)

Organosilanols, forming hydrogen-bonded complexes, 419
Organosilicon linear polymer with octaorganooctasilsesquioxanes (T_8) in main chain
  alternative approach to synthesis of T_8, 215, 217
  approaches to obtaining target linear polymers containing T_8 in main chain, 218, 221
  differential scanning calorimetry curve for high molecular fraction of polymer, 221, 222f
  establishing structure of compounds synthesized, 218, 220f
  experimental, 223–224
  first members of polyhedral organosilsesquioxane homologous, 216f
  formation of T_8 structure during hydrolytic polycondensation, 217–218
  ^{1}H NMR spectra of 1,4-bis(2-chloromethylphenylsilylethyl)hexamethyloctasilsesquioxane, 220f
  ^{1}H NMR spectra of 1,4-bis(2-phenyldimethylsilylethyl)hexamethyloctasilsesquioxane, 220f
  hydrosilylation reaction of functionalized and nonfunctionalized derivatives of starting macromonomer, 218, 219
  instrumental methods, 223
  intermolecular reaction of hydrolytic polycondensation of hexafunctionalized organosiloxanes, 217
  molecules containing incompletely condensed fragments (T_8), 215
  polyhedral organosiloxanes, 214–215
  polymers with completely condensed fragments (T_8), 215
  possible routes of chemical transformation leading to macromonomer in reaction system, 217–218
  size exclusion chromatography data for high molecular fraction of polymer, 221, 222f
  structural isomers T_8 with two functional groups, 216f
  structural study of polymeric organosilsesquioxanes, 215
  synthesis of 1,4-bis(2-chloromethylphenylsilylethyl)hexamethyloctasilsesquioxane, 223–224
  synthesis of 1,4-bis(2-isopropoxymethylphenylsilylethyl)hexamethyloctasilsesquioxane, 223
  synthesis of polymers, 224
  thermodynamically favored cubic structure of octaorganooctasilsesquioxanes, 214–215
  thermogravimetric analysis data for high molecular fraction of polymer, 221, 222f
Organosilicon (OS) exteriors. *See* Radially layered copolymeric dendrimers
Organosilicon polymers
  hydrosilylation polymerization, 226–227
  research interest, 226
  See also Halogenated poly(carbosiloxane)s
Oxidation reactions, silicon oxide (SiOx) film deposition, 549
Oxygen gas barrier films
  properties of SiOx-deposited poly(ethylene terephthalate) (PET) films, 556, 558–559
  See also Silicon oxide (SiOx) coated polyester films
Oxygen permeability
  permeation rate through silicon oxide (SiOx) deposited poly(ethylene terephthalate) (PET) film, 546–547
  property for contact lens, 296–297
  See also Fluorosiloxane based hydrogels

P

Particles
  dispersion in polymer melts, 204
  See also Reinforced
    poly(dimethylsiloxane) melts
PDMS. See Poly(dimethylsiloxane) (PDMS)
Per-deuterated siloxanes
  ^{13}C NMR of small, 64
  catalytic preparation of both linear and cyclic, 72
  fractionation of deuterated cyclic PDMS, 73, 74*t*
  gas chromatography/ammonia chemical ionization mass spectroscopic analysis of small cyclics, 69, 70*f*, 71*t*
  gas chromatography/electron ionization mass spectroscopic analysis of small cyclics, 65, 66*f*, 67*f*, 68*t*, 69
  ^{1}H NMR of small, 63
  infrared (IR) of small, 64–65, 66*f*
  preparation of larger, 69, 72
  preparation of small cyclic, 61, 62*f*
  ^{29}Si NMR of small, 64
Perfluorocyclobutane (PFCB)
  chemistry, 309–310
  linear PFCB fluorosilicones, 313
  network PFCB fluorosilicones, 314–320
  PFCB silicones from phenols, 311–313
  siloxane-containing PFCB materials, 311–320
  traditional materials, 310
  See also Fluorosilicones with perfluorocyclobutene (PFCB) aromatic ether linkage
2-Perfluorooctyl)ethyl]trichlorosilane (FOETS)
  AFM and LFM images of carboxylated NTS(NTS$_{COOH}$)/FOETS mixed monolayers in air, 346*f*
  AFM and LFM images of NTS-NTS$_{COOH}$/FOETS ultrathin film, 349*f*
  AFM and LFM images of NTS/FOETS mixed monolayer by Langmuir method, 344*f*
  LFM image of NTSCOOH/FOETS mixed monolayer in vacuum, 347*f*
  mixed monolayer of NTS/FOETS (18-nonadecyltrichlorosilane), 343, 345
  monolayer preparation, 333–334
  NTS$_{COOH}$/FOETS (carboxylated NTS/FOETS) mixed monolayer, 345
  NTS-NTS$_{COOH}$/FOETS ultrathin film, 345, 348
  schematic of domain height control in NTS/FOETS mixed monolayer, 350*f*, 351*f*
  See also 18-Nonadecyltrichlorosilane (NTS) monolayer
Permeability
  dendrimer-based networks, 262
  siloxane polymers, 3
Phase behavior. See Multi-block copolymers
Phenol release. See Interpenetrating polymer networks (IPNs)
Phenyl-substituted cyclosiloxanes. See Photoluminescence of phenyl- and methyl-substituted cyclosiloxanes
Phenyl-terminated oligomers
  production, 75, 76*f*
  reaction leading to cyclic and linear siloxanes, 75, 76*f*
Phenylalanine
  differential scanning calorimetry results for amino acid functional siloxanes, 161*t*
  evaluating efficacy and generality of synthetic procedure, 141
  See also Amino acid functional siloxanes
Photocurable epoxy coating

formulations. See Modifiers for photocurable epoxy coating formulations
Photohydrosilylation, photolabile platinum catalysts, 171
Photoinitiation
cationic polymerization, 291
photopolymerization method, 295
See also Cationic polymerization of monomers with silsesquioxane (T_8) core
Photoluminescence of phenyl- and methyl-substituted cyclosiloxanes
class of stereoregularly-built phenylated trimethylsiloxycyclosiloxanes, 116
compounds under investigation, 116
cycloaddition reaction of dichloroneopentylsilene to tolane (tolane cycloadduct, TCA), 116
effect of ring size of phenyl-substituted cyclosiloxanes, 119, 122, 124
instrumentation, 117–118
phosphorescence excitation spectra of stereoregular phenylcyclosiloxanes along with linear tetraphenylsiloxanediol, 124, 125f
phosphorescence time dependence of emission from stereoregular phenylcyclosiloxanes along with linear tetraphenylsiloxanediol, 124, 126f
photoluminescence emission spectra at room temperature of differently silicon-substituted TCA compounds, 120f
photoluminescence emission spectra of bis-TCA-disiloxanediol at room temperature as solid, 120f
photoluminescence emission spectra of functionalized silaspirocycles, 121f
photoluminescence emission spectra of octaphenylcyclotetrasiloxane solid at liquid nitrogen temperature, 121f
photoluminescence emission spectra of stereoregular phenylcyclosiloxanes along with linear tetraphenylsiloxanediol, 122, 123f
photoluminescence excitation spectra of stereoregular phenylcyclosiloxanes along with linear tetraphenylsiloxanediol, 122, 123f
photoluminescence of cyclosiloxane, 119
photoluminescence of silacyclobutene derivatives, 118–119
TCA-containing cyclosiloxanes, 117
TCA derivatives, 116–117
Photoresist
evaluation of poly(siloxyethylene glycol), 374
sensitivity characteristic curve of film against photo irradiation, 375f
Plasma polymerization
choice of silanes for silicon oxide (SiOx) film deposition, 547
reactor chamber for silicon oxide deposition, 548f
reactor for silicon oxide deposition, 545–546
silicon oxide deposition procedure, 546
See also Silicon oxide (SiOx) coated polyester films
Platinum
hydrosilylation reaction, 17
Karstedt catalyst, 17–18
Platinum-curable silicones, scheme for conversion from sand, 12f
PMCS. See Cyclolinear polyorganosiloxanes (POCS)
POCS. See Cyclolinear polyorganosiloxanes (POCS)
Poly(4-vinylpyridine) (PVPy)
organic-inorganic polymeric hybrids, 429–430
See also Silanol-containing polymers
Poly(amidoamine-organosilicon) (PAMAMOS) copolymeric

dendrimers. *See* Radially layered copolymeric dendrimers
Poly(amidoamine) (PAMAM). *See* Radially layered copolymeric dendrimers
Poly(*n*-butyl methacrylate) (PBMA)
  miscible polymer blends, 425–426
  See also Silanol-containing polymers
Polycarbonate (PC)
  evaluation of Tospearl in PC film, 541*t*
  treatment with Tospearl, 540
  See also Interpenetrating polymer networks (IPNs)
Poly(carbosilane)
  control of molecular weight by phenyllithium, 466, 468
  control of molecular weight by transition metal complex, 461, 465
  ^{1}H and ^{13}C NMR spectra of polyTMSB by phenyllithium, 467*f*
  ^{1}H NMR spectra of olefinic region of model compounds and polyTMSB by Pt catalyst, 463*f*, 464*f*
  initiation reaction of TMSB by phenyllithium, 466
  polymer from 1,1,2-trimethylsilacyclobutane (TMSB), 461
  proposed polymerization mechanism of TMSB by Pt catalyst, 465
  ^{29}Si NMR spectrum of polyTMSB by Pt catalyst, 462*f*
  stereoregular and/or optically active, 468–472
  See also Silicon-containing polymers
alt-Poly(carbosilane/siloxane)
  monomer preparation by Pt-catalyzed hydrosilation reactions, 434
  preparation by Grignard reaction, 433
  ruthenium catalyzed syntheses, 434–443
  unsaturated polymers by acyclic diene metathesis (ADMET) reactions, 434
  See also Ruthenium catalyzed synthesis of *alt*-poly(carbosilane/siloxanes)

Poly(carbosiloxane)s
  stereoregular and/or optically active, 468–472
  See also Halogenated poly(carbosiloxane)s; Silicon-containing polymers
Poly(dimethylsiloxane) (PDMS)
  acrylamide substituted, 171
  catalyzed ring-opening polymerization of octamethylcyclotetrasiloxane (D$_4$), 15
  contact angles for PDMS film adsorbed onto rigid substrate, 323–324
  contact angles of crosslinked PDMS on flexible substrate, 324
  contact angles on preformed elastomer surfaces, 324–325
  copolymerization with tougher materials, 270–271
  fillers, 16
  flexibility, 2–3
  glass transition, 2
  insertion of silphenylene group into backbone, 2–3
  permeability, 3
  preparation of Si–H on chain polymer, 16
  surface and interfacial properties, 4
  synthesis of aldehyde-functional polysiloxanes, 446
  unusual properties, 3
  vinyl-stopped, 15, 16
  See also Block copolymers by free radical polymerization; Cyclic poly(dimethylsiloxanes); Gelation studies; Hybrid polymers from atom transfer radical polymerization (ATRP) and poly(dimethylsiloxane) (PDMS); Multi-block copolymers; Reinforced poly(dimethylsiloxane) melts; Silicone–urethane copolymers; Thin film poly(dimethylsiloxane) (PDMS)
Polyethoxysiloxane. *See* Hyperbranched polyethoxysiloxane; Silicas

Poly(ethylene oxide) (PEO). *See* Multiblock copolymers
Poly(ethylene terephthalate) (PET). *See* Silicon oxide (SiOx) coated polyester films
Polyhedral organosiloxanes
range of individual compounds of type, 214, 216f
thermodynamically favored cubic structure of octaorganooctasilsesquioxanes, 214–215
See also Organosilicon linear polymer with octaorganooctasilsesquioxanes (T_8) in main chain
Poly(hydroxymethylsiloxane) (PHMS)
organic-inorganic polymer blends, 429–430
Tg values of hybrids, 430t
See also Silanol-containing polymers
Polymer melts
dispersion of aerosil silica particles, 206, 210
dispersion of polysilicate particles, 210
terminally attached and alkylated silicas in, 210, 213
See also Reinforced poly(dimethylsiloxane) melts
Polymer-modified glasses, silica-type materials, 6
Polymeric hybrids
organic-inorganic, 429–430
See also Hybrid polymers from atom transfer radical polymerization (ATRP) and poly(dimethylsiloxane) (PDMS); Silanol-containing polymers
Polymerization, ring-opening, preparation of siloxane polymers, 1–2
Poly(methyl methacrylate) (PMMA)
evaluation of Tospearl in PC film, 541t
See also Block copolymers by free radical polymerization

Polyols. *See* Modifiers for photocurable epoxy coating formulations
Polyorganosiloxanes, cyclolinear. *See* Cyclolinear polyorganosiloxanes (POCS)
Polyphosphazenes, desirable in hybrid materials, 270
Poly(propylene) (PP)
coextruded film with Tospearl, 540
evaluation of Tospearl in coextruded PP film, 540t
evaluation of Tospearl in single PP sheet, 540t
Tospearl additive, 539
Polysilicate particles
dispersion in polymer melts, 210
See also Reinforced poly(dimethylsiloxane) melts
Polysiloxanes
anionic polymerization of strained-ring cyclotrisiloxanes, 20
cationic polymerization of cyclosiloxanes, 20–21
interest in controlled synthesis, 20
use as segments of block and graft copolymers, 20
See also Block copolymers by free radical polymerization; Cationic ring-opening polymerization of cyclotrisiloxanes; Interpenetrating polymer networks (IPNs)
Poly(siloxyethylene glycol) (PSEG)
analytical measurement methods, 362
change in transmittance of PSEG aqueous solution in phosphate buffer, 371f
chemical structures of PSEG(m/n) and poly(divinylsiloxyethylene glycol) [PVSE(m/n)], 363
controlling mobility of PSEG series, 365
differential scanning calorimetry profiles for PSEG(1/9) and PVSE(1/7), 368f
dry etching resistance of PSEG, 374, 376

electroconductivity, 376
etching characteristics of PVSE(1/7) and diazonaphthoquinone/novolac resin film in oxygen plasma, 377f
evaluation as electron-beam resist, 374
evaluation as photoresist, 374
experimental materials, 360
glass transition temperatures (Tg) and melting points of PSEG series and PVSE(1/7), 370t
gel permeation chromatography profiles of PSEG(1/7), PSEG(2/7), and PVSE(1/7), 364f
^{1}H NMR spectra of PSEG(1/7), PSEG(2/7), and PVSE(1/7), 366f
hydrolytic stability in aqueous media, 369, 372
hydrolytic stability test method, 360–361
ion conductivity characteristics as function of temperature, 380f
ionic conductivity method for PSEG gel, 362
lower critical solution temperature (LCST), 369
negative tone images of PVSE resist with electron beam EB-exposure, 379f
negative tone images of PVSE resist with UV-exposure, 378f
normalized attenuance equation, 361
O_2 RIE resistance test method, 361–362
physicochemical properties, 365, 369
plots of LCSTs of PSEG versus silicon content in polymers, 373f
plots of Tg versus silicon content, 370f
polymer synthesis method, 360
preparation of PSEG gels and their ion conductivity, 376
PSEG as negative working resist, 372, 374
resist processing method, 361

schematic of polycondensation reactions for PSEG synthesis, 363
sensitivity characteristic curve of PVSE(1/7) film against electron-beam exposure, 375f
sensitivity characteristic curve of PVSE(1/7) film against photo irradiation, 375f
synthesis and characterization, 362, 365
synthetic results of PSEG and PVSE series from polycondensation of bis(diethylamino)siloxane (BAS) and oligo(ethylene glycol) (OEG), 367t
time course of normalized attenuance of polymers in phosphate buffer, 373f
Polysilylenes, desirable in hybrid materials, 270
Poly(styrene) (PS). *See* Block copolymers by free radical polymerization
Poly(tetramethylene oxide) (PTMO) based segmented polyurethanes. *See* Silicone–urethane copolymers
Polyurethanes. *See* Silicone–urethane copolymers
Poly(N-vinylpyrrolidone) (PVPr)
miscible polymer blends, 426–428
organic-inorganic polymeric hybrids, 429–430
See also Silanol-containing polymers
Potassium hydroxide (KOH). *See* KOH-catalyzed ring-opening polymerization of cyclic siloxanes
Preparative gel permeation chromatography (GPC)
fractionation of deuterated cyclic PDMS, 73f, 74t
instrument, 45, 47
method, 45, 47
schematic of instrument, 46f

R

Radially layered copolymeric dendrimers
  amphiphilic poly(amidoamine-organosilicon) (PAMAMOS) copolymeric dendrimers as inverted micelles, 266
  architecturally driven property differences between dendrimers and corresponding linear polymers, 247t
  C1s high resolution ESCA spectrum of PAMAMOS dendrimer-based network surface, 265f
  confined nanoscopic poly(amidoamine) (PAMAM) domains, 267
  copolymeric dendrimers, 248
  dendrimer-based molecular sponges, nanoscopic reactors, and inorganic-organic composites, 266
  dendrimers, 242–243, 248
  effect of curing time at room temperature on glass transition (Tg) of PAMAMOS dendrimer-based network, 259f
  effect of generation of PAMAM dendrimer reagent on synthesis rate and composition of resulting PAMAMOS dendrimer product, 253f
  effect of solvent on PAMAMOS dendrimers synthesis, 252f
  efficient curing of PAMAMOS dendrimer precursor, 260
  examples of PAMAMOS dendrimers, 249, 254f
  Langmuir trough isotherms, 255
  PAMAMOS, 242
  PAMAMOS dendrimer-based networks containing hydrophilic and hydrophobic nanoscopic domains, 257, 260
  permeability of PAMAMOS dendrimer-based networks, 262
  preparation in variety of different chemical compositions, 266
  preparation of poly(amidoamine-organosilicon) (PAMAMOS) dendrimer, 249
  scanning electron microscopy (SEM) image of cross-cut of film of PAMAMOS dendrimer-based network, 263f
  schematic of Michael addition reaction of silylated acryl ester, 251
  schematic of reiterative sequence of excess-reagent divergent growth method, 250
  schematic of water hydrolysis and subsequent condensation of silanol intermediates into siloxane interdendrimer bridges, 257, 258
  schematic representation of homopolymeric dendrimers, 244f
  schematic representation of radially layered and segmented ordered copolymeric dendrimers, 245f
  selected molecular characteristics of ethylenediamine core PAMAM dendrimers, 246t
  SEM/EDS spectrum of PAMAMOS dendrimer-based network surface, 264f
  surface pressure measurements for PAMAMOS dendrimer with PAMAM interior and two exterior layers of OS branch cells, 255, 256f
  surface properties of PAMAMOS dendrimer-based networks, 262
  surface properties of PAMAMOS dendrimers, 255, 257
  tapping mode atomic force microscopy (TMAFM) image of PAMAMOS dendrimer-based network, 261f
  thermal properties of PAMAMOS dendrimer-based networks, 260, 262
  typical procedure for network preparation, 257, 260
Radiation curable silicones
  quest for faster curing silicones, 170
  See also UV curable silicones
Random copolymers, polysiloxanes, 5

Reactive homopolymers
  block copolymers, 4
  cyclic trapping, 5
  elastomeric networks, 4–5
  types of reactions, 4
Reactive modifiers. *See* Modifiers for photocurable epoxy coating formulations
Refractive index, hydrogenated cyclic and linear PDMS, 54, 55*f*
Regulated structure. *See* Silicon-containing polymers
Reinforced poly(dimethylsiloxane) melts
  deconvolution of relaxation decay for PDMS 32K in 10.1% w/w aerosil, 209*f*
  dispersion of particles in polymer melts, 204
  effect of adsorption at interface on spin-spin relaxation times, 204
  effect of various surface-modified silicas on PDMS relaxation times, 212*f*
  experimental system, 205
  experimental transverse magnetization decays from aerosil silica in 32K polymer, 207*f*
  fitting relaxation decay by distribution of relaxation times dependent on volume fraction profile, 204–205
  nuclear magnetic resonance (NMR) method, 205
  PDMS samples, 205*t*
  polysilicate particles in polymer melts, 210
  relative relaxation times for two series of samples using 32K polymer with aerosil A200, 208*f*
  silicate samples, 206*t*
  terminally attached and alkylated silicas in polymer melts, 210, 213
  variation in proportions of components of T_2 for blends of 10K PDMS with polysilicate R3, 211*f*
  variation in T_2 relaxation time for blends of 10K PDMS with polysilicate R3, 211*f*
Relaxation times. *See* Reinforced poly(dimethylsiloxane) melts
Relay substitution reactions, mechanism of silane formation, 177
Resist
  evaluating poly(divinylsiloxyethylene glycol) (PVSE) as photoresist, 374
  evaluating PVSE as electron-beam resist, 374
  poly(siloxyethylene glycol) (PSEG) as negative working resist, 372, 374
  processing method, 361
Resistance
  dry etching of poly(siloxyethylene glycol) (PSEG), 374, 376
  etching characteristics of PVSE(1/7) and diazonaphthoquinone/novolac resin film in oxygen plasma, 377*f*
  negative tone images of PVSE resist with UV- and EB-exposure, 378*f*, 379*f*
  O_2 RIE resistance method, 361–362
Ring macromolecules
  reasons for studying, 41–42
  structural comparison of linear and cyclic polymers, 42*f*
Ring opening metathesis polymerization (ROMP)
  reactions using Ru catalyst, 435
  See also Ruthenium catalyzed synthesis of *alt*-poly(carbosilane/siloxanes)
Ring opening polymerization
  preparation of siloxane polymers, 1–2
  See also Cationic ring-opening polymerization of cyclotrisiloxanes; KOH-catalyzed ring-opening polymerization of cyclic siloxanes
Rubbery organic networks
  base-catalyzed one-stage polymerization, 487, 491
  change in Guinier radius and mass

fractal dimension of fractal structures during one-stage polymerization, 489f
clusters largest by one-stage polymerization, 491
clusters (siloxane-silica) by sol-gel process from tetraethoxysilane (TEOS), 486
dynamic mechanical analysis (DMA) method, 487
dynamic shear modulus of hybrid networks as function of temperature, 496f
epoxide network of diglycidyl ether of Bisphenol A (DGEBA) and poly(oxypropylene)diamine (Jeffamine D2000), 486
epoxy-silica hybrids syntheses procedures, 486
equivalent box model, 499
evolution of SAXS intensity during one-stage polymerization of hybrid DGEBA–D2000–TEOS, 490f
evolution of SAXS profiles during crosslinking of endcapped prepolymer SEC2000 by sol-gel process, 493f
evolution of SAXS profiles during one-stage polymerization of hybrid DGEBA–D2000–TEOS, 488f
evolution of SAXS profiles during two-stage polymerization of hybrid DGEBA–2000–TEOS, 492f
evolution of silica by small-angle X-ray scattering (SAXS), 487
experimental, 486–487
independent polymerization of organic and inorganic systems, 501
Kerner–Nielson model, 495
loss factor tan δ as function of temperature at 1 Hz, 495, 497f
mechanical properties, 495, 499
model by Davies, 499
network from endcapped prepolymer SED2000, 491
phase separation and rate during polymerization, 501
procedures to synthesize DBEGA–D2000–TEOS hybrid, 486
reaction mechanism of sol-gel process, 501
reinforcement of rubbery epoxide network with in situ formed silica by DMA, 495
relationship between structure and morphology of epoxy-silica hybrids, 501–502
relative modulus of organic-inorganic hybrid as function of effective volume fraction of hard phase, 500f
relative modulus of organic-inorganic hybrid as function of silica phase volume fraction, 495, 498f
SAXS intensity profiles of hybrid networks, 494f
silica structure evolution, 487, 491
structure and morphology of epoxy-silica hybrid, 491
theoretical bicontinuous models describing mechanical behavior of reinforced systems, 501–502
two-stage polymerization using prehydrolysis of TEOS, 491
Ruthenium catalyzed synthesis of *alt*-poly(carbosilane/siloxanes)
activated complex characterization by NMR and X-ray crystallography, 437–438
acyclic diene metathesis (ADMET) polymerization, 435, 436f
addition of ortho acetylstyrene yielding 1:1 complex [Ph$_3$P]$_2$RuCO, 437, 438f
alt(carbosilane/siloxane) copolymers incorporating crown ethers, 439
catalytic cycle involving coordination of unsaturated Ru center, 436
complex of [Ph$_3$P]$_2$RuCO with *N*-benzylideneaniline, 438
converting Si-vinyl groups, 439–440
copolymerization of aromatic ketones and 1,3-divinyltetramethyldisiloxane, 434–435

direct preparation of cross-conjugated copoly(arylene/1,1-vinylenes), 441–442
end group analysis by NMR spectroscopy, 437
fluorescence spectra of cross-conjugated materials, 442
polycyclic aromatic ketones copolymerizing with 1,3-divinyltetramethyldisiloxane, 437
preparation of hyperbranched material, 441
reactive terminal C=C double bonds, 439
reactivity with ferrocenyl ketones, 440–441
Ru catalyst isomerizing terminal α,ω-diene to internal dienes, 436–437
Ru catalyst preparation, 435
Ru catalyzed reaction between benzophenone and 1,3-divinyltetramethyldisiloxane, 440
structure of crystalline complex [Ph$_3$P]$_2$RuCO, 438f
utility in ring opening metathesis polymerizations (ROMP), 435

S

Sand
  carbo-electro reduction process, 11, 13
  scheme for conversion to Pt-curable silicones, 12f
  See also Chemistry of silicones
Self-assembly. See Cyclolinear polyorganosiloxanes (POCS)
Selfbias
  etching reactions for silicon oxide (SiOx) film deposition, 549
  See also Silicon oxide (SiOx) coated polyester films
Self-condensation. See Silanol-containing polymers
Semi-interpenetrating polymer networks (semi-IPNs)
  inter-polymer hydrogen bonds in copolymer blends, 429
  preparation, 421
  See also Silanol-containing polymers
Silacyclobutene derivatives, photoluminescence, 118–119
Silanol-containing polymers
  composition dependence of glass transition temperature of blends, 426–427
  composition dependence of Tg of poly(4-vinylphenyldimethylsilanol)/poly(N-vinylpyrrolidone) (PVPDMS-18/PVPr) blends, 428f
  composition dependence of Tg of poly(4-vinylphenylmethylphenylsilanol)/PVPr (PVPMPS–18/PVPr) blends, 427f
  composition dependence of Tg of styrene–VPMPS (ST–VPMPS) copolymers, 428f
  copolymerization of 4-vinylphenyldialkyl/arylsilane with styrene, 420
  Fourier transform infrared (FT–IR) spectroscopy method, 421
  homopolymerization of 4-vinylphenyldialkyl/arylsilane, 420
  inorganic siloxane-based silanol polymer or copolymers, 424
  interpolymer complexes and semi-IPNs, 429
  infrared (IR) frequency shifts in poly(styrene-co-4-vinylphenylmethylphenylsilanol) (ST–VPMPS) and poly(styrene-co-4-vinylphenyldimethylsilanol) (ST–VPDMS) with poly(n-butyl methacrylate) PBMA blends, 426t
  IR frequency shifts in ST–VPMPS and ST–VPDMS with PVPr blends, 426t
  monitoring conversion of ≡Si–H to

≡Si–OH bonds by IR, 422, 423*f*
organic-inorganic polymeric hybrids, 429–430
polymer modification of ≡Si–H containing precursor polymers with dimethyldioxirane solution in acetone, 421
preparation of blends, semi-interpenetrating polymer networks (semi-IPNs), and organic-inorganic polymeric blends, 421
preparation of dimethyldioxirane solution in acetone, 420
properties of siloxane-based silanol polymer and copolymers dependent on silanol composition, 424–425
selective oxidation of precursor polymers with dimethyldioxirane solution, 421–422
silanol hydrogen bonds in miscible polymer blends, 425–428
stability of 4-vinylphenyldialkyl/arylsilanol polymer and styrene copolymers, 422
strength of self-associated silanol hydrogen bonds by IR frequency shifts, 423, 424*f*
synthesis and characterization of polymers and copolymers, 421–425
Tg values of 50/50 w/w PHMS/PVPr and PHMS/PVPy hybrids, 430*t*
thermal analysis method, 421
Silanol functional groups, reactive intermediates in silicon chemistry, 419
Silanol intermediates, production, 75
Silica sol. *See* Hyperbranched polyethoxysiloxane; Silicas
Silicas
  alternative approach to synthesis with tailored properties, 504
  analytical methods, 504–505
  atomic force microscopy (AFM) measurements, 513–514
  behavior of molecular particle converting from polyethoxysiloxane to silica, 511
  blocking functional groups using trimethyltrifluoroacetoxysilane, 509–510, 511
  carbo-electro reduction process, 11, 13
  common interests with polymer chemistry, 504
  condensation method of polyethoxysiloxane hydrolysate, 506
  experimental materials, 504
  factors controlling formation of silica particles, 504
  gas-liquid chromatography (GLC) methods, 504–505
  gel permeation chromatography (GPC) method, 505
  high surface area, trichlorosilane (TCS) reaction, 16
  hydrolysis and subsequent intramolecular cyclization, 510
  hydrolysis method of polyethoxysiloxane, 506
  hyperbranched polyethoxysiloxane synthesis, 506–509
  infrared (IR) spectra of triethoxysilanol and polycondensation product, 507*f*
  IR absorption spectra of system, 507–508
  IR spectrum of silica sol dried in vacuum, 510*f*
  Kratky-plot of scattering curves of hyperbranched polyethoxysiloxane, trimethylsilyl derivative, and silica sol, 512*f*
  link between starting hyperbranched polyethoxysiloxane and products, 512–513
  log-log plot of experimental scattering curves of hyperbranched polyethoxysiloxane, trimethylsilyl derivative, and silica sol, 512*f*
  method of small-angle X-ray

scattering (SAXS), 505
molecular silica sols synthesis, 509–511
possible uses of silica sols, 514
SAXS measurements, 511–513
scanning force photomicrograph of silica sol on mica, 513f
^{29}Si NMR spectra of triethoxysilanol and hyperbranched polyethoxysiloxane, 508f
size exclusion chromatography (SEC) of trimethylsilylated hyperbranched polyethoxysilane and polystyrene standard, 509f
synthetic method for hyperbranched polyethoxysiloxane, 505–506
terminally attached and alkylated in polymer melts, 210, 213
tetraethoxysilane derivatives as building blocks of hyperbranched structure, 507
variety of chemical processes, 503
See also Hyperbranched polyethoxysiloxane; Reinforced poly(dimethylsiloxane) melts
Silicate particles
samples, 206t
See also Reinforced poly(dimethylsiloxane) melts
Silica-type materials
polymer-modified glasses, 6
sol-gel ceramics for in situ precipitations, 6
Silicon
formation of elemental, 11, 13
See also Chemistry of silicones
Silicon-29 nuclear magnetic resonance (NMR)
chemical shift for hydrogenated cyclic and linear PDMS, 56
hydrogenated cyclic PDMS fractions, 49–50
small per-deuterated dimethylsiloxanes, 64
spectrum of polymer from cationic ring-opening polymerization of model monomer, 24, 25f
Silicon-containing polymers
anionic ring-opening polymerization of optically active five-membered silicon monomer, 472
^{13}C NMR spectra of Si(CH$_3$)$_2$ of poly[{(1S)-1-(1-naphthyl)-1-phenyl-3,3-dimethyldisiloxane-1,3-diyl}ethylene] from racemic and optically active monomer, 473f
control of molecular weight by phenyllithium, 466, 468
control of molecular weight by transition metal complex, 461, 465
control of molecular weight of polycarbosilanes, 461–468
diad sequences of poly[{(1S)-1-(1-naphthyl)-1-phenyl-3,3-dimethyldisiloxane-1,3-diyl}ethylene], 472, 474f
diastereomeric splitting of methyl groups in isopropyl of methyl(1-naphthyl)phenyl-(-)-menthyloxysilane, 469f
^{1}H and ^{13}C NMR spectra of poly(1,1,2-trimethylsilacyclobutane) (polyTMSB) by phenyllithium, 467f
^{1}H NMR spectra of olefinic region of model compounds and polyTMSB by Pt catalyst, 463f, 464f
^{1}H NMR spectra of poly(methylphenylsilylenetrimethylene), 471f
HPLC separation of methyl(1-naphthyl)phenyl-(-)-menthyloxysilane, 469f
initiation reaction of TMSB by phenyllithium, 466
methyl(1-naphthyl)phenyl-(-)-menthyloxysilane as starting synthesis material, 468
optically active five-membered cyclic silicon compound synthesis by intramolecular hydrosilylation, 470, 472
polymerization of silacyclobutanes by

transition metal catalysts, 461*t*
polymerization of TMSB by alkyllithium, 466*t*
polyTMSB, 461
proposed polymerization mechanism of TMSB by Pt catalyst, 465
^{29}Si NMR spectrum of polyTMSB by Pt catalyst, 462*f*
stereoregular and/or optically active poly(carbosilane)s and poly(carbosiloxane)s, 468–472
synthesis and determination of optical purity of optically active vinylhydrodisiloxane monomer, 470
synthesis of optically active allylmethylphenylsilane, 468
Silicon oxide (SiOx) coated polyester films
atom composition of deposited SiOx films from silanes, 548*t*
atom composition of deposited SiOx films from tetramethoxysilane (TMOS)/O$_2$ mixtures, 550*t*
atom composition of SiOx deposited films under selfbias, 555*t*
atomic composition of deposited SiOx films, 547, 549
chemical composition of deposited SiOx films, 549, 551, 556
chemical composition of SiOx films under selfbias deposition by X-ray photoelectron spectroscopy (XPS) analysis, 551, 556
choice of silanes for SiOx film deposition, 547
etching reactions by selfbias for SiOx film deposition, 549
experimental materials, 545
gas barrier property, 544
method for XPS of deposited SiOx films, 546
oxidation reactions for SiOx film deposition, 549
oxygen gas barrier properties of SiOx-deposited poly(ethylene terephthalate) (PET) films, 556, 558–559
oxygen permeability coefficient equation, 547
oxygen permeation coefficient for SiOx films, 558*t*
oxygen permeation rate through SiOx-deposited PET film, 546–547
oxygen permeation rate through SiOx-deposited PET films, 558*t*
plasma polymerization reactor for SiOx deposition, 545–546
possible plasma chemical vapor deposition (CVD) method, 545
possible processes for elimination of carbonaceous compounds, 545
scanning electron micrographs (SEM) of SiOx-deposited films under selfbias and under no selfbias, 557*f*
schematic of reaction chamber for SiOx deposition, 548*f*
selfbias voltage at constant oxygen concentration in TMOS/O$_2$ mixture as function of RF power, 552*f*
selfbias voltage at constant RF power as function of oxygen concentration in TMOS/O$_2$ mixture, 550*f*
SiOx deposition procedure by plasma polymerization, 546
synthesis by sol-gel or plasma chemical vapor deposition, 544
two layers model, 547
XPS (C1s) spectra of deposited SiOx films as function of oxygen concentration in TMOS/O$_2$ mixture, 552*f*
XPS (C1s) spectra of SiOx deposited films under selfbias as function of oxygen concentration in TMOS/O$_2$ mixture, 554*f*
XPS (Si2p) spectra of deposited SiOx films as function of oxygen concentration in TMOS/O$_2$ mixture, 553*f*
XPS (Si2p) spectra of SiOx deposited films under selfbias as function of oxygen concentration in TMOS/O$_2$ mixture, 555*f*

Silicone industry, direct process of
   silicon to methylchlorosilanes, 11,
   12f
Silicone surfaces
   preparation, 325, 327
   See also Thin film
      poly(dimethylsiloxane) (PDMS)
Silicone–urethane copolymers
   characteristics of siloxane–urethane
      and polyether–urethane segmented
      copolymers, 402t
   comparison of tensile behavior of
      poly(dimethylsiloxane) (PDMS) and
      poly(tetramethylene oxide) (PTMO)
      based segmented polyurethanes,
      405f
   comparison of tensile properties of
      siloxane–urethane and polyether–
      urethane segmented copolymers,
      404t
   compositions of silicone–urethane and
      polyether–urethane reactions, 400t
   conventional polyether and polyester
      based polyurethane elastomers, 396
   differential scanning calorimetry
      (DSC) thermograms for model 4,4'-
      isocyanatocyclohexylmethane/1,4-
      butanediol (HMDI-BD) based
      polyurethane and PSU-4, 402, 403f
   dynamic mechanical thermal analysis
      (DMTA) curves for PSU-4, 403f
   effect of backbone composition on
      elastomeric properties, 401
   experimental materials, 397
   formation of two-phase morphologies,
      402
   ^{1}H NMR spectrum of α,ω-
      hydroxyhexyl terminated PDMS
      oligomer with Mn=900 g/mol, 398f
   investigation of tensile behavior of
      copolymers, 402, 404
   lamellar structure by strain induced
      crystallization, 405, 406f
   polymer syntheses, 397, 400
   polyurethanes from 4,4'-
      diphenylmethanediisocyanate (MDI)
      and mixed soft segments, 396
   preparation of high molecular weight
      silicone–urethane copolymers, 401
   preparation scheme of PDMS–
      urethane segmented copolymers,
      399
   product characterization, 400–401
   range of well-defined PDMS-
      containing multiphase copolymers,
      396
   reasons for failure in synthesizing high
      molecular weight silicone–urethane
      copolymers, 396
   stress-strain results, 404–405
Silicones
   surface applications, 322–323
   See also UV curable silicones
Siloxane-containing modifiers. *See*
   Modifiers for photocurable epoxy
   coating formulations
Siloxane-silica fillers. *See* Rubbery
   organic networks
Siloxane-type polymers
   applications, 5–6
   block copolymers, 5
   block copolymers from end-
      functionalized polymers, 4
   copolymers, 5
   cyclic trapping with reactive
      homopolymers, 5
   elastomeric networks, 4–5
   flexibility, 2–3
   formation from methylchlorosilanes,
      14–16
   homopolymers, 2–4
   medical applications, 5
   non-medical applications, 5–6
   permeability, 3
   preparation, 1–2
   random copolymers, 5
   reaction types in reactive
      homopolymers, 4
   reactive homopolymers, 4–5
   surface and interfacial properties, 4
   unusual properties of
      poly(dimethylsiloxane) (PDMS), 3

Siloxanes
  preparation of polymers containing pendant and terminal amino acids, peptides, and polypeptides, 129
  reaction mechanism proposal, 73, 75–76
  surfactant properties, 129
  surfactants, 128
  synthesis of linear polymers, 82
  See also Amino acid functional siloxanes; Poly(dimethylsiloxane) (PDMS)
Silphenylene group, insertion into backbone of poly(dimethylsiloxane), 2–3
Silsesquioxanes
  cubic siloxane molecules, 271
  initiators for atom transfer radical polymerization, 278, 279
  interest in functionalizing, 284
  See also Cationic polymerization of monomers with silsesquioxane (T_8) core; Hybrid polymers from atomic transfer radical polymerization (ATRP) and poly(dimethylsiloxane) (PDMS)
Silyl triflates, production, 75
Small-angle neutron scattering (SANS), dimension of rings and chains in dilute solution, 52
Small-angle X-ray scattering (SAXS). *See* Rubbery organic networks; Silicas
Soft contact lens, poly(dimethylsiloxane) (PDMS), 3
Sol-gel process
  commonality between polymer and silica chemistry, 504
  reaction mechanism, 485, 501
  silica-type materials, 6
  siloxane-silica clusters from tetraethoxysilane (TEOS), 486
  See also Rubbery organic networks; Silicas
Solvation model
  calculations, 94
  gas-phase reaction path of KOH-catalyzed ring-opening polymerization, 85
Star polymers, atom transfer radical polymerization (ATRP), 278
Stereoregular poly(carbosilane)s and poly(carbosiloxane)s. *See* Silicon-containing polymers
Structural organization. *See* Cyclolinear polyorganosiloxanes (POCS)
Structure, comparison of linear and cyclic polymers, 42f
Styrene polymerizations
  atom transfer radical polymerization (ATRP), 275–276
  ATRP of vinyl monomers from PDMS macroinitiators, 277t
  initiation by PDMS containing in-chain silylpinocolate moieties, 272
  kinetic plot for ATRP of styrene from difunctional PDMS macroinitiator, 275f
  kinetic plot for ATRP of styrene initiated from benzyl chloride-POSS (polyhedral oligomeric silsesquioxane), 280f
  molecular weight plot for ATRP of styrene from difunctional PDMS macroinitiator, 276f
  molecular weight plot for ATRP of styrene initiated from benzyl chloride-POSS, 281f
  polymerization from benzyl chloride-POSS, 278
  SEC traces of ATRP initiated from tetrafunctional cyclotetrasiloxane, 279f
  See also Hybrid polymers from atom transfer radical polymerization (ATRP) and poly(dimethylsiloxane) (PDMS); Silanol-containing polymers
Surface adsorption, hydrogenated cyclic and linear PDMS, 57
Surface properties
  dendrimer-based networks, 262

polysiloxanes, 4
See also Multi-block copolymers; 18-Nonadecyltrichlorosilane (NTS) monolayer; Thin film poly(dimethylsiloxane) (PDMS)
Surface treatment
fillers, 16
Tospearl, 541–542

T

Telechelic oligomers, silicon terminated
acyclic diene metathesis (ADMET) polymerization viability, 408
effect of chain limiter on molecular weight, 412
gel permeation chromatography (GPC) of PDMS–POCT–PDMS [polyoctenamer (POCT)] and PDMS macromonomer, 416f
^{1}H and ^{13}C NMR spectra of bis(5-hexenylchlorodimethylsilane) polyoctenamer, 414f
^{1}H and ^{13}C NMR spectra of bis(5-hexenylmethoxydimethylsilane) polyoctenamer, 411f
^{1}H NMR spectrum of copolymer, 415f
molecular weight analyses of block copolymer, 417
relationship between measured and calculated Mn (from ratio of monomer to chain limiter), 413f
synthesis of block copolymers, 412, 417
synthesis of chlorodimethylsilane terminated telechelic POCT oligomers, 412
synthesis of copolymer PDMS–POCT–PDMS, 415f
synthesis of methoxydimethylsilane terminated telechelic POCT oligomers, 410, 412
synthesis using ADMET chemistry, 409f
Telechelic siloxanes

blending compound 11 with commercial novolac-based cyanate ester resin (CER), 168
deprotection of tetrahydropyranyloxyphenyl (THP)-protecting group, 166
homopolymerization of compound 11 by curing, 167–168
infrared (IR) method, 165
nuclear magnetic resonance (NMR) method, 165
scanning electron micrographs (SEM) method, 165
SEM photograph of fracture surface of telechelic siloxane 12 in commercial CER after curing, 168f
SEM photograph of fracture surface of telechelic siloxane 13 in commercial CER after curing, 169f
synthesis of α,ω-(p-tetrahydropyranyloxyphenyl)-*oligo*-(dimethyl-*co*-diphenylsiloxane) (8), 166
synthesis of α,ω-(p-tetrahydropyranyloxyphenyl)-*oligo*-(dimethylsiloxane) (7), 165–166
synthesis of bis[1,3-(4-cyanatophenyl)]-1,1,3,3-tetramethyldisiloxane (11), 166
synthesis of bis[1,3-p-tetrahydropyranyloxyphenyl]-1,1,3,3-tetramethyldisiloxane (5), 165
synthetic scheme, 167
thermal analysis method, 165
thermogravimetric analysis of blends with novolac-based CERs, 168
Tensile properties
elongation of siloxane-modified epoxy coatings, 522, 526f
modulus of siloxane-modified epoxy coatings, 522, 523f, 524f, 527, 528f
strength of siloxane-modified epoxy coatings, 522, 524f, 525f
See also Modifiers for photocurable epoxy coating formulations
Tethered oils

foul release performance, 186–187
incorporation into silicone network, 185
mechanisms for oil retention, 183–184
synthesis of diphenyldimethylsiloxane oil, 184f
See also Biofouling release coatings
Tetraethyl orthosilicate (TEOS)
possible crosslinking reactions of TEOS functionalized polyols with cycloaliphatic epoxide under cationic UV conditions, 527
proposed condensation reactions of TEOS functionalized polyols under cationic UV conditions, 522
synthesis and characterization of reactive diluents, 518–519
See also Modifiers for photocurable epoxy coating formulations
Tetramethoxysilane (TMOS). *See* Silicon oxide (SiOx) coated polyester films
Thermal transitions. *See* Cyclolinear polyorganosiloxanes (POCS)
Thermoplastics, reinforcement with inorganic filler, 485
Thin film poly(dimethylsiloxane) (PDMS)
adsorption of PDMS films on solid substrate, 323–324
advancing and receding contact angles of *n*-hexadecane on poly(methylhydrogensiloxane)/PDMS (PMHS/PDMS), 330
advancing and receding water contact angle with molecular weight of PDMS, 327
characterization of undecenylsiloxane monolayers, 326t
contact angle characterizations of PDMS, 323–325
contact angle measurement technique, 327
contact angles of water on PDMS films end-grafted onto Si/SiO$_2$, 328t
contact angles on PMHS/PDMS film grafted onto Si/SiO$_2$, 328t

copolymer PMHS/PDMS, 325, 327
critical surface tension of wetting plot of PMHS/PDMS copolymer grafted on Si wafer, 329f
crosslinked PDMS coatings on flexible substrates, 324
diversity of application, 322–323
freshly cleaved muscovite mica, 325
hysteresis of PDMS-grafted silicon wafer, 328
low-hysteresis model for contact angle and surface studies, 330–331
PDMS elastomer surfaces, 324–325
plasma polymerization of hexamethyldisiloxane (HMDS), 330
preparation of silicone surfaces, 325, 327
SiH functional PDMS polymers grafted onto silicon wafer, 326t
Thiol-ene crosslinking, mercaptosiloxanes and vinylsiloxanes, 170–171
Tolane cycloadduct (TCA). *See* Photoluminescence of phenyl- and methyl-substituted cyclosiloxanes
Tone polyols. *See* Modifiers for photocurable epoxy coating formulations
Tospearl particles
applications, 538–542
application versus particle size, 538f
characteristics of commercial grades, 536t
coextruded polypropylene (PP) film, 540f
dispersion viscosities in polar and nonpolar solvents, 538t
effect of MeSi(OMe)$_3$/water ratio on particle size, 535f
effect of stirring rate and ammonia concentration on particle size, 534f
electron micrographs, 537f
evaluation in coextruded PP film, 540t
evaluation in polycarbonate (PC) film, 540, 541t
evaluation in poly(methyl methacrylate) (PMMA) sheet, 541t

evaluation in single PP sheet, 539, 540t
function and applications, 539t
hydrolysis and condensation of alkyl trialkoxysilanes, 533
kettle, 534f
particle size distribution for Tospearl 120, 536f
preparation, 533–535
process for product isolation, 535f
properties, 535–536
results of treating weatherstripping with, 539t
surface treatment with aminoethylaminopropyltrimethoxysilane, 542
surface treatment with hexamethyldisilazane (HMDZ) or trimethylchlorosilane (TMSCl), 541
thermogravimetric analysis of Tospearl 120 and organic resin fine particles, 537f
Triblock copolymers
motivation for synthesis, 272
organic/inorganic hybrid polymers from atom transfer radical polymerization and PDMS, 272, 275–276
See also Hybrid polymers from atom transfer radical polymerization (ATRP) and poly(dimethylsiloxane) (PDMS)
Trifluoromethanesulfonic (triflic) acid
condensations leading to phenyl-terminated linear oligomers, 76f
production of silyl triflates and silanol intermediates, 75f
reaction mechanism proposal, 73, 75–76
reaction of phenyl-terminated oligomers leading to cyclic and linear siloxanes, 76f
removing phenyl groups from dimethyldiphenylsilane, 59–60
Trifluorovinyl ether (TFVE). See Fluorosilicones with perfluorocyclobutene (PFCB) aromatic ether linkage
Trimethyl orthoformate (TMOF)
alternate methoxy source, 479–480
reaction method, 484
See also Bis-(3-trimethoxysilylpropyl)fumarate
1,1,2-Trimethylsilacyclobutane (TMSB)
^{1}H and ^{13}C NMR spectra of olefinic region of model compounds and polyTMSB by Pt catalyst, 463f, 464f
^{1}H and ^{13}C NMR spectra of polyTMSB by phenyllithium, 467f
^{29}Si NMR spectrum of polyTMSB by Pt catalyst, 462f
initiation reaction by phenyllithium, 466
monomer for poly(carbosilane)s, 461
polymerization by transition metal catalysts, 461t
proposed polymerization mechanism by platinum catalyst, 465
See also Silicon-containing polymers

U

Ultraviolet (UV) coatings
principal modes, 516
See also Modifiers for photocurable epoxy coating formulations
Urethanes. See Silicone–urethane copolymers
UV curable silicones
acrylated silicones, 171–172
acryloxymethyldimethylacryloxysilane (I) preparation, 172–173
condensation reaction between silane I and model compound, 173
Dow Corning's preparation of acrylamide terminated PDMS, 171
from acryloxymethyldimethylacryloxysilane, 172–177

further studies with silanol terminated PDMS, 173–174
Gol'din's synthesis of acrylate functional disiloxane, 173
hydrolysis of first intermediate silane establishing its identity, 175, 176
mechanism of formation of silane I and its reaction with silanol, 177
model silanol reaction with silane I, 173
neutralization of liberated hydrochloric acid, 172
nucleophilic substitution reactions, 175
reacting silane I in excess methanol, 177
reactions of silane I and related compounds, 175–177
reactivity of acetoxysilanes in condensation reactions with silanol, 173
relay substitution of chloromethyldimethylchlorosilane, 177
scrambling of acryloxy and acetoxy groups during preparation of mixed silanes, 175, 176
Shin–Etsu's preparation with multiple acrylate terminal groups, 172f
^{29}Si NMR chemical shifts of tetra- and penta-coordinate silanes, 174f
^{29}Si NMR studies, 174
surprising reactivity of silane I toward silanol, 174
transition states of silanol substitution and group transfer polymerization, 178f
trans-silylation of bis(trimethylsilyl)acetamide, 175
See also Curable silicones

V

Valine
DSC results for amino acid functional siloxanes, 161t
evaluating efficacy and generality of synthetic procedure, 141
See also Amino acid functional siloxanes
4-Vinylphenyldialkyl/arylsilane. *See* Silanol-containing polymers
Viscosity, bulk, hydrogenated cyclic and linear PDMS, 53
Volatile organic compound (VOC) emphasis on low, 516
See also Modifiers for photocurable epoxy coating formulations

W

Weatherstripping, treatment with Tospearl, 539
Wilkinson's catalyst, hydrosilation of 4-vinylcyclohexene oxide with silsesquioxane, 285

X

X-ray photoelectron spectroscopy (XPS)
method for deposited SiOx films, 546
See also 18-Nonadecyltrichlorosilane (NTS) monolayer; Silicon oxide (SiOx) coated polyester films
X-ray reflectivity (XR). *See* 18-Nonadecyltrichlorosilane (NTS) monolayer

Highlights from ACS Books

Desk Reference of Functional Polymers: Syntheses and Applications
Reza Arshady, Editor
832 pages, clothbound, ISBN 0-8412-3469-8

Chemical Engineering for Chemists
Richard G. Griskey
352 pages, clothbound, ISBN 0-8412-2215-0

Controlled Drug Delivery: Challenges and Strategies
Kinam Park, Editor
720 pages, clothbound, ISBN 0-8412-3470-1

Chemistry Today and Tomorrow: The Central, Useful, and Creative Science
Ronald Breslow
144 pages, paperbound, ISBN 0-8412-3460-4

Eilhard Mitscherlich: Prince of Prussian Chemistry
Hans-Werner Schutt
Co-published with the Chemical Heritage Foundation
256 pages, clothbound, ISBN 0-8412-3345-4

Chiral Separations: Applications and Technology
Satinder Ahuja, Editor
368 pages, clothbound, ISBN 0-8412-3407-8

Molecular Diversity and Combinatorial Chemistry: Libraries and Drug Discovery
Irwin M. Chaiken and Kim D. Janda, Editors
336 pages, clothbound, ISBN 0-8412-3450-7

A Lifetime of Synergy with Theory and Experiment
Andrew Streitwieser, Jr.
320 pages, clothbound, ISBN 0-8412-1836-6

Chemical Research Faculties, An International Directory
1,300 pages, clothbound, ISBN 0-8412-3301-2

For further information contact:
Order Department
Oxford University Press
2001 Evans Road
Cary, NC 27513
Phone: 1-800-445-9714 or 919-677-0977
Fax: 919-677-1303

Bestsellers from ACS Books

The ACS Style Guide: A Manual for Authors and Editors (2nd Edition)
Edited by Janet S. Dodd
470 pp; clothbound ISBN 0–8412–3461–2; paperback ISBN 0–8412–3462–0

Writing the Laboratory Notebook
By Howard M. Kanare
145 pp; clothbound ISBN 0–8412–0906–5; paperback ISBN 0–8412–0933–2

Career Transitions for Chemists
By Dorothy P. Rodmann, Donald D. Bly, Frederick H. Owens, and Anne-Claire Anderson
240 pp; clothbound ISBN 0–8412–3052–8; paperback ISBN 0–8412–3038–2

Chemical Activities (student and teacher editions)
By Christie L. Borgford and Lee R. Summerlin
330 pp; spiralbound ISBN 0–8412–1417–4; teacher edition, ISBN 0–8412–1416–6

Chemical Demonstrations: A Sourcebook for Teachers, Volumes 1 and 2, Second Edition
Volume 1 by Lee R. Summerlin and James L. Ealy, Jr.
198 pp; spiralbound ISBN 0–8412–1481–6
Volume 2 by Lee R. Summerlin, Christie L. Borgford, and Julie B. Ealy
234 pp; spiralbound ISBN 0–8412–1535–9

The Internet: A Guide for Chemists
Edited by Steven M. Bachrach
360 pp; clothbound ISBN 0–8412–3223–7; paperback ISBN 0–8412–3224–5

Laboratory Waste Management: A Guidebook
ACS Task Force on Laboratory Waste Management
250 pp; clothbound ISBN 0–8412–2735–7; paperback ISBN 0–8412–2849–3

Reagent Chemicals, Eighth Edition
700 pp; clothbound ISBN 0–8412–2502–8

Good Laboratory Practice Standards: Applications for Field and Laboratory Studies
Edited by Willa Y. Garner, Maureen S. Barge, and James P. Ussary
571 pp; clothbound ISBN 0–8412–2192–8

For further information contact:
Order Department
Oxford University Press
2001 Evans Road
Cary, NC 27513
Phone: 1-800-445-9714 or 919-677-0977